Grain Marketing

SECOND EDITION

Grain Marketing

EDITED BY

Gail L. Cramer
and Eric J. Wailes

UNIVERSITY OF ARKANSAS

Routledge
Taylor & Francis Group

NEW YORK AND LONDON

First published in 1993 by Westview Press

Published in 2021 by Routledge
605 Third Avenue, New York, NY 10017
2 Park Square, Milton Park, Abingdon, Oxon OX14 4RN

Routledge is an imprint of the Taylor & Francis Group, an informa business

Library of Congress Cataloging-In-Publication Data
Grain marketing / edited by Gail L Cramer and Eric J. Wailes. — 2nd ed.
    p. cm.
    Rev. ed. of: Grain marketing economics / edited by Gail L Cramer, Walter G. Heid. New York: Wiley, c1983.
    Includes bibliographical references and Index.
    ISBN 0-8133-1796-7
    1. Grain trade—United States. 2. Grain trade. I. Cramer, Gail L. II. Wailes, Eric James. III. Title: Grain marketing economics.
HD9035.G73   1993
380.1 '4131 '0973—dc20   93-20432
                                                                            CIP

ISBN 13: 978-0-3670-0750-8 (hbk)
ISBN 13: 978-0-3671-5737-1 (pbk)

To the following outstanding instructors:

The late Blaine C. Hardy, Wapato, Washington
Eldon E. Weeks, Pullman, Washington
Albert H. Harrington, Pullman, Washington
Lester V. Manderscheid, East Lansing, Michigan
The late Albert N. Halter, Corvallis, Oregon
William G. Brown, Corvallis, Oregon
Kenneth L. Robinson, Ithaca, New York
Stanley R. Thompson, Columbus, Ohio
James D. Shaffer, East Lansing, Michigan

Contents

Tables and Figures

Figures

Preface

Grain has an ever-present influence in our daily lives. It is consumed at every meal or snack, either directly or indirectly. It is a part of your breakfast cereal, toast, eggs, bacon, and rolls; your lunch of soup, sandwich, and milk; your dinner—red meat, poultry, seafood, pasta; and appetizers, cocktails, and after-dinner drinks. Grain is a versatile commodity as well as the world's most important diet staple.

This revised edition gives the student an opportunity to learn some basics about the present world grain industry and its operation. The U.S. grain industry is a central part of the world grain industry, which includes a series of important components as grain is moved from the "farm gate" to final consumption. Government policies, the farm and trade policies of the United States and other countries that trade in the world market, affect both the world grain industry and the ability of the United States to compete in it. Trade policy issues have become a greater concern in the 1990s as a result of major changes in the world's political structure and the growing prominence of international trade associations. The changes in the former Soviet Union and Eastern Europe have created new market opportunities. However, trade continues to be restricted in Japan and the European Community.

The U.S. grain industry services markets throughout the world. It faces stiff competition that often is motivated by a different set of values and trading strategies than those experienced within the U.S. system. To further complicate the trade relationships, dominant grain marketing firms operating in the United States are, in many instances, multinational corporations. A large share of their transactions often are with a single state agency or board that serves as a nation's sole buyer of grains and grain products. The complexity of this situation makes it difficult to administer national trade policy. Students are encouraged to study the ramifications of these trade relationships and to compare the effectiveness of alternative marketing systems.

The volume of grain handled by the U.S. grain industry will continue to escalate in the coming decades as world population, incomes, and livestock inventories increase. This scenario, of course, assumes a continuation of the ability to improve crop yields and to maintain competitiveness in the growing international market. Evolving lifestyles and diets, developing foreign agricultural systems, and the economic well-being of certain grain-importing countries such as those in the former Soviet Union will also exert their influences.

For the student interested in understanding how the grain marketing system functions, and who may be employed somewhere within the system, it is important not only to understand the marketing process but also to be aware of the system's ancillary services. This latter dimension, sometimes overlooked by students, includes all the activities involved in financing grain inventories, advertising and promotion, research (both public and private), education, public policy efforts of commodity associations, methods of handling risk and uncertainty (insurance and hedging), and the development and execution of government policies. Students should also be particularly aware of the vast amount of administrative work that must be performed as grains and their products flow through the system.

Although the United States has a free-enterprise system whereby individual grain firms may privately sell to foreign buyers—usually state trading agencies—these firms are required to meet certain federal regulations, such as reporting large-volume sales, honoring embargoes, and complying with other domestically imposed requirements. Grain firms must also negotiate with labor unions, be aware of diplomatic relations, and handle special shipments involving grain moving under government programs. Activities of the U.S. grain industry are also affected by such agencies as the Federal Grain Inspection Service, Environmental Protection Agency, Occupational Safety and Health Administration, and Federal Trade Commission, to name a few. To meet the standards of these regulations is time-consuming, costly, and disruptive for decisionmakers, but it is also deemed necessary; these standards have been developed and imposed by the democratic process.

The complexity of our grain-trading system far exceeds the lines of a flow diagram or even the discussion in the following chapters. Therefore, this book can only whet the student's quest for knowledge. Serious students should be encouraged to take field trips and educational tours through grain-related facilities. At some universities, extra credit may be earned if the student participates in a field trip to a major grain-trading center. Field trips enable students to gain firsthand knowledge and a chance to observe the real world in action. On such field trips, visits often are arranged by or with a board of trade, grain terminals, grain brokers, flour mills, baking plants, or other processors. Students farther away from major trade centers may take advantage of other grain-related facilities, visiting port terminals, quality-testing laboratories, or United States Department of Agriculture research centers. Other students may have the opportunity to travel to foreign countries to study the operations of other marketing systems. At the same time, a large number of foreign students and grain-trade teams visit the United States each year to learn more about the U.S. system.

This book blends descriptive, institutional, and analytical approaches. The overall plan allows the student to follow the movement of the major grains grown in the United States—corn, wheat, and soybeans—from farm production to final consumption. This flow includes all intermediate steps: assembly, storage, grading, transportation, processing, and merchandising channels, including international trade channels. Accompanying these chapters are treatments of supply and demand relationships, pricing, futures trading, sources of market information, and government policy.

Three chapters have been added to this second edition. A new introductory chapter describes marketing in terms of time, form, and space utility. Chapter 8 is devoted to options trading because of the increasing interest and complexity of this specialized market for commodities. A new chapter (Chapter 13) has been added on the European Community, a trade group that has become a major grain exporter and competitor with the United States in the world market. It has been ten years since the first edition of this book was published; thus tables and figures related to grain trade and policy changes have been thoroughly updated.

The coeditors of this book greatly appreciate the dedicated and cooperative spirit of the contributing authors. Also, we appreciate the assistance of Dr. Kenneth B. Young, Dana Hees, and Victoria-Sue Lombari for their proofreading and editing. Finally, we wish to express our appreciation to Sylvia Porter and Darlene Riese for typing the final manuscript.

Gail L. Cramer
Eric J. Wailes
University of Arkansas

1

Introduction

Donald W. Larson

Grain Marketing

The grain industry is an important part of the U.S. economy. Each year billions of bushels of grain are produced and marketed to domestic and foreign destinations. Millions of dollars are spent on the farm inputs needed to produce and market this grain, and hundreds of thousands of people are employed in the production and marketing of grain. The economic health of many farms, agribusinesses, rural communities, and consumers depends upon the efficiency, prosperity, and vitality of the grain industry. The purpose of this book is to provide a comprehensive understanding of the U.S. grain-marketing system so that students and other readers with interest in marketing may use this information to achieve a more efficient production and marketing system for U.S. grains. Also, the grain marketing systems in Canada and the European Community are described and analyzed because they have a significant impact on world grain markets.

In this text, grain marketing is defined as the performance of all business activities that coordinate the flow of goods and services from grain producers to consumers and users. Grain marketers usually classify grains into three broad categories (food, feed, and oilseeds) based upon their end use.[1] Food grains are those used for direct human consumption and generally include wheat and rice. Feed grains, or coarse grains, are used primarily for livestock feed and include corn, barley, sorghum, oats, and rye. Oilseeds, including crops such as soybeans, sunflowers, rapeseed (canola), and flax, are grown for their oil and protein meal content. The categories of food, feed, and oilseeds must be considered as general categories because some of the commodities, such as wheat, may be used primarily as food grains but at times may be also used as feed grains. Some grains, such as corn, have industrial uses. New domestic uses for corn, such as ethanol as an octane enhancer and corn

1

sweeteners, grew rapidly in the 1980s. Barley is used primarily as a feed grain but is also used in the brewing industry as malting barley.

During the 1980s, the U.S. grain production and marketing system was affected by a number of issues. These issues include the loss of export markets for U.S. grain, which may be caused by a number of factors such as the strong value of the U.S. dollar, protectionist trade policies in importing countries, increased self-sufficiency in food production among importing countries, and large foreign debt and high interest rates among developing countries. In addition, grain quality problems, grain transportation problems, increased price risk, costly government farm programs, and high U.S. support prices and grain embargoes adversely affected grain exports. In domestic markets, demand for grains has slowed because of increased consumer preference for white meats (production of white meat requires less grain per unit of meat) rather than red meats.

U.S. agricultural exports surpassed $42 billion in the 1980-1981 fiscal year, then declined steadily to less than $30 billion in the 1985-1986 fiscal year, and subsequently recovered to nearly $40 billion in 1990-1991. Grains, feeds, oilseeds, and oilseed products accounted for about two-thirds of U.S. agricultural exports in the early 1980s, but this share has decreased to about 50 percent in the early 1990s. Sharply lower prices and reduced export quantities have contributed to the overall decline in agricultural exports and the reduced share for grain products. In addition, subsidies on exports among competing producing countries to capture and recapture foreign markets have resulted in large government expenses in Europe and the United States. The United States and the European Economic Community (EC) compete forcefully in a near trade war on food products by trying to see which one will pay the bigger subsidy to sell farm products to a particular foreign market. Highly subsidized wheat sales to Egypt from the United States and the EC in the past illustrate the intensity of this food trade conflict.

The poor quality of U.S. grain has frequently been mentioned as an issue contributing to the loss of export markets. Importers have complained loudly about the poor quality of U.S. grain. Problems of short weighting in export shipments and grain grades below that specified in export contracts have been cited in recent years. Changes in the inspection service for export grain as well as grain grades and standards have been made in part to solve these problems.

High interest rates were a burden for grain producers and grain handlers throughout most of the 1980s. High interest rates in combination with low farm prices and declining land values contributed to the most severe farm financial crisis since the Great Depression. The large number of farm and grain elevator bankruptcies caused widespread concern throughout the U.S. economy. These bankruptcies as well as changes in the transport system led to a major restructuring of the grain production and marketing system. A much smaller number of larger, more specialized firms has emerged from this restructuring.

The large foreign debt of developing countries such as Brazil, Mexico, and Venezuela as well as most of the countries in Eastern Europe and the former Soviet Union has made it more difficult for them to import grain products because the foreign exchange to pay for the imports is not available. Much of the foreign exchange earned by these countries is needed to pay the interest and principal on their foreign debt, leaving little money to import grain products.

The 1981 Farm Bill with its high support prices is frequently mentioned as a factor contributing to the overproduction of grains and the resulting low farm prices and high inventories of these commodities. In addition, these high support prices may be an important reason for the lack of competitiveness of U.S. grains in export markets. One objective of the 1990 Farm Bill was to restore the competitive position of the United States in world markets by lowering farm price supports.

U.S. Production

Even when considering all the above problems, there are many reasons for one to be optimistic about the future of the grain production and marketing system in the United States. First, the United States has an abundant supply of highly productive cropland that responds well to commercial fertilizers and other capital inputs. Although land in farms has declined steadily since 1970 because of industrial and urban development and highway expansion, the area harvested for grains and soybeans has expanded in this same period (Table 1.1). Over 220 million acres (89 million hectares [ha]) of grains and soybeans were harvested annually in the 1980s compared to about 200 million acres (81 million ha) harvested in the early 1970s.

The rapid expansion of production during the 1970s was due to very favorable prices and profitability because of strong world markets for grain products. The area harvested for these crops, however, declined in the late 1980s because of low farm prices, government set-aside programs, and large crop inventories. Corn-for-grain acreage normally exceeds that for wheat and soybeans; however, soybean acreage has expanded rapidly since 1970 and has surpassed wheat and/or corn at various times (Table 1.2).

A second reason for optimism is the favorable climate for crop production. The temperate climate of the Northern Hemisphere is conducive

TABLE 1.1   U.S. Land in Farms and Acres Harvested: Feed and Food Grains, 1970-1990 (in millions of acres)

	Land in Farms[a]	Feed Grains	Soybeans	Food Grains	Total Grains and Soybeans
1970	1,102.7	99.2	42.2	46.8	188.2
1971	1,097.3	106.0	42.7	51.5	200.3
1972	1,093.0	93.7	45.6	50.1	189.6
1973	1,089.5	101.9	55.6	57.2	214.8
1974	1,087.7	99.7	51.3	68.6	219.7
1975	1,059.4	104.6	53.6	73.0	231.3
1976	1,054.1	106.2	49.4	74.1	229.7
1977	1,047.7	108.6	57.8	69.6	236.0
1978	1,044.7	105.7	63.6	60.3	229.7
1979	1,042.0	102.5	70.3	66.1	239.0
1980	1,038.8	101.3	69.9	75.0	246.4
1981	1,034.1	106.6	67.5	85.1	259.3
1982	1,027.7	106.1	70.8	81.8	258.8
1983	1,023.4	80.2	63.7	64.4	208.4
1984	1,017.8	106.6	67.7	70.7	245.0
1985	1,012.0	111.7	63.1	67.9	242.7
1986	1,005.3	101.5	60.4	63.7	225.6
1987	998.9	86.8	58.1	58.9	204.0
1988	994.5	80.4	58.8	56.6	195.9
1989	991.1	91.1	59.5	65.3	215.9
1990	987.7	89.4	56.5	72.5	218.5

[a]Redefinition of "farms" in 1975 reduced number of farms by about 9 percent. Figures before and after 1975, therefore, are not strictly comparable.

SOURCES: Land in Farms: National Agricultural Statistics Service series (1970-1974) *Agricultural Statistics 1976*; (1975-1986) *Agricultural Statistics 1986*; (1990) *Agricultural Statistics 1990*.

Feed Grains and Food Grains: (1970) *Agricultural Statistics 1980*; (1971-1985) *Agricultural Statistics 1986*; (1986) calculated from *Crop Production 1986 Summary*, p. A4; (1980-1989) *Agricultural Statistics 1990*; (1990) *Crop Production 1990 Annual Summary*, p. A4.

Soybeans: (1970) *Agricultural Statistics 1990*; (1971-1985) *Agricultural Statistics 1986*; (1986) *Crop Production 1986 Summary*, p. A4; (1980-1988) *Agricultural Statistics 1990*; (1989-1990) *Crop Production 1990 Annual Summary*, p. A4.

TABLE 1.2   U.S. Acreages Harvested and Production of Corn for Grain, Wheat, and Soybeans, 1971-1990 (in millions)

	Corn for grain		Wheat		Soybeans	
	Acres	Bushels	Acres	Bushels	Acres	Bushels
1970	57.3	4,152	43.5	1,352	42.2	1,127
1971	64.1	5,646	47.6	1,619	42.7	1,176
1972	57.5	5,580	47.2	1,546	45.6	1,271
1973	62.1	5,671	53.8	1,711	55.6	1,548
1974	65.4	4,701	65.6	1,782	51.3	1,216
1975	67.5	5,829	69.6	2,122	53.5	1,547
1976	71.3	6,266	70.7	2,142	49.3	1,288
1977	71.6	6,505	66.6	2,046	57.8	1,767
1978	71.9	7,268	56.4	1,776	63.6	1,869
1979	72.4	7,928	62.4	2,134	70.3	2,261
1980	72.9	6,639	71.1	2,381	67.8	1,798
1981	74.5	8,119	80.6	2,785	66.1	1,989
1982	72.7	8,235	77.9	2,765	69.4	2,190
1983	51.4	4,175	61.3	2,420	62.5	1,636
1984	71.9	7,674	66.9	2,595	66.1	1,861
1985	75.2	8,877	64.7	2,425	61.5	2,099
1986	69.2	8,250	60.7	2,092	58.3	1,940
1987	59.2	7,072	56.0	2,107	57.0	1,923
1988	58.2	4,929	53.2	1,812	57.4	1,549
1989	64.7	7,525	62.2	2,037	59.5	1,924
1990	67.0	7,933	69.4	2,739	56.5	1,922

SOURCES: (1970-1976) *Crop Production 1979 Annual Summary*; (1977-1986) *Crop Production 1986 Annual Summary*; (1986-1987) *Crop Production 1988 Annual Summary*, pp. A4-A5; (1988-1990) *Crop Production 1990 Annual Summary*, pp. A4-A5.

to high crop yield.   The growing season is typically long enough to support complete development of the plants, and the amount and timing of the annual rainfall fits the needs of the growing and maturing plant in terms of high yields and quality of product. In areas where rainfall is not adequate, crops requiring less rainfall, such as sorghum, wheat, and barley, are widely grown.   In addition, many farms have irrigation systems to supplement the natural rainfall in these drier areas.

A third reason for optimism is the biotechnology of agriculture. Research and development of new technology have provided U.S. farmers with increased output for many years, and this system has been a model for agricultural development in most of the world. The U.S. Department of Agriculture (USDA), land grant colleges and universities, and private firms have developed and disseminated a wide range of technologies for grain farmers. A wide range of hybrid and improved seeds, chemical fertilizers, pest management, and farm machinery have contributed to larger yields per acre as well as output per unit of labor (Table 1.3). These technological advances result from a strong

TABLE 1.3   Average Yield for U.S. Grains, 1970-1990 (in bushels per acre)

	Corn	Wheat	Soybeans
1970	72.4	31.0	26.7
1971	88.1	33.9	27.5
1972	97.1	32.7	27.8
1973	91.2	31.7	27.7
1974	71.4	27.4	23.2
1975	86.2	30.7	28.8
1976	87.4	30.3	25.6
1977	90.8	30.7	30.6
1978	101.0	31.4	29.4
1979	109.5	34.2	32.1
1980	91.0	33.5	26.5
1981	108.9	34.5	30.1
1982	113.2	35.5	31.5
1983	81.1	39.4	26.2
1984	106.7	38.8	28.1
1985	118.0	37.5	34.1
1986	119.3	34.4	33.3
1987	119.4	37.7	33.7
1988	84.6	34.1	27.0
1989	116.3	32.7	32.3
1990	118.5	39.5	34.0

SOURCES: (1970-1976) *Crop Production 1976 Annual Summary*; (1977-1986) *Crop Production 1986 Summary*; (1986-1987) *Crop Production 1988 Annual Summary*, p. A5; (1988-1990) *Crop Production 1990 Annual Summary*, p. A5.

commitment to and strong support for applied and basic research in these institutions. This commitment and support must continue in order to maintain productivity in agriculture and will likely lead to further technological advances in the future.

The fourth reason for optimism is the high level of technical and managerial skill of U.S. farmers. Their ability to adopt new technology and combine all the resources of production in the most cost-effective manner results in an efficient grain production system. For this reason, the United States has been able to maintain a leadership position in the production of basic grains for world markets.

The fifth reason is that the United States has the infrastructure and institutions to respond to increased international grain requirements. The U.S. marketing system has been responsive to world grain needs so that over the past half-century no other country in the world can match the record of the United States as a consistently reliable supplier of grain and grain products at reasonable prices.

A unique feature of the U.S. grain-marketing system not found in many other grain-producing countries is the high degree of competition and freedom from government intervention. For example, farmers in other countries are often forced to sell their crops to the government at a regulated price. U.S. farmers or any other subsequent owners of grain are free to sell it to whomever they wish and whenever they wish for the best price they can get. Any person or firm with the necessary capital has free entry into any phase of the grain business or may be involved in several phases at the same time, such as grain production, elevator storage, and processing. This feature of the U.S. grain industry also contributes to the efficiency and flexibility with which the system adjusts to changing market conditions.

World Production

Perhaps more than any other product, grains and oilseeds must compete in a global marketplace. Competitiveness in world markets is a major issue. World production and trade patterns have changed dramatically since the energy and drought crises of the early 1970s. Countries such as Argentina and Brazil emerged as major producers and exporters of soybeans in the 1970s. The EC countries changed from net importers to net exporters of grains. Rapid rates of economic growth and increasing demand for animal products from consumers with higher incomes led to rapid growth in demand for grain imports among newly industrialized countries such as Korea and Taiwan. World production and trade patterns will continue to adjust to these changing markets. The leading countries in production of wheat, corn, coarse grains, and

TABLE 1.4 Leading Grain-producing Countries of the World, 1991-1992

	Wheat			Corn	
	Million[a] Metric Tons	% of World		Million[a] Metric Tons	% of World
China	96.0	17.6	United States	189.9	39.3
EC-12	90.4	16.6	China	95.0	19.7
Former USSR	78.0	14.3	E. Europe	30.9	6.4
India	54.0	9.9	EC-12	26.6	5.5
United States	53.9	9.9	Brazil	26.0	5.4
E. Europe	39.2	7.2	Mexico	14.5	3.0
Canada	32.8	6.0	Former USSR	11.0	2.3
Australia	10.0	1.8	S. Africa	8.0	1.7
Argentina	8.5	1.6	Argentina	7.6	1.6
Others	82.9	15.1	Thailand	3.7	0.8
			Others	70.1	14.3
World	545.6	100.0	World	483.4	100.0

	Coarse Grains			Soybeans	
	Million[a] Metric Tons	% of World		Million[a] Metric Tons	% of World
United States	218.5	27.1	United States	54.04	51.3
China	110.6	13.7	Brazil	17.50	16.6
EC-12	88.8	11.0	Argentina	10.50	10.0
Former USSR	85.5	10.6	China	10.10	9.6
E. Europe	61.4	7.6	EC-12	1.68	1.6
Canada	22.7	2.8	Paraguay	1.60	1.5
Argentina	11.0	1.4	Others	9.85	9.4
S. Africa	8.6	1.1			
Australia	6.4	0.8			
Thailand	4.0	0.5			
Others	188.7	23.4			
World	806.0	100.0	World	105.26	100.0

[a]One metric ton of wheat or soybeans contains 36.7437 bushels, and one metric ton of corn contains 39.368 bushels.

SOURCES: Wheat, Corn, and Coarse Grains: *World Grain Situation and Outlook* (USDA, Foreign Agricultural Service), January 1992; and Soybeans: *World Oilseed Situation and Outlook* (USDA, Foreign Agricultural Service), January 1992.

soybeans are shown in Table 1.4 for 1991-1992. The United States is the largest producer of corn, coarse grains, and soybeans, and China is the largest producer of wheat.

The Competitive Market

The competitive market of economic theory assumes a large number of buyers and sellers; freedom of entry and exit; knowledge of demand, supply, and prices; and rational behavior among participants. In this kind of a market, a uniform price will prevail among buyers and sellers of a particular grain that is traded at a particular place and time. These three dimensions of the competitive market concept—the place or location, the time period, and the form of a commodity—are an appropriate economic framework in which to analyze grain markets. The remainder of this chapter will describe in more detail these three dimensions of the competitive market concept.[2] These three dimensions may be considered as the utilities of place, time, and form.

Place Utility: Production and Consumption Regions

As can be seen from Figures 1.1, 1.2, and 1.3, U.S. grain production is concentrated in the Great Plains and Cornbelt states. The primary wheat

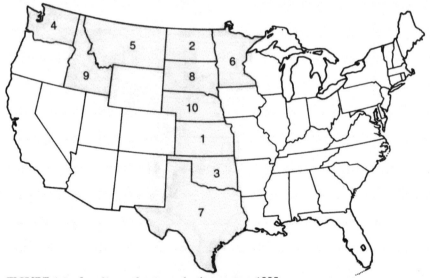

FIGURE 1.1   Leading wheat-producing states, 1990.

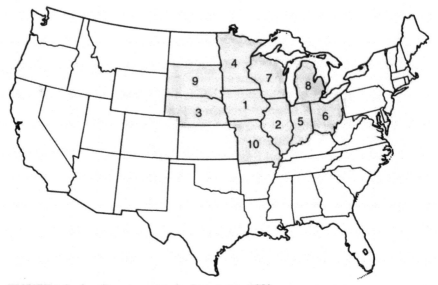

FIGURE 1.2   Leading corn-producing states, 1990.

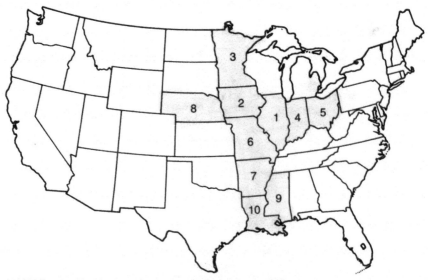

FIGURE 1.3   Leading soybean-producing states, 1990.

production regions are the central and northern Great Plains and Pacific Northwest (Figure 1.1). Corn and soybean production is concentrated largely in the Cornbelt states, which include Indiana, Illinois, Iowa, Minnesota, Missouri, Nebraska, and Ohio (Figures 1.2 and 1.3). During the 1970s, some of the states in the southeastern United States, such as Arkansas, Mississippi, Louisiana, and Tennessee, also expanded rapidly into soybean production because of attractive prices on world markets. Because of this expansion in the South, soybean production is less concentrated in the Cornbelt than is corn production. Soybean expansion in the Southeast slowed dramatically in the 1980s, and some of these states may reduce soybean acreage in the 1990s.

The location of wheat, corn, and soybean production in the Cornbelt and Great Plains is quite distant from the major domestic population centers on the East Coast and West Coast and from the export ports along the Great Lakes, East Coast, Gulf Coast, and West Coast. Because of the substantial distances between the location of production and the location of domestic markets and export sites, the marketing system must create important place utility to transfer these products from the production areas to the domestic and export markets.

Within the concept of a competitive market, one would expect that prices at market destinations compared to prices at the sources of production should not differ by more than the costs of transferring the commodity from the point of production to the market destination. If the price differential exceeds transfer costs, entrepreneurs will ship more products from the point of production to the market destination. This competitive action will increase prices at the point of origin and reduce prices at the market destination, leading to an eventual equilibrium position in which the difference in market prices will equal the cost of transferring the product from origin to destination. In a similar manner, if the price differential is less than the cost of transfer, entrepreneurs will reduce the movement of commodities from a particular origin to a particular destination, leading to pressures that reduce prices at the origin and increase prices at the destination. This process will occur until the price differential again is equal to the cost of transfer.

These price differences over space are illustrated in Figure 1.4, which shows the average price received by farmers for corn by state in 1990. Prices tend to be lowest in the major producing states of the Cornbelt where excess supply is greatest, and prices tend to be highest at major processing and utilization centers and export ports where the deficit is greatest. These deficit areas tend to be in the East, West, and South. These price differences are needed to pay for the costs of transporting grain from the production areas to the market destinations.

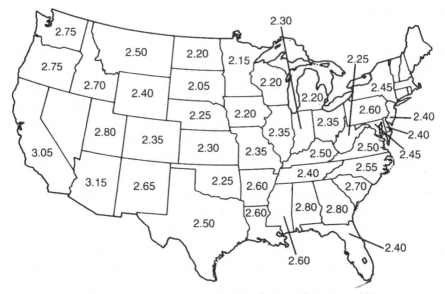

FIGURE 1.4   State average prices received by farmers for corn, 1990 crop.

Time Utility

The marketing system creates time utility by storing products from surplus supply periods such as the harvest period to supply future deficit periods when that production is needed. This concept is particularly important in grain marketing because production of grain is concentrated in a relatively short period of time (the harvest season) while the demand for grain products occurs at a fairly even rate throughout a calendar year. The timely provision of goods when needed is a very important function of marketing. Because of this disparity between production and consumption, prices at different times must differ by the amount needed to reward entrepreneurs who are willing to assume the risks and costs of storage to make the product available at a later period when it is needed.

In a competitive market equilibrium the price for a commodity such as wheat at some future time minus the price for the commodity today must equal the carrying charges (storage, interest, and insurance). If the price at some future date exceeds today's price by more than the carrying charges, entrepreneurs will have the incentive to hold the commodity today for sale at a future period. This increase in the amount of the commodity held for later sale will put upward pressure on today's cash prices and downward pressure on prices in a future period. This process will continue until prices in the future period minus today's prices are equal to the carrying charges. In a similar manner, if the price at some future period minus today's price is less than the carrying charge, entrepreneurs will not be willing to hold the commodity in storage for later sale. This difference will put downward pressure on today's price and upward pressure on the price for some future period. This process will continue until the price differential once again equals the cost of holding the commodity from today's date to the particular future period. The more distant the future period, the higher the price should be to cover these carrying charges. That is, the price of a commodity four months into the future will be higher than today's price by the carrying charges for four months, and the price of a commodity eight months into the future must be higher than today's price by the cost of holding the commodity for eight months.

In the real world, changes in interest rates, insurance, and storage charges are the main factors that affect the cost of holding inventories, and increases in any of these factors would then require an increase in the differential between today's price and the price at some future time period. Likewise, a reduction in any of these factors will lead to a decrease in the differential between today's price and some future price.

Futures markets represent an important tool that is used by grain traders and grain producers to assess the amount that the marketplace is willing to pay entrepreneurs to hold the commodity from one period to another. Concepts of futures markets, hedging, and other marketing alternatives will be discussed in considerable detail in later chapters.

In the U.S. grain-marketing system, farmers and grain firms provide a major part of this storage function. As can be seen in Table 1.5, farmers and grain firms both can store large amounts of grain products. Storage capacity and the amount of stocks held indicate that farmers generally store as much as grain firms. Since 1987, however, firms are holding larger amounts in storage and for longer periods.

Form Utility

The third dimension of the competitive market concept is the product form. The form of the commodity, as indicated by the raw product as compared to a processed product or as indicated by different qualities of the same product, must be consistent with the cost of transformation. In the competitive market concept, the price of the processed product minus the price of the raw product adjusted for any physical losses in trans-formation must equal the cost of processing. For example, the price of wheat flour, including the value of the by-products minus the price of wheat, must equal the cost of wheat milling in a competitive market situation. In a similar way, the price of the oil and meal from soybeans

Donald W. Larson

TABLE 1.5   U.S. Off-Farm Grain Storage Capacity and On-Farm Grain Stocks, 1970-1990 (in billions of bushels)

	Rated Off-Farm Storage Capacity	On-Farm Stocks As of December 1[a],[b]
1970	5.6	5.6
1971	5.7	5.8
1972	5.7	5.6
1973	5.8	5.7
1974	5.9	5.2
1975	5.9	4.1
1976	6.1	5.1
1977	6.3	5.1
1978	6.6	6.3
1979	7.0	7.0
1980	7.1	7.5
1981	7.2	6.3
1982	7.3	7.7
1983	8.0	9.4
1984	8.1	5.4
1985	8.1	6.8
1986	8.3	8.5
1987	9.6	8.2
1988	9.6	5.9
1989	9.4	6.4
1990	9.1	6.9

[a]Includes wheat, rye, corn, oats, barley, sorghum, and soybeans.

[b]On-farm storage capacity was first reported in 1978. These data are presented as a proxy for minimum on-farm storage capacity.

SOURCE: U.S. Department of Agriculture, *Grain Stocks*, various issues.

minus the price of the raw soybeans must equal the cost of soybean processing (Table 1.6).

The price differential between various qualities of the same product must also equal the cost of transforming the product from one grade or quality to another in this competitive market situation. For example, the price differential between No. 2 yellow corn and No. 3 yellow corn must be equal to the cost of changing from a No. 3 yellow corn to a No. 2

TABLE 1.6   Value of Processed Soybean Meal and Oil Compared to the Raw Soybean Value and the Costs of Processing Per Bushel, 1990-1991[a]

Item	Amount
Soybean Oil Value	
11.1 lbs. x 20 cents/lb.	$2.22
Soybean Meal Value	
47.5 lbs. x $180.00 per ton/2000	4.28
Total Value of Oil and Meal	$6.50
Less processing cost	–.35
Purchase Price of Raw Soybeans	$6.15

[a]The extraction rate for oil and meal can vary for different production regions and from year to year.

yellow corn.   If this differential is small, neither producers nor merchandisers will have the incentive to produce and sell No. 2 yellow corn as compared to corn of some lower quality.

As illustrated above, the competitive market provides an appropriate framework in which to analyze the grain-marketing system.   This competitive market structure results in a marketing system that provides the right commodity at the right place at the right time and guarantees that the market will provide the incentives for the creation of the time, place, and form utilities demanded by consumers and that price differentials will be adequate to cover the cost of providing these utilities.

Summary

Grain marketing is defined as the performance of all business activities that coordinate the flow of goods and services from grain producers to consumers and users.   Grains are classified as food, feed, or oilseeds, depending on their end use.   The U.S. grain production and marketing system was affected by a number of important issues in the 1980s that caused a decline in exports; however, grain exports in value terms have increased in 1990-1991 to near the peak level of 1980-1981.   Grain exports currently constitute about half of the total agricultural exports.   The United States has continued to be engaged in a near trade war with the

EC and has provided export subsidies on grain to maintain its competitive position in the world market.

There has been a major restructuring of the U.S. grain production and marketing system since the early 1980s because of a farm financial crisis, bankruptcies of both farmers and grain elevators, and changes in the grain transportation system. A much smaller number of larger, more specialized firms has emerged from this restructuring. Farm policy was also changed in the 1985 and 1990 Farm Bills to improve the competitiveness of the United States in world markets by lowering farm price supports.

The future of grain production and marketing in the United States looks promising because of the abundant supply of highly productive cropland, the favorable climate for crop production, the continued technological advances in agriculture, and the skills of U.S. farmers. This country continues to be the largest producer of corn, coarse grains, and soybeans in the world market; China is the largest wheat producer.

The competitive market is an appropriate economic model to analyze grain markets. The concept of place utility in theory explains the differences in price needed to pay the costs of transferring grain from production areas to market destinations. The concept of time utility explains the differences in prices needed to pay carrying charges for holding grain from harvest time to the time when it is required by the final user. The concept of form utility explains the differences in prices needed to pay for processing grain from the raw form to a finished form, such as wheat to flour. Because grain marketing has a competitive marketing structure in the United States, the market provides adequate incentives to guarantee that the above important utility functions are performed and that the price differentials are adequate to cover the cost.

Notes

1. Oilseeds are included under grains because they are handled by the same facilities and by the same firms.

2. The U.S. marketing system is not a perfectly competitive market as it does not meet all the requirements of economic theory; however, the market structure is competitive in terms of most of the conditions specified in theory.

Selected References

U.S. Department of Agriculture, *Agricultural Statistics*, various years.
U.S. Department of Agriculture, *Crop Production*, various issues.
U.S. Department of Agriculture, *Grain Stocks*, various issues.

2

Grain Supply and Utilization

Walter G. Heid, Jr.

Grain distribution in the United States has evolved from a simple system, in which grain was produced and fed on the farm, into a complex marketing system. A large percentage of farm production is now marketed off farms. Some of this grain is processed into food and feed for domestic uses, while other grain is exported as whole grain, primarily in overseas markets. With the growth in importance of foreign demand, the U.S. market for corn, soybeans, and wheat has assumed a worldwide dimension. An understanding of this emerging grain supply and utilization situation as it relates to traditional marketing institutions and practices should provide a better understanding of the total U.S. grain-marketing system.

The specific numbers shown in this chapter are not as important to the student of marketing for their value per se as they are to show how dynamic the U.S. grain industry must be to keep abreast of change. To be competitive in such a dynamic market requires a good grasp of marketing tools; sound intelligence; quick, accurate decisions; and a certain amount of good luck.

The U.S. marketing system seems to be in a constant state of adjustment. U.S. exports of a specific grain may increase or decrease as much as 50 percent or more from one year to the next because of the great variability of foreign demand. Such fluctuations in demand are largely due to the vagaries of weather and to the import policies of a number of countries throughout the world. The servicing of export markets places much more stress on the U.S. marketing system than when the system was primarily serving the domestic market. Both the U.S. production policy and the marketing institutions have attempted to adapt to this expanded market. Nevertheless, the marketing system has experienced difficulties in responding to the variability of foreign demand.

Foreign demand for grain and oilseeds has grown tremendously since World War II. In 1950, U.S. exports were less than 15 percent of production of wheat, soybeans, and corn. Forty years later, 45 percent of the total U.S. output of those crops was distributed overseas.

With the growth in foreign markets, estimating foreign demand for U.S. grain has become a complex task; demand cannot be forecast on the basis of traditionally predetermined variables. Total demand for these commodities is now determined by trade agreements as much as by disposable income; by tariffs as much as by population; by export enhancement programs (EEPs) as much as by a country's ability to pay cash; and by diplomatic relations as much as by preference. Occasionally, poor weather conditions overseas cause temporary spurts in the demand for U.S. grain. This situation can be either positive or negative. Also, unfavorable economic conditions and the trend toward food self-sufficiency tend to have a negative effect on the amount of U.S. grain exports.

Supply and Distribution

In the marketing process, buyers and sellers are brought together for the purpose of negotiating transactions, with price as the confirming mechanism. At times when the quantity supplied greatly exceeds the quantity demanded, various government programs may complement cash markets to move products. For example, wheat may be distributed through cash sales, short-term credit at commercial rates, long-term concessional credit, or Public Law (PL) 480 donations.

Supply includes annual production and carry-over from previous crops, plus imports, if any. Thus, the term *supply* refers to the quantity of an economic good available per unit of time for distribution in the market at alternative prices, all other things being equal.

· *Demand* is defined as the amounts of a product that buyers are willing to purchase at alternative prices per unit of time, all other things being equal.

According to economic theory, demand indicates the differing amounts of a product that will be purchased at differing prices. However, in a world filled with hungry people this statement does not adequately describe the basis for distributing the supplies of grains and oilseeds. Even a more precise term, *effective demand*, does not totally explain what takes place in the market. Effective demand is the desire of the consumer, backed with purchasing power. Effective demand only explains the quantity that people are able to acquire by spending a proportion of their disposable income. Although most commodities are purchased at a price, some are distributed as a donation, on credit, or through other arrangements. Thus, true food demand includes both what some consumers are able to purchase and what others obtain through various forms of charity.

The people of Egypt, for example, both need and desire many things—more wheat, better clothing, better homes, and so on. However, the Egyptian demand for all these goods is very limited because they do not have the purchasing power to satisfy their needs, let alone their wants. Therefore, part of the Egyptian needs are met through receipts of wheat and flour from the United States under an export enhancement program. Many other countries receive similar aid from the United States as well as economic aid from other developed countries.

Supply Factors

Many factors cause supply to increase and decrease. Some factors, such as production decisions or weather, may affect total supply. Other factors, such as embargoes, may affect the available supply for general export or to a specific country.

With a large percentage of U.S. grain production finding its way into foreign markets, the world supply has become an important factor affecting U.S. prices. Knowledge of world supplies is often less than perfect, making marketing and government policy decisions tenuous.

Grain supply is affected by both nature and man. It is affected by government action and by the independent decisions of several million farmers. The supply is also affected by the buying and selling decisions of those who take title to it in the marketplace.

So many factors influence supply that its control has heretofore eluded the policymakers' best efforts to maintain a balance between supply and demand. In the United States, as well as worldwide, grain supplies historically have oscillated between periods of shortages and periods of abundance.

Nature. Perhaps weather more than any other single factor explains the problem of year-to-year variation in total output. Weather may affect planting intentions, yields, and harvest. Droughts, hail, and floods often claim large acreages of crops. Conversely, ideal weather conditions often contribute to bumper crops. In addition to weather, insects and diseases may affect production. In 1970, for example, the U.S. corn supply was hit by the southern leaf blight. Yields were reduced by 13.5 bushels per acre, or 17 percent.

The larger the crop yield, the larger the variation in total production from year to year. Usage of commercial fertilizer and newly improved seeds have greatly increased yields, especially of corn and wheat in the

United States as well as rice in other parts of the world. Weather, crop diseases, and insects have no respect for large yields. The history of U.S. wheat production serves to illustrate that losses in high-yielding fields are often much greater than losses in low-yielding fields. Yields increased from less than 15 bushels per acre in the early 1900s to nearly 40 bushels by the late 1980s. U.S. wheat yields deviated from the trend line by less than 2 bushels 88 percent of the time from 1900 to 1932. However, from 1933 to 1977, actual yields varied 18 percent of the time from the expected by 3 bushels or more. If one assumes 70 million harvested acres, this degree of variation results in a 210-million-bushel production variation in about one year out of five. Wheat yields continued to increase in the late 1980s, and with higher yields forcast, even greater variations in production can be expected in the years ahead.

Acreage Control and Subsidies. Government programs have affected wheat and feed-grain production levels since the Great Depression. In 1932, realized net farm income dropped to less than one-third of the 1929 net, and farm prices fell more than one-half. In an effort to correct farm economic problems associated with the depression, the Agricultural Adjustment Act (AAA) of 1933 was passed. Among other provisions, the act authorized the secretary of agriculture to secure voluntary acreage reductions for basic crops, including corn and wheat. Except during World War II, the Korean War, and in much of the 1970s, most farm legislation has been designed to limit grain supplies.

The AAA of 1933 and successive farm legislation that introduced the term *parity* sought to bolster farm prices as well as to control supplies. Ironically, while the provisions that strengthened farm grain prices actually encouraged larger supplies, other provisions of the farm programs were designed to reduce acreages.

Except during World War II, and the 1970s after the large grain sale to the former Soviet Union, overproduction has been a major farm problem. Beginning in 1955, the United States commenced a long period of national wheat-acreage allotments. The 55-million-acre wheat allotment in 1955 reduced wheat acreage 7 million acres below the 1954 level.

The Agricultural Act of 1956 established a conservation reserve and an acreage reserve system. The purpose, again, was to adjust supply. Under the conservation reserve, known as the Soil Bank, farmers designated certain cropland for the reserve and put the land into a conservation use—usually some type of grass. Many farmers placed their entire farms in the Soil Bank. By July 15, 1960, there were 28.6 million acres of cropland under contracts for a maximum of 10 years.

Under the acreage allotment program, farmers reduced the acreage of basic crops to an established allotment, conserving their remaining acres and receiving payment for the diversion.

Except for a brief period in the mid-1970s when "fence-to-fence" production was encouraged, farm programs throughout the 1960s and 1970s were designed to reduce corn and wheat supplies and to support prices. By contrast, the government has never attempted to control the supply of soybeans to maintain a reasonable balance between supply and demand. By the late 1980s and early 1990s, the government farm program was characterized by "runaway public costs"; policymakers were exploring ways to invoke mandatory acreage controls.

Embargoes. A grain embargo may effectively limit export supply for various reasons. In the 1970s, the U.S. government imposed several embargoes. The first, President Richard Nixon's 1973 embargo on exports of soybeans and related products, attempted to check U.S. food price inflation. President Gerald Ford imposed the second on Poland in 1975 to prevent transshipment of grain (reselling imported grain) to the Soviet Union. The third embargo, imposed by President Jimmy Carter in 1980, sharply reduced the sale of corn and wheat to the Soviet Union in protest of that country's involvement in Afghanistan.

Embargoes impose a financial hardship on farmers, who are the first to feel the depressing effect of the sudden loss of a market. Also affected are grain merchants and suppliers of farm inputs.

The effects of embargoes are felt for a long time, especially in the country imposing the embargo. An embargo creates an immediate shock and then shock waves much like an earthquake. The effects of the 1980 embargo on livestock production in the Soviet Union and on U.S. farm input purchases came months after the initial shock.

Within the international grain-trading system, leakages occur in two ways. First, transshipment of U.S. grain exports to the embargoed nation can occur through third countries. Second, the actions of multinational trading firms cannot be completely controlled. Perhaps an even greater reason for the long-run ineffectiveness of embargoes is other countries who have no quarrel with the embargoed nation and also wish to expand their own export markets.

Embargoes are trade sanctions. When trade policy is politicized, trade relationships are restructured. The 1973 soybean embargo, for example, opened the door for Brazil to enter as an alternative supplier of soybean meal and oil. As the result of this embargo, the Brazilian government, which was just in the process of making heavy investments in its soybean industry, took advantage of the situation. Less than a decade later, with its soybean industry flourishing, Brazil had become a major exporter.

After the 1980 embargo, the United States not only lost much of the Soviet Union's important market, but other markets as well. The Soviet Union and other countries found satisfactory, dependable suppliers of grain elsewhere. As a result of this embargo, the United States not only lost part of its foreign market, but it also gained the image of an unreliable supplier. Overcoming such a reputation is, at best, slow when the world supply-demand situation signifies a buyer's market—one in which the quantity supplied far exceeds the quantity demanded.

Grain as a Weapon. Grain is increasingly being used as a political weapon. Embargoes are only one example. "Food for Crude" became a popular slogan in the 1970s when the United States was experiencing shortages of crude oil. Withholding grain supplies from developed countries and other major importing countries is, in some respects, a harmless war game. But withholding grain supplies from under-developed countries can have dire consequences. Limiting or denying food supplies to some underdeveloped nations, in some situations, can quickly result in starvation. Poor nations may have neither the capability nor the economic ability to move rapidly enough to locate and obtain supplies from alternative sources before starvation begins. Yet, it is becoming increasingly common for food suppliers to keep the political persuasion of developing countries linked to food aid needs.

Grain Reserves. Grain held in reserve for use during critical shortages in world production may reduce the effective supply. Grain designated for use under the Food for Peace program, also known as PL-480, is an example. If, by law, a quantity of grain is set aside for donation, it is, for all practical purposes, removed from the normal market, and the reserved grain is no longer responsive to the pricing system. Usually, however, such reserves account for only a token percentage of the total supply.

Bilateral Supply Agreements. Unlike politically motivated grain distribution, bilateral supply agreements are long-term commitments. World food insecurity produces bilateral agreements, opportunities to capture or lock in a long-term market. Proponents claim these long-term agreements introduce market stability, enhancing supply management at the production level. Opponents discount the importance of stability and suggest that these agreements take away the system's flexibility in responding to short-term needs. They contend that one bilateral agreement opens the way for others, and soon the situation arises where the governments control demand. Opponents of bilaterals also claim that these agreements erode confidence in our system of projecting a reasonable return on commercial investments, whether the proposed investment is for production inputs—fertilizers and agricultural chemicals—or market activities such as investments in storage and transportation.

During a seller's market in an 18-month period, 1980 to mid-1981, the United States announced 19 long-term agreements, nearly all for corn, soybeans, or wheat. These agreements, running three to five years, involved many tons of grain, and some of these agreements continue to be renegotiated as they expire or as major world events affect the negotiating countries.

As world population, livestock, and poultry numbers increase worldwide and the supply for grain and oilseeds tightens, more importing countries can be expected to request bilateral agreements, becoming steady or preferred customers of exporting countries. Countries that are considered "most favored nations" and those that can best afford grain will be first in line for available supplies. However, in times of plentiful grain supplies, bilateral agreements quickly find disfavor. Buyers become very price conscious, desiring to shop for the best buy. This comparison shopping, of course, makes supply management in major producing countries quite difficult. By the mid-1980s, the most-favored-nation program backfired on the United States somewhat as cash customers began asking why they did not receive the same buying advantages.

Planning Horizons. Grain supplies are also affected by decisions made by farmers and members of the grain trade. These decisions—crop production, allocation of supplies through alternative marketing channels, and distribution for specific uses—are based largely on the economic theory of resource allocation.

Time plays an important role in these decisions. Regardless of the planning horizon (short run versus long run), decisions made on the farm and in the marketplace affect both the volume of supply and its availability in terms of place, time, form, and price. Specifically, time-based decisions can affect supplies as follows:

1. In the short run, storage costs, the seller's need for cash, the price offered, and the market outlook may influence sellers to sell or hold their product.
2. In the intermediate period, farm production costs may change, thus causing, for example, a cutback in commercial fertilizer usage and, in turn, lower yields. Or in the marketplace, changes in grain price levels or competition may affect the volume of supply.
3. In the long run there may be changes in production facilities as well as changes in all of the factors mentioned in (1) and (2). For example, there may be new agricultural development or marginal land may be brought into production. Additional processing capacity may be added, or changes in technology may occur.

Distribution Factors

In the U.S. economy, the pertinent marketing question has always been "how much will be bought at a specified price?" and not "how much will be needed or desired?" However, as producers, policymakers, and social scientists begin to think of the world as one large market, they give more attention to humanitarian concerns. Especially since World War II, total demand for U.S. farm products has reflected greater concern with the economic well-being of people in foreign countries. Distribution programs such as the Marshall Plan, PL-480, Cooperative for American Relief Everywhere (CARE), and the Catholic Relief Services (CRS) are examples. U.S. involvement in relief-type programs has varied. Commodity Credit Corporation (CCC) credit and PL-480 programs accounted for 78 percent of total U.S. wheat distribution in foreign markets in 1964-1965, for 20 percent in the late 1970s, and for a larger percentage by the mid-1980s and into the 1990s. Over a recent 25-year period, 30 percent of the U.S. wheat exports were shipped under various government programs. More commodities tend to be distributed under relief programs when U.S. supplies are large than when they are short.

Just as the existence of charitable forms of distribution blurs the use of recognized economic theory to explain and predict utilization patterns, so have they clouded the role of the United States in feeding the world. Some countries view the United States as the grocery store of the world—the world's breadbasket. Although it is true that the United States, *physically*, has a production capacity far exceeding domestic needs, it does not *economically* have an unlimited capacity to feed the world. Without sufficient return to cover their production costs, farmers cannot endlessly produce. Similarly, the movement, processing, and storage of products at all stages of the marketing channels entail costs; and the commodities cannot be given away freely. The need for charitable types of distribution has persisted and will likely increase in the future. The marketing system, handling volumes of grain distributed under relief programs in addition to normal commercial activities, must be compensated just as farmers must be paid for their performance.

The debate concerning the morality of attempting to feed the world is in its infancy. In the quarter century from 1950 to 1975, the United States contributed $25 billion in global food aid. This charitable gesture met with a diversity of opinion. On one side, some said the United States should have given even more food aid to poor nations. On the other side, others complained that too much food was being given away. The debate centered around the U.S. taxpayers' burden, the effects on domestic and foreign food prices, a desire for higher farm prices, and the U.S. balance of payments versus humanitarian concerns. In essence, the effects of food aid on the U.S economy, not the hunger needs per se, proved the major focus of debate.

Opponents of food aid adopted the philosophy that efforts by rich nations to feed the world's poor are self-defeating because such efforts were believed to compound the problems of the Malthusian equation on food and population by keeping the poor alive only to spawn more people who, in turn, make increasing demands on the world's food supply. Based on this philosophy, termed *lifeboat ethics*, opponents of food aid suggest that rather than continually trying to save the world's poor from starvation, the United States and other food-exporting nations should tend to matters in their own "lifeboat" and let famine, floods, and pestilence keep the world's population under control. Other groups of would-be benefactors advocate contraception aid and education to decrease population growth and thereby starvation.

Another philosophy, the theory of triage, was suggested as a means of selectively distributing food. *Triage* refers to the World War I medical practice of dividing the wounded into three categories: (1) those who would survive without treatment, (2) those who would die even if given treatment, and (3) those who would survive with but die without treatment. Thus, when applied to feeding the hungry, the triage method would concentrate limited food resources on those who could truly benefit from them and ignore those who would likely die even with food. Proponents of food aid counter these theories by claiming that it is well within the power of humanity to produce and distribute enough food to prevent masses of people from starving, regardless of existing or projected circumstances.

Unfortunately, this food distribution debate has no simple solution. The rich cannot afford to give food away, and the poor cannot afford to pay. The debate will continue to surface whenever short food supplies or high prices cause widespread hunger or economic depression.

It is the task of those engaged in marketing to move products to the right place, at the right time, and in the right form. *Marketing*, in the traditional sense, is defined as *the performance of all business activities involved in the flow of goods and services from the initial producer to the final consumer.*[1] For marketing wheat, corn, and soybeans, this task means that crops harvested in a 2- to 3-month period must be marketed over a 12-month period. This task not only means matching a given supply with demand, but also involves storage, transportation and handling, processing, manufacturing, wholesaling and retailing as well as financing, risk-bearing, researching, and advertising—the ancillary services that accompany the marketing process. To accomplish these tasks, people in charge of marketing must know the factors affecting demand—population, incomes, prices, and quantities—because this information provides the basis for projecting future demand trends. It serves as the basis for production and marketing decisions and related agricultural policies.

U.S. Population. When explaining food utilization, population is by far the most important factor. The Bureau of the Census projects the U.S. population to be somewhere between 260 and 280 million by the year

TABLE 2.1 Per Capita Civilian Consumption of Wheat, Corn, and Soybean Products, United States, 1975 and 1980 to 1990

	Wheat			Corn							Soybeans		
	Total Consumed (million bushels)	Per Capita Consumption (pounds)		Total Consumed (million bushels)	Per Capita Consumption (pounds)						Total Crush (million bushels)	Per Capita Consumption (pounds)	
		Flour	Cereal		Flour & Meal	Hominy & Grits	Syrup	Sugar	Starch			Margarine	Shortening
1975	534	119	2.7	367	7.7	2.7	28.8	5.4	2.1		865	11.1	17.1
1980	610	117	2.7	504	7.4	2.8	35.2	3.8	2.7		1,020	11.4	18.2
1981	602	116	2.9	418	7.7	2.7	40.0	3.5	2.9		1,030	11.3	18.5
1982	616	117	3.0	462	8.0	2.9	44.7	3.5	2.9		1,108	11.1	18.6
1983	643	117	3.2	528	8.4	3.0	49.2	3.5	3.3		983	10.4	18.5
1984	651	119	3.4	584	9.4	3.1	55.4	3.5	3.5		1,030	10.5	21.3
1985	674	125	3.7	618	10.2	3.2	63.0	4.2	3.7		1,053	10.9	23.0
1986	712	126	3.9	650	11.9	3.3	63.9	4.2	4.2		1,179	11.5	22.2
1987	721	130	4.2	678	13.6	3.3	65.6	4.3	4.2		1,174	10.6	21.5
1988	726	130	4.3	691	14.0	3.4	67.4	4.2	4.4		1,058	10.4	21.6
1989	753	129	4.5	711	14.0	3.4	67.9	4.4	4.1		1,146	10.2	21.5
1990a	796	136	4.6	728	14.0	3.4	68.7	4.5	4.3		1,187	10.9	22.3

aPreliminary.

SOURCES: U.S. Department of Agriculture, Agricultural Statistics 1992, and data compiled by the Economic Research Service.

2000. Although the annual rate of increase has declined from earlier years, the total increase from 1980 to 1990 alone was about 10 percent. An assessment of the U.S. population structure shows several trends besides an increase in numbers that will affect future food consumption. These trends include:

1. Migration to warm climate states
2. Shift in racial and ethnic groups
3. Increasing urbanization
4. Increasing number of families with single head of house
5. Shift in age groups
6. Increasing number of meals eaten away from home.

As a consequence of these trends, consumers may not increase their intake of total food, but their tastes and preferences for specific foods do change over time. As shown in Table 2.1, per capita consumption of most grain and oilseed food products is increasing or at least is higher than it was in 1975.

Disposable Income Spent for Food. As disposable income increases, people tend to demand more expensive foods. In the United States, the consumer has gone two steps further: first, demanding prepared foods, and second, demanding the ultimate—restaurant-prepared foods. Away-from-home food expenditures changed little until the inflationary period of the middle to late 1970s. Increases in away-from-home food expenditures reflect a viable economy. A slowing economy would likely shift to greater home preparation of food. Poverty-level incomes encourage the use of more grain products and fewer livestock products.

In the United States, the wealthiest spend less than 5 percent of their disposable income for food, while the poorest spend well over 50 percent. Food expenditure data show that, on the average, U.S. consumers have spent 15 percent or less of their disposable income for food since 1970. By 1984, food expenditures as a percentage of disposable income had dropped to 12.4 percent (Table 2.2). By 1991, this expenditure stood at 11.6 percent. Of this amount, 37.5 percent was spent in away-from-home sites. According to at least one study, as much as 40 percent of the U.S. family's food dollar went toward dining out. In most foreign countries, over 25 percent of disposable income is spent for food, with nearly 100 percent of disposable personal income being spent for food in some developing countries. U.S. families with average or above average incomes have a great deal of choice as to the type of food item purchased—a choice based on desires and willingness to pay. In contrast, poorer U.S. consumers based their choice more on necessity—on what they most need and on what they are able to pay.

TABLE 2.2   U.S. Food Expenditures in Relation to Disposable Income, 1970 and 1975 to 1991 (in billions of dollars)

	Disposable Personal Income	Food Bought for Use at Home[a]		Food Bought Away from Home[b]		Total Food Expenditures	
		Amount	Percent of Income	Amount	Percent of Income	Amount	Percent of Income
1970	722	74.2	10.3	26.4	3.7	100.6	13.9
1975	1,151	115.1	10.0	45.9	4.0	161.0	14.0
1976	1,264	122.9	9.7	52.6	4.2	175.5	13.9
1977	1,391	131.6	9.5	58.6	4.2	190.2	13.7
1978	1,568	145.0	9.2	66.8	4.3	211.7	13.5
1979	1,753	161.8	9.2	76.9	4.4	238.7	13.6
1980	1,952	178.5	9.1	85.4	4.4	263.9	13.5
1981	2,175	190.4	8.8	95.9	4.4	286.2	13.2
1982	2,320	197.8	8.5	104.6	4.5	302.3	13.0
1983	2,494	207.8	8.3	114.3	4.6	322.1	12.9
1984	2,760	219.1	7.9	122.6	4.4	341.7	12.4
1985	2,943	228.4	7.8	124.5	4.4	357.9	12.2
1986	3,132	236.4	7.5	133.6	4.4	359.6	11.9
1987	3,290	244.9	7.4	147.4	4.5	392.3	11.9
1988	3,548	256.7	7.2	158.1	4.5	414.8	11.7
1989	3,788	274.9	7.3	165.9	4.4	440.7	11.6
1990	4,059	297.3	7.3	177.3	4.4	474.6	11.7
1991	4,218	304.6	7.2	182.9	4.3	487.5	11.6

[a]Includes purchases for off-premise consumption and that produced and consumed on farms. Excludes government-donated foods.

[b]Includes food furnished to commercial and government employees and purchased meals and beverages. Excludes food paid for by government and business.

SOURCE: *Food Consumption, Prices, and Expenditures, 1970-90*, Economic Research Service, USDA Statistical Bulletin 840, August 1992.

World Population.   The growing world population has a very pronounced effect on the demand for U.S. grains and oilseeds. Beginning in 1960, more wheat was exported than utilized within the country. Most of the U.S. wheat exports as well as some corn and soybean products were exported for human consumption. By 1980-1981, foreign demand for wheat accounted for approximately 65 percent of total disappearance. After that marketing year, several factors, including foreign exchange rates, mounting foreign debt, and a general lack of hard currency, forced many importing countries to adopt new food policies. A distinct trend toward self-sufficiency in grain production began to curtail the growth in U.S. grain exports. In 1990-1991, for example, exports accounted for only about 44 percent of total U.S. wheat disposition.

On January 1, 1990, the world population stood at about 5.3 billion. The continued growth of world population will increase the potential for foreign markets in the future, and may exceed 6 billion by the year 2000, possibly reaching 8 billion by the year 2020 (Figure 2.1). Because most of the population growth is expected in the developing countries, some increase in food production will also occur in these countries, but this increase will not likely keep pace with population growth. Even if the rate of food production parallels the rate of population growth, the potential for famine exists in many countries.

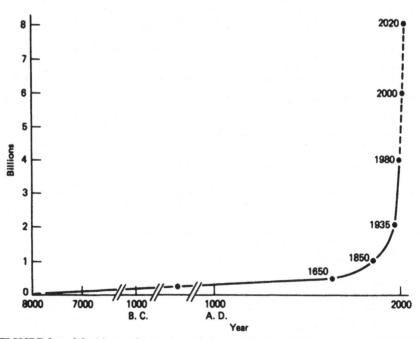

FIGURE 2.1 World population growth—an indicator of future grain and oilseed needs.

SOURCE: Adapted from Norman Borlaug, "The Human Population Monster," presented at the Landon Lecture Series, Kansas State University, Manhattan, Kansas, March 20, 1979.

Optimistically, one hopes the world's people will be fed. There are millions of acres of land in the United States and in other parts of the world that might be suitable for cropland but that are not presently cropped. In many areas, land must be cleared or drained before it can be cropped. Also, the technical knowledge necessary to obtain big increases in yields is being spread to all corners of the earth. Thus although the Malthusian theory—that population is capable of increasing faster than the means of subsistence—should not be ignored, it would appear that with better communication and cooperation, more ingenuity and the will to survive, world leaders can and will take the necessary steps to follow a rational food production and trade policy in the foreseeable future. Steps taken by many developing countries, the former Soviet Union, and China, to become food self-sufficient support this conclusion. At the same time, weak export markets caused a serious agricultural recession in the United States in the mid-1980s. Given this situation, U.S. agricultural policymakers were faced with such diverse choices as idling production capacity or obligating huge grain-storage costs, and the challenge of devising new marketing strategies.

Livestock and Poultry Numbers and per Capita Consumption. Changing numbers of livestock and poultry have a major effect on the demand for feed grains and high-protein meal. Changing feed conversion ratios also may affect demand although they change little in the short run. In the 1970s, these ratios varied from just over two pounds for broilers up to five pounds or more of feed per pound of gain for beef production. In the 10-year period from 1980 to 1990, U.S. livestock and broiler production changed as shown in Table 2.3. Little further gains in efficiency are likely in the twentieth century.

TABLE 2.3   U.S. Livestock and Broiler Production, 1980-1990 (in millions)

Type	1980	1985	1990
Cattle (beef)	100	100	98
Hogs	57	54	54
Sheep	13	11	11
Dairy	11	11	10
Broiler hatch	4,280	4,802	6,315

SOURCE: Economic Research Service, U.S. Department of Agriculture.

In the 1970s, no significant trend developed in the per capita consumption of red meat. By the 1980s, however, per capita red meat consumption began declining. At the same time, domestic consumption of poultry (broilers and turkeys) increased rapidly (Table 2.4). From the end of World War II to 1990, per capita poultry consumption more than tripled.

In view of this trend in livestock and poultry consumption, the future domestic market for feed grains and high protein meal may be reduced because a smaller quantity of grains is required to produce a pound of poultry than a pound of red meat. Nevertheless, the long-run domestic market for feed grains and high-protein meal will be determined largely by three factors: U.S. population growth, the increasing domestic demand for poultry products, and the steadily growing foreign demand for U.S. red meat and broilers. In the short run, the prices of feed grains and high-protein meal are the biggest factors. The domestic feed use of corn, for example, may be reduced by as much as 10 percent in one year if its price is high relative to the price of feeders and finished cattle.

Worldwide, numbers of both livestock and poultry are increasing as economic conditions improve. However, since increases in numbers are

TABLE 2.4 Per Capita Consumption of All Red Meat and Poultry, 1980-1990 (in pounds)

	All Red Meat	*Poultry*[a]
1980	180.8	39.7
1981	178.4	40.9
1982	171.0	41.8
1983	176.5	42.6
1984	176.0	43.7
1985	177.4	45.2
1986	174.8	47.1
1987	170.6	50.7
1988	173.8	51.7
1989	168.4	53.6
1990[b]	163.2	55.9

[a]Includes chickens and turkeys.
[b]Forecast by the U.S. Department of Agriculture.

SOURCES: U.S. Department of Agriculture, *Agricultural Statistics 1992*, and Economic Research Service, *Livestock and Poultry Outlook and Situation*, various issues.

not always accompanied by sufficient increases in feed grain production, deficit situations result. For example, the rapid growth of Japan's livestock and poultry industries in recent years led to that country's increased imports of feed grains and soybean meal. This scenario is common in several developing countries. At times, substantial quantities of low-protein U.S. wheat as well as corn and soybeans will also be demanded by these foreign countries to support their livestock industries. The U.S. share of these markets will depend on its ability to compete for these markets as well as on political considerations.

Supply-and-Demand Relationships

Changes in supply and demand result in equilibrium price changes. Figure 2.2 shows the effect on equilibrium price of four possible changes in supply and demand.

In each of these cases, either supply or demand is held constant while the other is allowed to change. An example of a supply decrease and demand unchanged situation is a government program that limits crop acreage in some manner, as shown in Figure 2.2D. Such programs should theoretically result in less production and higher farm prices. Whether farmers are better or worse off with an acreage limitation program depends on the elasticity of the demand curve. If the demand for the grain products affected by the program is inelastic rather than elastic, then a crop reduction program will result in higher total receipts to farmers.

Many more complex cases might be illustrated. One could diagram the following situations and list the factors that could cause these shifts.

1. Supply and demand change in *opposite* directions.
   a. Supply increases and demand decreases.
   b. Supply decreases and demand increases.
2. Supply and demand change in the *same* direction.
   a. Both supply and demand increase.
   b. Both supply and demand decrease.

Elasticity of Demand

Elasticity measurements can serve as decisionmaking tools for individual firms and for those people involved in government policy. Price elasticity of demand is computed as the percentage change of quantity demanded divided by the percentage change in price. The computed coefficient thus measures consumers' responsiveness of quantity demanded to a price change.

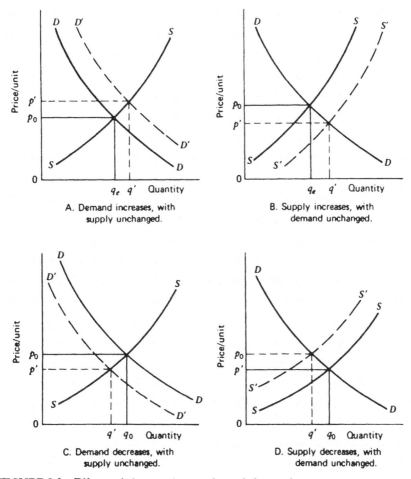

FIGURE 2.2   Effects of changes in supply and demand.

Income elasticity is computed as the percentage change in quantity purchased divided by the percentage change in income. Measurement of income elasticity requires knowledge of the relationships between changes in the potential buyer's income and the quantity taken.

For most purposes, both price and income elasticity are measured between two points on a curve. Such measurements are termed *arc elasticity*. This type of elasticity is generally used when researchers or policymakers are concerned with farm programs designed to affect the price of agricultural products. Elasticity can also be measured at a specific point on a curve, thus the term *point elasticity*.

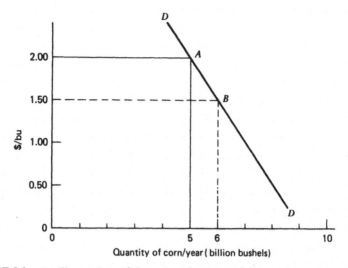

FIGURE 2.3   An illustration of the price elasticity of demand.

Price Elasticity.   Arc price elasticity, as previously defined, may be computed using the following formula:

$$E_d = \frac{(Q_1 - Q_2)/(Q_1 + Q_2)}{(P_1 - P_2)/(P_1 + P_2)}.$$

The numerator of this formula represents the percentage change in quantity demanded. The denominator represents the percentage change in price.

Figure 2.3 assists in illustrating this method of calculating price elasticity. By substituting the values shown in the figure into the formula we can compute the price elasticity as follows:

$$E_d = \frac{(5 - 6)/(5 + 6)}{(2.00 - 1.50)/(2.00 + 1.50)} = \frac{-1/11}{.50/3.50} = \frac{-1}{11} \times \frac{3.50}{.50} = -.64.$$

Thus, the price elasticity of demand between points *A* and *B* in Figure 2.3 is −0.64. The elasticity of any straight line demand curve will vary according to the segment of the curve being measured.

For most goods, the elasticity coefficient sign will be negative because price and quantity demanded change in opposite directions. When the absolute value of the elasticity coefficient is less than one, as in the preceding example, demand is termed *inelastic*. When the coefficient absolute value is greater than one, the demand is termed *elastic*. When the coefficient absolute value is exactly one, demand is termed *unitary*.

Price elasticities of demand vary greatly, both at the farm level and at different stages of the marketing process. For example, the approximate price elasticity of demand at the farm level for cattle is –0.68; for hogs, –0.46; for eggs, –0.23; and for fluid milk, –0.14.

The elasticity of demand at the farm is usually less than the elasticity at the retail level. The price elasticity of demand for cereal (baking products) at the retail level is about –0.15, whereas at the farm level it is generally about –0.02 for wheat and about –0.07 for corn.

Farmers generally receive far less than 50 percent of the consumer's dollar for their grain products, so the percent of price change at the farm level (the denominator in the elasticity formula) is greater than at the retail level. Thus, the elasticity of demand at the farm usually is less than at the retail store. In other words, the wider and more stable the margin between farm prices and retail prices, the less elastic is the demand at the farm compared to the demand at the retail store.

Factors influencing elasticity include: (1) the availability of close substitutes for a commodity, (2) the number of uses to which a commodity can be put, and (3) the price of a commodity relative to the consumer's disposable income.

An example of the elasticity of substitution is related to classes of wheat. The elasticity of substitution between soft red winter and hard red winter wheat, when compared with that for soft red winter and white wheat, suggests that soft red winter can substitute for hard red winter more easily than for white wheat.

Income Elasticity. Buyers will purchase more of most goods as their incomes increase, but not necessarily in direct proportion with the income increase. In the case of food, the quantity purchased increases as income rises, but at a decreasing rate (Figure 2.4A). Conversely, in the case of most nonfood goods, the quantity increases at an increasing rate as income rises (Figure 2.4B).

Arc income elasticity, as previously defined, may be computed using the following formula:

$$E_d = \frac{(Q_1 - Q_2)/(Q_1 + Q_2)}{(I_1 - I_2)/(I_1 + I_2)}.$$

Just as for price elasticity of demand, the numerator in this formula represents the percentage change in quantity demanded. However, in this formula the denominator represents the percentage change in income along a segment of the curve.

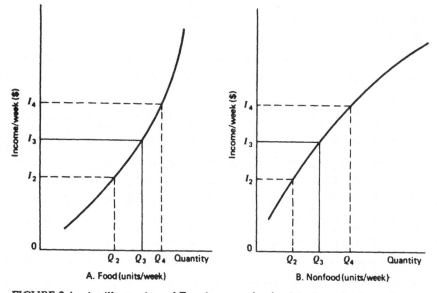

FIGURE 2.4   An illustration of Engel curves for food and nonfood goods.

Figure 2.4A provides an illustration for calculating the income elasticity of demand coefficient using the consumption of food as an example. With I_2, I_3, and I_4 representing weekly incomes of $200, $300, and $400, respectively, and Q_2, Q_3, and Q_4 representing 28, 38, and 45 units of food purchased at those three income levels, the income elasticity of demand between weekly incomes of $300 and $400 is measured as follows:

$$E_i = \frac{(38 - 45)/(38 + 45)}{(300 - 400)/(300 + 400)} = \frac{-7/83}{-100/700} = \frac{7}{83} \times \frac{700}{100} = .59.$$

In this hypothetical example, the income elasticity of demand is 0.59, meaning that a 100 percent increase in income results in a 59 percent increase in the quantity of food purchased. If the income elasticity of demand had been measured at a lower level of income, say a change from $200 to $300 per week, the elasticity would have been 0.76.

Cross-Elasticity. Economists use a concept called the *cross-elasticity of demand* to measure the extent to which commodities are related to each other. One might consider, for example, the commodities X and Y. The cross-elasticity of X with respect to Y equals the percentage change in quantity of X taken divided by the percentage change in the price of Y.

This can be expressed mathematically by:

$$E_{xy} = \frac{(Q_{x1} - Q_{x2})/(Q_{x1} + Q_{x2})}{(P_{y1} - P_{y2})/(P_{y1} + P_{y2})}.$$

If the computed cross-elasticity coefficient is positive, the two commodities are termed *substitutes*. Butter and margarine are examples of close substitutes. If the price of one increases, the demand for the other increases and vice versa. High cross-elasticities indicate close relationships—substitute goods in the same industry. If the computed cross-elasticity coefficient is negative, the two commodities are termed *complementary*. Canning jars and lids are examples of complementary products. If the price of jars increases, the quantity of jars demanded decreases and the demand for lids also decreases. High negative cross-elasticities indicate strong complementary commodities. Low cross-elasticities, coefficients close to zero, indicate remote relationships.

Elasticity of Supply

The concept of supply elasticity is similar to that of demand elasticity. The computed coefficient is always positive because a change in price will bring about a change in quantity in the same direction when the supply curve slopes upward and to the right.

Commodities that are very responsive to price changes have elastic supply curves, and those that respond little to price changes have inelastic supply curves. For grain products, time is an important factor in determining supply elasticity. After crops are seeded, the quantity cannot be changed greatly regardless of price changes; the supply is largely fixed, depending on the weather and certain other factors. However, in the longer run, grain supplies are responsive to price and price forecasts as farmers respond to changes in the relative prices and production costs of alternative crops.

Corn Supply and Utilization

The United States, the major corn-producing country, produces about 40 percent of the world's total. In world trade, the United States accounted for about 75 percent of the total in the 1980s, except for a decrease between 1985 to 1987. In 1988 the country resumed its previous market share, which continued into the 1990s.

World corn production generally has maintained a safe margin over consumption. However, after stocks reached uncomfortably low levels for three consecutive years, 1972 to 1974, production began to exceed consumption, and world stocks grew to very high levels.

Supply

Except for occasional years of dry weather or widespread crop disease that limited production, U.S. production of corn rose steadily from 1940 to the 1980s. Total supply (current production plus carry-over) exceeded 10 billion bushels in 1985-1986. This quantity is placed in perspective by comparing it to production, carry-over, and utilization.

Production. U.S. production of corn for grain reached new record levels of 4 billion bushels in 1963, 5 billion in 1971, 6 billion in 1976, 7 billion in 1978, 8 billion in 1981, and nearly 9.5 billion bushels in 1992. Part of these production increases can be attributed to acreage shifted to corn, and part to yield increases. In 1985, corn was harvested from a record 75 million acres. Average yields reached 100 bushels per acre for the first time in 1978, and continued upward to 130 bushels per acre in 1992.

Carry-over. Historically, corn carry-over has been highly variable both in terms of total volume and relative to total supply. The largest carry-over in history occurred in 1985-1986, when it reached an estimated 4.9 billion bushels, or 40 percent of the total supply.

In the late 1950s and 1960, when the Commodity Credit Corporation (CCC) owned over a billion bushels of corn annually, the market was glutted. Then in the 1960s, supply control measures, coupled with the opening of new export markets, reduced the burdensome carry-over to manageable proportions. At the same time, CCC activity lessened. A decade later, the government no longer took ownership, and stored no corn; the industry had entirely assumed the carry-over function.

During the 1970s, carry-over of corn was much lower, ranging from a high of about 1.1 billion bushels on October 1, 1972, to a low of 361 million bushels just three years later. Then, in the 1980s, carry-over set an all-time record, reaching 4.9 billion bushels on September 1, 1987. This record U.S. carry-over represented 59 percent of total U.S. corn production in 1986. In comparison, carry-over accounted for only 15 percent of U.S. corn production in 1991-1992. Such a buildup of grain stocks as occurred in 1986, and the inability of the United States to maintain a supply-demand balance, indicates the absence of a production policy that is responsive to world demand conditions and causes serious marketing problems. Such volatility is typical of export markets. There are major differences in demand variability between domestic and export markets.

Utilization

In the quarter century from 1955 to 1980, total demand for U.S. corn more than doubled, increasing from about 3 billion bushels to about 7.3 billion bushels. With export demand declining, total disappearance dropped to about 6.5 billion bushels by 1985, but increased to approximately 8.4 billion bushels in 1992.

Corn is used primarily for domestic purposes, including livestock and poultry feed, wet- and dry- processed products, alcoholic beverages, and seed. Exported corn is largely in the form of whole grain, although lesser volumes of corn meal and other products are also shipped overseas. In 1991, exports accounted for approximately 20 percent of total disappearance, nearly 4 percent less than the average for the period between 1985 and 1990.

Feed. The largest use of corn is for livestock and poultry feed, as it has been throughout U.S. history. However, tremendous changes have occurred in corn marketing methods.

In the early 1940s, less than 25 percent of production was sold off farms. The primary demand was for livestock feed on the farm where it was grown. By the 1950s and 1960s, about 30 to 40 percent of production was sold off farms, although, a large percentage of this volume eventually made its way back to farms or feedlots. By the 1980s, about two-thirds of production entered marketing channels, primarily for feed uses and exports. Only a small percentage was returned to farms. In 1990, due to a large level of production, feed uses accounted for only 60 percent of production.

The figures in Table 2.5 show the total volume of corn produced and utilized domestically for feed from 1950 to 1990. The feed market for corn represents the most stable outlet for farmers and a steady dependable business for feed manufacturers. However, farmers, corn merchandisers, and feed manufacturers constantly must be aware of trends or changes in the industry. Such factors as changing price relationships between feed grains and between corn and wheat can effect the feed demand for corn. The demand by foreign buyers is another major factor that may affect the supply-demand relationship for a feed ingredient, such as corn. Also, marketing channels and marketing opportunities may change over time as the location of feed manufacturing moves, say, from centralized points to decentralized feedlots. Thus, while the demand may appear fairly steady or a trend fairly consistent, dynamic forces in the marketplace require a constant monitoring effort and frequent changes in marketing strategies.

Food. Domestic food uses increased nearly 50 percent during the 1970s and continued to grow in the 1980s. By 1990, this market for corn products accounted for about 9.4 percent of total disappearance.

TABLE 2.5   Corn Produced and Utilized Domestically for Feed, 1950-1990 (in millions of bushels)

	Produced	Used for Feed	Percent Fed
1950	2,764	2,482	90
1960	3,907	3,092	79
1970[a]	4,152	3,593	87
1975	5,841	3,570	61
1980	6,639	4,133	62
1985	8,876	4,114	46
1986	8,226	4,669	57
1987	7,131	4,798	67
1988[b]	4,929	3,941	80
1989	7,526	4,389	58
1990	7,934	4,669	59

[a]Year of abnormally low production because of corn blight.
[b]Year of abnormally low production because of severe drought experienced in the Midwest, in particular, the big corn-producing states.

SOURCE: U.S. Department of Agriculture, *Agricultural Statistics, 1992*, and *World Grain Situation Outlook*, FG 1-93, Jan. 1993.

Corn used for food is either wet or dry processed. Wet-processed products include starch, dextrin, syrup, and various other by-products. Other corn is processed into alcohol and distilled products. Dry-processed products include corn oil, corn meal, and grits.

Per capita consumption of corn syrup increased rapidly, nearly tripling in the 1970s, and continued to grow at a phenomenal rate in the 1980s. Correspondingly, the wet mill grind increased about 60 percent from 1970 to 1980 and continued to follow a sharp upward trend in the 1980s. Per capita consumption of syrup increased from 15.8 pounds in 1970 to 68.7 pounds by 1990, as shown in Table 2.1. The primary cause of this change was the high-fructose corn syrup (HFCS) industry, which came into prominence seemingly overnight. That industry managed to condense 50 years of normal product growth into a decade or two. U.S. sugar programs stimulated growth and development of the HFCS industry, despite a declining world sugar price, and kept the domestic refined price artificially high.

This case, of course, is a classical example of changing price relationships resulting in new marketing opportunities. Artificially high sugar prices stimulated or triggered the substitution of HFCS for sugar in many food and beverage products. When the wet corn industry determined that a large untapped market potential existed for corn sweeteners, it committed millions of dollars to new HFCS plants and new technology, assuring its customers that high fructose would be available in the quantity and quality needed.

By 1980, HFCS accounted for 33 percent of the total sweetener market. By 1984, the major soft drink manufacturers approved the use of HFCS at 100 percent levels. By the mid-1980s, the industry had captured 50 percent of the total market and was still growing. HFCS, by then, was well accepted not only by soft drink manufacturers but also by the baking, canning, dairy products, and confectionery industries.

The market for dry corn milling products also became explosive by the mid-1980s. Per capita consumption, which stood at 7.4 pounds in 1970, reached 14 pounds in 1990. This increase in dry milled corn demand was largely accounted for by growing consumer preferences for masa flour products, including tortillas and tortilla chips. Retail sales of corn snack foods and tortilla chips increased almost 70 percent between 1980 and 1985, and the industry speculates that these corn products may eventually replace potato chips as the snack food leader.

Alcohol and Distilled Spirits. The utilization of corn by the alcohol and distilled spirits industries accounted for just over 1 percent of total disappearance annually in the 1970s. The use of corn for distilled liquors declined while its use for fermented malt liquors increased. Corn products—grits, flakes, and refined starch—are used as supplementary carbohydrates in the manufacture of beer. Within certain quality tolerances, these products can be substituted for malt. Until the popularization of light beers, the brewing industry had steadily increased the use of grains in its brewery operations. Just over 10 pounds of corn were used per barrel of beer in the period from 1951 to 1976. Since the late 1970s, the introduction of light beer has resulted in less demand for corn products. Instead, carbohydrates (other than malt) are being supplied by dextrose rather than by grains.

Ethanol is another alcohol product that is gaining in prominence. Since the days of high gasoline prices in the mid-1970s, ethanol production has grown substantially. Accordingly, the demand for corn for this use increased from only 4 million bushels in 1978 to nearly 400 million bushels by 1990.

Seed. Seed corn accounts for less than 1 percent of total utilization. However, hybrid seed should not be considered along with other uses when studying the marketing system. Hybrid seed is not handled in the same marketing channels as other corn dispositions. Hybrid seed is a farm input distributed by hybrid seed companies to their farm customers through a network of hybrid seed dealers.

Exports. U.S. corn exports grew steadily after World War II but increased most rapidly in the 1970s. Exports accounted for 33 percent of the total utilization in 1980 compared to less than 12 percent in 1970. U.S. corn exports passed one billion bushels for the first time in 1972 and have exceeded that amount in each succeeding year. By 1980, U.S. corn exports reached 2.4 billion bushels. Then, by 1985, this volatile market was back to 1.2 billion bushels. However, from 1985 to 1989, the corn export market grew steadily:

1940	15	million bushels
1950	117	
1960	292	
1970	517	
1980	2,408	
1985	1,227	
1986	1,493	
1987	1,716	
1988	2,026	
1989	2,368	
1990	1,725	

Like markets for other grains and oilseeds, the world demand for corn was affected by worldwide economic conditions and renewed efforts for food self-sufficiency. World corn production showed remarkable growth in the 1970s and early 1980s, increasing from 244 million metric tons (mmt) in 1969 to 480 mmt in 1985-1986. U.S. exports, as a percentage of total world trade, changed from 44 percent in 1970 to 78 percent in 1980, and then back to 58 percent in 1985. By 1988, the United States had regained a market share of 79 percent and stayed near that level through 1990. The drop in export demand during the 1980s largely explains the massive buildup of inventory.

The eight major importers of U.S. corn, shown in Table 2.6, have not changed significantly since the early 1970s, although the volume demanded has changed. The European Community (EC) countries, for example, took over 15 mmt in the early 1970s, but by the mid-1980s, U.S. shipments to EC countries had dropped to 5 mmt or less. Most other major markets have increased their corn purchases. The eight major importers accounted for 84 percent of the total U.S. overseas corn market in the late 1980s. The U.S. share of these eight markets averaged 86 percent as compared to 80 percent for all markets. Thus the United States continues to dominate the growing world corn trade.

TABLE 2.6 Annual Three-Year Average Imports of Corn, Showing the Eight Largest U.S. Markets and the U.S. Market Share, 1987-1988 to 1989-1990 (in thousand metric tons)[a]

| Countries | Rank | Average Annual Imports | | U.S. Percentage |
		U.S. 3-Yr Avg	World 3-Yr Avg	
Japan	1	13,911	16,267	86
USSR	2	12,361	14,533	85
Korea, Rep.	3	4,896	5,600	87
Taiwan	4	4,269	4,567	93
Mexico	5	3,622	3,800	95
EC-12	6	2,717	3,800	72
Eastern Europe	7	1,701	1,733	98
China	8	220	233	94
Total, 8 Countries		43,697	50,533	86
Other Countries		7,992	14,067	57
Avg. Total World Exports		51,689	64,600	80

[a]There are 39.368 bushels of corn in one metric ton.

SOURCES: Foreign Agricultural Service, *World Grain Situation & Outlook*, FG-11-91, Nov. 1991, and Supplement 14-88, Dec. 1988; FATUS (ERS), 1990 Supplement.

Soybean Supply and Utilization

Until the 1970s, the United States and mainland China produced most of the world's soybeans. In 1975 these two countries produced nearly 90 percent of the total. However, a decade later, the situation had changed. Brazil and Argentina accounted for over 20 percent of total world production; the United States about 60 percent; and China about 8 percent.

The United States competes in a rapidly growing world oilseed market in which soybeans account for almost one-half the total production. In 1990 this country produced 50 percent of all the world's soybeans. Government statistics, which indicate the supply and utilization situation for soybeans and other oilseeds, normally break soybeans into three groups: whole beans, meal, and oil.

Supply

Compared to corn and wheat, soybeans are a new crop in the United States. Production was less than 100,000 bushels prior to World War II. After 1950, soybean production increased faster than that of any other major grain or oilseed. Total production reached 1 billion bushels in 1966. Thirteen years later, in 1979, it peaked at 2.3 billion bushels. In 1985, generally in response to world supply-demand conditions, production had backed off to about 2.0 billion bushels. Total supply remained fairly constant at about 2 billion bushels from 1978 to 1985, but dropped below the 2 billion bushel level in the late 1980s.

Production. The volume of U.S. soybean production increased largely because of increased acreages. From 1967 to 1984, for example, soybean acreage increased from 39.8 to 66.1 million acres. By 1979, soybean acreage had surpassed 70 million acres. However, soybean yields remained fairly constant from 1985 to 1990.

In a world market situation where total oilseed production increased about 50 percent from the mid-1970s to the mid-1980s, the U.S. soybean supply has been well managed. However, factors looming on the horizon in the late 1980s could cause serious oversupply problems for U.S. soybean producers and the domestic market. As the United States accounted for less and less of total production in the 1980s, its share of world trade also declined. Thus, by the late 1980s, it appeared as if the nation might find itself in a soybean situation of oversupply, much as it faced for corn and wheat.

Carry-over. Soybean carry-over was of little consequence prior to the 1950s, and in the 34-year period from 1950 to 1984, it exceeded 300 million bushels only five times. Each time, these larger-than-normal volumes of carry-over followed years of exceptionally large production. Each time, the soybean industry was able to handle those minor gluts without the need for government supply management programs. Then in 1985 and 1986, inventories jumped to over 500 million bushels, indicating that farmers were not responding to the reduction in export volume that began in 1982. However, steady export demand coupled with low production in 1988 helped to lower carry-over by 1990.

Utilization

Whole soybean disappearance, led by domestic uses, increased as follows after 1950:

1950	298	million bushels
1960	580	
1970	1,258	

1975	736
1980	1,834
1985	1,878
1986	2,040
1987	2,057
1988	1,673
1989	1,869
1990	1,840

About 96 percent of all soybeans are either crushed domestically or exported as whole beans. The remainder is used for seed or fed as whole beans.

Crushings. Soybean crushings account for almost 65 percent of total disappearance. Crushings yield about 11 pounds of crude soybean oil and 48 pounds of 44 percent protein meal per bushel. Of the meal produced, about 80 percent is utilized domestically and 20 percent exported. This ratio is compared to 74 and 26 percent, respectively, during the 1970s. Of the oil produced, 86 percent is used domestically, and 14 percent exported.

Soybean meal is used in the U.S. feed industry, especially for poultry and hogs. Increases in U.S. livestock and poultry-feeding rates are boosting soybean meal consumption, although in the case of livestock, this rate increase is partially offset by decreasing numbers. During the early 1970s, protein feed accounted for about a tenth of all concentrate feed. However, rising feed costs encouraged the industry to adopt more efficient feeding practices. By the 1980s, protein supplements accounted for 14 percent. In 1990, about 26 mmt of soybean meal were processed into feed.

Soybean oil is used largely in foods and competes with other oils such as canola, corn, coconut, palm, and peanut. Despite competition from these oils, consumption of soybean oil by U.S. food manufacturers has remained relatively constant. In 1990, soybean oil accounted for about three-fourths of all oils and fats utilized in food products. Over 93 percent of domestic soybean oil production was used in the manufacture of shortening, margarine, cooking and salad oils, and other edible products. The remainder was used in such nonfood products as paint and varnish, resins, plaster, and other oil-based products.

Exports. U.S. soybean export volume peaked in 1981 when 45 percent of the year's production was channeled into foreign markets. (See Table 2.7.) After the 1981-1982 marketing year, the U.S. export volume of whole beans, meal, and oil has trended downward. In the five-year period from 1986-1987 through 1990-1991, exports of whole beans averaged 35 percent of production.

TABLE 2.7   U.S. Exports of Soybeans, 1950-1990 (in million metric tons)

	Whole Beans[a]	Soybean Meal	Soybean Oil
1950	0.8	0.2	0.22
1960	3.5	0.5	0.32
1970	11.8	4.2	0.81
1980	19.7	6.7	0.74
1981	25.3	6.9	0.94
1982	24.6	7.1	0.92
1983	20.2	5.4	0.83
1984	16.3	4.9	0.75
1985	20.1	6.0	0.57
1986	20.6	7.3	0.54
1987	21.8	6.9	0.85
1988	14.4	4.9	0.75
1989	16.9	4.8	0.61
1990[b]	15.2	5.0	0.35

[a]There are 36.7437 bushels of whole beans in one metric ton.
[b]Forecast by the U.S. Department of Agriculture.

SOURCE: U.S. Department of Agriculture, *Agricultural Statistics, 1992.*

The chief reason for the rapid growth in exports until the early 1980s was the fast-increasing demand for high-protein meal for livestock feed in Western Europe, Eastern Europe, Japan, and other parts of Asia. These areas imported whole beans and processed them into meal and oil. Conversely, until the 1980s, developing countries generally lacked soybean-crushing capabilities and therefore imported processed meal and oil. However, by the late 1980s, more and more of these countries were investing in soybean-crushing capacity or becoming self-sufficient in other substitutable oilseeds.

The chief reason for the decline in U.S. soybean exports following 1981 was not world economic conditions but rather the growth in world production. World oilseed production literally exploded after the mid-1970s. Increases in both yield and area seeded contributed to this rapid rise in production. These trends show no immediate signs of change.

The worldwide drive toward self-sufficiency in oilseed production also appears quite successful.

Soybean marketing is not as provincial as that of corn and wheat. Instead, students of soybean marketing must think beyond the commodity in question. Soybean exporters not only compete with other exporting countries but also with other oilseeds such as cottonseed, peanut, sunflowerseed, rapeseed (canola), flaxseed, copra, palm kernel, and palm oil. All are more or less substitutable notwithstanding consumer preferences due to perceived health benefits. Edible use of corn and canola oil has increased substantially since the mid-1980s. Countries in which major increases in oilseed production can be expected over the next few decades include China, India, Argentina, and Brazil. Lower costs of production and processing in these countries will present a serious challenge to the U.S. marketing system.

The U.S. soybean market is affected by still another unique factor not experienced to such a degree by corn and wheat markets. That factor is the competition and trade of processed products in world markets. In addition to the crushing by Argentina, Brazil, and China and their subsequent export of the meal and oil, the United States must compete with the European Community. The 12 EC countries, some major U.S. customers for whole beans (Table 2.8), process imported soybeans for resale. In 1990, for example, the 12 EC countries exported 75 percent of the volume of meal as the United States did, and they led the world in soybean oil exports, although they did not produce many of these beans.

Although the United States still dominates the world soybean trade, South American countries are fast becoming major players in the world market. Brazil and Argentina together accounted for 28 percent of world whole-bean exports in 1990. They also accounted for 51 percent of the world soybean meal exports and 49 percent of the world soybean oil market, two important value-added markets.

Brazilian policy is to crush soybeans at home and export the products—mainly meal. Correspondingly, only about 13 percent of Brazil's crop was exported as beans in 1985-1986. Since the 1970s, Brazil has significantly increased its crushing capacity. By 1980, trade estimates placed the country's processing capacity at 19 million tons, and the country crushed more than 85 percent of its soybean production. Conversely, Argentina still exported about 40 percent of its supply as whole beans in the same year.

Major markets for U.S. whole beans are Japan, the Netherlands, Taiwan, and Spain. These four countries are all long-term customers for unprocessed U.S. soybeans. They, alone, account for almost 56 percent of the U.S. export market (Table 2.8). The 15 largest markets for soybeans take about 15 million tons, accounting for 91 percent of the total U.S.

TABLE 2.8   Annual 5-year Average Imports of Soybeans[a], Showing the 20 Largest U.S. Markets and the U.S. Market Share, 1987-1988 to 1990-1991 (in thousand metric tons[b])

| Countries | Rank | Average Annual Imports | | U.S. Market Share Percentage |
		U.S. 4-Yr Avg	World 4-Yr Avg	
Japan	1	3,557	4,550	78
Netherlands	2	2,901	3,598	81
Taiwan	3	1,780	1,979	90
Spain	4	1,334	2,317	58
Mexico	5	1,143	1,161	98
Korea, Rep.	6	909	1,049	87
Germany	7	899	2,749	33
Belgium-Luxembourg	8	595	1,162	51
The former Soviet Union	9	442	858	52
United Kingdom	10	419	717	58
Portugal	11	400	832	48
Israel	12	397	400	99
Eastern Europe	13	326	664	49
France	14	201	356	57
Romania	15	186	346	54
Canada	16	175	192	91
Other West Europe	17	167	520	32
Indonesia	18	143	521	27
Italy	19	130	569	23
Yugoslavia	20	107	219	49
Total, 20 Countries		16,209	24,757	65
Other Countries		756	1,620	47
Average Total World Exports		17,079	26,489	64

[a]Does not include meal and oil.
[b]There are 36.7437 bushels of soybeans in one metric ton.

SOURCE: *World Oilseed Situation & Outlook*, FOP 11-89, FOP 11-91, Nov. 1989 and 1991. FOP-89, Nov. 1989, and FOP 11-91, Nov. 1991.

export market. These same 15 countries receive almost 80 percent of their total supply of beans from the United States. This percentage, of course, is much higher than for wheat, indicating that the United States through the mid-1980s has had less competition for world soybean markets than it faced for wheat exports.

The former Soviet Union ranked 16th during the 1980-1981 to 1984-1985 marketing period. However, the former Soviet Union, unlike the top 15 countries, imported only 24 percent of its demand for whole soybeans from the United States. During that five-year period, the former Soviet Union imported a larger total volume of beans than did Mexico.

Major markets for soybean meal in the early 1980s were West Germany, Italy, the Netherlands, the United Kingdom, Venezuela, and Canada. These six countries accounted for almost 70 percent of the total U.S. meal export trade. However, in contrast to its market share for whole beans, the United States had only 60 percent of these six major soybean meal markets.

Major markets for soybean oil in the early 1980s were Pakistan and India. These two countries purchased an annual average tonnage of about 350 thousand metric tons in the early 1980s. No other country took over 60 thousand tons. After the 1984-1985 marketing year, the United States lost much of its soybean oil market to its competitors.

Thus, the U.S. soybean export market, like those for corn and wheat, remains largely one for the unprocessed commodity. In 1986, for example, whole beans accounted for nearly 80 percent of the tonnage of exported soybeans and soybean products. Also, by the mid-1980s, it became clear that the United States was not only losing its share of the total world soybean market, it was losing at a faster rate its market for processed soybean products.

As the U.S. soybean industry entered the 1990s, world competition was increasing on at least four fronts: (1) increased world production, (2) increased processing capacity in foreign countries, (3) growing competition from substitute oilseeds, and (4) increased price competition in world markets. In the late 1980s, the U.S. government used credit programs in an attempt to reverse a declining market share.

To further increase the utilization of soybeans, the American Soybean Association conducted promotional activities in the United States and in over 70 other countries using funds received from a voluntary producer check-off program. The types of programs undertaken by this association included:

- Programs to improve soybean oil quality and increase sales in Venezuela, South Korea, Mexico, India, Pakistan, Japan, Portugal, Egypt, Italy, Greece, United Kingdom, France, and West Germany;
- Use of soybean oil as a pesticide carrier in agricultural and forest chemicals;
- Use of soybean oil to reduce grain dust at commercial elevators;

- Programs to expand soybean meal sales for poultry in Nigeria, China, Mexico, Portugal, Turkey, and Yugoslavia, and for fish production in South Korea, Spain, and Finland.

Such promotional efforts may be futile as long as costs of production, processing, and marketing per ton of soybeans or soybean products remain above the level in Argentina or Brazil, and these and other countries continue to expand their oilseed production. Country competition for world markets is no different than firms competing for domestic markets. Pricing is the number one factor.

Wheat Supply and Utilization

The United States accounted for about 10 percent of total world wheat production in the early 1990s, down about 6 percent from the 1970s level. However, the United States often accounts for a much larger share of the total supply because of its capacity to store wheat coupled with its policy of paying farmers and grain merchandisers to hold wheat in storage.

By the end of the 1970s, exports accounted for approximately 65 percent of total U.S. wheat disappearance. The United States was the major wheat supplier in the world trade, accounting for about 40 to 45 percent of the total traded. World wheat and flour trade continued to increase in volume through the late 1980s, but by 1990 the U.S. market share had dropped to 30 percent. Exports, by the early 1990s, accounted for only a little over half of total disappearance, and most of those were made possible only through the use of costly subsidy programs.

In the four decades following World War II, the U.S. wheat industry endured previous periods of market surpluses, but none that equaled the market glut that occurred in the mid-1980s.

Supply

The total supply of U.S. wheat rose steadily, tripling from 1949 to 1984. Total U.S. wheat supply surpassed the 3 billion bushel mark for the first time in 1977 and reached 4 billion bushels in 1984 and again in 1986. The most recent records, of course, reflected the country's loss of world market share, the resulting increase in stocks, and the export programs designed to reduce those stocks.

Production. After reaching a record 75.9 million in 1949, the wheat acreage harvested was sharply reduced over the next quarter century. Nevertheless, market prices continued to reflect a general glut on the market. Between 1949 and 1975, wheat production increased 62 percent even though total acres harvested were reduced about 8 percent. In other words, although acreage harvested was lower, production remained at record highs.

On one side of the aisle, policymakers frantically attempted to control supply by acreage reduction programs, while at the same time, plant breeders made rapid advances in yield per acre. For example, yield per acre was only 14.5 bushels in 1949, but in 1975 it was 30.5, or more than double. In fact, by the mid-1980s, wheat yields of 35 to 40 bushels per acre were common. Wheat yields averaged 37.5 bushels from 1983 to 1987, dropped to only 32.8 bushels in 1989, and increased to 39.5 bushels in 1990.

Events in the early 1970s triggered wheat-marketing problems that lasted into the late 1980s. Back-to-back poor world harvests caused extremely short world supplies in 1972. Worldwide, wheat stocks declined by almost 2 billion bushels from 1969 through 1972. In response to world demand, U.S. stocks dropped from 983 million bushels in 1970 to only 340 million bushels in 1974.

With their first opportunity since the end of World War II to test their capacity to produce, U.S. farmers responded. Harvested acres increased from 43.6 million in 1970 to 65.6 million in 1974. In the succeeding years they have not fallen below 60 million, although yields caused huge increases in production, and a massive buildup in stocks occurred by the mid-1980s. Thus, the production response far exceeded the wheat industry's needs. In 1986, production totaled 2.1 billion bushels, and over 1.9 billion bushels of old wheat were carried into the new marketing year. This situation, however, changed dramatically by the late 1980s. In 1991, due to strong exports and bad weather that affected yields, production was reported at 2.0 billion bushels and beginning stocks at 0.86 billion.

The preceding scenarios typify the conditions under which the U.S. wheat industry has performed throughout much of the time since World War II. (Students of grain marketing need to study this highly unstable production situation very carefully.)

The U.S. wheat-marketing system is unique: whether or not there is a market, wheat producers are subsidized, and grain merchandisers and farmers are paid to store excess supplies. Thus, incentives that should normally be expected through the pricing system do not apply to the U.S. wheat-marketing system. In the U.S. system, production is influenced by government programs as well as supply and demand, and government compensation for storage is often preferable to merchandising margins.

Until the 1980s, excess stocks proved to be a temporary phenomenon though a challenge for the marketing system to find sufficient storage space for both existing supplies and new production. However, reversal of the 1980s oversupply situation is now less likely than in the past. The success of plant breeders around the world unleashed technology. In the decade of the 1980s, world wheat production increased about 25 percent. Thus, the upper limits of U.S. wheat production remained untested as production continually outpaced demand.

Further reductions in acreage harvested and production were in order because of reduced sales caused by high cost of production, poor quality, a loss of image as a dependable supplier, stiff competition from other exporters, and a trend toward food self-sufficiency by importing countries. Yet, no immediate reversal of this situation was in sight.

To policymakers, the U.S. wheat industry, and students of grain marketing these are challenging conditions. Economic efficiency is difficult to maintain when conditions change from a seller's market to a buyer's market. Farmers want to use their production capacity, and the grain industry abhors unused capacity as well. Yet, there continues to be tremendous variability in foreign demand. As a result, the ability of the United States to compete in world trade will determine what happens to U.S. production and stocks.

U.S. wheat production is, therefore, highly dependent upon two policies. First is the policy directly related to supply. Second is the policy related to maintaining a presence in the world market. Both policies must work in harmony.

The easiest factors for policymakers to rely upon with regard to establishing production controls are population (world consumption) and trends in yields. The world demand for wheat and wheat products grew at an annual rate of 4.5 percent from 1960 to 1990. Almost certainly, this demand will continue to grow at least in proportion to world population growth. However, as mentioned, a part of this growing demand is being met by increased production by wheat importing countries. How to tap the remaining market is the big question.

Because the world's supply of cropland is limited, in the long run, yields must be improved if production is to keep pace with demand. However, improvements in crop yields must be kept in harmony with growth in demand.

Carry-over. Excess supplies have often been the center of agricultural policy debate in the U.S. Congress, especially when taxpayers were called on to pay the cost of storing excess supplies. Stocks exceeded the volume of wheat used domestically in 12 consecutive years in the 1950s and 1960s, reaching more than double the domestic use level from 1959 to 1963. Stocks were at a record high in 1961. At that time, they amounted to 1.4 billion bushels, an equivalent of 231 percent of domestic needs. Then in 1986, the carry-over reached 1.9 billion bushels, approximately equivalent to total utilization for an entire year. In contrast, the volume of carry-over was low in both 1952 and 1974, less than 50 percent of our domestic needs.

These extremes in carry-over reflect continual stresses placed on the wheat industry and grain markets in general. In times of rapid stock draw-down, the industry often operates well above normal transportation and handling capacity. In periods of stock buildup, storage capacity may be fully utilized, but day-to-day operations are sometimes slow.

Policymakers not only need to regulate production to prevent large variations in carry-over, they also need to be concerned with the situation by class of wheat. Yet, U.S. policymakers have always ignored this distinction in their supply management deliberations. Hard, bread-type wheats are not the same commodity as soft wheats. Durum is also different from all other classes. Merchandisers do not treat all wheat in the same manner. Each class is priced separately and marketed class identity is preserved. When abnormally large stocks accumulate, an analysis by class will often show that the excess supply problem does not apply across the board. For example, when carry-over reached 1.9 billion bushels on June 1, 1986, a year and one-half supply of hard red spring wheat was in storage, but soft red winter wheat was in short supply.

Utilization

The utilization of wheat consists of two major components, domestic and export. Domestic uses include flour products, breakfast cereals, livestock feed, seed, and industrial uses. Exported wheat is largely in the form of whole grain, although lesser volumes of flour and bulgur are also shipped overseas. In the quarter century from 1959 to 1984, total demand for U.S. wheat more than doubled. Then, due to a sudden decline in exports, total demand dropped back to late 1970 levels. In the early 1990s, U.S. exports accounted for only about 50 percent of total disappearance whereas in the previous decades they had accounted for over 60 percent of total utilization.

Food. Flour is the major product derived from wheat. About 45 pounds of flour are extracted from each bushel of milled wheat. Wheat flour is used to manufacture bread, cookies, crackers, cake mixes, macaroni and spaghetti, and other products.

Each class of wheat has its own distinct characteristics in terms of milling quality and end-use value. Flour milled from hard red spring wheat is high in protein and is used primarily for baking bread and rolls. This class of wheat is often blended with lower protein hard red winter wheats for making bread. Soft red winter and soft white wheats are lower in protein content and are used primarily in the manufacture of crackers, biscuits, and cakes; some of these wheats are used as feed in

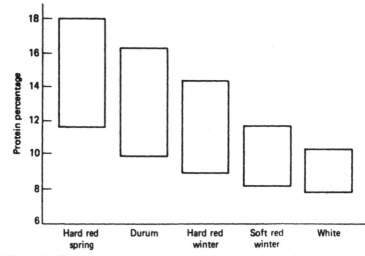

FIGURE 2.5   Protein range for flour uses of major wheat classes.

years of excess supply and low prices. Durum, a high-protein wheat, is generally processed into semolina, which is then used to produce macaroni, spaghetti, and other pasta products (Figure 2.5).

In 1980-1981 and 1990-1991, the USDA estimated the breakdown in utilization by class (Table 2.9). Note how these percentages by class vary greatly from year to year. This variation usually depends on world production, carry-over, and other factors that affect relative prices. In

TABLE 2.9   Domestic Usage and Exports of Wheat by Class, 1980-1981 and 1990-1991 (in percentages)

| | 1980-1981 | | 1990-1991 | |
Class	Domestic	Exported	Domestic	Exported
Hard Red Winter	33	67	65	35
Hard Red Spring	48	52	54	46
Soft Red Winter	34	66	54	46
White	20	80	33	67
Durum	47	53	59	41

SOURCE: U.S. Department of Agriculture, *Agricultural Statistics, 1992.*

TABLE 2.10   Wheat by Class, Percentage Utilized, and Percentage in Carry-over, 1985-1986

Class	Percentage Utilized	Percentage in Carry-over
Hard Red Winter	48	45
Hard Red Spring	16	32
Soft Red Winter	21	6
White	12	11
Durum	3	6

SOURCE: U.S. Department of Agriculture, *Agricultural Statistics, 1992.*

1985-1986, the loss of some major markets greatly affected utilization by class (Table 2.10).

Some substitution among classes is possible when supplies and prices dictate. However, the amount of substitution for domestic food uses is limited because of end-product quality considerations. Along with decisions to substitute one class for another, wheat marketing in the future will be further complicated because of the difficulty in distinguishing between classes of wheat, especially the hard red winter, hard red spring, and soft red winter classes. The complication arose when plant breeders started crossing the parentage of one class with another in an attempt to produce improved varieties.

Per capita consumption of wheat products steadily declined for many years, reaching a low of 110 pounds per capita in 1972. Then, due to diet changes, per capita consumption rose rapidly, reaching 140.6 pounds in 1990. Nevertheless, U.S. consumers were still expressing their preference for prepared products and non-wheat-based foods as opposed to flour itself.

The amount of wheat used for domestic food has increased steadily since 1940:

1940	492	million bushels
1950	493	
1960	497	
1970	517	
1980	611	
1985	674	
1986	712	
1987	721	

1988	726
1989	749
1990	791

In 1990, an estimated 796 million bushels of wheat were used for domestic food purposes. Food uses accounted for about 58 percent of annual disappearance. Over time, food uses exhibit one of the most stable, but growing, aspects of the entire grain industry. Thus, the grain processors' ability to plan needed capacity and their ability to fully utilize this capacity has been exceptionally good.

Seed. Seed use varies with planted acreage which, in the case of wheat, is often affected by farm program changes. Seed use in the 1980s ranged from 85 to 113 million bushels or about 4 percent of total utilization.

Feed. The demand for feed wheat is quite variable. It was generally higher in the 1980s than at any time since World War II, accounting for about 10 to 11 percent of total disappearance. The practice of feeding wheat to livestock in the United States is largely related to price, but is gaining in general acceptance. Compared with corn, the feeding value of wheat is slightly higher, but the amount of wheat that can be included in feeding rations is more limited. Usually the price of good quality wheat precludes its use as feed. In the past, it was most often fed on the farm where produced. However, feed wheat began entering U.S. marketing channels in significant amounts in the 1980s, and in 1984-1985, over 400 million bushels of wheat were fed to livestock in the United States.

The use of wheat as a feed grain in foreign countries is also cutting into the demand for corn as it has to some extent in the United States. Major countries that commonly use wheat for feeding include the former Soviet Union, the Philippines, South Korea, and Mexico.

Industrial Uses. Hard and durum wheats account for about 85 percent of the milled wheat products used for industrial purposes. Normally, the industrial use of wheat is limited to second clears, a by-product of flour milling. Also, some whole wheat is wet processed to produce industrial starch and gluten products. Specific industrial uses include plywood and composition board, laminating adhesives, industrial starch for laundries, textiles, billboard and wallboard pastes, paper additive, and alcohol. Wheat is also used in making whiskey, beer, cosmetics, fertilizer, paving mixes, and certain polishes. Historically, industrial and alcoholic beverage uses account for only about 0.5 percent of total wheat utilization. In the future, however, the percentage could increase, depending largely on the economic feasibility of gasohol, a fuel product that can be derived from grain. Given current prices, barley, milo, and corn would probably be used to make gasohol rather than hard wheats.

Exports. The volume of wheat exports expanded following World War II, but did not show phenomenal growth until the 1970s. Then, in the 1980s, U.S. wheat exports averaged 1.3 billion bushels. Much of this volume, however, was shipped under government export enhancement and credit programs.

1940	41	million bushels
1950	366	
1960	654	
1970	741	
1980	1,514	
1981	1,771	
1982	1,509	
1983	1,426	
1984	1,421	
1985	909	
1986	999	
1987	1,598	
1988	1,419	
1989	1,233	
1990	1,068	

During the 1980s, the United States exported wheat and flour to about 120 countries. However, 25 countries accounted for more than 90 percent of the total (Table 2.11). The U.S. share of these 25 markets averaged 49 percent, compared with 41 percent for all world markets. This U.S. wheat share is a much smaller share of the world market than is its share for corn or soybeans. Competition became quite stiff by the late 1980s, making future expansion of world wheat markets by the United States uncertain. There appears to be greater market potential for expanding some of the country's larger wheat markets—like the former Soviet Union, Iraq, Italy, Colombia, or Bangladesh—rather than smaller developing markets.

Summary

The U.S. grain-marketing system does not stop at the country's borders. It extends worldwide. Production uncertainty in foreign countries and, in turn, variability in foreign demand has led to periods of large surpluses followed by periods of short supply. These surpluses and shortages place much greater stress on the U.S. marketing system than if the system were to serve only the domestic market. Nevertheless, the U.S. grain

TABLE 2.11 Average Annual Imports of Wheat and Wheat Flour, Showing the
25 Largest U.S. Markets and the U.S. Market Share, 1987-1988 to
1990-1991 (in thousand metric tons)[a]

| Countries | Rank | Average Annual Imports | | U.S. Market Share Percentage |
		U.S. 4-Yr Avg	World 4-Yr Avg	
Former Soviet Union	1	5,772	16,600	35
China	2	5,206	13,250	39
Egypt	3	3,398	6,732	50
Japan	4	2,894	5,572	52
Korea, Rep	5	1,796	3,373	53
Algeria	6	1,283	4,190	31
Philippines	7	1,052	1,254	84
Pakistan	8	984	1,480	66
Morocco	9	956	1,464	65
Taiwan	10	773	893	87
EC-12	11	746	2,150	35
Iraq	12	695	2,425	29
Bangladesh	13	688	1,798	38
India	14	553	640	86
Venezuela	15	523	1,054	50
Sri Lanka	16	509	683	74
Eastern Europe	17	503	2,229	23
Israel	18	501	640	78
Mexico	19	475	665	71
Colombia	20	442	695	64
Tunisia	21	409	924	44
Jordan	22	379	591	64
Ecuador	23	366	373	98
Peru	24	340	849	40
Yemen	25	258	1,084	24
Subtotal		31,502	71,604	44
Other Countries		4,292	26,626	16
World Total		35,794	98,230	36

[a]There are 36.7437 bushels of wheat in one metric ton.

SOURCES: *World Grain Situation & Outlook*, FG-10-91, Oct. 1991; Circular Series,
Export Markets for U.S. Grain & Products, EMG-9-91, Sept. 1991.

production and marketing system has performed exceedingly well when called on to increase its capacity for storage, transportation, and handling, even on short notice.

The demand for corn, soybeans, wheat, and their products grew tremendously in the 1960s and 1970s and then tapered off in the 1980s. Exports are an important channel of trade for U.S. grain and oilseeds. They offer the greatest challenge to the U.S. grain-marketing system and to policymakers. They will likely be of even greater importance in the longer planning horizon. As of the late 1980s, the success of evolving self-sufficiency policies and the ability of developing countries to reverse declining economic conditions leaves considerable uncertainty as to the opportunity for U.S. grain exports in the 1990s. In all likelihood, a sizeable market will exist if the United States can compete pricewise and qualitywise.

Projected world human population and livestock numbers suggest that the world demand for all grain will continue to grow. To meet this growing demand, some additional cropland will be brought into production. There is concern that this added cropland, especially in developing countries, may be less productive than that presently in production. At the same time, there is concern over the loss of cropland presently in production due to erosion, industrial expansion, and urban sprawl. In the long run, increased yields must be forthcoming to prevent world hunger and starvation. The greatest grain production capacity will remain geographically apart from the greatest growth in demand for food.

Projected trends in both human population and livestock numbers suggest a challenge to the U.S. grain-marketing system and an opportunity for it to serve these growing markets for food grains, feed grains, and oilseeds. By the year 2020, according to some population projections, the world may be inhabited by 8 billion people. It is hoped that production can be doubled to meet this anticipated demand. If so, then the marketing system must be expanded accordingly to handle the distribution of supplies. In view of this outlook, challenges abound for plant breeders, farmers, grain merchandisers, grain processors, and all others associated with the grain trade. What appear as problems necessitating market growth can be turned into opportunities for improvements in marketing channels, facility design, trading procedures, pricing methods, and the balancing of supply and demand.

Notes

1. Richard L. Kohls, *Marketing Agricultural Products*, Macmillan Co., New York, 1955, p. 7.

Selected References

Bakken, Henry H., *Theory of Markets and Marketing*, 1st ed., Mimir Publishers, Inc., Madison, WI, 1953.

Baumol, William J., *Economic Theory and Operational Analysis*, 4th ed., Prentice Hall, Inc., Englewood Cliffs, NJ, 1977.

Brandow, George E., *Interrelation Among Demands for Farm Products and Implications for Control of Market Supply*, Agricultural Experiment Station Bulletin No. 680, Pennsylvania State University, University Park, PA, August 1961.

Cramer, Gail L., and Clarence W. Jensen, *Agricultural Economics and Agribusiness*, 2nd ed., John Wiley & Sons, Inc., New York, 1982.

Heid, Walter G., Jr., *U.S. Wheat Industry*, Economics, Statistics, and Cooperatives Service, USDA, Agricultural Economic Report No. 432, Washington, DC, August 1979.

Leath, Mack N., and others, *U.S. Corn Industry*, Economic Research Service, USDA, Agricultural Economic Report No. 479, Washington, DC, February 1982.

Leftwich, Richard H., *The Price System and Resource Allocation*, 6th ed., Holt, Rinehart and Winston, New York, 1976.

Whittaker, Edmund, *A History of Economic Ideas*, 4th ed., Longman, Green, and Co., New York, 1947.

3

Marketing Channels and Storage

Robert L. Oehrtman and L. D. Schnake

Individual grains are produced seasonally and are harvested over a relatively short period in locations that generally are quite far removed from the consuming areas. As grains and their products move through the market channels, they are bought and sold many times. In the handling of these goods, a variety of important economic functions are performed by the marketing system. Grain and grain products are moved from the raw products producers, processed into desired forms, delivered to the desired locations, and, through the storage function, made available to consumers at the desired time.

Marketing Channels

Grain-marketing channels are the various agencies or institutions through which grain products flow from the producer to the final consumer. In these marketing channels, decisionmakers solve the problems of time, form, and location through storage, processing, and transportation, respectively.

When grain producers harvest their crops, they are faced with making a decision between two alternatives: (1) store the grain for use or sale in the future or (2) sell the grain immediately to a marketing agency. Producers who decide to store their grain may store the grain on the farm and incur costs of doing so or pay a storage charge to a commercial facility for storing the grain. If the decision is to sell, the producer must decide whether to sell the grain to a local farmer, country elevator, subterminal elevator, terminal elevator, processor, or an export-marketing agency.

Agencies in the Marketing Channel

Before going further in the discussion of marketing channels, we need to define several terms used to describe agencies in the marketing channel.

Off-farm sales include all whole grain sold from farms. These sales of grain may be to other local farms, country elevators, subterminal elevators, terminal elevators, processors, or export-marketing agencies.[1]

Country elevators refer to grain establishments that receive a majority of their whole grain from farmers. Grain is normally delivered by the producer to local country elevators by farm truck or by commercial trucking services. Country elevators fall into three general classes:[2] (1) independent or privately owned elevators, (2) farmer owned elevators, and (3) line elevators.

Independent, or privately owned, elevators are single-unit firms owned and operated by individuals who provide their own capital and usually operate the elevators themselves. Elevators owned by farmers include two types of organizations. The first is a cooperative elevator organization established by its farmer members, has its own capital structure, and returns savings to the member patrons as dividends on the basis of their patronage. The second type of farmer-owned elevator is governed by a directorate elected from and by the stockholders with profits distributed as dividends on stock ownership rather than on the basis of patronage.

Line elevators are multiunit chains of elevators with two or more establishments owned by a grain company, usually located along a railroad system. Line-elevator operators receive instructions regarding buying prices and operating practices from the parent company, a corporation, a cooperative, or an independent firm. These may be flour mills, grain processors, or grain import/export firms.

Subterminal elevators are establishments that receive over one-half of their supply of grain from other elevators, usually country elevators.[3] Subterminal elevators are typically located near metropolitan areas. They often are the only large grain-handling facility in the immediate vicinity that does not buy the majority of its grain directly from the farmer. The manager of a subterminal elevator merchandises grain directly to terminal elevators, processors, and exporters rather than selling to interior dealers or a commissioned broker.[4]

Subterminal elevators primarily, and to some extent river elevators, serve as *surge tanks* in locations outside the traditional terminal elevator markets. They provide functions such as blending to achieve greater uniformity of grain and to facilitate marketing the grain. Storage at these intermediate points in the marketing channel facilitates the ability to move grain quickly to the final position when needed. Thus, subterminal and river elevators have a year-round demand for their services.[5]

Terminal elevators are establishments that receive over one-half of their supply of grain from other elevators. They are located in major terminal markets and railroad centers such as Chicago, Kansas City, Minneapolis, Hutchinson, Des Moines, Enid, and Fort Worth. Certain of these markets, such as Chicago, Minneapolis, and Kansas City, provide a futures market as well as a cash market for grain.[6] Terminal elevators provide many types of marketing functions, such as assembly, storage pricing, grading, conditioning, and financing. They vary in capacity from a few hundred thousand bushels to more than 50 million bushels. However, it is the location of the elevator rather than its physical size that determines its classification.

Terminal elevator operators buy grain from cash grain merchants, subterminal elevators, river elevators, and country elevators. Depending on the location and facilities of the terminal elevator, grain may be received by truck, rail, barge, or boat. Grain is sold from the terminal elevator to processors, millers, distillers, feed manufacturers, exporters, and occasionally to elevators in other parts of the country.[7]

Like country elevators, terminal elevators are classified according to ownership: independent, cooperative, and integrated. Independent and cooperative elevators are similar in organization to country elevators. An integrated terminal elevator company may own line country elevators, river houses, and subterminals and be engaged in exporting and leasing rail equipment; it may also own barges and ships. Although overall policy, such as financing of the integrated terminal company, may be provided by the home office, day-to-day merchandising activities are carried on by the individual manager at each terminal location.[8]

Terminal agencies are business organizations that facilitate the assembly and distribution of grain but do not have physical facilities for handling grain. Terminal agencies may or may not take title to grain. Included among these agencies are wholesale dealers, car lot receivers, commission merchants, and brokers.[9]

Port terminals are establishments that receive over half of their supply of grain from other elevators and have facilities to load oceangoing ships. The major marketing functions performed by port terminals include buying, selling, short-time storage, financing, risk taking, and assembling and blending grain in shipload volumes to accommodate the specification of foreign grain buyers. Examples of typical locations of port elevator facilities are Chicago, Duluth-Superior, Toledo, Philadelphia, Baltimore, Norfolk, Mississippi River sites, Gulf Coast sites, Columbia River, Puget Sound, San Francisco-Oakland, and Stockton.

Economic Integration of Marketing Agencies

Marketing channels have changed in recent years because of economic integration. *Economic integration* is the process of combining the management function of two or more economic activities and may be thought of as a fusion or coordination of business units. Integration tends to reduce the length of marketing channels and to increase the economic control, or marketing power, of those firms involved. Market power is achieved by producers or marketing agencies banding together through economic integration for the purposes of achieving a common objective. Like most other processes, economic integration exists in varying degrees.

Integration may take place in the form of ownership or contractual agreements. It may be established by acquisition, merger, or consolidation, and by coordination, cooperation, or contract. The degree to which economic integration exists depends on the advantages of integration and the human motivation of management.

Vertical integration occurs when a firm combines activities that are unlike but are sequentially related to those that it currently performs.[10] Vertical integration can then be interpreted as the process of combining management functions by moving forward or backward in the marketing channel.

Forward integration is a business tie with another firm that is located one step closer to the final marketing stage; *backward integration* is a business tie with another firm that is located one step closer to the source of the raw product or a business tie with a farm operation itself.[11]

Horizontal integration occurs when a single management gains control either through voluntary contract or through ownership over a series of firms performing similar activities at the same level in the production or marketing sequence.[12] Horizontal integration, then, is the process of combining the management function of two or more firms at the same level of the marketing channel. For example, if a processor wants to expand but does not want to take on additional marketing functions, another processing firm may be purchased. This integration may be accomplished by negotiating a contract to combine the management functions of these two or more firms without changing the legal ownership of the firm.[13]

In some cases, firms will be involved in both horizontal and vertical integration, or *dual integration,* a network or combination of both vertical and horizontal integration within one company. Decisions leading to involvement in vertical integration, horizontal integration, or both, may be based on expected technological gain. Decisions to integrate may also be based on increased efficiency in coordinating market decisions, the gain of market supply, improved financial position, greater use of managerial know-how, or simply the drive for power and control.

Cash Grain Merchandisers

A cash grain merchandiser, or cash merchant, is any person or firm dealing in the buying and selling of a cash commodity. A cash grain merchandiser operates in the cash market, acting as both a buyer and a seller of grain, with margins or spreads in the price or basis generating that person's income.[14]

The cash grain merchant may locate in a deficit grain-producing area, or in a surplus grain-producing area, buying grain in the surplus area to supply to the deficit area. The cash grain merchandiser in surplus areas operates between the country elevator, exporter, processor, miller, and other buyers of grain.

One of three types of cash grain merchandisers is the *country merchandiser*. A country merchandiser is generally located in a rural area, buying grain from country elevators and selling it to terminal markets, to grain processors, and to exporters. The country merchandiser takes title to the purchased grain but does not operate storage facilities. Shipping instructions are given to the country elevator for shipment of the grain. The operation consists primarily of making bids and offers and giving shipping instructions by telephone. A merchandiser seldom sees the grain that is bought and sold.

Another type of cash grain merchandiser is the *terminal market merchandiser*. This merchandiser is located in terminal market areas, buying grain from country positions to be shipped to terminal markets such as Portland or Minneapolis. Like the country merchandiser, the terminal market merchandiser takes title to the grain that is bought but does not operate storage facilities.

The third type of cash grain merchandiser is the *terminal elevator company*, involved in similar merchandising operations as other cash grain merchandisers, which also operates terminal and subterminal elevators for storage as well as merchandising.[15]

Commission merchants are also merchandisers, but they differ from cash grain merchandisers in that they do not take title to grain. They function strictly as agents, bringing the buyer and the seller of the grain together. When terms and conditions of the sale are mutually agreed on, the brokerage service has been completed. Commission fees earned by the broker are usually paid by the seller of the grain. Commission brokers are important in the marketing of fats and oils (lard, soybean oil, cottonseed oil, and so on), mill feeds, protein meals, and by-products of various processing operations.[16]

Country elevators must compete with other country elevators for farmers' grain, so the most important service performed by the cash grain merchandiser is in locating the highest bidder for country grain. Country elevators find it useful to utilize the services of cash grain merchandisers to compare bids of alternative outlets for grain, even though some country elevators sell directly to processors or terminal elevators.[17]

The cash grain merchandiser specializes in meeting the needs of country elevator managers. This effort includes providing shippers with information on the availability of grain, arranging shipment of grain, arranging financing, and analyzing the outlook for cash and futures prices. The cash grain merchandiser may also handle futures market transactions for country elevator managers.[18]

Grain Storage

Individual grains are harvested during a relatively short period but are consumed at a rather uniform rate throughout the year. It is storage that makes grain available at the desired time. Storage is the marketing function that matches production patterns with consumption patterns over time.

The Location and Types of Storage Facilities

Storage facilities are located at all stages in the grains complex: on farms; off farms at country, subterminal, terminal, and port elevators; and at milling and processing plants.

Grain is temporarily stored in piles on the ground when adequate local storage facilities are not available to hold the grain. The grain is kept on the ground until adequate transportation facilities are available to move it to storage at other points in the grain-marketing system. The losses from short-term ground storage are usually small but nevertheless greater than if proper storage were available.

On-farm facilities include woodbins, round metal bins, specially lined silos for storage of high-moisture grain, corncribs, and flat storage facilities, which may include machinery sheds for temporary storage. On-farm storage serves two purposes: It preserves grain for later use as feed on the farm where produced, and it holds the grain for later sale. Most wheat and soybeans fall into the latter practice. Corn and other feed grains are stored on farms for both purposes.

Off-farm storage facilities are usually referred to as elevators whether they are country elevators, subterminal or terminal elevators, or storage facilities of grain processors. Elevators are so called because the grain is elevated and poured into bins or transport vehicles.

Elevator facilities include older, steel-covered wood country elevators, large round steel bins, silos of steel or concrete construction, and flat storage. Flat storage facilities are built wider and lower than conventional silos with minimal handling equipment to provide low-cost storage space for grain. Sometimes flat storages are simply a shed-type roof over a floor connected to the side of an elevator. Flat storage facilities are built to provide relatively low-cost storage space during periods of grain surplus and expected low turnover rate. Power shovels, augers and/or mobile power loaders may be required to move the grain from flat storage, so such storage space is used to provide safe storage for grain not expected to be handled or moved immediately.

Grain Storage Capacity

Yields of corn, wheat, and sorghum have increased significantly and rapidly since World War II. The period of harvesting has been shortened with adoption of high-speed harvesting equipment. These developments have called for increased storage volume not only for increased production but also to meet peak demand.

Higher moisture harvesting of corn, sorghum, and soybeans and field shelling of corn since World War II have led to the development of storage-drying-cleaning systems. Storage of ear corn in cribs has greatly decreased, and storage of shelled corn in bins has greatly increased.

Grain storage capacity is needed for each year's production and carry-over stocks. In addition, storage space is needed for "pipeline" stocks of processors, and storage space is needed by grain handlers for receiving, blending, cleaning, conditioning, and loading operations. This latter space is often termed *working space*.

Historically, grain storage capacity in the United States has adjusted with the trend of increased production. Changing volumes of carry-over have, at times, resulted in either shortages of space or excess space. Government programs have played a significant role in the steady expansion of storage capacity, especially at the farm level. They have been used since 1933 to stimulate expanded storage in conjunction with crop loan programs. In 1938, the Commodity Credit Corporation (CCC) acquired its first storage bins to handle corn in default of government loans. Farm storage was encouraged in 1941 by an advance CCC payment of seven cents per bushel to farmers to construct or repair facilities for storage of grain placed under loan.

Government programs also played a significant role in providing off-farm capacity during and after World War II. CCC-owned bin capacity reached 300 million bushels in 1942 with the construction that year of prefabricated wooden bins with a capacity of 100 million bushels. This capacity was reduced to 45 million bushels during World War II by sales to farmers and by use as building materials to relieve a general shortage. Bins were purchased by the CCC again from 1949 to 1957 except for 1952 and 1953. By June 20, 1957, CCC-owned bin capacity peaked at 990 million bushels. All CCC-owned bins were sold to farmers or private individuals by the mid-1970s. Most of the bins remained in use for grain storage.

Storage facility loans to increase storage capacity were made available to farmers by the CCC, at the direction of Congress, starting in 1949. This program continued until 1962. In 1977, the government again initiated storage facility loans to farmers. Favorable interest rates were offered to encourage program participation, and the programs were well received by farmers.

In the 1950s government programs to increase storage capacity were directed toward private investment in off-farm commercial storage facilities. Off-farm storage capacity expanded significantly during this period. Government actions to stimulate building off-farm storage capacity included providing occupancy guarantee programs for commercial grain facilities, amending the Farm Credit Act to permit the Bank for Cooperatives to loan up to 80 percent of the cost of newly constructed agricultural storage facilities by farmer cooperatives, authorizing Small Business Administration loans for the purpose of construction of grain storage facilities, and amending the Internal Revenue Code of 1939 to permit rapid depreciation (20 percent per year) of newly erected storage facilities.

At times, the government has also provided temporary storage for grain: from 1949 to 1951 and 1954 to 1962 through CCC lease of Maritime Commission ships, and in 1950 and 1951 through leasing of military buildings.

Government action to stimulate increased grain storage capacity has not been the only factor in expanded U.S. grain storage capacity. World population continues to increase. Many developed countries are continuing to increase meat consumption and decrease cereal consumption by humans. Net gains in grain consumption have occurred from animal feeding. In addition, there have been substitutions among cereal grains, for example, the dietary shift from rice to wheat in Japan. Developing country governments, in many instances, have taken action to improve the diets of their population through the use of cereals. These factors and others have increased world demand for grains. In response to these new changing demands, storage capacity was increased accordingly.

The export market has become a major market for U.S. grains. Efficient operators in the U.S. system have seen profit opportunities in this development and have responded to the economic incentive. Economic incentives associated with the export market have led to the expansion and greater utilization of grain storage space required for orderly foreign marketing. The majority of U.S. grain storage capacity is in areas of concentrated grain production, particularly in the Cornbelt states and wheat and sorghum states in the Great Plains. Since the dramatic expansion of exports in the early 1970s, the greatest expansion in storage capacity has occurred in areas of heaviest production density, particularly the Cornbelt states of Iowa, Indiana, and Illinois.

Government actions, such as the soybean embargo of 1973 and the partial embargo of U.S. grains to the former Soviet Union in 1980, create marketing uncertainties that have also influenced some farmers to build additional storage to "weather unexpected marketing storms."

The size of farming operations has increased significantly in the United States since World War II. Many operators, in their desire for more control over the physical product, have taken on increased marketing responsibility. Farm storage capacity has been increased to gain the control desired.

Typically, storage capacity of grain processors has tended to increase at the rate of industry growth. Most processors have storage space and working space to accommodate a 30- to 45-day inventory of raw materials. Most processors also have space, usually flat storage, for processed products. This space, however, is not included in government grain storage statistics. Yet, it is a part of the total pipeline, serving the function of storing grain and grain products as they move through the grain-marketing channels. In recent years, storage capacity of wet corn processors has increased faster than for most other processors. This unique trend is expected to continue for several years because of technological changes in corn sweetener production.

Although grain storage capacity has expanded at all stages in the U.S. grain complex in recent years, the greatest growth in capacity has been near the point of production. The closer that storage is to the point of production, the greater is the flexibility in making final market choices.

Structural Changes of the 1980s

Structural changes in the U.S. grain-marketing system have been more extensive and far reaching in recent years than in any time in history. The 1980s can best be characterized as a period of consolidation and increased concentration in grain marketing. To understand the economics of these changes, one has to look at the stimulus to investment in marketing infrastructure resulting from the grain export boom of the 1970s.

U.S. grain exports more than tripled in the 1970s reaching an all-time record in 1980. This increase in exports stimulated investments in rail cars, barges, storage, and port facilities, much of which did not come on line until the 1980s when grain exports began an extended period of decline.

U.S. grain exports declined to 3.0 billion bushels in 1986 from their record high of 5.0 billion bushels in 1980. Competition for the reduced volume drove marketing margins down; at the same time, the new surplus infrastructure investments became burdensome. The industry continues to be plagued by excess capacity and associated narrow marketing margins. These structural adjustments in the system, which were extensive in the 1980s, continue in the 1990s.

Changes in Grain Facilities and Structure

The 10 largest U.S. grain companies, as listed in the *1991 Grain Guide*, operated 857 grain facilities with aggregate storage capacity of 1.579 billion bushels. Subterminal elevators have increased in both number and importance in response to special unit-train rates offered by the railroads. This trend will likely continue as more grain moves directly from country gathering points to ports or to domestic processors without moving through terminal markets.

Privately owned grain companies and regional grain-marketing cooperatives expanded rapidly through acquisitions during the decade of the 1970s until the early 1980s. However, in 1981, regional grain-marketing cooperatives began downsizing their operations through joint ventures with major multinational corporation mergers between regionals.

The cooperative grain-marketing system in 1990 is vastly different from that of a decade earlier when U.S. grain exports peaked. The downsizing of interregional and regional grain-marketing cooperatives during the decade was necessitated by heavy investment in grain-marketing infrastructure during the grain export boom. Increased competition and reduced marketing margins on the smaller volume of grain exports in the 1980s resulted in reduced revenue and operating difficulties that necessitated structural adjustments.

Changes in U.S. Grain Export Market Structure. A major government study in 1982 categorized the market structure of the U.S. grain export system into four groups: (1) major multinational corporations, other than Japanese; (2) Japanese-owned or affiliated firms; (3) farmer-owned cooperatives; and (4) all other exporting firms.[19]

Data on changes in U.S. grain export market shares during the decade of the 1980s are not available. However, they would probably show that the share of U.S. grain exports handled by farmer-owned cooperatives has declined. The share of total grain-export elevator storage capacity controlled by cooperatives declined from 21 percent in 1981 to 15 percent in 1989 (Table 3.1). Most of this capacity is now located in the Great Lakes, the ports through which the smallest amount of U.S. grain exports flow. Cooperatives no longer control export space at the Mississippi Gulf through which the largest share of grain exports leave the United States.

The share of port storage capacity held by the five major multinational grain-exporting firms also declined from 50 percent to 46 percent during the same period. On the other hand, the share of port capacity held by "other" firms increased from 28 to 39 percent. Two of the largest grain-exporting firms in the "other" category include the Archer Daniels Midland Company and ConAgra, Inc. Both have expanded their grain operations in recent years. They would probably be classified as major multinational grain exporters in 1991.

In fact, excess capacity in the system today has squeezed marketing margins and intensified competition. The surplus in grain-exporting capacity has been estimated at close to 50 percent.[20] In addition, more of our grain exports today are dependent upon favorable credit terms or subsidies provided by the U.S. government.

Grain Storage Capacity Increases. The first national survey of grain storage facilities in the United States was made in 1978. It showed aggregate farm and off-farm storage capacity at nearly 17 billion bushels,

TABLE 3.1   Total Export Elevator Capacity Controlled by Exporter Group, 1981 and 1989 (in percentages)

Exporter Group	1981	1989
Five Major Multinationals[a]	50.3	46.0
Farmer-Owned Cooperatives	21.4	15.3
Others[b]	28.3	38.7
Total	100.0	100.0

[a]Includes Cargill, Continental, Bunge, Dreyfus, and Garnac.
[b]Includes public elevators and elevators operated by port authorities.

SOURCES: 1981 figures, Neilson C. Conklin and Reynold P. Dahl, "Organization and Pricing Efficiency of the U.S. Grain Export System." *Minnesota Agricultural Economist*, Agric. Ext. Service, University of Minnesota, No. 635, May 1982, p. 3. 1989 figures, Export Elevator Directory, U.S. Dept. of Agric., Federal Grain Inspection Service, January 1989.

made up of 10 billion bushels of storage on-farm (59 percent of the total) and 7 billion in off-farm facilities (41 percent of the total) (Tabel 3.2). This capacity was equivalent to a full year and one-half of grain production in the United States, which was about 12 billion bushels per year in 1978.

As grain exports declined in the 1980s, stocks accumulated despite sizeable acreage idled under federal farm programs. Grain stocks reached an all-time high of 8.4 billion bushels at the end of the 1986-1987 marketing year (Figure 3.1). Most of these stocks were stored under government programs such as the farmer-owned reserve, regular price support loan, and CCC ownership.

Grain storage capacity increased in response to the stock buildup, reaching a record 22.9 billion bushels on December 1, 1988, an increase of 36 percent from 10 years earlier. The total of on-farm storage capacity of 13.3 billion bushels (58 percent of the total) and off-farm capacity of 9.6 billion bushels (42 percent of the total) could handle about two years of U.S. grain production.

TABLE 3.2   Grain Storage Capacity in the United States, On-Farm and Off-Farm, by State (in millions of bushels)

	April 1, 1978			December 1, 1990		
	On-Farm	*Off-Farm Commercial*	*Total*	*On-Farm*	*Off-Farm Commercial*	*Total*
Iowa	1,492	635	2,127	1,900	1,011	2,911
Illinois	1,154	787	1,941	1,150	1,212	2,362
Minnesota	1,192	368	1,560	1,400	564	1,964
Nebraska	833	488	1,321	1,120	802	1,922
Kansas	370	831	1,201	440	892	1,332
Texas	264	838	1,102	220	924	1,144
N. Dakota	691	142	833	860	230	1,090
Indiana	507	283	790	690	387	1,077
Wisconsin	437	130	567	500	180	680
Missouri	347	210	557	400	287	687
Others	2,637	2,275	4,912	3,720	2,616	6,336
Total	9,924	6,987	16,911	12,400	9,105	21,505

SOURCE: *Grain Stocks*, National Agricultural Statistics Service, USDA, January 1991.

FIGURE 3.1   U.S. grain exports and ending stocks.

Nearly six out of ten bushels of U.S. grain storage capacity represents farm storage, reflecting the steady expansion of these facilities in recent years under farm program incentives. Farmers found it advantageous to have farm storage in order to participate in the regular nine-month farm price support program. Having their own storage gives farmers more flexibility in grain marketing.

As grain stocks accumulated in the 1980s under federal farm programs, the grain trade derived more income from storage and handling grain for the government. The income from such operations offset, in part at least, declines in income associated with reduced grain exports and marketing margins. But the precipitous drop in grain stocks as a result of the 1988 drought resulted in excess grain storage capacity and reduced storage income for the grain-marketing system.[21]

Transportation Economics Induce Structural Change. Structural change in the U.S. grain-marketing system has also been induced by the changing economics of grain transportation. Transportation cost is the largest single component of grain-marketing costs because grain has a high bulk relative to its value. Hence, grain-marketing firms must be alert to available opportunities to minimize transportation costs if they are to remain competitive. Changes in transportation costs as impacted by intermodal competition and railroad deregulation over the past two decades have been important factors inducing structural change in the grain-marketing system at both the country and terminal market levels.

Changes in transportation technology and costs accelerated the decline of cash grain trade at grain exchanges in terminal markets following World War II.[22] First, the increased volume of grain shipped by truck bypassed terminal rail markets and was not traded at grain exchanges whatsoever. Second, new multicar rates offered by the railroads to compete with increased truck-barge competition were point-to-point rates that did not include the transit privilege. Transit privileges were an integral part of the railroad rate structure under which grain could be stopped at an intermediate point between its origin and its final destination for inspection, storage, or processing without additional charge. The through rate applied under transit billing. As more multicar rates were offered by the railroads, the transit privilege was eroded and virtually eliminated. The demise of the transit privilege and deregulation of the railroads as authorized by the Staggers Rail Act of 1980 sharply reduced the flow of grain from country points to grain exchanges in terminal markets for resale.

Decentralization of Cash Grain Trade. Most grain now moves directly from gathering points in the country to domestic users such as flour mills or to export elevators without moving through a terminal market such as Minneapolis, Kansas City, or Chicago for resale.[23] Grain merchants are still located at grain exchanges in these markets, but trading in individual cars, or unit trains, is most likely to occur near origin points in the country rather than by sample on the grain exchange floor.[24]

In addition to diminishing the role of grain exchanges in the marketing of cash grain, railroad deregulation has diminished the role of terminal elevators at these markets, particularly terminal elevators built many years ago to handle rail grain. Many of these elevators are now obsolete for grain merchandising and are suitable only for long-term storage, primarily of government-owned grain.

Cash grain marketing has become more decentralized, with subterminal elevators located in the country and shipping grain in unit trains taking over the functions formerly held by many older terminal elevators. Subterminals are also likely to replace many country elevators, which will continue to decline in number.[25]

Costs of Storage

The costs of grain storage, regardless of the position of the storage in the marketing channel, may be classed as fixed and variable. *Fixed costs* must be met whether or not grain is stored. *Variable costs* are incurred only if grain is stored. Fixed costs are related to the storage facilities themselves as a result of ownership. Each item included in this category relates to the storage facilities and not to the grain or oilseed in storage. These items include depreciation, maintenance, interest on capital investment, insurance on structures and equipment, and property taxes.

Variable costs are related to the grain stored. They include: insurance and taxes on grain stored; loss from quality deterioration; costs of quality maintenance; interest income on the grain investment foregone while the grain is in storage; labor, transportation, and handling expenses due to the storage operation; and costs of shrinkage.

The costs of grain storage vary depending on position in the grain-marketing channel, the individual storage operation, the particular year, and general business conditions. Economies of size favor commercial off-farm elevators in grain drying and storage; however, on-farm drying and storage capacities are expanding more rapidly. An explanation of this phenomenon, which seemingly contradicts the least cost principle of economics, encompasses several factors that are mentioned in detail in the following section.

Producer Decisionmaking

Farmers' decisions concerning grain storage are basically: (1) whether or not to store, (2) how much to store, (3) where to store, and (4) how long to store.

Seasonal price changes provide the incentive for storage; thus, the first question involves a comparison of costs of storage with possible gains from a price rise later on. Differences in seasonal prices, however, present problems in selecting a marketing strategy. The uncertainty of predicting future short-term price disturbances and market response limits the success of any strategy to minimize risk and income variability or to maximize average value received per bushel. Factors that influence how much to store are needs for on-farm processing and feeding, distance to the elevator, and income tax management.

Where to store is a question with no definitive answer. Profitability of grain storage is not the same for all individuals in the grains complex. Each individual faces a different set of circumstances. A farmer who does not already have farm storage capacity, has a weak equity position, and has high capital requirements elsewhere in the operation would be better off not investing in on-farm storage.

On-farm storage affords the producer direct control over the physical commodity and gives farmers greater flexibility. It frees them from having to deal with only one local elevator, giving them time to shop around for the best bid. In some production areas, farmers can avoid waiting in line to unload at harvest time. If grain is placed in farm storage bins, the speed of harvest can be increased so that expensive farm machinery and equipment are not idled and valuable time is not wasted.

Government programs that offer farmers the option to store grain on the farm with specified storage rates, possibly in addition to other payments, may influence on-farm storage decisions. In some instances, off-farm storage charges may be higher than on-farm costs. This difference may be because of elevator operator uncertainty as well as incomplete information on farmer response to price changes and the investment in on-farm facilities based on past elevator charges.

How long to store would normally be based on marketing strategy, and whether the added benefits exceed added storage costs. However, this marketing strategy may be influenced by government grain loan or reserve programs. If grain is stored under a government program, the conditions for release from storage will be specified; thus, the storage time may be fixed or variable, depending on the specific program.

Costs of Holding Grain

Grain producers are constantly faced with the decision of when to sell their grain.[26] During periods of violent price changes, choosing the time to sell is a difficult task.

If a grain producer decides to delay the sale of her or his grain until a later date, she or he must decide whether to store the grain in a commercial elevator and pay storage costs or store the grain on the farm.

Commercial storage has several advantages for the producer. These may include the following:

1. There are no shrinkage or grade changes to be absorbed.
2. The grain is handled only once by the producer.
3. No investment in facilities is required.
4. The grain can be sold any time after delivery.

The disadvantages of commercial storage for the producer may be as follows:

1. Fixed storage costs per day must be paid.
2. A producer cannot choose a different location to sell his or her grain without a penalty.

Storing grain on the farm is not free of cost and, in fact, once the decision is made to construct grain storage on the farm, it must be utilized to keep the storage cost per bushel to a minimum.

There are several advantages to having grain storage facilities on the farm. These include the following:

1. The producer can choose the location and the buyer of the grain.
2. On-farm storage may be more convenient during the harvest period.
3. The farmer does not have to pay commercial storage rates.
4. There is less concern about the lack of available commercial storage.
5. In the case of wheat, seed is readily available from the grain in storage.

Some disadvantages also accompany producer grain storage facilities. These are as follows:

1. Grain may get out of condition, and the producer will receive a lower price.
2. The producer must absorb all shrinkage in weight.
3. Grain must be handled twice by the producer.
4. The producer assumes risk from natural disasters such as flood, fire, and wind.

Major Costs

There are two major costs in holding grain for sale at a later date: the storage costs and the opportunity costs of the value of the grain. The opportunity cost for stored grain will be different for every grain producer and depends greatly on her or his financial condition.

Table 3.3 shows the opportunity costs of capital in holding grain for six months after harvest, in cents per bushel. The opportunity cost figures are based on annual rates of interest ranging between 6 and 10 percent, which are applied to grain prices ranging from $3.00 to $5.00 per bushel.

Opportunity cost of capital and storage costs for holding grain need to be considered together. If the storage cost of grain is 1.5 cents per bushel per month or 9.0 cents per bushel for six months, then this amount added to the opportunity cost of capital indicates the approximate cost of holding grain for the later sale date. This storage cost for six months added to the opportunity cost of capital for six months at various interest rates for grain prices between $3.00 and $5.00 per bushel is presented in Table 3.4. For example, if the price of grain is $4.00 per

TABLE 3.3 Opportunity Cost of Capital to Hold Grain for Six Months (in cents per bushel)

Grain Price ($/bushel)	Rate of Interest				
	6%	7%	8%	9%	10%
3.00	9.0	10.5	12.0	13.5	15.0
3.25	9.8	11.4	13.0	14.6	16.3
3.50	10.5	12.3	14.0	16.3	17.5
3.75	11.3	13.1	15.0	16.9	18.8
4.00	12.0	14.0	16.0	18.0	20.0
4.25	12.8	14.9	17.0	19.1	21.3
4.50	13.5	15.8	18.0	20.3	22.5
4.75	14.3	16.6	19.0	21.4	23.8
5.00	15.0	17.5	20.0	22.5	25.0

TABLE 3.4 Opportunity Cost of Capital at Different Interest Rates Added to the Storage Cost to Hold Grain for Six Months (in cents per bushel)

Grain Price ($/bushel)	Rate of Interest				
	6%	7%	8%	9%	10%
3.00	18.0	19.5	21.0	22.5	24.0
3.25	18.8	20.4	22.0	23.6	25.3
3.50	19.5	21.3	23.0	25.3	26.5
3.75	20.3	22.1	24.0	25.9	27.8
4.00	21.0	23.0	25.0	27.0	29.0
4.25	21.8	23.9	26.0	28.1	30.3
4.50	22.5	24.8	27.0	29.3	31.5
4.75	23.3	25.6	28.0	30.4	32.8
5.00	24.0	26.5	29.0	31.5	34.0

bushel at harvest and the opportunity cost of capital is 10 percent, the producer must acquire an additional 29 cents per bushel, or a price of $4.29, to hold his or her grain for six months and attain a break-even price compared to harvest. Thus, the producer must realize a higher price for his or her grain the longer the sale is postponed beyond harvest. The only exception is if there are greater income tax benefits for sales made in the next tax year.

Quality Maintenance of Grain in Storage

Grain can be durable or highly perishable while it is being held in storage. Durability of the stored grain can be promoted by using proper storage facilities and good storage practices. For quality maintenance, on-farm storage structures should do the following:[27]

- Hold the grain without loss from leaks and spills;
- Prevent rain, snow, or soil moisture from reaching the grain;
- Protect grain from rodents, birds, objectionable odors, and theft;
- Provide safety from fire and wind damage;
- Permit effective treatment to prevent or control insect infestation;
- Provide headroom over the binned grain for sampling, inspection, and ventilation;
- Be easy to clean and inspect;
- Be equipped with an aeration system to cool the grain, minimize moisture migration, facilitate fumigation, if needed, and limit insect development.

The ease or difficulty in achieving quality maintenance may begin with the farmer's selection of a given grain variety. Some varieties are more resistant to insects while in storage than others.

Upper limits on storage quality may be set during harvest. Inspection and cleaning of harvesting equipment prior to harvest will eliminate residues that contaminate newly harvested grain with insects and microorganisms. Chaff, weeds, weed seeds, and other foreign matter that affect grain drying and provide a good environment for insect and mold development in stored grain can be eliminated by proper combine adjustment. Proper combine adjustment will also minimize cracked and broken kernels that contribute to insect and mold problems, the dustiness of grain, and ultimately to grain grade reduction.

Control of temperature and moisture in stored grain is a prerequisite to grain quality management. Harvesting grain as close as possible to a safe moisture level minimizes drying costs and storage losses. Harvesting grain when it is too wet provides an environment for insect development and fungal invasion and development. Fungal spores can be distributed throughout the stored grain by migrating insects.

Management techniques that can reduce grain moisture for conventional storage include drying with or without heat, aeration, stirring, or cooling. Acid preservatives, high-moisture storage facilities, and plastic covers may be used for high-moisture grain preservation.

Maximum moisture content for safe storage up to one year for aerated good-quality grain is considered to be 14 percent for corn, 13 percent for wheat, and 12 percent for soybeans. For longer storage, the moisture content should be lower.

The proper application of insecticide during farm binning operations or later fumigation in conjunction with other good storage practices is an effective management practice to protect grain from insect damage for about one storage season.

Proper preparation of on-farm storage bins for newly harvested grains is a necessity for grain quality maintenance. All leftover grain should be removed; walls, ceilings, sills, ledges, and floors should be thoroughly swept and the sweepings destroyed; needed repairs should be made; walls should be treated inside and outside with an approved residual insecticide; and trash and litter should be removed from outside the bin area. After binning is completed, the grain should be leveled and bin surfaces treated with an approved insecticide to help prevent insects from entering or feeding on the grain surface. Because grain temperature is a good indicator of grain condition, the temperature should be checked at least once a month and more often during summer and early fall months when temperatures are conducive to insect development.

Temperature sensor cables installed in the storage bin provide an excellent means of temperature monitoring. A thermometer attached to a stick and inserted into the grain can also be used to measure temperature in farm bins.

Proper control of the two life-sustaining elements, temperature and moisture, is fundamental for stored grain quality maintenance. When grain temperature is over 60°F, checks should be made for insects and mold. If checks also reveal high moisture, grain may require drying to control or prevent mold growth. However, in both cases, control of temperature and moisture by aeration may be all that is necessary. If checks indicate the presence of insects, fumigation may be necessary. Fungi increase both temperature and moisture content by their growth. Temperatures of more than 60°F are required by most stored-grain insects to develop damaging populations. Many species require 70°F or higher temperatures.

Applicators of grain fumigants are required to take special training and to be certified before they can purchase and use chemicals for grain fumigation because the chemicals are classified as restricted pesticides by the Environmental Protection Agency. As a result of government regulation, it may be safer and more economical for some farmers to hire commercial applicators to fumigate.

The quality of grain received by a country elevator determines where it will be binned at the elevator. Moisture content, test weight, and dockage are usually checked before binning. Checks are also made for insect damage and live, stored-grain insects.

Grain may be sampled by several methods, not all of which are officially accepted by the USDA. Some elevator operators take samples by simply obtaining a random "coffee can" of grain from the top of the grain load or by inserting the can in the grain stream emptying from the delivery vehicle; some operators probe the grain using a mechanical or hand-operated probe; some use Pelican, Ellis cups, or Woodside mechanical samplers; and some use diverter-type mechanical samplers installed to traverse the grain stream in the elevator. If diverter samplers are used, they must be approved and operators licensed before the sample can be considered an official sample for official grades. The coffee-can method and certain probing devices are not approved to obtain samples for official grades; yet, these methods are often used both to determine the bin in which to store a lot of grain and as a basis for paying the farmer.

Infrared reflectance devices became generally available in the 1970s and made possible for the first time rapid determination of protein, moisture, and oil content of grains. Elevator operators can use these devices advantageously in binning grain into homogenous lots excluding moisture variation and in other grain-handling and load-out operations.

Moisture is highly variable, but not a serious problem as long as the moisture content is not too high and the grain is turned often or aerated. Moisture can vary as much as 4 percent within a few feet in a grain field. Such moisture variations continue to exist in any lot of grain, whether it is in rail car, truck, or barge. Thus, it is difficult to obtain a precise moisture content reading for a large lot of grain because the moisture content of individual kernels in a sample varies, and precise segregation of grain by moisture content at the elevator is impossible because no elevator has enough bins. As a result, an elevator bin may contain truckloads of grain of various moisture contents. If the grain is properly handled, this variation presents no problem as the wet and dry grain will reach a moisture equilibrium within two or three days after being placed in storage. Temperature differences of the various loads will cause air currents and moisture movements in the grain. Special care should be taken where spoutlines develop when grain is poured into an elevator.

A spoutline is a core of fine particles that accumulates at the peak of the grain pile and fills the spaces between kernels as a bin fills. Whole kernels slide down the incline of the pile. The diameter of the core is proportional to bin width and the percentage of fines in the grain. The center of the core is a dense mass that impedes or may prevent air circulation or escape of heat in the grain.

Temperature cables or probes are necessary for regular checks on grain temperature in large commercial bins. Devices exist that allow readings to be taken at various points in the grain mass and recorded automatically in the elevator office.

Prior to 1950, and the adoption of aeration, it was common practice in the elevator industry to "turn" grain as a part of quality maintenance programs. Turning is a process of emptying a grain bin, elevating the grain, rebinning, and possibly blending with another bin. In the turning process there was a certain amount of ventilation of the grain as well as dilution or dispersal of trouble spots. This process was expensive in terms of equipment operation (particularly energy) and maintenance. It also increased the number of broken kernels and the dust content of grain. Turning creates dust, a concern in the grain industry because of its role in elevator fires and explosions and its link to lung diseases in elevator employees. Consequently, turning grain as a routine quality-control measure is now practiced mainly at older facilities where aeration systems have not been installed.

Once grain loses quality, it can never be improved. However, poor-quality grain and superior-quality grain can be blended to raise the overall grade of the lots of grain. Whenever possible, elevator operators will blend so that the lots of grain meet the minimum of specific official grades. This practice permits such grain to move through marketing channels to end uses where quantities of lower quality grain are not objectionable. For the grain elevator manager, it is a profitable practice. However, this practice necessitates grain quality surveillance at all points in the grain-marketing system to prevent quality decline, particularly from hidden insect infestation and storage fungi.

Finally, grain can be damaged in handling due to elevator equipment design and operation and speed of operations. Artificially dried corn is more subject to breakage in handling due to brittleness and stress cracks. The more times grain is handled, the more breakage occurs and the more susceptible the grain is to deterioration.

Warehouse Regulations

Little thought to factors other than shelter from the elements may be given by farmers who store grain at public warehouses. Perhaps of greater importance, too little thought may be given to the qualifications of people who weigh, sample, or determine quality factors on which price will be determined. The farmer may have little, if any, knowledge about the financial responsibility of the warehouse manager. Questions should be answered to the farmer's satisfaction concerning warehouse storage. Why are the products being placed in storage? What is expected to be returned? Is the warehouse manager financially responsible? Does the warehouse manager have adequate storage and handling facilities and competent employees to care for the grain while it is in storage? Will a warehouse receipt be given and, if so, what will be its terms? Will it show the correct weight and grade? Will the warehouse receipt constitute a definite, enforceable contract? If the warehouse manager does not want to give a warehouse receipt, should grain be entrusted to that person? How much will it cost to store? What are the terms of the warehouse manager's tariff? Is the warehouse manager bonded, by whom, for the benefit of whom, in what amount, and with whom is the bond filed? Is the warehouse manager subject to supervision, and the warehouse and contents subject to inspection by the appropriate government agency? What is the supervision?[28]

The need to protect the interests of producers and other grain owners who store grain with public warehouse managers was the driving force that brought state and federal government units to regulate grain warehouses many years ago. States have been regulating public grain storage for over 130 years, and federal regulations have been in place since 1916.

Warehouse Licensing

The regulation of public grain storage begins with operating license requirements. The license is granted to a firm with its owners and responsible officers identified; thus, a transfer of ownership of storage facilities requires a new license to operate the same facilities.

There are two types of licenses: state and federal. Many states have laws requiring licenses; federal licenses are voluntary.

License requirements vary from state to state. Generally, states require new application for licensing each year. Bonding and insurance requirements vary from state to state. Inspection of facilities prior to licensing depends on the state. Reports required and reporting frequency are also variable. Inspection of operations after licensing depends on state regulations.

To qualify for a federal license under the U.S. Warehouse Act, a warehouse manager must have a suitable and properly equipped warehouse, determined to be so by a review of facts gathered on an inspection by a federal warehouse examiner. The warehouse manager must submit a complete current financial report on a prescribed form and have a good business reputation. Financial requirements for a person to be federally licensed as a warehouseman mandate that the person have allowable net assets equal to the greater of $50,000 or 25 cents multiplied by the warehouse capacity in bushels. The Act also requires that the warehouseman file a bond with the secretary of agriculture, to be renewed each year. Such bond must be in the amount required by the secretary and approved by him or his designated representative.[29]

Qualified personnel with knowledge of how to weigh, inspect, and grade grain are required. Inspectors and weighers must be licensed under the Warehouse Act. Their licenses are valid only for carrying out the requirements of the U.S. Warehouse Act, not for complying with the U.S. Grain Standards Act.

Warehouse Receipts

Federally licensed warehouse managers must have adequate equipment to properly weigh and grade grain. They must post a tariff (schedule of charges) for receiving, loading out, storage, insurance, conditioning, and all other warehouse services. The tariff must be furnished to the USDA and is subject to disapproval. Initial inspection and license fees must be paid at the time of application.

Federally licensed warehouses are monitored by unannounced warehouse examinations as often as twice a year. Such examinations are also performed in some states by state inspectors. The office, books, records, papers, and accounts relating to the warehouse as well as the contents of the warehouse are examined. When minor discrepancies or adverse conditions are found, the warehouse manager is required to bring operations into compliance within a specified period. Serious violations may lead to license suspension. Improprieties in connection with warehouse receipts, inspection, or weighing carry penalties of imprisonment and fines.

Grain in storage is much like money in a bank. The depositor has a legal right in both cases to get back in kind and value the things deposited. If grain is stored with identity to be preserved, the owner is entitled to get back the identical product stored.

When money is deposited at a bank, the bank is required to give the depositor a deposit slip. And if the bank teller fails to give a deposit slip to the depositor, the depositor will not likely forget to demand a copy. However, when farmers deliver grain to an elevator, they may not be offered a warehouse receipt for their grain, and they may not think about the receipt unless they want to borrow money on the grain; then it may be too late. As a matter of good business practice, warehouse receipts should be demanded any time the ownership of delivered grain has not been transferred. Also, the owner of the grain should see a copy of the warehouse tariff.

A warehouse receipt is formal acknowledgment by a warehouse manager of grain received for storage. A scale ticket, which lists weight, and may also provide such other information as test weight, moisture, protein, and grade, is not formal acknowledgment of receipt for storage and is not a legal document of title. Scale tickets (Figure 3.2) and warehouse receipts (Figure 3.3) are not synonymous and should not be

FIGURE 3.2 A typical scale ticket.

FIGURE 3.3 A warehouse receipt.

confused. The following items are typically stated on the face of a warehouse receipt: location of the warehouse in which the grain is stored, date of issue of the receipt, the rate of storage charges, description of the grain (quantity, grade, kind), facts of ownership, any lien(s) that the warehouse manager claims, consecutive number of the receipt, and the signature of the warehouse manager.

The warehouse receipt may provide for delivery to: (1) a specified person, (2) a specified person "or that person's order," or (3) to the "bearer." The latter two forms are considered negotiable and can be transferred with all rights of property from one party to another without notice to the warehouse manager. All rights to property of the original depositor go with the transfer.

The Uniform Commercial Code (UCC), adopted by the American Bar Association in 1952 and by all states except Louisiana, supersedes the Uniform Warehouse Receipts Act that was adopted voluntarily in all states to facilitate the easy transfer of warehouse receipts, which in turn facilitates financing and risk-bearing functions in grain marketing. Under the UCC, a warehouse receipt is a certificate of title for agricultural products stored in a warehouse. Some states require registration of warehouse receipts in lieu of or in addition to sending a copy of the warehouse receipt to state authorities.[30]

The contents of the federal warehouse receipt are essentially those set out in the former Uniform Warehouse Receipts Act and now covered in the UCC. In markets requiring registration of warehouse receipts, all federally licensed warehouses in that market comply with the local regulation. The federal warehouse receipt is uniformly dependable and acceptable in financial circles as reliable collateral for loans. Bankers who deal with warehouse receipts from several states prefer receipts from federally licensed facilities over state-licensed facilities because of uniform administration of federally licensed warehouses.

Commodity Credit Corporation Influence on Warehouse Managers

Government influence on warehouse managers includes requirements of CCC storage contracts with warehouse managers. The statutory obligation of the CCC, as owner of and lender on large quantities of grain from time to time, has led to a body of rules that must be met by warehouse managers storing grain under CCC ownership or loan.

The contractual arrangement which the warehouse manager enters into with CCC to store grain is known as a Uniform Grain Storage Agreement (UGSA); rice is covered by the Uniform Rice Storage Agreement (URSA). The CCC does not license warehouses that store CCC-managed grain, but it does require compliance with its contract terms that are comparable to those for licensing in many states. A warehouse must be examined by a person designated by CCC before being approved by CCC for storage or handling of grain.

The basic standards for CCC warehouse approval require a net worth that is the greater of $50,000 or 25 cents per bushel times the maximum storage capacity that the elevator can accommodate in the customary manner.[31] If the calculated net worth requirement is greater than $25,000, any deficiency between $25,000 and the calculated minimum can be satisfied by bonds acceptable to and meeting CCC requirements, or by cash and negotiable securities, or with a legal liability insurance policy, providing the policy contains a clause or rider making the policy payable to CCC.

The warehouse manager must also have sufficient funds available to meet ordinary operating expenses, maintain accurate records of inventory and operations, and use only prenumbered warehouse receipts and scale tickets. In addition, a work force and equipment must be available to complete loadout within approximately 60 working days for that quantity of grain for which the warehouse is or may be approved under a UGSA, and within approximately 90 working days for that quantity of rice for which the warehouse manager is or may be approved under the URSA. The warehouse manager, officials, or supervisory employees in charge of the warehouse operations must have experience, organization, technical qualifications, and skills to provide proper storage and handling services. These personnel must also have satisfactory records of integrity, judgment, and performance and no prior suspensions or debarment under CCC regulations.

The warehouse must be of sound construction, in a good state of repair, and adequately equipped for grain handling and storage operations. The warehouse must be under the control of the contracting warehouse manager at all times, and not subject to greater than normal risk of fire, flood, or other hazards. Adequate and operable firefighting equipment for the type of warehouse and grain must be available.

Also, a warehouse manager with a Certificate of Competency issued by the Small Business Administration will be accepted by CCC for basic standards pertaining to net worth, operating funding, adequate work force and equipment, qualification of personnel, and suitability of the storage facilities. If a warehouse manager with the required minimum $50,000 net worth fails (or if the warehouse fails) to meet one or more of the CCC standards, the warehouse may be approved if CCC determines the warehouse services are needed and satisfactory protection can be provided for the grain. Additional bond coverage will be prescribed by CCC in such cases that must be made by CCC-approved instruments.[32]

Payment for storage and handling services of CCC grains was made according to set schedules of payment until July 1975. These schedules were adjusted periodically to reflect changes in costs. Since July 1975, CCC has used an offer-rate system that introduced flexibility into payment for storage and handling services.

CCC requires that U.S. grain standards be used to establish the quality of grain received. States generally require warehouse managers to accept grain in storable condition up to licensed house capacity. Under the UGSA, however, the warehouse manager is not obligated to accept any specific quantity of CCC-managed grain.

Summary

Storage is the marketing function that matches production patterns with consumption patterns over time.

Grain storage facilities are located on farms and off farms. The types of on-farm grain storage facilities include woodbins, round metal bins, specially lined silos used for high-moisture grain storage, corncribs, and flat storage facilities that may include machinery sheds when needed. Off-farm storage facilities, referred to as elevators, include steel-covered wood elevators, large round steel bins, silos of steel or concrete construction, and flat storage.

Grain storage capacity is needed for each year's production and for carry-over stocks. Storage space is also needed for working stocks and for receiving, blending, cleaning, conditioning, and loading operations. U.S. grain storage capacity has increased with the trend of production. Government programs have been used quite effectively to stimulate increases in both on-farm and off-farm capacity.

A well-built, well-prepared storage facility does not guarantee sound-quality grain. Grain must be properly conditioned and checked regularly to maintain its quality. Temperature and moisture are basic elements that must be controlled in stored grain to prevent loss of quality from insects and molds.

A sound storage facility may protect grain from physical deterioration or loss, but safe grain storage is more than protection of the physical commodity. Safe storage also protects the owner's investment in the grain. The warehouse receipt, like a bank deposit slip, is formal acknowledgment of receipt of a deposit. Warehouse receipts, other than those made out for delivery to a specified person, are considered negotiable and can be transferred with all rights of property from one party to another. Warehouse receipts should always be demanded when grain is deposited in a public warehouse. Similarly, warehouse receipts should not be confused with scale tickets, which are not negotiable, are not formal acknowledgment of receipt, and do not transfer any property rights.

As farming operations continue to increase in size, and farmers become more marketing conscious, on-farm storage capacity will increase relative to off-farm storage capacity.

Finally, it is likely that the need for grain storage in the United States will increase as the world population grows and the demand for cereals increases.

Notes

1. Walter G. Heid, Jr., "Grain Marketing—A General Description," *Marketing Grain*, Proceeding of the NCM-30 Grain Marketing Symposium at Purdue University, Lafayette, IN, North Central Regional Research Publication No. 176, January 1968, p. 16.

2. Lloyd Besant, Dana Kellerman, and Gregory Monroe, eds., *Grains*, Board of Trade of the City of Chicago, 1977, p. 58.

3. Heid, op. cit., p. 17.

4. Besant, et al., op. cit., p. 66.

5. L. D. Schnake and James L. Driscoll, *Number and Physical Characteristics of Grain Elevators*, USDA, Economics, Statistics, and Cooperatives Service-22, May 1978, p. 2.

6. Walter J. Wills, "Organization of the Grain Industry," *An Introduction to Grain Marketing*, The Interstate Printers and Publishers, Inc., Danville, IL, 1972, p. 29.

7. Besant, et al., op. cit., p. 75.

8. Ibid.

9. Heid, "Grain Marketing—A General Description," p. 17.

10. John W. Goodwin, *Agricultural Economics*, Reston Publishing Company, Reston, VA, 1977, p. 285.

11. Heid, op. cit., p. 44.

12. Goodwin, *Agricultural Economics*, p. 285.

13. Omri N. Rawlins, *Introduction to Agribusiness*, Prentice-Hall, Inc., Englewood Cliffs, NJ, 1980, p. 152.

14. Besant, et al., op. cit., p. 69.

15. Ibid.

16. Ibid.

17. Ibid.

18. Ibid.

19. GAO Staff Study, *Market Structure and Pricing Efficiency of the U.S. Grain Export System*. GAO/CED-82-61, June 15, 1982.

20. "Facing Up to Terrible Dilemma in Grain Trade," *Milling and Baking News*, July 2, 1991.

21. Reynold P. Dahl, "Structural Change and Performance of Grain Marketing Cooperatives," *Journal of Agricultural Cooperation*, 5, 1991: 66-80.

22. Today most cash grain is traded by telephone. Merchants and processors telephone bid prices each day to country elevators, usually for forward delivery. Forward selling enables country elevators to fix the price as they purchase grain from farmers and have time to schedule load-out and shipping without assuming a price risk.

23. Dahl, op. cit., pp. 66-80.

24. "Changing Face of Breadstuffs," *Milling and Baking News*, 1983.

25. "Grain Terminal Must Adapt to New Role," *Milling and Baking News*, June 5, 1984.

26. Gary M. Mennem and Robert L. Oehrtman, *Costs of Holding Wheat*, OSU Extension Facts, No. 428, Oklahoma State University, Stillwater, OK. September 1976.

27. Charles L. Storey, Roy D. Spiers, and Lyman S. Henderson, *Insect Control in Farm-stored Grain*, USDA, Science and Education Administration, Farmers' Bulletin No. 2269, December 1979.

28. U.S. Department of Agriculture, *The Yearbook of Agriculture—1954*, Marketing.

29. Warehouse Division (Agricultural Marketing Service), U.S. Warehouse Act, as Amended—Regulations for Grain Warehouses, USDA, October 1987.

30. The bankruptcy proceedings in 1981 of an Arkansas firm with grain elevators in Arkansas and Missouri brought national attention to grain warehouse receipts. Delay in delivery of grain, or payment for grain stored in a Missouri elevator owned by the firm, was clouded by two issues. The primary issue was conflict in jurisdiction: Did a state regulatory agency in Missouri or the bankruptcy court in Arkansas have jurisdiction? The ruling was that the bankruptcy court had jurisdiction. A subsidiary issue was, "Was a warehouse receipt a title document?" Legislation has been proposed by both houses of Congress (S-1365, passed by the Senate) with the main intent to establish that grain held in storage for somebody else is not a part of an elevator's assets and should be promptly distributed to the owners in species or in kind.

31. The National Grain and Feed Association, *The NGFA Grain Book*, Washington, DC, October 1991, p. 183.

32. U.S. Department of Agriculture, *Standards for Approval of Warehouses for Grain, Rice, Dry Edible Beans, and Seed*, Commodity Credit Corporation Handbook I-IM, Part 1421—Grain Warehouse Standards, Washington, DC.

Selected References

Besant, Lloyd, Dana Kellerman, and Gregory Monroe, eds., *Grains*, Board of Trade of the City of Chicago, Chicago, IL, 1977.

Carr, Camilla A., Keith E. Jackson, Darlene Logsdon, and David R. Miller, *Grain Elevator Bankruptcies in the U.S.: 1974 Through 1979*, Illinois Legislative Council Memorandum File 9-197, March, 1981.

Christensen, Clyde E., *Storage of Cereal Grains and Their Products*, American Association of Cereal Chemists, Inc., St. Paul, MN, 1974.

Goodwin, John W., *Agricultural Economics*, Reston Publishing Company, Reston, VA, 1977.

Harris, Troy G., and John Minor, *Grain Handling and Storage*, Ag Press, Manhattan, KA, 1979.

Heid, Walter G., Jr., "Grain Marketing—A General Description," *Marketing Grain*, Proceedings of the NCM-30 Grain Marketing Symposium at Purdue University, Lafayette, North Central Regional Research Publication No. 176, January, 1968.

Hill, Lowell D., and P. J. Van Bickland, "Grain Marketing," *Advances in Cereal Science and Technology*, American Association of Cereal Chemists, Inc., St. Paul, MN, 1976.

Mennem, Gary M., and Robert L. Oehrtman, *Costs of Holding Wheat*, OSU Extension Facts, No. 428, Oklahoma State University, Stillwater, OK, 1976.

National Grain and Feed Association, *The NGFA Grain Book*, Washington, DC, October 1991.

Rawlins, N. Omri, *Introduction to Agribusiness*, Englewood Cliffs, NJ, Prentice-Hall, Inc., 1980.

Schnake, L. D., and James L. Driscoll, *Number and Physical Characteristics of Grain Elevators*, ESCS, USDA, Agricultural Economics Report No. 22, Washington, DC, May 1978.

Sinha, R. N., and W. E. Muir, eds., *Grain Storage: Part of a System*, AVI Publishing Company, Inc., Westport, CT, 1973.

Sorenson, L. Orlo, and Donald Anderson, *The Marketing Operations of the Commodity Credit Corporation Through 1962*, North Central Regional Research Bulletin No. 167, North Dakota State University, Fargo, September 1965.

Storey, Charles L., Roy D. Spiers, and Lyman S. Henderson, *Insect Control in Farm-stored Grain*, USDA, Science and Education Administration, Farmer's Bulletin No. 2269, Washington, DC, December 1979.

U.S. Department of Agriculture, *Standards for Approval of Warehouses for Grain, Rice, Dry Edible Beans, and Seed*, Commodity Credit Corporation Handbook I-IM, Part 1421—Grain Warehouse Standards, Washington, DC.

———, *Yearbook of Agriculture—1954*, Marketing, U.S. Government Printing Office, Washington, DC, 1954.

———, Warehouse Division (Agricultural Marketing Service), *U.S. Warehouse Act as Amended—Regulations for Grain Warehouses*, Washington, DC, January 1980.

U.S. General Accounting Office, *More Can Be Done to Protect Depositors at Federally Examined Grain Warehouses*, CED-81-112, Washington, DC, June 19, 1981.

Wills, Walter J., *An Introduction to Grain Marketing*, Interstate Printers and Publishers, Inc., Danville, IL, 1972.

Working, Holbrook, "The Theory of Price of Storage," *American Economic Review*, Vol. 39, No. 5, December 1949.

4

Grain Transportation

L. Orlo Sorenson

Transportation is an important and integral part of the marketing system for grain. Prices paid to farmers for corn, soybeans, or wheat frequently are determined by the price at a central market (e.g., Gulf export price) less the cost of transporting grain to that market. Costs of transporting grain help determine geographic price relationships; availability of markets; and locations of production, processing, and storage within the very complex, total system of grain production and distribution.

The transportation system is used to move goods from surplus areas to deficit areas for consumption. Demand for transportation services is derived from the demand for commodities at locations other than where they are produced. The market price at the destination must be enough higher than at the origin to offset the cost of transportation for the movement to be rational in a profit-maximizing market economy. Because several supply areas normally compete for sales, relative transport rates from various surplus areas to a deficit location are very important in determining which production area will supply that market and at what price.

Transportation creates both place and time utility. Greater usefulness of the product, as reflected by a higher price at the destination, is the way in which place utility is created. Creation of time utility is also credited to transportation because of elapsed time in moving from origin to destination. However, time utility frequently is merely a by-product of transportation. Static storage costs are not incurred when grain is in transit. Time in transit thus may be a factor in selection of type of transport.

The Domestic Transportation System

The U.S. domestic transportation system includes five major modes or types of transport: railway, highway, inland waterway, pipeline, and airway. Grain and grain products move principally by railway, highway, and inland waterway. Each mode of transport has unique cost and service dimensions, largely determined by various characteristics of system components, that is, differences in way (highway, water, railway), in terminals, and in vehicles (trucks, barges, or trains).

System Components

Ways. Highways are the most ubiquitous of the three principal grain transportation modes. There are approximately 3.9 million miles of roads and streets in the United States, including the 42,500-mile national system of interstate highways (Figure 4.1). The quality of roads varies throughout the system. Rural roads may be low surface (unpaved) with low weight-limit bridges, allowing passage only of small trucks or farm-trailer loads. High-surface (paved) roads allow larger loads and continuous vehicle movements where highway entry and exit are controlled in a way that avoids intersecting traffic. All highways have size and weight limits that control maximum vehicle dimensions and vehicle weights. Weight density of whole grain and carrying capacities of trucks normally permit grain loads that exceed legal weight limitations of highways.[1]

Railway mileage operated by Class I railroads in the United States has declined from a maximum of 254,000 miles in 1916 to 132,000 miles in 1990, and continued decline is likely as unprofitable lines are abandoned. Railways are not of uniform quality, with significant portions of rail lines having serious weight and speed limitations. Portions of the system were not built to carry gross rail-car weights as great as fully loaded covered-hopper cars. Other portions of rail line will carry trainloads of hopper cars at high speeds without difficulty.

Waterways, of course, are limited to developed and maintained navigation channels on rivers, canals, and intracoastal waterways. The total length of navigable waterways in the United States (excluding open water navigation on the Great Lakes) with a 6- to 9-foot channel depth provide an additional 3,516 miles of navigable water (Figure 4.2).

Waterway depth and navigation characteristics cause waterway carrying capacity to vary significantly by waterway segment. Standard grain barges used on the Mississippi system (195 by 35 feet) can load approximately 5,485 bushels per foot of grain depth. At 7-foot grain

FIGURE 4.1   The national system of interstate highways.

FIGURE 4.2   Waterways of the United States.

depth, a standard barge will carry 38,395 bushels. A barge with 10-foot grain depth will load 54,850 bushels, the equivalent of 16.5 covered hopper rail cars or 64 trucks. Self-propelled barges used on the Columbia-Snake river system are larger and frequently carry loads up to twice that of individual barges used in tows on the Mississippi River system.

The St. Lawrence River-Great Lakes Seaway system permits ocean vessels to load grain at Great Lakes ports as far west as Duluth, on the western tip of Lake Superior—a distance of about 2,200 miles from open water at the mouth of the St. Lawrence River. The major restriction in the seaway system is at the Welland Canal between Lake Erie and Lake Ontario. The canal has a depth limit of 27 feet, which is too shallow for many fully loaded ocean vessels. Larger vessels load grain to maximum Welland Canal depth and "top-off" at Montreal or other St. Lawrence River loading points before entering open water.

Terminals. Transportation terminals are provided for loading, unloading, storing, transferring, assembling, classifying, grading, and routing shipments of grain and grain products. Terminal activities are significant time and cost factors in rail transportation of grain, especially when grain is moved in single-car rail shipments requiring frequent reassembly into new trains or transfer to connecting railroads. Grading of grain in rail cars also causes delays at terminals. Frequently, grain is unloaded for storage or processing at terminal sites and reloaded for subsequent rail movement to consumers. Rail-carrier terminal costs arise from: (1) investment in terminal facilities, (2) terminal operation, and (3) delays of road equipment.

Terminal activity for barges is primarily at loading and unloading points but may also occur at "fleeting" points, where barges are reassembled into larger or smaller tows as required by channel conditions. On the Mississippi River system, barges are frequently reassembled where river tributaries enter the main stem of the river.

For trucks, terminal activities normally are limited to loading, sampling for grade either at origin or destination, and unloading. Terminal delays are minimal except in periods of high-volume movement, when truck unloading delays may occur.

Vehicles. The third major component of each system is the vehicle. Vehicles include power units (locomotives, towboats, and highway tractors) and mobile containers for grain or grain products (rail cars, barges, and highway trailers). Only in the cases of straight trucks and self-propelled barges used on the Columbia-Snake system are the container and the power source incorporated in a single unit. Proper and expeditious assembly of container units and power units is important in minimizing transport costs and transit time for rail carriers and barge line operators. Trucks normally move from origin to destination without intermediate stops for assembly or equipment exchange.

Capital Investment: Private Versus Public

All forms of transportation involve large fixed investment in ways; hence, all are capital intensive when all costs (public and private) are considered. Truck and barge transportation is less capital intensive than rail in terms of private investment because of greater public investment in ways and terminals. Public investment in highways and waterways is justified by: (1) general public benefits (national defense, interregional communication, and so on), (2) the extreme lumpiness of the investment, and (3) a long gestation period between investment and payoff. The level of investment in ways would likely be inadequate if left totally to the private sector.

With minor exceptions, railroad tracks and terminals as well as operating equipment are privately owned. Railroads in the United States reported operating revenues in 1990 of $28.4 billion and a net investment of $48.1 billion—a capital turnover ratio (operating revenue to capital investment) of 0.59. Hence, rail costs encompass a major element of fixed costs associated with large private capital investment that significantly affects rate making and service behavior of the carrier.

Waterways and river terminals have been provided at public expense—waterways and navigation aids normally are provided by the federal government and port facilities by state and local governments. For many years, barge operators were not required to pay user charges for river navigation facilities. However, since 1980, barge operators have been required to pay a specific tax on the fuel (10 cents per gallon in 1992) used in river operations. Revenues from this tax are used to pay part of the public cost of maintaining and operating river navigation facilities.

The largest public investment in any mode of transportation is in highways. Highway users, including truckers, provide funds for highway construction and maintenance through gasoline taxes; license fees; excise taxes on vehicles, tires, and parts; and other special assessments. Such charges, however, convert the user's highway investment and maintenance responsibility from fixed to variable costs. Because they do not invest in specific roadways and their vehicles are adaptable to different types of products, motor carriers can adjust capacity quickly as changes in demand for services occur, avoiding prolonged excess capacity or shortages that sometimes exist in the railroad industry. With elastic supply conditions in trucking and in the absence of economic regulation, truck rates strongly reflect operating costs.

Grain Transportation Requirements

Grain is transported almost exclusively in bulk in the United States. Although semiperishable, grain does not require a highly controlled environment during transport. However, it does require protection from rain and from contamination by insects or by inert foreign materials while in transit.

The volume of grain movement varies among years depending on market conditions and yields per acre in producing areas. The demand for services from an individual transport mode also varies in response to supply changes by competing modes. Short-run availability of grain transport services is influenced by: (1) river navigation conditions, (2) variation in grain and nongrain transport demands for rail freight cars and locomotives, and (3) changes in demand for truck transportation for nongrain uses. Short-run demand for grain transportation is affected by: (1) quantities of grain and soybeans produced; (2) the volume of production sold off farms and, hence, entering the commercial marketing system; and (3) the amount of grain and soybeans sold for export (Table 4.1).

Grain movements are seasonal. Grain increasingly moves from farms to country elevators at harvest time. An estimated 60 percent of corn sold off farms in corn surplus states is marketed from October to

TABLE 4.1 Average Annual Production and Estimated Sale Off-Farms of Major Grain and Soybean Crops in the United States, 1980 through 1990 (1,000 bushels except rice)

Crop	Production	Percent Sold	Quantity Sold	Exports
Wheat	2,877,909	93	2,211,455	1,352,182
Corn	7,223,636	60	4,334,182	1,861,545
Sorghum	745,182	60	447,109	248,727
Oats	430,909	60	258,545	2,000
Barley	481,000	75	360,750	82,091
Soybeans	1,890,750	95	1,796,212	735,570
Rice (cwt.)	144,382	95	137,163	75,045
Total	13,293,768	72	9,545,416	4,357,160

SOURCE: *Production and export data*: Various outlook and situation reports, USDA. Sold off farms: Author's estimates.

December. Eighty-five to 90 percent of the wheat marketed by farmers is delivered to the elevator in the harvest period. From local elevators, grain for domestic use is transported to processors, livestock feeders, millers, feed manufacturers, and elevators in other areas. Export grain may move directly from local assemblers to export elevators for loading on ocean vessels. Larger quantities of grain for ultimate export may move through inland terminals at locations such as Minneapolis, Omaha, Kansas City, Chicago, Enid, Hutchinson, Lewiston, and others, or through subterminals in producing areas where equipment and transportation arrangements for multiple carload or unit train shipments have been established.

Grain for export moves from surplus producing areas to ports by those means of transportation that represent lowest cost to the shipper. The proximity of surplus producing areas and the uniqueness of water transport channels provide a general guide to patterns of export flow. The Mississippi River system, the Great Lakes-St. Lawrence Seaway system, and the Columbia-Snake River system are important inland waterway systems for movement of grain to ports. Exports of grain and soybeans by port areas in 1982 and 1987 are shown in Table 4.2.

The data in Table 4.2 also indicate the distinct importance of Gulf of Mexico ports in exporting grain. In 1990, 52 percent of all U.S. wheat, 77 percent of corn, 81 percent of sorghum, and 93 percent of soybean exports moved through gulf ports. Specialization among types of grains occurs at Gulf of Mexico ports. Mississippi River and east gulf ports export relatively large volumes of soybeans and corn, and north and south Texas gulf ports export relatively greater amounts of wheat and sorghum.

Operating Cost Economics

Fixed, Variable, Joint, and Common Costs. The various modes or types of transportation have been described as being composed of ways, terminals, and vehicles. Regardless of type of carrier, each element of a system generates costs to the operator. Cost behavior among modes of transport varies, however, depending on the nature and behavior of relevant cost elements.

Fixed costs, those that do not vary with output, are greater for railroad operations than for trucks and barges because rail carriers own the roadway, whereas trucks and barges pay user charges for use of highways and waterways.

Variable costs, those that vary with output, are a greater part of operating costs for trucks and barges than for railroads. Variable costs are incurred in handling specific traffic. For truckers, variable costs

TABLE 4.2 Selected Grains and Soybeans Inspected for Export by Port Area, 1982 and 1987 (1,000 bushels and percentages)

Port Area	Wheat		Corn		Sorghum		Soybeans		Total	
	1982	1987	1982	1987	1982	1987	1982	1987	1982	1987
Lakes	92,658 (6.5)	76,222 (4.9)	128,769 (6.5)	66,348 (4.4)	— (0.0)	— (0.0)	50,789 (5.5)	46,407 (6.0)	272,216 (6.0)	188,977 (4.7)
Atlantic	67,886 (4.7)	27,101 (1.7)	389,662 (19.8)	79,368 (5.2)	— (0.0)	— (0.0)	88,819 (9.7)	23,039 (3.0)	546,367 (12.0)	129,417 (3.2)
Gulf	816,315 (56.9)	1,004,279 (64.2)	1,316,480 (66.9)	1,122,717 (74.0)	181,756 (76.0)	160,238 (84.4)	748,248 (81.4)	658,916 (85.5)	3,062,799 (67.2)	2,946,150 (72.9)
Pacific	456,776 (31.9)	454,292 (29.1)	130,219 (6.6)	214,264 (14.1)	54,276 (22.7)	16,014 (8.4)	29,229 (3.1)	18,248 (2.4)	670,500 (14.7)	702,818 (17.4)
Interior	— (0.0)	2,098 (0.1)	2,881 (0.2)	34,288 (2.3)	3,245 (1.3)	13,953 (7.4)	2,624 (0.3)	23,783 (3.1)	8,750 (0.1)	74,122 (1.8)
Total	1,433,635 (100.0)	1,563,901 (100.0)	1,968,011 (100.0)	1,516,485 (100.0)	239,277 (100.0)	190,295 (100.0)	919,709 (100.0)	770,393 (100.0)	4,560,632 (100.0)	4,041,482 (100.0)

SOURCE: Agricultural Marketing Service, *Grain Marketing News*, USDA, 1982, 1987.

include fuel, drivers' wages, and road tolls as well as wear and tear on the trucks and tires that would not have occurred if specific traffic had not been carried. Barges lie in a more intermediate position because the investment in vehicles (barges and towboats) is somewhat greater than the investment in trucks by motor carriers.

Two other cost features occurring commonly in transportation also are significant. These are joint costs and common costs. *Joint costs* result when provision of one product or service automatically results in production of another. When wheat is milled into flour, mill feeds also are produced. Milling costs are thus joint costs of producing wheat flour and wheat by-products. In transportation, when service from locations A to B results in service from locations B to A (backhaul), costs are joint between primary haul and backhaul.

Common costs are of the same nature, except that products do not need to be produced in fixed proportions. Coal and wheat may be transported in the same train, thereby incurring common costs, but there is no physical or technical requirement that both services are produced at the same time or in fixed proportions. Joint and common costs and fixed and variable costs are not mutually exclusive categories. Either fixed or variable costs also may be joint or common costs.

It is difficult to assign fixed costs to specific shipments or to specific classes of shipments. It is equally difficult to assign joint or common costs to specific shipments. Where large elements of fixed, joint, and common costs occur, as is the case with railroads, significant areas of pricing discretion may be used beyond the assignment of direct variable costs in arriving at transport market rates for particular services. In the case of trucks, a much larger portion of costs is normally directly assignable to variable costs, hence less pricing discretion exists, except in cases of backhaul.

Distance-Cost Relationships

The relationship between cost and length of haul also is governed by technical relationships in producing the service. Three types of distance-transfer cost relationships are illustrated in Figure 4.3. Intercept values T_a, T_b, and T_c represent an initial charge to cover terminal charges, loading costs, and other charges not associated with line haul costs. The cost function progression beyond the intercept represents line-haul costs per unit (ton, hundredweight, or bushel) as distance increases. Type A function is linear, illustrating a uniform and equal progression of costs as distance increases. Type B represents a function in which costs increase by some quantity at an en route point. This cost increase may occur for truckers at the point where it is necessary to change drivers or to park and rest for a substantial period. It also may occur for railroads at points

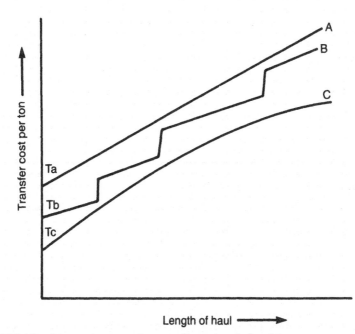

FIGURE 4.3 Alternative transfer cost-distance relationships.

where en route terminal stops occur and for barges at fleeting points along the route.

Type C, like A, is a continuous function, but carrier costs increase at a decreasing rate as length of haul increases. This decrease results from operating economies, such as those associated with efficient equipment use and advantageous assignment of personnel.

Rail shipment and barge transportation generally cost less than truck transportation on longer hauls. General cost relationships between rail and truck movement of grain as related to length of haul are illustrated in Figure 4.4. Costs of trucks reflect a low intercept value because of low fixed costs and low terminal costs but a relatively steep increase in costs as length of haul increases. Railroads represent a higher intercept cost and a lower cost progression as length of haul increases. This relationship suggests a cost-minimizing use of trucks on short hauls (D_1) and railroads or barges on longer hauls (D_3), with cost-equalizing at some intermediate point (D_2). The length of haul associated with D_2 varies significantly, depending upon specific transport conditions. For large trucks (e.g., 850 to 950 bushel loads), the intersection point (D_2) frequently occurs between 250 and 300 miles. Increased energy costs tend to shorten the length of haul over which trucks are the lower cost mode.

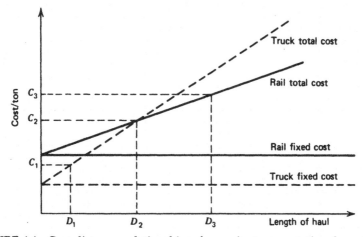

FIGURE 4.4   Cost-distance relationships for grain transportation by rail and truck.

If a continuing shipment is transferred from truck to rail at point D_2 in Figure 4.4, a reduced rate of increase in the cost function occurs beyond point D_2. This type of cost surface may also occur with transfers from trucks to barges or from small trucks to larger trucks, from single-car to multiple-car train shipment, or from a small assembly of barges to a larger one. Less-than-proportional rate increases with increased distance commonly describe commercial grain transportation rates. Intermode transhipments and changed configurations of vehicles within any transport mode are part of the cause.

Barges have a cost pattern over differing lengths of haul that is similar to that of railroads, although port (terminal) costs normally are higher than for rail, and line-haul costs are lower. Barge cost levels and cost configurations vary greatly among river segments, making it difficult to generalize about cost relationships between rail and barge in long-haul traffic. Similar adaptation of vehicles to distance hauled, as illustrated in Figure 4.4, applies to truck movement alone because of flexibility in size and design of trucks.[2]

Conditions of Efficient Performance

Regardless of type of carriage, certain loading and transport conditions influence unit vehicular costs. To reduce private transportation cost, or to negotiate effectively with for-hire carriers, grain and grain products shippers can benefit by an understanding of physical conditions affecting shipment cost. Some conditions influencing efficient resource use in transport are:[3]

1. Product weight per unit of space
2. Optimal units of cargo
3. Optimal vehicle size
4. Continuous flow
5. Direct routing
6. Adaptation of power unit to power needs.

Product Weight and Density. Weight of product per unit of space affects the ability to load to the rated capacity of the vehicle. Transport cost per unit of weight is lower when product density permits loading to maximum allowable weight before the volume limit of the vehicle is reached. Reaching the volume limit is not normally a problem with whole grains but may be a problem with milled grain or grain by-products. For example, weight per cubic foot of wheat flour is approximately 80 percent that of whole wheat; hence, a lower maximum weight of wheat flour than of wheat can be loaded in a vehicle of a given size.

Optimal Units of Cargo. Transportation cost is also high when the volume of shipment is less than the capacity of the vehicle in use. Carrier costs are associated with the capacity of vehicles moved whether fully loaded or not. If vehicles travel empty, or if power units are underutilized, costs for the carrier are nearly as much as when all space is filled and power units are working at rated capacity.

Optimal Vehicle Size. If less-than-carload or less-than-truckload shipping costs are high and size of shipment cannot be adjusted to the vehicle, it is important to match the vehicle capacity to the size of the load. When shipment size can be adjusted, unit cost economies normally occur with larger vehicles. Economies occur from savings in operator labor and from a higher ratio of payload to gross weight. Reducing vehicle weight also will permit heavier payloads when weight restrictions occur on the roadway. For example, use of aluminum instead of steel in covered-hopper rail cars reduces car weight to approximately 63,000 pounds, permitting grain loads up to 200,000 pounds without exceeding a 263,000-pound weight restriction on a sizable portion of the nation's rail line. The economic incentive for larger vehicles is present in all modes of transportation.

Continuous Flow. Transport that maintains continuous movement can be undertaken at lower cost than when delays and speed variations occur. For railroads, delays occur when cars are stopped for grade sampling, when grain movement is stopped for storage or processing under transit provisions, when individual cars are classified and assembled into road trains, and when interconnecting railroads exchange freight and equipment. Further delays occur when reduced speed or stopping is required in congested areas or at highway grade crossings.

A steel wheel on a steel rail produces relatively low amounts of friction but also provides little traction, and hence, has difficulty in creating speed. Because of relative traction conditions, trucks can overcome grades with less additional effort than locomotives. According to McElhinney, a truck requires only a little more than twice as much tractive force to go up a 2 percent grade than to run on a horizontal surface, but a train requires five times as much.[4]

Expeditious loading and unloading, including access to terminals without delays and flow-through terminals that permit short turnaround time, are important to all three major modes in grain transportation. Terminal delays reduce efficiency of rail freight car use and of trucks and drivers in motor transportation. In barge transportation, delays of towboats and their crews are especially costly.

Direct Routing. Distance of travel is minimized when grain or grain products move as directly as possible from origin to destination. Out-of-line movement, permitted without extra charge by some carriers, can only increase cost to carriers. Differences in directness or minimum mileage movement exist among types of carriers in point-to-point movement. Movements between the same two points may result in a shorter distance for trucks than for rail or barge where circuity is normally involved. Point-to-point rail distances are frequently 10 to 15 percent greater than those for trucks, and barge distances may be as much as 30 percent greater than those for rail.

Adaptation of Power Unit to Power Needs. The efficient use of the power source requires that the locomotive, towboat, or highway tractor be adapted to the size of load. Maximum use of power sometimes is difficult when multiple containers are needed in a single haul. If towboats are built to handle eight-barge tows, it is inefficient to move only two barges per tow. If railroad locomotives are built to produce the power to pull a 65-car train over normal grades, substantially fewer cars per train is inefficient. Many highway tractors generate more than sufficient power to carry loads up to the maximum permitted by highway load restrictions.

Many structure, conduct, and performance attributes as well as trends in transport services available to grain shippers are guided by the above characteristics of carrier costs. For example, the increasingly popular unit train in grain transport combines economies of larger-sized shipments, appropriate matching of power units and load, and continuous movement to achieve lower carrier costs than are possible with single-car movements. With increased intensity of competition among carriers for grain traffic in the 1990s, the role of carrier costs in determining transport rate systems will increase.

Carrier Organization

The differing cost characteristics among transport modes are significant in determining organization and economic structure, and hence, pricing and service differences among modes. Organization and economic structure, in turn, help determine the institutionalized rules under which each type of transportation operates.

Railroads. Railroad costs include large elements of fixed costs associated with investment in tracks, terminals, and rolling stock. Investments in trackage and terminals characteristically involve excess capacity. The indivisible nature of such investment means that traffic rarely will be matched with investment in a way that even approaches minimum average total cost, as illustrated by the U-shaped, short-run average cost curve.

Class I railroads were defined in 1987 as those with $94.4 million or more annual operating income. In 1987, there were 16 Class I railroads in the United States, accounting for about 91 percent of total freight revenues.[5]

Because of the small number of firms in the rail industry and the tendency for excess capacity in major capital investment (mainly tracks), many shippers are served by only one railroad. However, important competition exists among railroads in grain-gathering areas because of the opportunity to truck grain to a shipping point on a competing rail line. Grain also may be trucked directly to an inland terminal point, where service from more than one railroad is available. Additionally, railroads serving the same traffic origin areas (e.g., western Nebraska), but different market destinations (e.g., Gulf vs. West Coast export points) compete for traffic from the origin area. Railroads may also compete for traffic from different origins to a specific destination. For example, a railroad serving eastern Nebraska may compete for delivery of corn to Kansas City with a railroad wishing to satisfy demand for corn at Kansas City from origins in central Iowa. Thus, important competition for rail grain traffic exists within the rail system, even though the local elevator-shipper is served by only one railroad. Trucks and barges also provide a competitive, although not completely substitutable, service for many grain shippers.

Motor Carriers. In contrast to railroads, the trucking industry approaches pure competition. Independent truck operators serving grain industries number in the thousands. Few economies of scale for trucking firms have been identified, and the cost of entry for truckers is relatively low under unregulated entry conditions. A trucking firm's capacity can be expanded or contracted in relatively small, discrete units (a vehicle and a driver). Local elevators frequently operate their own trucks for local deliveries or for relatively short-haul movements to terminal or subterminal transshipment points. These conditions normally provide an elastic supply of trucking services. Grain trucking is a very competitive industry.

Inland Water Carriers. Barge cost patterns are similar to those of trucks in many respects. Historically, barge operators have not been required to provide fixed investments for development or maintenance of waterways or for navigation improvements. Terminal facilities are provided by public investment through local or state governments or by shippers.

Fixed costs for barge operators are those associated with ownership of vehicles (barges and towboats). Although barging equipment requires higher levels of investment than trucks, its capacity is adjusted to demand over time through investment and disinvestment in barges and towboats. Like trucking, the barge industry has a competitive structure that approaches pure competition. However, the supply of barge services is substantially less elastic in the short run than truck services, resulting in highly volatile barge rates over short-run cycles in demand for barge transport of grain.

Transportation Regulation

Economic regulation of transportation firms has British common law antecedents dating back to 1670. Many of the common law regulatory features were incorporated into the U.S. Interstate Commerce Act of 1887—an act written to provide for specific economic regulation of the railroads. Farmers and farm groups were instrumental in bringing about regulation to control alleged unethical practices of the railroads. Specific grievances of midwestern farmers included severe rate discrimination among shippers based on differences in transport options available to the shipper. Farmers also complained about high rates and heavy taxation to provide subsidies to railroads, about stock watering, and about the transportation industries' attitude of indifference to the shipping public.

Regulatory legislation for rail common carriers prescribed: (1) the use of just and reasonable rates, (2) nondiscriminatory rates, (3) no undue preference or prejudice among individual shippers or shipper communities, and (4) mandatory publishing of all rates and charges. Announced charges carried the force of law until such rates were abolished or superseded. In subsequent federal legislation between 1887 and 1980, railroad regulation became more detailed and was extended to other types of transportation.

In the decades of the 1960s and the 1970s, it became evident that economic regulation of transportation, as prescribed by statute, was not working well. Regulated carriers had difficulty competing against elements of the for-hire transport system that were unregulated and against private (shipper-provided) transportation. Many carriers were not doing well financially, especially the railroads. Other agencies (especially trucking) appeared to suffer from limited entry and less-than-optimum competitive service. Deregulation in transportation began with reduced restrictions on barges in 1973. Reduced regulation of freight and passenger airlines was legislated in 1977 and 1978, respectively, and of trucks and railroads in 1980.

For the barge industry, existing skeletal rate and service regulation was virtually eliminated in 1973. Barge grain tariffs have not been issued since deregulation.

Deregulation did not remove all regulations of a specific nature applying to trucks and railroads. For trucks, a major feature of deregulation was an easing of requirements for entry into interstate trucking. It also liberalized contract authority, resulting in greater use of shipper/carrier contracts and less dependence on common-carrier authority. The impact of truck deregulation for grain firms was mainly in interstate movement of processed products. Interstate truck movements of unprocessed grains never were regulated. In addition, increased freedom to contract for backhaul loads provided greater opportunity for two-way hauls and reduced costs and rates.

In the case of railroads, deregulation: (1) increased rail rate flexibility, (2) provided for almost unlimited confidential shipper/carrier contracts, (3) limited regulatory control over joint rates and routes, and (4) limited participation of rate bureau members in rate proceedings to those railroads directly involved in movement of the shipment for which a rate was to be determined.[6]

A substantial portion of railroad grain and grain product movement is under private contract, for which many common-carrier obligations do not apply. Rail carriers and many motor carriers also retain operating authority as common carriers and are subject to common-carrier obligations on noncontract traffic. Common carriers must serve all who request service within the range of common-carrier services normally supplied. Rates are subject to Interstate Commerce Commission surveillance for control of undue preference or prejudice to shippers. Publication of tariffs is required for services provided under common-carrier obligation. Carriers are responsible for safe delivery of all goods entrusted to them for transport.

Freight Rate Factors

Commercial transport rates are subject to many factors involving transport supply and cost conditions, transport demand and competitive conditions, product transport characteristics, and characteristics of routes over which transport vehicles will travel.

A listing of factors to be considered in transport pricing is provided by Harper under commodity cost, commodity demand, route cost, and route demand factors as follows:[7]

Commodity cost factors include:

1. Loading characteristics
2. Susceptibility to loss and damage
3. Volume of traffic
4. Regularity of traffic
5. Type of equipment required.

Commodity demand factors include:

1. Value of the commodity
2. Economic conditions in the user industry
3. Rates on competing commodities.

Route cost factors include:

1. Distance
2. Operating conditions
3. Traffic density.

Route demand factors include:

1. Competition with other carriers
2. Production point competition
3. Market competition
4. Traffic density.

The diversity of factors that affect freight rates results in great complexity in rates. Contract rates, in addition to the above cost and demand factors, will vary depending upon relative negotiating strengths among shippers and between shippers and carriers. Backhaul opportunities for carriers also may be highly significant in determining specific rates.

Common-carrier rates reflect traffic demand conditions that vary with products transported, competition from other carriers, and other demand features. Hence, products with similar transport characteristics (wheat and corn, for example) transported from point A to point B may have different rates. In the case of railroads, with a high proportion of fixed, joint, and common costs, marginal costs of transporting specific commodities between two points may be quite low. If a given rate is sufficient to cover marginal costs associated with its transport and provides any additional revenue to help cover committed costs, it pays for the carrier to take the additional traffic.

Demand-based pricing (value-of-service pricing) is especially applicable to railroad rate-making because of rail cost characteristics. According to Johnson and Wood:[8]

> The complexity of the freight rate structure is related to the extensive use of "value-of-service" pricing as opposed to full-cost pricing. Full-cost pricing refers to setting a price that covers both fixed and variable cost plus a margin for profit. Value-of-service pricing, which is also called differential pricing, discriminatory pricing, or charging what the traffic will bear, involves using variable cost only to establish a floor below which rates normally will not go. The objective is to set rates that will maximize the contribution received over and above the variable costs incurred for carrying each shipment. The result is that a ton of steel, a ton of gravel, a ton of canned goods, a ton of liquor, and a ton of furniture, each moving from city A to city B, pay different rates. In fact, one commodity may just barely cover variable costs involved, while another product pays as much as 100 to 200 percent more than the full costs of carriage.

Rate differentials responding to value-of-service conditions occur in grain transportation. Rail shipments that parallel low-cost barge routes along the Mississippi River may be near railroad variable-cost levels, whereas long-distance shipments originating in the Great Plains or in intermountain states (e.g., Montana) may have rates that greatly exceed variable costs.[9] Except for backhaul considerations, value-of-service pricing has little application in truck transport. Rate levels 'for barge movement of grain may vary in time from carrier variable-cost levels to above full costs, depending upon short-run, relative, supply-and-demand conditions for barge movement.

Shipper Determination of Rates

Shipper determination of rates may involve: (1) selection of the single appropriate tariff rate for a specific shipment, or (2) negotiation of transportation rates with a carrier under individualized contracts. Barge and truck rates are predominantly negotiated rates. However, common-carrier trucking firms still publish tariffs, especially for intrastate hauls, that designate rates for common-carrier service. For-hire barge rate and service conditions arise through carrier/shipper negotiations.

Rail tariffs on grain and grain products provide the source of rate determinations for many shippers of grain and grain products. In cases of contract rates, carrier rate quotations are frequently stated as a percent of a specific tariff. Shipper knowledge of tariffs, therefore, is important in contract discussions with carriers. Barge rate quotations are still made as a percent of Waterways Freight Bureau Freight Tariff No. 7, which was published in 1968.

Tariffs. Grain and grain product freight tariffs are published by carriers and carrier associations with a major objective of recording the prices that are charged for common-carrier service in the transportation of those commodities. Thousands of grain tariffs are in force in the United States, with numerous alternative rates available between a given set of origins and destinations. Mistakes in assessing rates often are made because of the large number of alternatives in routing, shipping configuration, intermediate and en route services available, and rate variations based on ultimate disposition of the grain. Expertise in the use of freight tariffs requires study and an understanding of the methods and purposes of tariff construction as well as an understanding of rules and procedures guiding tariff publication and use. Tariffs normally are not difficult to use after the rules are learned. However, staying current with tariff changes and applications of tariff rules is tedious and may require substantial time and effort.

Rate tariffs are more than price lists. Tariffs contain rules, obligations, and service limitations applied to specific movements. For common carriers, the Interstate Commerce Commission (ICC) requires that rate tariff circulars contain the following divisions: (1) title page; (2) table of contents; (3) list of participating carriers; (4) alphabetical list of commodities covered; (5) alphabetical list of stations; (6) geographical list of stations; (7) explanation of symbols, reference marks, and abbreviations; (8) list of exceptions; (9) rules and regulations; (10) statement of rates; and (11) routing provisions.

Contracts. Contracts for specific services negotiated between shippers and carriers have existed in the barge industry for decades. Because terms of contracts are private, analytical studies of contract terms are not available. Truckers were permitted to operate as contract carriers prior to 1980, but each carrier was limited to contracts with a maximum of eight shippers. Since 1980, the "rule of 8" has been eliminated, and greater contracting freedom is allowed.

The use of shipper/carrier contracts in rail transportation was authorized by the Stagger's Rail Act of 1980. Contract arrangements have provided advantages for both shippers and carriers above those provided in tariff conditions and have become very prevalent in rail transport of grain and grain products.

The terms of 127 wheat contracts reviewed in a 1985 study indicate that contracts specify many transport conditions in addition to rates.[10] These included volume commitments, multiple origins and destinations, minimum shipment size, refunds and allowances, provisions for rate adjustments, transit provisions, switching charge absorption, routing provision, and several others.

Private Transportation

Private transportation results when the user of a transportation service and the owner or operator of vehicles are the same. For-hire transportation dominates in grain and grain products industries, but situations exist in which private transportation is desirable.

Private transportation is perhaps most common in trucking. Private trucks usually are used in local delivery and collection systems. In other cases, large tractor-trailer combinations may be used in longer-range movements, especially when favorable backhaul conditions exist. On rivers, barge ownership by shippers is common, with towing service also provided privately or by a towing company on a contract basis. Grain shippers also are owners of railroad freight cars that are used to supplement the supply of cars provided by railroad carriers. Railroads reported 1,339,000 rail freight cars in service in 1986, of which 762,000 were owned by railroads and 450,000 were owned by car companies (for lease) or by shippers.[11]

Shippers may elect to use private transportation either for cost savings or for service improvement. In some circumstances, especially where specialized equipment is required, a shipper may be able to provide transportation at lower cost than for-hire carriers. Where regulatory restrictions apply to for-hire carriers and not to private carriers, cost advantages may accrue to the private carriers.

If private transportation complements other elements of a logistic system and service is improved, shippers may choose private transport even at a higher cost. Several types of service improvement may occur with private transportation. Faster delivery may be possible. This improvement may result from the benefits of direct routing over circuitous routing or interchange of freight that may occur with for-hire carriers. Special transport requirements, such as precise timing of delivery or special handling, may be facilitated with private transport.

Loss and damage in transit may be reduced. Fast, prompt delivery in private transport may reduce inventory costs. Perhaps most important, availability of service is assured, at least to the extent of private transport capability.

Traffic conditions of the firm are important in determining entry into private transport. Shippers may not want to provide private vehicles if they will be used only in peak demand seasons, as could be the case with rail freight cars. With firms using trucks, total shipment needs to be large enough to keep the truck busy throughout the year. Return haul loads are also an important characteristic of traffic pattern if privately owned facilities will be used.

A firm may prefer to stay out of the transportation business because it does not want to dilute its administrative resources or its capital. Operation of a transport service requires heavy commitments in each area. Specialized labor force requirements also accompany entry into transportation service. All things considered, shippers may frequently find that specialized, for-hire transporters operating competitively provide service at least as satisfactory and of a similar cost as that provided by private carriage.

Shipper-Carrier Relationships

Dealing with for-hire carriers establishes a contractual relationship between the shipper and the carrier. Procedural relationships that are dictated by trade practices accompany contracts and the negotiation of contracts. Performance rules for shippers and carriers also are specified by law or by regulatory commissions.

Ordering Service

For domestic transportation modes, ordering service by a shipper can normally be done by telephone when shipments are planned. Lead time in ordering equipment is important. Telephone orders normally are followed and verified by memo to record the equipment order and to avoid misunderstanding or miscommunications in the verbal order.

Barge operators frequently operate under annual or seasonal contracts with major grain shippers located at barge-loading points. Where contract arrangements exist, shippers may guarantee a minimum annual volume or a minimum percentage of total shipment to river port destinations. In return, the barge operator agrees to provide minimum quantities of equipment on a schedule that fits the shipping patterns of the grain elevator and/or at rates below existing single-barge rates.

Greater preshipment planning and earlier contracting are required for ocean shipping. Initial contact with carriers, usually through brokers, to determine availability of service is desirable. For grains, a shipping contract may be for a specified voyage (called a *voyage charter*) or for a specified period (called a *time charter*). If ship arrangements can be worked out, the carrier will send the shipper a freight contract (called a *charter party*), which specifies the origin and destination of shipment, quantity to be shipped, loading, sailing, and port arrival dates, along with other detailed arrangements attendant to the specific contract. A written record of arrival for loading and sailing times is important. Delays, either by the shipper or carrier, can be costly. Demurrage claims for delayed ships incur financial liability at a rapid rate. Undue ship delays also may cause congestion at the shipper's port facilities, resulting in subsequent damage claims against the carrier.

Shipping Contracts

A contractual relationship exists between a shipper and a carrier on all shipments. In the case of common carriers, the terms of the contract are spelled out on the reverse side of the bill of lading. Contracts for barge or motor carrier service specify shipper and carrier responsibilities in a long-term contract, using trip-ticket or load manifests to record individual shipments. Unregulated truckers frequently use trip tickets in lieu of formal contracts.

The Bill of Lading. Since 1919, a uniform bill of lading has been prescribed for common carriers regulated by the ICC. The bill of lading specifies the obligations of both carrier and shipper; it constitutes a receipt for goods assigned to a shipper and is a contract for carriage. Liability for goods in transit cannot be contracted away by the carrier, except in special cases where "release rate" contracts may be approved by the ICC. Exceptions to common-carrier liability exist in limited circumstances. Although uniform among carriers within a mode, bills of lading for railroads differ slightly in specified liability of the carrier from that for ocean shipping.

Uniform Order Bill of Lading. In truck, rail, and barge transportation, there is a negotiable form of the bill of lading called the Uniform Order Bill of Lading, making possible the transfer of title to goods after they have been shipped. The shipper may consign goods to himself or herself, holding title to the goods until they are consigned to someone else, if sold while in transit. The order bill of lading allows the shipper to collect for the goods shipped before the consignee takes delivery. The order bill may be transferred through the banking system and may be used as collateral for a loan. The consignee must surrender the order bill to the carrier before taking delivery of grain.

Ocean Bill of Lading. In the case of ocean shipping, the bill-of-lading conditions are governed by rules established by leading maritime nations meeting at The Hague in 1921. Each convention participant passed national legislation to implement actions taken at The Hague. Both international law and the ocean bill of lading give substantial exemptions from liability. Ocean transport companies are essentially liable for their own negligence, which is substantially less than the bill-of-lading liability of domestic surface carriers. The Harter Act of 1893 requires that U.S. carriers "exercise due diligence to make the ships seaworthy, properly manned, and fit for cargo."[12]

Loss and Damage Claims

Shippers' claims for en route loss and damage to grain in transit are frequent. If destination weights indicate loss in transit, carriers are responsible for the loss in value of the grain, but responsibility must be documented. The shipper may be at fault if: (1) losses occur at the loading point in reelevating after weighing, (2) losses result from leaking spouts between scales and rail car, (3) the distributor is improperly set or leaking, or (4) dust collectors or auxiliary blowers remove weight after weighing for loading. If loss has occurred because cars have not been properly prepared for loading, the shipper is also responsible.

Losses in quantity of grain or deterioration in quality of product en route are normally the responsibility of the carrier. Exceptions exist for losses that may occur as a result of "acts of God," such as tornado or flood conditions in which a diligent carrier could not avoid damage, or if quality deterioration resulted from a preexisting condition of the product (e.g., insect infestation of the grain when received for shipment). Under usual conditions, the carrier is expected to deliver the quantity and quality of product entrusted to him by the shipper.

Freight Charge Claims

Freight shipments by common carrier are entitled to the lowest applicable rate. Because many tariffs specifying different shipping conditions and various routings exist for rail shipments of grain, errors can be made in the selection of the legally applicable rate. When overcharges occur, shippers may file claims with the carriers for return of the overcharge. Auditing of freight bills either internally or by an external auditor is common practice. Interest may be claimed on the amount of the overcharge, if held for over 30 days.

Trends in Transportation Service

Changes in demands for transportation service by grain shippers; changes in real costs of providing service arising from improved technology; changes in relative prices of inputs; and changed patterns of regulation all have contributed to changes in the pattern of service offered by carriers in the current decade.

Expanded market movements of grain have increased the volume to be transported. Continued prominence of export markets results in a longer average distance of shipment. Increased competition for traffic and reduced regulation has forced greater attention to carrier cost patterns in establishing rate and service conditions.

Large-volume shipments to export destinations provide an opportunity for reducing costs of transportation through barge and trainload shipments. Repetitions of trainload movements from the same origin to the same destination permit cost reduction through better equipment utilization. In grain transportation, the result has been the development of subterminals in producing areas, especially in corn, wheat, and soybean areas, and a shift in the pattern of local assembly of grain to larger shipping points.

Concurrent developments in costs and rate practices have reduced the general level of traffic on feeder rail lines, resulting in either abandonment of rail service in many local communities or in formation of shortline, regional railroad companies. From 1980 to 1988, the total mileage operated by short-line and regional railroads increased by almost 62 percent, to 33,645 miles.[13] Investments in highways and improvements in trucks have caused shifts in grain and nongrain freight service in local communities to greater use of trucks. Rail service is no longer profitable on some local lines in rural communities. Railroads abandoned road miles at an accelerated rate in the 1980s compared with previous decades.[14]

Cost and competitive conditions have resulted in the following major trends in grain transportation in the 1980s:

1. Increased rail-rate flexibility has made railroads competitive in some markets where they previously were not. The same rate flexibility and accompanying uncertainty for shippers has created an increased need for risk-reduction strategies, planning, and market awareness.
2. Private contracts have permitted a better matching of service need of shippers and service supply conditions of carriers through private negotiations.
3. Increased used of barges has accompanied increased export movement and technological improvements in tow boats and navigation systems.
4. Pricing innovations in grain transport have modified transportation markets. In 1978, the Merchants' Exchange of St. Louis began offering an open auction for settling rate and service contracts with carriers. In early 1988, the Burlington Northern Railroad offered

negotiable "certificates of transportation" for sale, which could be a forerunner of a forward pricing system for grain transportation by rail.

5. Changes in railroad service patterns in rural communities have increased reliance on trucks and short-line railroads for local shippers. Short-line and regional railroads have emerged in the 1980s as a significant element in the grain transportation system. Short-line and regional roads compose 18 percent of total railroad mileage in 1988.

These recent trends pose many issues concerning: (1) transportation for rural communities, (2) the pricing of transportation services, (3) level access to transport services for all shippers, and (4) the competitiveness of U.S. products in world markets. Transportation will continue to be a dynamic element in grain marketing in years to come.

Summary

Transportation adds place utility to commodities. It involves moving goods from a location of less intensive wants to another where greater satisfaction of wants can be obtained from those goods.

The transportation network in the United States includes five major modes or types of transport: railway, highway, inland waterway, pipeline, and airway. Grain and grain products are transported primarily by rail, truck, and barge.

Transportation rates help define the market area for surplus producing regions. Relative transport rates are highly significant to individual shippers in determining markets in which they can be competitive.

In two major types of grain transportation, rail and truck, rules governing performance (regulation) have changed dramatically in the 1980s. Deregulation has resulted in a more volatile, more flexible, and more dynamic system. More transport rates are determined through negotiated contracts, which require more knowledge and negotiating expertise on the part of the shipper.

It is important for grain merchandisers, processors, and handlers to understand carrier cost and competitive conditions in a deregulated transportation environment as well as rules and procedures associated with regulated common carriers. Some shippers use common-carrier service frequently. Contracts commonly use tariffs as a base from which to specify rates and services.

In a relatively deregulated environment, grain transport rates have been reduced in some areas, have stayed the same in others, and perhaps have increased in others. Good management requires increasing amounts of information and understanding of transportation events and environment.

Notes

1. Weight limits vary significantly by type of road and type of vehicle and by regulatory jurisdiction. Common weight limit for a 4-axle semi on the national interstate highway system is 72,000 pounds gross, but this limit also may vary by state. Legal limits on the interstate system will permit loads of 900 to 950 bushels of corn or soybeans or 850 to 900 bushels of wheat.

2. Raymond G. Bressler and Richard A. King, *Markets, Prices and Interregional Trade*, John Wiley & Sons, Inc., New York, 1970, p. 115.

3. Marvin L. Fair and Ernest W. Williams, *Economics of Transportation and Logistics*, Business Publications, Inc., Dallas, TX, 1975.

4. Paul T. McElhinney, *Transportation for Marketing and Business Students*, Littlefield, Adams & Co., Totowa, NJ, 1975.

5. Association of American Railroads, *Railroad Facts*, 1987.

6. For discussion of the role of rate bureaus in rate making, see Donald V. Harper, *Transportation in America, Users, Carriers, Government*, 2nd edition, Prentice-Hall, Inc., Englewood Cliffs, NJ, 1982, p. 210.

7. Ibid., pp. 188-193.

8. James C. Johnson and Donald F. Wood, *Contemporary Physical Distribution and Logistics*, 3rd edition, 1986, p. 177.

9. For other examples see: Orlo Sorenson, Dale G. Anderson, and David C. Nelson, *Railroad Rate Discrimination: Application to Great Plains Agriculture*, Great Plains Agricultural Council Publication No. 62, University of Nebraska, Lincoln, 1973.

10. Keith A. Klindworth, L. Orlo Sorenson, Michael W. Babcock, and Ming H. Chow, *Impacts of Rail Deregulation on Marketing of Kansas Wheat*, Office of Transportation, U.S. Department of Agriculture, 1985.

11. Association of American Railroads, op. cit.

12. Donald V. Harper, *Transportation in America, Users, Carriers, Government*, 2nd edition, Prentice-Hall, Inc., Englewood Cliffs, NJ, 1982.

13. Association of American Railroads, *On Track, A Rail Industry Report*, Vol. 2, No. 17, September 1-20, 1988.

14. Association of American Railroads, *Statistics of Regional and Local Railroads*, Washington, DC, 1988.

Selected References

Association of American Railroads, *On Track, A Rail Industry Report*, Vol. 2, No. 17, Washington, DC, September 1-20, 1988.

———, *Railroad Facts*, Washington, DC, 1987.

———, *Statistics of Regional and Local Railroads*, Washington, DC, 1988.

Bressler, Raymond G., Jr., and Richard A. King, *Markets, Prices and Interregional Trade*, John Wiley & Sons, Inc., New York, 1970.

Harper, Donald V., *Transportation in America, Users, Carriers, Government*, 2nd ed., Prentice-Hall, Inc., Englewood Cliffs, NJ, 1982.

Johnson, James C., and Donald F. Wood, *Contemporary Physical Distribution and Logistics*, 3rd ed., Macmillan Publishing Company, New York, 1986.

Klindworth, Keith A., L. Orlo Sorenson, Michael W. Babcock, and Ming H. Chow, *Impacts of Rail Deregulation on Marketing of Kansas Wheat*, Office of Transportation, U.S. Department of Agriculture, Washington, DC, 1985.

McElhinney, Paul T., *Transportation for Marketing and Business Students*, Littlefield, Adams & Co., Totowa, NJ, 1975.

Sorenson, Orlo, Dale G. Anderson, and David C. Nelson, *Railroad Rate Discrimination: Application to Great Plains Agriculture*, Great Plains Agricultural Council Publication No. 62, University of Nebraska, Lincoln, 1973.

5

Grain Grades and Standards

Lowell D. Hill

Grain grades, discounts, and complaints about poor quality have generated heated debate among farmers, exporters, importers, and government agencies. In an attempt to recover lost export volumes Congress has amended the U.S. Grain Standards Act, mandated studies, held numerous hearings, and pressured industry to alter handling practices. Evaluating the diverse and often antagonistic positions of the different participants requires that students identify the issues and apply economic principles to the proposed solutions.

The Issues

Most complaints by foreign buyers have focused on the difference in quality between that stated on the certificate and that received in the processors' plants. Mold damage, excessive broken kernels, insect infestation, and low yields of processed products are the most frequent complaints. Although U.S. exports have sometimes declined following a year of an unusually large number of complaints, the correlation between the number of complaints and any economic variables related to exports provides no evidence of a cause-and-effect relationship between quality and market share. Still, the probability that customer dissatisfaction influences future purchases has justified numerous congressional actions to impose changes in grades on a reluctant industry.

The changes proposed have been of four types: (1) redefining grade factors, (2) adding new grade factors, (3) changing the factor limits, and (4) regulating marketing and handling practices. Change in factor definitions is illustrated by the debate in the early 1990s about treating broken corn and foreign material (BCFM) as one grade factor or two. The issue was studied extensively in the 1930s, 1970s, and 1990s, but the

121

studies were not sufficiently conclusive to support separation of BCFM into two categories.

Requests for additional grade factors are generally motivated by the search for indicators of product yields. In the 1986 Grain Quality Improvement Act, Congress inserted value in end use as a purpose of grades and required future grade changes to take end use value into account. The Federal Grain Inspection Service (FGIS) responded in 1991 by proposing that oil and protein contents be reported on all export certificates for soybeans. Debate continues about which uses shall be reflected in grades (e.g., protein content of corn is important information for livestock feeders, but grades provide no measure of chemical composition) and which readily measurable characteristics are correlated with each end use.

Factor limits have been changed many times since grades were promulgated. None generated a more heated response than the 1991 Federal Register proposal to reduce the limits on foreign material for No. 1 and No. 2 soybeans by 50 percent. Farmers and grain handlers demanded proof that the positive effect on exports would justify the increased costs and the loss of farm income. The status quo has strong support, and economic benefits from a change in factor limits have been very difficult to prove.

In frustration over the failure of buyers in the market channel to alter practices through economic incentives, Congress turned to regulation. The 1986 legislation prohibited any elevator from reintroducing dust or foreign material that had been removed from the grain. This regulation was intended to reduce levels of dust and foreign material in exports, but in fact it had little effect on delivered quality. Grain with high levels of foreign material was still part of blending and handling practices, and export elevators continued to load vessels with grain containing amounts of foreign material close to the maximum allowed by the contract. Where regulation runs counter to economic incentives, grain handlers often find marketing strategies that capture the economic advantages.

Evaluating the issues in grain grading requires a knowledge of its history, because various actions in the past created the status quo that has been so difficult to change. Understanding the role of grades and standards in a competitive market is also essential because it has been argued that the availability of private brands with certain reputations for quality make government grades unnecessary. For example, brands with certain reputations for quality may be used as alternatives to prescribed government grades and standards.

Changes in grades and standards must be examined from a base of knowledge about grading and handling practices at country and export elevators. A set of economic principles is needed to provide the tools with which to evaluate the costs and benefits of alternative solutions. The facilitating functions of a market include the standardization and description of the commodity to maintain a competitive and efficient pricing system.

Grades and standards are essential to the efficient operation of a competitive market that handles generic commodities. Grades provide the language essential for communicating information about value, and they encourage price competition in the market.

To facilitate exchange in a modern, industrialized society, government agencies and trade organizations in most countries have developed rules and procedures for determining the quality of goods in open trade. These rules can be simple or complex, ranging from visual inspection of the commodities at a neighborhood fruit stand, and subsequent acceptance or refusal to purchase the goods, to very technical contracts and standards specifying the processing characteristics of a material whose terms are worked out only through extensive negotiation.

One of the requirements for a perfectly competitive market is a homogeneous product to facilitate price comparisons by buyers. The grain market approaches this requirement by providing buyers with a standardized product interchangeable in market transactions. If the characteristics of different qualities are such that their relative value in use can be established as a fixed, known ratio then the homogeneity criterion of perfect competition can still be met despite any variation in quality among sublots.

Uniformity of terminology and measurement of important characteristics are essential for establishing relative values among the various forms or qualities of the product. In the case of grain grades and standards, this uniformity or standardization serves four basic purposes: (1) it permits buying and selling products by description rather than by inspection of each lot offered for sale; (2) it permits commingling of grain from many sources into a few categories or grades having uniform characteristics, thereby facilitating the marketing function; (3) it describes characteristics of grain so that buyers and sellers can estimate value for use in marketing and processing; and (4) it provides the tools whereby the market may communicate preferences and generate incentives for the improvement of quality.

History of Grades and Standards

The first grades and standards were those established for wheat in 1856 by the Chicago Board of Trade. These were a simple designation of white, red, and spring prime quality wheat. In 1857, the Chicago Board of Trade authorized the appointment of grain inspectors, and added numerical values to the class designation of wheat, creating club spring, No. 1 spring, and No. 2 spring. Grades were also assigned to corn, oats, and barley in that year. In 1858, the first official grain inspectors were appointed by the boards of trade in Chicago and Detroit, followed by Milwaukee in 1859.

Exchanges in other cities quickly followed Chicago's lead in developing grain standards and establishing inspection points. Between 1858 and 1865, grades were adopted and inspectors were appointed in Chicago, Milwaukee, Detroit, St. Louis, Cleveland, and Toledo. The New Orleans Board of Trade did not adopt grading and inspection until 1881. Despite annual revisions in grades and inspection procedures, frequent complaints of unfair and inaccurate grading were voiced in Chicago as well as in other markets where grain was graded.[1]

In an attempt to provide a more uniform system, individual states developed grading and inspection regulations on a statewide basis. Illinois was the first to provide inspection under the control of the Railroad and Warehouses Commission in 1871. Minnesota established state grades in 1885, and several states followed its lead. State control of grading and inspection failed to resolve inequities and complaints, however, because each state adopted its own grades and terminology. These grades and terms varied from state to state and from year to year, creating confusion and dissatisfaction among producers as well as fostering complaints from customers in other countries.

By 1906 the diversity among grades had made grades an impediment rather than a facilitator to efficient marketing. A survey by the Uniform Grades Congress of the National Grain and Feed Dealers Association reported the number of grades in use as "104 for wheat, 59 for corn, 57 for oats, 45 for barley, 10 for rye, 1 for no grade general rule, and 1 for no established grade."[2]

Despite many attempts by state agencies and grain trade organizations to improve the system, the inequities became increasingly more evident to producers and the grain trade. Frequent attempts to institute federal regulations during the late 1800s met with considerable resistance. Most of the trade associations continued to search for compromises among various factions in the trade that would permit uniform grades to be voluntarily adopted throughout the country. Congressmen and their farm constituents were divided between support for the voluntary use of federal grades and the more restrictive position of federal inspection as well as grades enforced throughout the entire market by federal employees. After many years of debate and frustration, a compromise amendment attached to the 1917 agricultural appropriations bill established the Grain Standards Act. The bill was signed into law as PL-190 on August 11, 1916. PL-190 provided for official federal grades with federal supervision of private inspection agencies and required that all interstate and export shipments be inspected and graded according to the official U.S. grain standards.

The secretary of agriculture was authorized to issue a license for inspecting and grading grain to anyone who was competent. Anyone not holding a federal license was forbidden to issue a U.S. grade certificate. The regulation provided for licensed inspectors wherever the need arose, and for an appeal system to be handled by federal employees. Federal employees were not permitted to make original inspections, but were only allowed to handle appeals following an original grade by a licensed inspector employed by state agencies, boards of trade, or private firms.

Current grading and inspection procedures still follow the basic philosophy of the 1916 act, but changes have been made. Numerous regulatory changes were made in the grade standards between 1917 and 1976 that did not require legislation. Most of the changes involved only nomenclature or small numerical changes such as raising the allowance for heat-damaged kernels or raising or lowering the limits for test weight. Several changes were also made in inspection procedures.

The Grain Standards Act was amended five times between 1917 and 1975. Only two of these amendments had economic significance. The first, made in 1940, transferred authority for soybean grades from the Hay, Feed, and Seed Division to the Grain Division. The 1968 amendment included removal of the mandatory inspection for all grain moving in interstate commerce. This change permitted interstate shipments between firms on the basis of their own evaluation of quality and eliminated the inefficiency of channeling rail cars through inspection yards for the sole purpose of official inspection.

With the uncovering of numerous abuses of the grain standards in 1975, a thorough investigation of grain quality and inspection procedures was conducted by the General Accounting Office, congressional committees, and agencies of the USDA. Several grain firms and inspection agencies were convicted of fraud, bribery, and misgrading. The result was a major revision of the Grain Standards Act. Two of the more important changes introduced were:

1. Inspection agencies at all of the major export points in the United States were to be operated by the FGIS, a new agency established within the USDA. This inspection service had the option of authorizing state agencies to perform the inspections where they thought this was in the best interest of the industry. However, grading at many of the ports was immediately taken over by

federal inspectors, and the private agencies and the agencies operated by boards of trade were eliminated from grading export grain in these ports.
2.   Official weights were placed under the jurisdiction of the inspection division.

The USDA was also charged with responsibility for performing the research and developing information to evaluate the present grades and standards and to determine whether changes were needed to maintain equity and the U.S. competitive position in world markets.

The broad-based review of inspection procedures, a major rewriting of the Grain Standards Act, and creation of a separate agency to administer the act under the 1975 amendment failed to solve the problems cited in complaints of foreign buyers about accurate representation of grain quality. In 1976 another amendment made significant changes in inspection procedures but failed to resolve many of the issues that first prompted the investigation. In the 1980s, with declining exports and a resurgence of formal complaints about quality, the media and Congress renewed efforts to obtain changes in grain grades. A flurry of legislative bills focused on regulating practices that were thought to reduce export quality. A task force was created under the leadership of the North American Export Grain Association (NAEGA) to bring together members from all segments of the industry, including producers, to discuss solutions and arrive at a consensus as to what should be changed to better measure quality for maintaining marketing efficiency. The report of this task force (*Commitment to Quality*) was submitted to Congress in June 1986.[3]

In November 1986, H.R. 4613 was signed into law as the Grain Quality Improvement Act, PL-99-641. This act established a new set of criteria by which to judge the adequacy of grades and standards. It prohibited the reintroduction of dust and foreign material once it had been removed from the grain stream. This legislation provided the basis for several regulatory changes in grain grades and inspection procedures.

The Grain Standards Act of 1916 and all of its revisions give the USDA two types of responsibilities: (1) establishing uniform grades and standards for measuring quality in domestic and international trade, and (2) supervising licensed inspectors to verify the accuracy and equity between buyers and sellers. This responsibility also includes providing regional and federal inspection appeals procedures and, following the 1976 amendment, original inspection of exported grain.

The results still did not satisfy the critics, and the 1990 Farm Bill contained a section on grain quality. Among the provisions of Title XX of the Food, Agriculture, Conservation, and Trade Act of 1990 was the requirement for a study to determine the cost of delivering cleaner grain to enhance the U.S. competitive position and a requirement that grades reflect value in end use.

Inspection Procedures

Three types of grading certificates are issued by the FGIS or its licensed representatives. One of the three is the *white certificate,* issued when the sampling and grading have been performed by licensed employees of an official inspection agency. The white certificate attests to the accuracy of both the sample and the grading, thereby indicating that the results of the sample are representative of the lot from which it was taken. This white certificate is required for all grain sold for export and for any sale that designates an official U.S. grade.

A *yellow certificate* (also called a *warehouseman's certificate*) is issued on samples provided by a licensed warehouseman from an identified lot of grain using an approved diverter-type mechanical sampler. The sample is submitted to an the FGIS office and receives an official grade. The licensed warehouseman and the approved sampling device add credibility to the representativeness of the sample; the FGIS attests to the accuracy of the grade for the sample that was submitted. The yellow certificate cannot be used for exports.

The third is the *pink certificate,* often referred to as a submitted sample. Usually there is no "third party" involved in obtaining the sample, and the official inspection agency cannot guarantee the sampling method or that it represents the lot from which it was taken. Therefore, the pink certificate attests only to the quality and grade of the grain in the sample.

Anyone who questions the accuracy or representativeness of the sample may request an appeal inspection by one of the field offices of the FGIS. The appeal may take either of two forms. The field office may be requested to resample the grain, assuming the rail car or barge lot is still intact, and issue a certificate indicating that it was an appeal certificate on that particular lot of grain. The second alternative is for the field office to grade the file sample. Inspection agencies are required to maintain a file sample of approximately 2,000 grams of the grain on every certificate issued for a specified period of time—30 to 90 days depending on the type of inspection. This file sample is obtained by dividing the original sample, using one-half for grading and storing the other half.

On an appeal inspection, the inspection agency issues a new certificate that shows both the original grade and the appeal grade. If there is a difference in the result for any grade factor between the original inspection and the appeal inspection, the results of the appeal supersede the results of the original inspection. As a result of random sampling variation, identical results from each half of the sample are unlikely. Grade changes during an appeal may be the result of three causes: (1) error of interpretation by inspectors, especially on subjective grade factors; (2) error of measurement by mechanical equipment, such as improperly calibrated moisture meters; and (3) random variation. If grain is blended close to the grade limit on any factor, there is a high probability that the original sample and the file sample will be certified as being different grades.

The random variation due to sampling can be easily demonstrated with a simple experiment, as illustrated by the following classroom example. Thoroughly mix 1 pound of soybeans with 49 pounds of corn. Take 10 random samples of about 100 grams from the mixture, separate the soybeans from the corn, and weigh and calculate the percent of soybeans and the percent of corn. Next, construct a histogram of the results. Finally, calculate the probabilities that any one sample is exactly 2 percent; less than or equal to 2 percent; between 1.5 and 2.5 percent.

This statistical relationship can be illustrated by using the factor of heat damage in corn, where the No. 1 grade limit is 0.1 percent, the No. 2 grade limit is 0.2 percent, and No. 3 limit is 0.5 percent. If a 1,000-gram sample has exactly 0.2 percent heat-damaged kernels, the probability that a 250-gram subsample will grade No. 1 is 0.14; No. 2 is 0.44; and No. 3 is 0.42.[4] The variation in grading results due to random sampling error follows known probability distribution and statistical properties, and the probability that any one grade will be assigned can be mathematically calculated. Operating-characteristic curves have been developed for each grade factor and are used by the FGIS in evaluating changes in sampling procedures and grade limits.[5] (As a supplemental exercise, students may calculate an operating characteristic curve under estimated values of x and s^2. Recalculate the curve allowing an appeal inspection using the file sample.)

Most of the official grades other than for export grain are determined by licensed inspectors hired by private or by state agencies. The rates charged for inspection by private agencies are determined by the individual inspection agencies. Although the FGIS supervises the rates being charged and has regulations concerning discriminatory practices, it does not establish the rates, and rates may differ among inspection agencies. Each inspection agency is allowed a particular region within a state or across states over which it has sole jurisdiction. This jurisdiction prevents inspection agencies from competing on price; more importantly, it prevents them from competing on the grade of grain recorded on the certificate. Rates for inspection services performed by the FGIS are established at the national level for all inspection points. In recent years, Congress has required the FGIS to move toward user fees to cover all operating costs. Debate over the impact of this philosophy revolves around public goods definition and market response to higher fees.

The Standards for Grain

Grades and standards for all grains are based on numerical values for a set of factors selected to reflect the quality of each type of grain. Quality characteristics may be included as grade factors, where factor limits determine numerical grade; as standards where values do not affect numerical grade but must be reported on every certificate; or as official criteria where values are determined and reported only when requested.

The number of grade factors differs among grains. Factors common to all grains are measures of purity, density, and damaged kernels. In the case of corn, purity is indicated by the factor of BCFM. This factor is defined as whole kernels and pieces of kernels as well as all matter other than corn that will pass readily through a 12/64-inch round hole sieve, and all matter other than corn that remains in the sieved sample. Soybean grades have a similar factor, also called foreign material, with a definition similar to corn except that it specifies an 8/64-inch sieve. Wheat standards include both dockage, defined as all material other than wheat that can be removed by use of the approved sieve, and foreign material, defined as all matter other than wheat that remains in the sample after removal of dockage and shrunken and broken kernels.

Density is included in grades for most grains and is approximated by a factor called test weight, expressed in pounds per bushel. The approved devices include a brass quart bucket and a scale calibrated to convert the weight to the bushel equivalent. Test weight in each grade differs among grains and is not directly related to the bushel weight established for determining quantity in commercial trade. For example, No. 1 soybeans must have a test weight of at least 56 pounds (25.2 kilograms). However, a bushel of soybeans is defined as that quantity required to weigh 60 pounds (27 kg), regardless of volume or test weight. Commercial trade in the United States has over time universally accepted the weight of a bushel of each type of grain, even though no national standards exist.[6]

The definition of damaged kernels is similar for each grain and includes damage caused by molds, disease, heat, frost, and insects. Heat damage is usually treated as a subset of total damage. Moisture was removed as a grade-determining factor from wheat grades in 1934 and from corn, soybeans, and sorghum in 1985. The rationale for this decision was that moisture is a condition of grain that can be changed by environment and drying or aerating. Safe storage moisture or optimum moisture for processing is specified by all buyers, independent of the desired numerical grade.

Other factors are unique to the different grains. Corn has five numerical grades plus a sample grade determined by four factors (Table 5.1). Soybeans have four grades plus a sample grade determined by six factors (Table 5.2). Splits and color included in soybean grades have no comparable factor in corn or wheat. Grades for wheat are perhaps the most complex with five grades determined by eight factors plus "total defects" as the sum of three other factors (Table 5.3).

TABLE 5.1   Grades and Grade Requirements for Corn

Grade	Minimum Test Weight per Bushel (percent)	Maximum Limits		Broken Corn and Foreign Material (percent)
		Damaged Kernels		
		Heat-Damaged Kernels (percent)	Total (percent)	
U.S. No. 1	56.0	0.1	3.0	2.0
U.S. No. 2	54.0	0.2	5.0	3.0
U.S. No. 3	52.0	0.5	7.0	4.0
U.S. No. 4	49.0	1.0	10.0	5.0
U.S. No. 5	46.0	3.0	15.0	7.0

U.S. Sample grade is corn that (1) does not meet the requirements for U.S. Nos. 1, 2, 3, 4, or 5; or (2) contains 8 or more stones that have an aggregate weight in excess of 0.2 percent of the sample weight, 2 or more pieces of glass, 3 or more crotalaria seeds (*Crotalaria* spp.), 2 or more castor beans (*Ricinus communis* L.), 4 or more particles of an unknown foreign substance(s) or a commonly recognized harmful or toxic substance(s), 8 or more cockleburs (*Xanthium* spp.) or similar seeds singly or in combination, or animal filth in excess of 0.2 percent in 1,000 grams; or (3) has a musty, sour, or commercially objectionable foreign odor; or (4) is heating or otherwise of distinctly low quality.

SOURCE: Federal Grain Inspection Service, U.S. Department of Agriculture, *Official U.S. Standards for Corn*, Washington, D.C., 1992.

TABLE 5.2 Grades and Grade Requirements for Soybeans

| Grade | Minimum Test Weight per Bushel (percent) | Damaged Kernels | | Foreign Material (percent) | Splits (percent) | Soybeans of Other Colors (percent) |
		Heat-Damaged Kernels (percent)	Total (percent)			
U.S. No. 1	56.0	0.2	2.0	1.0	10.0	1.0
U.S. No. 2	54.0	0.5	3.0	2.0	20.0	2.0
U.S. No. 3a	52.0	1.0	5.0	3.0	30.0	5.0
U.S. No. 4b	49.0	3.0	8.0	4.0	40.0	10.0

Maximum Limits

U.S. Sample grade is soybeans that (1) do not meet the requirements for U.S. Nos. 1, 2, 3, or 4; or (2) contain 8 or more stones that have an aggregate weight in excess of 0.2 percent of the sample weight, 2 or more pieces of glass, 3 or more crotalaria seeds (*Crotalaria* spp.), 2 or more castor beans (*Ricinus communis* L.), 4 or more particles of an unknown foreign substance(s) or a commonly recognized harmful or toxic substance(s), 10 or more rodent pellets, bird droppings, or equivalent quantity of other animal filth per 1,000 grams of soybeans; or (3) have a musty, sour, or commercially objectionable foreign odor (except garlic odor); or (4) are heating or otherwise of distinctly low quality.

aSoybeans that are purple mottled or stained are graded not higher than U.S. No. 3.

bSoybeans that are materially weathered are graded not higher than U.S. No. 4.

SOURCE: Federal Grain Inspection Service, U.S. Department of Agriculture, *Official U.S. Standards for Soybeans*, Washington, D.C., 1992.

TABLE 5.3 Grades and Grade Requirements for Wheat (all classes of wheat except Mixed wheat)

Grade	Minimum Limits of Test Weight per Bushel		Maximum Limits						
	Hard Red Spring Wheat or White Club Wheat[a] (pounds)	All Other Classes and Subclasses (pounds)	Damaged Kernels		Foreign Material (percent)	Shrunken and Broken Kernels (percent)	Defects[c] (percent)	Wheat of Other Classes[d]	
			Heat-Damaged Kernels (percent)	Total[b] (percent)				Contrasting Classes (percent)	Total[e] (percent)
U.S. No. 1	58.8	60.0	0.2	0.2	0.5	3.0	3.0	1.0	3.0
U.S. No. 2	57.0	58.0	0.2	4.0	1.0	5.0	5.0	2.0	5.0
U.S. No. 3	55.0	56.0	0.5	7.0	2.0	8.0	8.0	3.0	10.0
U.S. No. 4	53.0	54.0	1.0	10.0	3.0	12.0	12.0	10.0	10.0
U.S. No. 5	50.0	51.0	3.0	15.0	5.0	20.0	20.0	10.0	10.0

U.S. Sample grade is wheat that (1) does not meet the requirements for the grades U.S. Nos. 1, 2, 3, 4, or 5; or (2) contains 8 or more stones or any number of stones which have an aggregate weight in excess of 0.2 percent of the sample weight, 2 or more pieces of glass, 3 or more crotalaria seeds (Crotalaria spp.), 2 or more castor beans (Ricinus communis L.), 4 or more particles of an unknown foreign substance(s) or a commonly recognized harmful or toxic substance(s), 2 or more rodent pellets, bird droppings, or equivalent quantity of other animal filth per 1,000 grams of wheat; or (3) has a musty, sour, or commercially objectionable foreign odor (except smut or garlic odor); or (4) is heating or otherwise of distinctly low quality.

[a]These requirements also apply when Hard Red Spring wheat or White Club wheat predominate in a sample of Mixed wheat.

[b]Includes heat-damaged kernels.

[c]Defects include damaged kernels (total foreign material and shrunken and broken kernels). The sum of these three factors may not exceed the limit for defects for each numerical grade.

[d]Unclassed wheat of any grade may contain not more than 10.0 percent of wheat of other classes.

[e]Includes contrasting classes.

SOURCE: Federal Grain Inspection Service, U.S. Department of Agriculture, Official U.S. Standards for Wheat, Washington, D.C., 1992.

The numerical grade is determined by the lowest quality of any of the factors. For example, if a sample of corn has a test weight of 58 lbs. per bushel, BCFM of 1 percent, and total damage of 1 percent, but heat damage of 0.2 percent, the sample would be graded No. 2 even though it was equal to or better than No. 1 on all factors but heat damage. A sample of corn containing 8 percent BCFM would be sample grade even though it was equal to No. 1 on all other factors (Table 5.1).

Discounts and Quality Pricing

Discounting grain of varying quality is a method for equating value and price. Each of the grade factors may be used as a basis for determining the final price for the grain. Prices are usually established for a base grade with discounts applied to each factor rather than a different price for each numerical grade. For example, corn with excessive BCFM is generally discounted at so many cents per bushel per percentage point above the No. 2 maximum of 3 percent. The discount on each grade factor is determined primarily by the market, which presumably reflects differences in value due to quality differences. However, many of the discounts appear to be related to tradition, and they remain relatively constant over many years and geographical regions despite changes in relative value.[7] Neither grade limits nor discounts are fixed by regulation for unofficial grades. Elevator managers are free to set any discount that competition will allow or to ignore some grade factors when testing corn delivered by farmers.

Although no longer a grade-determining factor, one of the more important discount factors is that of moisture. Price adjustments for differences in moisture have four purposes: (1) to compensate for the different proportion of water and dry matter, (2) to cover the cost of conditioning high-moisture grain so that it may be safely stored and merchandised, and (3) to adjust the quantity of drying capacity to the demand for drying. In addition, some managers include a fourth cost factor in the discount for higher moisture: (4) to cover the risk involved in handling and storing high-moisture grain.

Adjustments for excess moisture in grain take many forms: (1) a discount subtracted from the base price; either fixed cents per bushel per point or a graduated scale where the cents per bushel per point differ at different moisture levels; (2) a discount per point calculated as a percent of base price; and (3) a shrink plus drying charge. The trend in the industry has been toward the latter two alternatives.

Moisture discounts based on cents per bushel per percentage point of moisture above a base level confuse the four purposes because the discounted price is being paid for water as well as for grain. The value

TABLE 5.4 Bushels of Corn Remaining When 1,000 Bushels of Corn Are Dried to Selected Moisture Levels[a]

Beginning Moisture (percent)	Ending Moisture Levels (percent)							
	13.0	14.0	15.0	16.0	17.0	18.0	19.0	20.0
13.0	1,000							
13.5	989.3							
14.0	983.5	1,000						
14.5	977.8	989.2						
15.0	972.0	983.4	1,000					
15.5	966.3	977.6	989.1					
16.0	960.5	971.7	983.2	1,000				
16.5	954.8	965.9	977.4	989.0				
17.0	949.0	960.1	971.5	983.1	1,000			
17.5	944.3	954.3	965.6	977.1	989.0			
18.0	937.5	948.5	959.7	971.2	983.0	1,000		
18.5	931.8	942.7	953.8	965.2	976.9	988.9		
19.0	926.0	936.9	947.9	959.3	970.9	982.8	1,000	
19.5	920.3	931.0	942.1	953.3	964.9	976.7	988.8	
20.0	914.5	925.2	936.2	947.4	958.9	970.6	982.6	1,000
20.5	908.8	919.4	930.3	941.4	952.8	964.5	976.5	988.7
21.0	903.0	913.6	924.4	835.5	946.8	958.4	970.3	982.5
21.5	897.3	907.8	918.5	929.5	940.8	952.3	964.1	976.2
22.0	891.6	902.0	912.7	923.6	934.8	946.2	958.0	970.0
22.5	885.8	896.2	906.8	917.6	928.7	940.1	951.8	963.7
23.0	880.1	890.3	900.9	911.7	922.7	934.0	945.6	957.5
23.5	874.3	884.5	895.0	905.7	916.7	927.9	939.4	951.2
24.0	868.6	878.7	889.1	899.8	910.7	921.8	933.3	945.0
24.5	862.8	872.9	883.2	893.8	904.6	915.7	927.1	938.7
25.0	857.1	867.1	877.4	887.9	898.6	909.6	920.9	932.5
25.5	851.3	861.3	871.5	881.9	892.6	903.5	914.7	926.2
26.0	845.6	855.5	865.6	875.9	886.6	897.4	908.6	920.0
26.5	839.8	847.7	859.7	870.0	880.5	891.3	902.4	913.7
27.0	834.1	843.8	853.8	864.0	874.5	885.2	896.2	907.5
27.5	828.3	838.0	847.9	858.1	868.5	879.1	890.1	901.2
28.0	822.6	832.2	842.1	852.1	862.5	873.0	883.9	895.0
28.5	816.8	826.4	836.2	858.1	856.4	866.9	877.7	888.7
29.0	811.1	820.6	830.3	840.2	850.4	860.8	871.5	882.5
29.5	805.3	814.8	824.4	834.3	844.4	854.8	865.4	876.2
30.0	799.6	809.0	818.5	828.3	838.4	848.4	859.2	870.0

[a]Invisible shrink calculated at 0.5 percent.

SOURCE: Author.

of the water, the value of the dry matter, and the residual income to compensate for drying costs vary with grain prices and with the moisture content. If an elevator manager selects a discount rate per point of moisture such that it will just compensate for water loss and drying costs, that rate would have to be adjusted with every change in base price and moisture level. Such an adjustment would require detailed calculations and frequent changes in discounts.

There is an alternative to price discounts that reduces the confusion and complexity by adjusting the quantity of grain consistent with a base moisture and assessing a separate drying charge. Excess moisture in grain neither adds to nor detracts from the value of products such as starch, oil, meat, and flour that are derived from the dry matter. The quantity of dry matter per bushel or per ton determines the quantity of products obtained. The most accurate method for adjusting for excess moisture would be to purchase on the basis of the dry matter by adjusting the weight of the wet grain to the equivalent weight at some base moisture using existing shrinkage tables. A drying charge levied against the seller would be used to cover the cost of drying, to control supply and demand for drying, and to cover risks of quality losses due to moisture. To use this procedure for calculating price and value, it is important to understand the principles for calculating shrink—the loss of weight due to removal of water.

The shrink that occurs in drying high-moisture grain is the same regardless of the grain being considered. Therefore, it is possible to develop tables that show the shrink, or conversely, that show the bushels remaining when a certain quantity of grain is dried (Table 5.4). Table values can be calculated from a simple mathematical relationship that is independent of the condition of the grain, the moisture level from which it started, and the kind of grain being considered.

The calculation of shrink when drying corn illustrates the procedures and establishes the basic formula. One hundred pounds of corn with 25 percent moisture contains 75 pounds of dry matter and 25 pounds of water (Figure 5.1). During drying, water is evaporated, which reduces the amount of water and therefore the total weight of the original quantity of corn. Removing 10 pounds of water does not result in 15 percent moisture corn; the resulting 90 pounds would contain 16.67 percent moisture (15 pounds of water in 90 pounds of total weight remaining is 16.67 percent moisture). Because the total weight is changed during drying, 100 pounds of corn with 25 percent moisture corn must have 11.76 pounds of water removed to become 15-percent-moisture corn. (This result is easily verified by dividing 13.24 pounds of water remaining by 88.24 pounds of total weight remaining and multiplying by

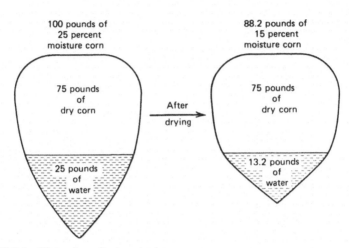

FIGURE 5.1   Water loss during drying.

SOURCE: Author.

100 to get 15 percent.) The formula for calculating shrink, remaining bushels, or moisture content, is based on a simple relationship:

$$DM_w = DM_d \tag{5.1}$$

where DM stands for quantity of dry matter in the corn, and the subscripts w and d stand for wet grain before drying and dry grain after drying. This relationship states that drying grain removes only water and does not affect the quantity of dry matter (an assumption that will be relaxed later to provide a more general formula). Because measurements are generally made on the percent of moisture rather than on the percent of dry matter, DM_w and DM_d can be represented by the equations:

$$(100 - \%M_w)Q_w = DM_w \tag{5.2}$$
$$(100 - \%M_d)Q_d = DM_d \tag{5.3}$$

where $\%M_w$ is the moisture percent of the wet corn, $\%M_d$ is the moisture percent of the dried corn, Q_d is the quantity (weight or bushels) of dry corn, and Q_w is the quantity (weight or bushels) of wet corn. Substituting Equations (5.2) and (5.3) into Equation (5.1) gives:

$$(100 - \%M_w)Q_w = (100 - \%M_d)Q_d. \tag{5.4}$$

To solve for the bushels remaining after drying (Q_d), the variables can be transposed to give:

$$\frac{100 - \%M_w}{100 - \%M_d}\ Q_w = Q_d. \tag{5.5}$$

In the previous example of 25-percent-moisture corn with 10 pounds of water removed, the remaining quantity of dry grain was calculated by Equation (5.5):

$$\frac{100 - 25}{100 - 15}\ (100\ \text{pounds}) = Q_d$$

$$\frac{75}{85}\ (100\ \text{pounds}) = 88.2\ \text{pounds of corn and water.}$$

The process of commercial drying often results in some loss of weight beyond the removal of water. This loss may be in the form of "bees' wings" blown into the air or small particles lost in handling. The loss is usually smaller than the sampling variance for moisture and cannot be readily verified by experimental methods. It also varies with management practices of the elevator. Consequently, a rule of thumb has been adopted that designates this "invisible" loss as equal to one-half of 1 percent of the wet weight. This factor is included in the Minary Charts and Table 5.4 and can be incorporated into Equation (5.5) by subtracting ($0.005Q_w$) from the remaining bushels.

As a shortened procedure for calculating shrink, many elevators have adopted what is called a *shrink factor*. This procedure consists of taking a rule-of-thumb value, such as 1.25 percent, and multiplying that percentage times the percentage points of moisture removed from each bushel. For example, corn dried from 25 percent moisture to 15 percent loses 10 percentage points. Points removed times shrink factor times bushels of wet corn gives bushels left during drying:

$$10 \times .0125 \times 1{,}000\ \text{bu} = 125\ \text{bu.}$$

Shrink (125 bu) subtracted from original wet bushels (1,000 bu) equals remaining bushels (875 bu) for which the farmer would be paid. In addition, the elevator would include a charge for drying, usually quoted in cents/bushel. (You may wish to calculate the effect of calculating the cost of drying against wet bushels versus dry bushels.)

As the beginning and ending moistures change, the mathematical value of the shrink factor also changes. Thus, a constant factor for all moisture levels cannot be used as an accurate substitute for the Minary Charts or the formula. The factor is approximately equal to 1.183 percent if the grain is dried to 15.5 percent moisture; 1.176 percent if ending moisture is 15 percent; 1.163 if ending moisture is 14 percent. (You can demonstrate an understanding of the principles by calculating the factor for 13 percent and the weight loss incurred when grain is overdried.) Table 5.5 compares two different shrink factors, the dry matter basis calculated on the formula, and the Minary Charts (formerly used in the grain trade), which include the formula plus 0.5 percent invisible shrink. Shrink factors above 1.25 are sometimes used as an indirect way of covering other costs, such as handling losses or creation of excess foreign material during drying. The use of shrink factors to conceal other charges is generally undesirable from the standpoint of pricing efficiency in a competitive market and does not permit price to serve the function of communicating differences in value to producers.

TABLE 5.5   Bushels of 15.0 Percent Moisture Corn Remaining from 1,000 Bushels of Corn at Various Beginning Moisture Levels, with Shrink Computed by Use of Four Different Adjustment Factors

Adjustment Factor	Beginning Moisture (percent)							
	16	18	20	22	24	26	28	30
Dry Matter Basis[a]	988	965	941	918	894	871	857	824
Minary Chart	983	960	936	913	889	866	842	819
Factor of 1.2[b]	988	964	940	916	892	868	844	820
Factor of 1.3[c]	987	961	935	909	883	857	831	805

[a]These values were obtained by dividing the percentage of dry matter in the corn at the beginning moisture level by the percentage of dry matter remaining at 15.0 percent moisture and multiplying this ratio by 1,000 bushels. No invisible shrink was included in the computation.

[b]These values were calculated by multiplying 0.012 × points of moisture removed × 1,000 bushels and subtracting from 1,000 bushels. The factor 1.2 × moisture removed gives the shrink per 100 bushels.

[c]These values were calculated by multiplying 0.013 × points of moisture removed × 1,000 bushels and subtracting from 1,000 bushels.

SOURCE: Author.

The choice between on-farm and off-farm drying must include returns to drying based on the difference in net payment between delivering 1,000 bushels of wet corn at a discounted price and delivering fewer bushels of corn remaining after drying at full base price. The net payment will depend on the elevator's discount schedule, shrink factor, drying charge, beginning and ending moisture level, and base price as well as the farmer's cost of drying. The farmer's calculation of drying cost must also include shrink, but it should be actual water loss, not the rule of thumb "pencil shrink" factor that often varies among elevators.

With the removal of moisture as a grade-determining factor in 1985, many elevators changed their base moisture from 15.5 percent to 15.0 percent. Many elevators dry the corn to 14 percent to provide a margin of safety for longer-term storage. When purchasing corn for storage, the elevator uses a shrink factor to calculate the quantity of corn remaining after drying. Surveys of discounting practices at country elevators indicate that many elevators use shrink factors greater than actual moisture loss.[8]

When calculating alternative drying, storage, and marketing strategies, farmers and elevator managers must calculate the differences between the amount of dry matter delivered and the dry matter for which payment will be received. There is a significant difference in net payment to a farmer who delivers 1,000 bushels of 25-percent-moisture corn that will be given a pencil shrink of 1.3 percent to 14.0 percent storage moisture, and the net payment if that same corn were dried on the farm to 15.0 percent moisture and sold for full weight and full price. The reader is encouraged to calculate the losses from several examples.

Failure to adjust all grain to the equivalent weight at some base moisture creates inequities in the market. Farmers delivering 17-percent-moisture corn receive a discount, but farmers delivering 13-percent-moisture corn receive no premium. The farmer delivering 1,000 bushels of 17 percent will be paid for 971.5 bushels of 15.0-percent-moisture corn (Table 5.4). When dried to 15 percent moisture, that 1,000 bushels will weigh out to be 976.5 bushels. If that same corn were dried to 13 percent moisture prior to delivery, the scales would show 53,424 pounds equal to 954.0 bushels. The farmer delivering 13 percent moisture would be penalized 22.5 bushels for delivering corn at a moisture content satisfactory for long-term storage. The elevator manager can blend the 13 percent and 16 percent to achieve 15 percent, eliminating the moisture discount and selling more total bushels than were purchased (954.0 + 971.5 = 1,925.5 bushels *purchased* compared with 1,000 + 954 = 1,954 bushels *sold*). The income from blending is a common source of gross income covering merchandising and operating costs. Which farmer in our example provided most of the elevator blending income?

Grading Practices at the Country Elevator

Official inspection is not required for grain moving in domestic market channels. Interfirm sales are often inspected by licensed private inspection agencies, but receipts from farmers are nearly always graded by elevator employees. Although some states have prohibited the use of certain sampling devices, the method of obtaining the sample from the farmer's truck is, in general, selected by the elevator personnel. Sampling is often the weakest link in the grading procedure. Methods vary from a coffee can from the top of the truck (a very inadequate method) to multiple probes with hand or hydraulic-operated equipment, to multiple cuts from the unloading stream of the truck, to mechanical samplers following the truck dump. Not all of the methods can be relied upon to provide a representative sample. The method selected should assure a sample that is random, unbiased, and representative. This requires multiple samples randomly distributed, with each kernel and location having an equal probability of selection.

Farm deliveries do not require official grades, so the elevator manager may analyze the sample for only those factors of concern. These factors may include grade as well as nongrade factors. For example, test weight of soybeans is often omitted in quality evaluations of farmer deliveries in years when average test weight is above the No. 1 limit of 56 pounds per bushel. Conversely, protein is often included in the analysis of farm-delivered wheat and discounts or premiums applied even though protein is not a grade-determining factor. The country elevator is free to choose those grade and nongrade factors that will be measured. Market conditions and competition generally determine the factors tested and the discounts applied. Producers usually have the option of requesting and paying for an official grade from a licensed inspector, but the buyer is not obligated to purchase on the basis of those results.

Most shipments from country and subterminal elevators are sold by grade—often determined by licensed inspectors. The sales contract may specify origin or destination grades. Interfirm transfers frequently move without official inspection although good management practices require accurate knowledge of quality at each point in the market channel. Discounts will be assessed against shipments when any grade factor exceeds the limit set by the contract, and buyer and seller must agree if the quality is to be determined at origin or destination. Origin and destination grades do not always agree. First, because probability plays a role in the sample obtained. Even with kernel characteristics uniformly distributed throughout the lot, the probability of every sample containing exactly the same percentage of every factor is almost zero. Factor values near the break point of the grade result in frequent changes of the grade due solely to random chance. In addition, grade changes can occur during transit. Handling increases breakage in corn and thus the percentage of BCFM; heat and biological activity may increase mold growth and the percentage of damaged kernels.

Grading Practices at the Export Elevator

Grain received at the export elevator arrives primarily by rail and barge. Inspection by elevator personnel or a licensed inspector is a common practice to establish quality at destination. Sales contracts specify either origin or destination grade for establishing discounts. All exported grain must be graded by the FGIS or its designated agency. A quality certificate is issued for the entire cargo or for portions of the cargo. A separate certificate may be issued for individual holds or for portions of a hold if the portions are separated within the hold. This is usually accomplished by plastic, plywood, burlap, or canvas "separations" placed across the hold when the first certificated quantity is loaded. The next portion is then loaded on top.

At the time that loading procedures are determined, the FGIS establishes sublot size (60,000 bushels is common for large vessels and high-speed loading), and the factor results for each sublot are recorded on the loading log. Statistical variability around the targeted level for each factor is accommodated by a loading plan—"cu-sum ship loading plan." Using standard deviations, "breakpoint" values are calculated for each factor. When the cumulative values of successive sublot samples exceed the grade or contract limit by more than the "breakpoint," the sublot is rejected. Either the buyer or the seller may request a reinspection or an appeal of the results. For those port elevators with shipping bins, the grain is held until the grade of the sublot is determined. If the reinspection fails to produce a value for that defect that is low enough to reduce the cu-sum below the breakpoint, the sublot cannot be loaded, and the grain is returned to the main storage bins for additional cleaning or blending. If the grain is already on the vessel, it must be removed or given a separate certificate of quality.

Nearly all grain is exported on grade and weight as determined prior to loading the vessel, and the contract specifies certificate final—meaning discrepancies in weight or quality at destination are not a legal basis for claims unless the buyer can prove that incorrect values were generated during loading. Many changes in the quality of the shipment may occur between the time the certificate is issued and the time the grain arrives in the plant of the foreign buyer. In addition to breakage during handling and biological activity during transit, most foreign processors receive only a small portion of the total vessel. Segregation during loading and unloading of a 50,000-ton vessel may result in wide variation

of quality among 50- or 100-ton sublots. Few importers regrade or blend
the contents of the vessel, and each sublot is sold on the basis of a single
certificate issued for the entire vessel at loading. Many complaints by
foreign buyers are the result of segregation among sublots and
inadequate sampling methods at destination.

Foreign complaints have stimulated many actions and reactions by
Congress, farmers, and industry groups. Despite recent publicity about
lost market share due to customer dissatisfaction, complaints by foreign
buyers have a long history. As early as 1857, the Chicago Board of Trade
received complaints from European importers regarding the quality of
grain. On January 1, 1899, the incoming president of the board reported
that merchants in England and France were threatening to boycott U.S.-
origin wheat unless changes were made in quality control.[9] A 1928
boycott against U.S. barley resulting from death of hogs (later proven to
be unrelated to the barley) elicited this response from U.S. grain traders:
"The plea of a hog's stomachache over being fed too expensive barley
cannot be allowed to cloud the issue."[10] The idea that complaints were
often an attempt to break an undesirable contract became a rallying cry
many years later when exporters claimed that price, not quality,
explained most complaints from foreign buyers.

The justification for changing grades takes many forms. Some of these
are based on incomplete information or unrealistic assumptions about
response of buyers and sellers to a change in information. Most of the
changes in grades have been preceded by a request for a study of the
potential impacts on export volume, farm income, and destination
quality.

Economic Principles for Evaluating Grain Grades

Impact on Exports

Export market share is determined by many factors; the quality of grain
is a minor factor in determining production and exports of other
countries. Unless changes in U.S. grades or quality induce other
countries to decrease their export volume or unless total world demand
increases, quality changes will have little effect on U.S. market share.
Price and quality are a package, and either one can be used by competing
countries to counter U.S. changes in grades. Specialized qualities or more
detailed and accurate descriptions of quality may gain U.S. producers
entry into specialized markets offering higher prices. Taking this market
away from competitors will force their grain into another market.
Increased market share for U.S. producers means a decreased market
share for another exporting country. In response to higher quality of U.S.
grains, which country will reduce production and why?

Farm Income

The value of each grain crop is determined by the value of products that can be derived minus the costs of transporting, marketing and processing, and competitively determined profit margins. Changing the nomenclature or classification of the crop after harvest will have little effect on any of the variables that determine farm income. If the quality or grade changes result in access to new or higher-priced markets, or if the changes produce an incentive to produce higher quality in subsequent crop years, there may be a positive effect on farm income. Grades alone have little direct effect on farm income except through their effect on marketing efficiency.

Improved Quality

Grades do not create quality; they only describe it. Changes in quality come about only through the actions and decisions of managers. Grades, in conjunction with market prices, can produce an incentive for change, but the effect depends on farmers and marketing firms making changes in their practices. Lowering grade limits does not force the sale of higher quality because importers are still free to choose quality in their contracts. For example, if the limits on foreign material (FM) for No. 2 soybeans were set at 1 percent instead of 2 percent, the importer could respond with a contract for No. 3 soybeans instead of No. 2. Changing grades does not automatically increase quality. Adverse weather conditions may lower average quality independent of factor definitions or grade limits. However, the market tends to rely on grade factors as a basis for discounts, and these discounts become the incentives that stimulate change in practices and quality.

Purposes of Grades

Grades can only be evaluated in the context of some set of performance criteria. The purposes for which grades were developed were not stated explicitly in the 1916 legislation. However, the 1986 Grain Quality Improvement Act established four purposes of grades and standards:

1. To define uniform and accepted descriptive terms to facilitate trade;
2. To provide information to aid in determining grain storability;
3. To offer end users the best possible information from which to determine end-product yield and quality;
4. To create the tools for the market to establish quality improvement incentives.

Each of the four objectives will be discussed below.

To Facilitate Trade. Almost any set of factors can meet this criterion. It is important that the grain can be grouped into relatively few categories to make buying, selling, and classification efficient and inexpensive. Trade is facilitated by a small number of grades determined by a minimum number of factors. Three numerical grades in the corn standards would be more desirable than five from the standpoint of simplicity of trading.

The factors in current grades are for the most part easily measured and provide a basis on which buyers and sellers can communicate price and discounts. Some of the factors have the additional advantage of being commonly used in international trade (for example, heat damage) and thereby serve to facilitate communication within the international market as well as the domestic market.

To Aid in Determining Grain Storability. The factors that influence storability of grain have been well documented by many scientists dating as far back as the early 1900s with the grain storage studies done by Dr. J.W.T. Duvel, of the USDA.[11] These factors are moisture, temperature, air flow, mold, and insect infestation. The length of time the grain has been held in storage and the condition under which it has been stored are also important factors.

Current grading standards provide little information on direct measures of storability. Moisture content is provided in terms of averages, but it has been demonstrated that the range of moisture among individual kernels may be more important than the average. The grades identify damage, but this is an arbitrary determination of a stage of mold development and storage deterioration. The total damage factor does not show how far deterioration has progressed or the rate of deterioration; it only records the percent of kernels where damage has progressed beyond an arbitrary minimum, which the official definition designates as "damage." Numerical grades alone provide no information on storability; the factors of BCFM (or splits and dockage in soybeans and wheat), test weight, damage, and average moisture are indirect indicators at best.

Changes in grading practices and grade standards have done little to improve the measurement of storability since the 1920s when Dr. Duvel suggested acidity as a measure of storage life and an indication of mixtures of old and new crop grain. (Incidentally, this recommendation was never adopted.) The 1986 changes in the interpretive line slides determining damage for soybeans did not improve the ability of the grades to predict storability. Interpretive line slides are photographs of the different damage characteristics of soybeans used by inspectors to judge the severity of deterioration. The correlation between the damaged kernels and the levels of free fatty acid was used by the FGIS to demonstrate the increased ability of grades to indicate oil quality; the relationship to storability was not demonstrated.

Most of the measures for storage life at the present time are based on laboratory procedures. No commercial test is available to provide a quick indication of the stage of deterioration in a lot of grain or the time remaining before it will go out of condition.

To Measure Value in Processing or Use. Because most grains are used for more than one purpose, it is not a simple matter to identify those characteristics that influence value. Some, such as oil and protein content of soybeans, are quite evident. The nutritional composition, the starch content and recovery rate in corn, and the baking characteristics of wheat are all end-use properties that are desired by processors. Few of these can be converted directly to measures of physical or chemical properties of the raw grain. Research in recent years has identified some factors that relate to value that can be measured in the commercial market channel. For example, hard-endosperm corn provides a higher yield of flaking grits.[12] Obviously, high-oil and high-protein soybeans provide higher yields of oil and soybean meal. Breakage susceptibility tests identify the ability of corn to withstand handling without increased breakage. Several tests are available that indicate baking and milling characteristics of wheat. These tests are frequently used in laboratories in the United States and Europe, for example, falling number, farinagraph, hardness, and baking tests. There is not complete agreement within the industry that these attributes are always a clear indicator of value.

Current grades do a very poor job of meeting the criterion of measuring end-use value. Soybean grades include little indication of the oil and protein content. New varieties of wheat have diminished the effectiveness of class and grade in evaluating flour and baking characteristics of wheat. Numerical grade on corn is often unrelated to its feeding value or yield of starch. In general, those characteristics of the raw grain most closely associated with value of products derived are not currently in numerical grades. Purity and cleanliness in terms of percent of foreign material or other grains provide one of the few indications of value in the current standards. Even here, there is a lack of clarity; the term *foreign material* means different things in each grain, and in most grains does not differentiate between broken kernels and impurities.

To Provide Market Incentives. Grade or quality factors that result in price differentials in the market provide an incentive for managers at each point in the market channel to make decisions to improve the quality and avoid the discounts. This incentive works back through the market channel to producers who may change harvesting, drying, and storing practices or may select different varieties. Preference for varieties with certain quality characteristics will, in turn, be reflected to plant breeders and generate better varieties in the future. In a competitive market, profit is a strong driving force in the decisions of firms, whether they are exporters, river elevators, producers, or plant breeders. Although the market itself sets the incentives in terms of prices and price differentials, the grades are not neutral in this scenario. For example, there is an incentive for farmers to select varieties of corn that will weigh at least 54 pounds per bushel to avoid discounts in the market. This incentive to the producer has been translated into incentives for the plant breeders who have spent significant research funds to develop varieties that will produce a high-test-weight corn under normal conditions. Test weight for No. 1 corn was an arbitrary number established in 1916 and has been changed at least three times since the grade standards were first introduced. The incentives for delivering corn with a test weight value of at least 54 pounds per bushel are the result of a combination of the limit set on a grade factor and a market response.

Current grades provide incentives in several ways. For example, with the market practice of basing corn prices on grade No. 2, which allows 3 percent BCFM, farmers have an incentive to incorporate 3 percent BCFM in their deliveries. By the same token, elevators have an incentive to clean or blend in such a way as to deliver 3 percent BCFM throughout the market channel. Any shipment containing less than 3 percent BCFM is a lost profit opportunity. Any shipment containing more than 3 percent BCFM is usually discounted. The step functions between the grades are automatic incentives for blending, even in the case of soybeans where the base price is for No. 1 quality. No. 1 grade soybeans allows 2 percent total damage. The incentive for blending is the opportunity to sell badly damaged, low-value beans for the price of No. 1 beans so long as they are incorporated in the shipment at less than the 2 percent rate. The wider the range between grades, the greater the number of factors, and the greater the number of grades, the greater is the incentive for blending.

The reverse side of the coin is the lack of incentive for improving quality on those characteristics omitted from the standards. With no price differential for soybeans with high oil and protein, farmers have had no incentive to select varieties that will represent greater value to the processor. Yield becomes the primary and in most cases the only criterion by which to select the variety to be planted. A second example is the drying temperature of corn. Although most corn processors object to the use of high temperatures during drying, the market does not differentiate between corn dried at high temperatures and corn dried at low temperatures. The premiums paid by a few dry millers for low-temperature-dried corn indicate that farmers do respond to these incentives when they are offered. At the present time, the incentives are not offered by means of grades but by means of contracts between processors and farmers in localized areas. It is especially difficult for buyers who are some distance from the production point to obtain the qualities they desire when those qualities are not incorporated in uniform standards, grades, and terminology.

The removal of moisture as a grading factor reduced the number of factors on which blending was required by grades, but the market still generates income from blending wet and dry corn. The change in rounding procedures for percent dockage in wheat removed the incentive to blend dockage just below the next higher break point. However, the number of grades and the steps between grades have not been significantly altered, and the incentives and disincentives still fall short of the ideal.

Economic Guide to Future Changes

The four purposes of grades do not lead directly to a system of grades and standards. An intermediate set of guidelines is required for each of the purposes identified in the 1986 Grain Quality Improvement Act. The four purposes and their requirements are as follows:

1. To define uniform and accepted descriptive terms to facilitate trade. This purpose requires a small number of categories established by clearly defined factors. The factors must be readily measured in commercial trade and objectively determined by technology that gives repeatable results at each point in the market channel. The grades and factors must be acceptable to, and used by, most participants in the market. Trade is also facilitated by stability and absence of change in the grades because any change results in uncertainty and adjustment.

2. To provide information to aid in determining grain storability. This purpose would be met by tests that reflect storage history as well as predicting remaining storage life before loss of value in use. Infestations by molds, fungi, and insects need to be accompanied by the extent of the development and deterioration. The kind of infestation is also an important measure of storability as a guide to the actions required to inhibit further deterioration. Damage should be differentiated at least into storage and field molds and fungi.

3. To offer end users the best possible information from which to determine end product yield and quality. The characteristics of raw grain that indicate the quality and quantity of processed

products differ among different industries. Factors selected for inclusion in grades should either be common to several industries or be important to an industry consuming a significant portion of the crop. The more directly the factor measures the desired end product, the more efficiently will the grades reflect value.

4. To create the tools for the market to establish quality improvement incentives. Incentives in grades are created in part by including factors that are economically important. To provide the market the maximum opportunity to establish price incentives, the grades should: (a) minimize the distance between factor limits for each grade; (b) report all values as accurately as measurement technology allows, using standard mathematical procedures for rounding to the nearest significant digit; (c) convey important economic information to producers enabling them to respond to producer preferences related to value.

The economic criteria developed for each of the four purposes of grades permit an evaluation of alternative grade structures, factors, and factor limits.

Choosing the Factors to Include in the Grades. Once it is accepted that grades are necessary to facilitate merchandising and communication in international and domestic markets, the choice of which factors shall be grade determining requires an examination of the second and third purposes of grades established by the 1986 Grain Quality Improvement Act. Each factor included must meet the same tests: (a) does it provide information on storage? (b) does it provide information indicating value to the end user? The factors related to value may change as research identifies new or improved measures of value. Many of these are already known but are not part of grades or standards. For example, acidity, oil, and protein are factors generally accepted as measuring value in whole soybeans, yet are not part of grades. However, effective September 4, 1989, the FGIS offered information on oil and protein contents when requested on export shipments. The FGIS has proposed that oil and protein contents be reported on all official export inspection certificates. This proposal is encountering opposition from several sources.

Grades, Standards, and Official Criteria. The factors selected as indicators of value may be included in the grades, in the standards, or in official criteria. Grade factors determine the numerical grade according to the factor limits established. Factors in the standards do not determine grade but must be included on the certificate as information whenever an official inspection is made. Official criteria are measured and reported on the certificate only when requested.

Assigning each factor to one of the three categories requires a guideline that can be objectively employed. Grades should serve the needs of a majority of users, and they should reflect value for those uses. These criteria suggest that grade factors are those factors related to cleanliness, purity, and soundness. Using this guideline, the grades would include factors such as impurities, fines, total damage, and heat damage. The lower the numerical values of any of these defects, the greater is the economic value of the grain. Zero damage is preferable to any larger number.

Physical and chemical attributes such as broken kernels, moisture, oil, and protein content influence economic value for most processing uses, but their importance and optimum levels differ among users. Higher or lower numerical values on these factors do not necessarily mean higher economic value over the entire range or for all uses. For example, the best level of protein in wheat depends on the ultimate product to be made from the flour. Lower moisture content means more dry matter per pound, but 5-percent-moisture corn is not generally of greater value than 12 percent because of the effects of overdrying. There is usually some optimum value for each of the factors in the standards, but this optimum varies with the use and location in the market channel.

Under Section 4 of the Grade Standards Act, factors in the grades must be measured whenever an official inspection is performed. However, factors considered as official criteria are measured only on request. One may debate the advisability of that particular part of the law, but in its present form it leads to the conclusion that only those nongrade characteristics of most importance to the largest number of users would be incorporated as standards. Those of lesser importance, or of importance to only a few users, would be official criteria available upon request to those buyers who need them. For example, moisture and chemical properties such as protein and oil content might be incorporated as standards for one or more grains. Breakage susceptibility and kernel hardness in corn, kernel size in soybeans, and falling numbers (measure of gluten strength) in wheat are examples of factors appropriate to be made available as official criteria. The list of official criteria might change over time.

The advantage of putting the major factors in as standards rather than official criteria is that the characteristics would be integrated into the market channel much more readily. Obligatory measurement throughout the market would spread the cost across the entire industry. The cost per unit would be insignificant, and therefore the information would be readily available as an incentive. A characteristic that must be specified by a separate request from each individual buyer increases the cost of information. For example, the true value of information on test weight is irrelevant under the present grades since all official inspections require that test weight be measured and recorded. The marginal cost to the buyer for that information is nearly zero. In contrast, if only one buyer has specified oil content in his export soybean contract, the cost of information would be quite high because the cost of measurement for oil throughout the market channel would be borne by the single buyer and would be spread over only those bushels that he purchased.

The four purposes of grades provide a basis for choosing factors and for dividing these factors between grades, standards, and official criteria. The next step is to set limits on each factor to designate separate numerical grades.

Setting the Grade Limits on Each Factor. It was previously discussed that the base value on the various factors was an automatic incentive throughout the market channel to add materials or to blend to reach that base limit. Blending damaged beans with good beans does not increase the *value* of the damaged beans, but it does increase the *price* of those damaged beans for they may now be sold as No. 1. The criteria for incentives for improving quality dictates that the base be set at zero. The argument that zero is unobtainable or that farmers may be unable to deliver perfection is irrelevant if we accept the four purposes of creating standards. Those four purposes do not include "minimizing farmers' discounts" or "describing the average quality delivered by good farmers." The objective of a grade is to "describe the value of the lot of grain being sold." Under current grades, the percent of damaged kernels in corn can change from zero percent to 3.0 percent, or from 5.0 percent to 6.9 percent without changing grade, so the numerical grade does not provide complete information on differences in value. Current grades for soybeans allow 1.0 percent foreign material with no discount, implying that any level between zero and 1.0 percent represents equal value. The first 0.5 percent of foreign material in a load of soybeans has no more real value than the third 0.5 percent of foreign material, even though the third is discounted and the first is not. Tighter limits on existing grade factors would reduce the incentives for blending and provide a more accurate measure of value. However, it is as difficult to justify an arbitrary limit of 0.5 percent as it is to justify an arbitrary limit of 1.0 percent. The only objective limit is zero on each grade factor, with discounts for any higher levels. The discount increments would be based on the accuracy of current sampling technology.

The base for the nongrade factors is, of course, immaterial because it is not grade determining, and the market is now free to choose what, if any, price adjustment is to be made for different levels of those factors.

The zero base concept is limited by the freedom of the market to respond. Unless (or until) export contracts and prices are established at zero base, merchandisers could start discounts at any level they desired, including the current factor levels for No. 2 corn and No. 1 wheat and soybeans.

Number of Grades. The final question in setting grades is the number of different grades required. The number of numerical grades differs among grains—malting barley has three; soybeans and sorghum have four; corn and wheat have five. Historical records provide no rationale for these numbers. The market seldom uses more than two of the numerical grades. The justification for the different numbers is not clear. The fewer numerical grades, the simpler is the marketing process, and the less space required to segregate the grades in storage and transport.

Few markets quote prices for No. 1, No. 2, and No. 3 corn, soybeans, or wheat. In nearly all cases, a base price is given on one grade and discounts or premiums attached to deviations on each factor from that base. For example, at the local elevator level, the price of corn is quoted on the basis of No. 2 grade, and farmers' prices are established by discounting each factor that falls outside of the No. 2 limits. Importers generally specify No. 2 soybeans or No. 3 corn, and prices in the contract are usually quoted for one grade with adjustments for deviations. A single grade for each grain would eliminate incentives for blending other than to meet the quality characteristics desired by the buyer. In addition, it would force the foreign buyer to specify the quality characteristics and the level of those characteristics that is desired. The importer would no longer receive 4 percent BCFM by default when ordering No. 3 corn. The buyer would be forced to specify the level of foreign material desired and would know in advance the trade-off with price.

The disadvantage of a single grade is that nearly every buyer must use one grade or specify levels on each factor. This situation would result in increased diversity of contracts, less opportunity to resell uniform lots, and increased transaction costs. The final number of grades to be established for each grain must be a compromise between the purposes of incentives, identifying value, and facilitating market transactions.

Deficiencies in Current Grades and Standards

The principles established in the preceding discussion permit a critique of current grades and suggest some needed changes.

1. The FM or BCFM factors in most grains classify broken pieces of grain, dirt, weed seeds, and other grains as having equal value. A method is needed for separating these materials. Changing screen size does not solve the problem, but only changes the size of dirt, weeds, or grain that will still be combined as FM.

2. The lowest factor approach to numerical grades places grain of widely different characteristics in the same grade even though

samples differ on all but one factor (e.g., FM). For instance, grain sorghum that is No. 1 on all factors except FM, but No. 5 on FM would grade the same as sorghum that is No. 5 on all factors including mold. Setting grades according to the lowest quality of any factor provides the incentive for blending every rail car or every vessel to the lowest quality on all factors.

3. The range of values between grades—especially in the case of heat damage—is too small to distinguish with our present sample size and sampling procedures. The probability of obtaining a correct grade for No. 2 corn on the basis of heat damage is less than 0.5. Factor limits should be established that are consistent with statistical properties and sampling methods.

4. Drying corn or soybeans below base moisture set by the market for reporting price results in a loss of weight that is not compensated by an increase in price when the grain is sold. Thus, a producer delivering 1,000 bushels of 9-percent-moisture soybeans into a market with a base price set at 13 percent moisture will incur a loss of 46 bushels of water that could have been sold at the full price of soybeans. In addition, the elevator receives a wide range of moisture levels but is selling into a market with a base moisture of 13 percent for soybeans and 15 percent for corn. This marketing practice creates a strong incentive for blending wet and dry grain to meet the base moisture level regardless of the effect on keeping qualities.

5. The present set of grades excludes economic factors of importance while including others of very limited economic importance to end users. The best example is test weight. Research at several universities has shown almost no relationship between test weight and feeding value of corn.[13] Test weight also appears to have no relationship to oil and protein content in soybeans, yet the price of grain continues to be discounted because of test weight, and incentives exist to seek higher test weight varieties without a reward of higher value but only higher price in the marketplace.

6. If broken corn is an important factor on which to grade (and discount) corn, then the price differential must be carried back to the point where harvesting and drying methods generated breakage-prone corn. Therefore, breakage susceptibility must be included in price differentials to the farmer to provide the economic incentive to dry at lower temperatures, harvest at lower moisture levels, design and adjust equipment to minimize stress cracks, and to plant varieties that are less likely to break during handling.

7. The debate of the 1800s over private funding vs. public funding of

grain grading has resumed in the 1990s. With the political emphasis on private enterprise and users paying the full cost of services, steps have been taken to require federal agencies to recover full cost of inspection and grading through user charges. Current proposals would require grain inspection fees to also cover costs of the research and standardization conducted by the FGIS.

The issue is one of identifying public versus private goods. Setting standards for weight and measurement technology and qualifications of inspectors benefits buyers, sellers, the industry, and consumers. The National Bureau of Standards, the Food and Drug Administration, and many other service agencies do not rely on user fees and private enterprise to set standards. The benefits of these services are too widespread and diverse to permit assignment of benefits on an individual use basis. Similarly, the benefits of grading food products such as meat or grain accrue more to the industry and consumers as a group than to individual users. The economic incentive for the individual grain handler under full-cost recovery strategy is to minimize grading and standardization to reduce costs. The use of private grades and unlicensed inspectors will also increase as official inspection charges rise. Private grades were the only alternative in 1900 when a survey revealed over 100 different grades for wheat in use in the market. The inefficiencies, confusion, and abuses of the system generated the demand for federal grades. When most of the benefits from uniform grading are aggregated across the industry and consumers, the cost of standardization must be paid from general revenues, or individual firms must be forced by law to use and pay for the service.

Quality Measurement in Other Countries

Grades and standards for grain have been developed in nearly all exporting and importing countries. The procedures and responsible agencies differ widely, from a few simple descriptive terms to very detailed definitions and limits.

Both buyer and seller of any lot of grain are interested in determining the value of the lot as a basis for establishing a price. The greater the distance between buyer and seller, and the more complex the marketing system, the more that buyer and seller must rely on grading standards and standardized terminology for a description of the characteristics of the grain.

Quality has been contractually specified in many different ways in the history of international grain trading.[14] Every major grain-exporting country uses some form of numerical grading for corn, soybeans, and wheat. There are several similarities. All forms include impurities and damage, but these are identified in many different ways. For example, in South African corn grades, the factor "defective kernels" includes mold damage and broken kernels; in U.S. corn grades, broken kernels and impurities are combined in the factor of "BCFM"; in Argentine grades, broken corn, impurities, and damage by mold are three separate factors. Moisture is not used as a grade-determining factor in any major exporting country,[15] but it is used for adjusting weight or price with some variant of the formula provided previously in this chapter. In many countries, grain exports are controlled or actually handled by government agencies, and the quality of exports is determined by government decision. The South African Maize Board is the sole exporter from that country and conducts inspections for export. Ceroil is the official exporter of corn and soybeans from China, and another government bureau conducts inspecting and grading. In Argentina, the Junta Nacional de Granos operated the inspection agency controlling quality from farmer to exporter and operated export houses alongside the private traders until 1991. A move to privatize industries put responsibility for quality control in private hands.

Summary

The ultimate goal of grades and standards for grain is to improve the efficiency with which buyers communicate their preferences to sellers. Grades do not establish value but merely standardize the terminology with which the market communicates information about value between buyers and sellers. Changes in standards should be made only if the value of the increased information is at least equal to the increased cost of implementing the change. In general, this principle requires that inspection and grading procedures be rapid and inexpensive, yielding estimates of quality that are efficient (i.e., of minimum variance) and unbiased.

In the short run, standards have little effect on quality or total value of the crop; they only describe the quality in terms accepted by the industry. In the longer run, the fact that the trade generally prices grain on the basis of established grade factors provides an incentive for changing production and marketing practices to alter the quality characteristics of the crop. Grades, in conjunction with prices and price differentials, communicate information through the market system. To the extent that managers are free to respond to economic incentives, information on value will increase the efficiency with which quality and quantity preferences of buyers are met.

Notes

1. Lowell D. Hill, *Grain Grades and Standards: Historical Issues Shaping the Future*, University of Illinios Press, Urbana-Champaign, April 1990.

2. C. Louise Phillips, *History of Grain Inspection in the United States, 1838-1936*, Washington, DC, USDA, 1936, p. 52.

3. North American Export Grain Association, *Commitment to Quality: A Consensus Report of the Grain Quality Workshops*, June 1986, Washington, DC.

4. T. E. Elam and Lowell D. Hill, "Potential Role of Sampling Variation in the Measurement of Corn Grading Factors," *Illinois Agricultural Economics*, Agricultural Experiment Station, University of Illinois at Urbana-Champaign, Vol. 17, No. 1, January 1977.

5. Operating-characteristic curves for grain provide an interesting application of statistical properties. Type I and Type II errors can be assigned relatively accurate values in a simulated problem of blending grain to grade limits. Targeted values set too low incur the cost of lost opportunity to sell low-quality grain at base price. Targeted values set too high will result in rejection of a proportion of the lots based on sample results and will incur the cost of off-loading the sublot or of negotiating a price discount with the buyer. The proportion of rejected lots can be estimated from statistical properties of density functions calculated from known standard deviations. (Bernard Ostel, *Statistics in Research*, Iowa State University Press, Ames, 2nd, ed., 1963, pp. 477-488).

6. Lowell D. Hill, op. cit.

7. Brian Anderson and Lowell D. Hill, "Corn and Soybean Price and Quality Discounts at Illinois Interior Grain Elevators During Fall 1989," Dept. of Agricultural Economics Staff Paper 90 E-449, University of Illinios, Urbana-Champaign, March 1990.

8. Aleksandar Bekric and Lowell D. Hill, "Corn and Soybean Prices and Quality Discounts at Illinois Grain Elevators," Dept. of Agricultural Economics, Publ. No. AE-4674, University of Illinois, Urbana-Champaign, June 1991.

9. Lowell D. Hill, op. cit., p. 26.

10. Lowell D. Hill, op. cit., p. 301.

11. Lowell D. Hill, op. cit., p. 185.

12. Lowell D. Hill, Marvin Paulsen, Aziz Bouzaher, Martin Patterson, Karen Bender, Allen Kirleis, *Economic Evaluation of Quality Characteristics in the Dry Milling of Corn*, Bulletin 804, Agricultural Experiment Station, University of Illinois, Urbana-Champaign, Nov. 1991.

13. Lowell D. Hill and A. H. Jensen, "The Role of Grades and Standards in Identifying Nutritive Value in Grains," *Feed Composition, Animal Nutrient Requirements and Computerization of Diets*, Fonnesbeck, Harris, and Kearle, eds., Utah Agricultural Experiment Station, Utah State University, Logan, 1977.

14. A good description of several of these contracts is provided by G. H. Morsink, "Comparative Quality of U.S. Corn for Feed, Food, and Oil in the Netherlands," in *Corn Quality in World Markets*, Lowell D. Hill, ed., Interstate Printers and Publishers, Inc., Danville, IL, 1975.

15. Lowell D. Hill, ed., Proceedings of "Uniformity by the Year 2000: An International Workshop on Maize and Soybean Quality," Department of Agricultural Economics, University of Illinois, Urbana-Champaign, 1991.

Selected References

Agricultural Research Service, *Grain Breakage Caused by Commercial Handling Methods*, Marketing Research Report No. 968, U.S. Department of Agriculture, Washington, DC, June 1973.

Anderson, Brian, and Lowell D. Hill, "Corn and Soybean Price and Quality Discounts at Illinois Interior Grain Elevators During Fall 1989," Department of Agricultural Economics, Staff Paper No. 90 E-449, University of Illinois, Urbana-Champaign, March 1990.

Bekric, Aleksandar, and Lowell D. Hill, "Corn and Soybean Prices and Quality Discounts at Illinois Grain Elevators, 1990," Department of Agricultural Economics, Publ. No. AE 4674, Agricultural Experiment Station, University of Illinois, Urbana-Champaign, June 1991.

Bermingham, Steve C., and Lowell D. Hill, *A Fair Average Quality for Grain Exports*, Department of Agricultural Economics, Publ. No. AE-4459, Agricultural Experiment Station, University of Illinois, Urbana-Champaign, July 1978.

Dorfman, Robert, and Peter O. Steiner, "Optimal Advertising and Optimal Quality," *American Economic Review*, Vol. 44, No. 5, December 1954.

Elam, T. E., and Lowell D. Hill, "Potential Role of Sampling Variation in the Measurement of Corn Grading Factors," *Illinois Agricultural Economics*, Vol. 17, No. 1, January 1977.

Farris, Paul L., "Uniform Grades and Standards, Product Differentiation and Product Development," *Journal of Farm Economics*, Vol. 42, No. 4, November 1960.

Federal Grain Inspection Service, *Historical Compilation of Changes in the Grain Standards of the United States*, U.S. Department of Agriculture, Washington, DC, July 1986.

———, *The Official U.S. Standards for Grain*, U.S. Department of Agriculture, Washington, DC, 1992.

Freeman, Jere E., "Quality Factors Affecting Value of Corn for Wet Milling," *Transactions of the American Society of Agricultural Engineers*, Vol. 16, No. 4, 1973.

Friedman, Milton, *Sampling Inspection*, H. A. Freeman, Milton Friedman, Frederick Mosteller, and W. Allen Wallis, eds., McGraw-Hill Book Company, Inc., New York, 1948.

Hill, Lowell D., and Gene C. Shove, *Drying Corn at the Country Elevator*, Extension Circular 1053, University of Illinois, Urbana-Champaign, March 1972.

Hill, Lowell D., Steve C. Birmingham, and Randall Semper, *Sampling and Measurement Problems in Grain Grading*, Department of Agricultural Economics, Publ. No. AE-4407, University of Illinois, Urbana-Champaign, October 1976.

Hill, Lowell D., and A. H. Jensen, "The Role of Grades and Standards in Identifying Nutritive Value in Grains," *Feed Composition, Animal Nutrient Requirements and Computerization of Diets*, Paul V. Fonnesbeck, Lorin E. Harris, and Leonard C. Kearle, eds., Utah Agricultural Experiment Station, Utah State University, Logan, 1977.

Hill, Lowell D., Marvin R. Paulsen, and Margaret Early, *Corn Quality Changes During Export*, Illinois Agricultural Experiment Station, SP-58, University of Illinois, Urbana-Champaign, September 1979.

Hill, Lowell D., Mack N. Leath, Odette Shotwell, Donald G. White, Marvin R. Paulsen, and Philip Garcia, *Alternative Definitions for the Grade Factor of Broken Corn and Foreign Material*, Bulletin 776, Illinois Agricultural Experiment Station, University of Illinois Urbana-Champaign, October 1982.

Hill, Lowell D., Eugene Kunda, and Clint Rehtmeyer, *Price Related Characteristics of Illinois Grain Elevators*, Publ. No. AE-4561, Department of Agricultural Economics, University of Illinois, Urbana-Champaign, September 1983.

Hill, Lowell D., "Effects of Regulation on Efficiency of Grain Marketing," *Case Western Reserve Journal of International Law*, Vol. 17, No. 3, Summer 1985, pp. 389-419.

———, *Principles for Use in Evaluating Present and Future Grain Grades*, Department of Agricultural Economics, Staff Paper No. 85 E-329, University of Illinois, Urbana-Champaign, September 1985.

———, *Removal of Moisture as a Determinant of Numerical Grade*, Department of Agricultural Economics, Staff Paper No. 85 E-330, University of Illinois, Urbana-Champaign, October 1985.

Hill, Lowell D., and M. R. Paulsen, *Maize Production and Marketing in Argentina*, Illinois Agricultural Experiment Station, Bulletin 785, University of Illinois, Urbana-Champaign, July 1987.

Hill, Lowell D., Wojciech J. Florkowski, and Julia P. Brophy, "Production Response of Illinois Farmers to Premiums for Low-Temperature Dried Corn," in *Agribusiness: An International Journal*, Vol. 4, No. 2, March 1988, pp. 197-209.

Hill, Lowell D., "Grain Grades: They Lack Economic Rationale," in *Choices*, published by the American Agricultural Economics Association, Herndon, VA, Vol. 3, No. 1, First Quarter, 1988, pp. 24-27.

———, Video, "Soybean Quality: Meeting the Challenge," Department of Agricultural Economics, University of Illinois, Urbana-Champaign, 1989.

———, Video, "Quality in the U.S. Grain Marketing System," Department of Agricultural Economics, University of Illinois, Urbana-Champaign, February 1989.

Hill, Lowell D., *Grain Grades and Standards: Historical Issues Shaping the Future*, University of Illinois Press, Urbana-Champaign, April 1990.

Hill, Lowell D., and Marvin R. Paulsen, *Changes In Corn Quality During Export from New Orleans to Japan*, University of Illinois Agricultural Experiment Station, Bulletin 788A, University of Illinois at Urbana-Champaign, June 1990.

———, *Changes in Corn Quality During Export from New Orleans to Japan*, University of Illinois Agricultural Experiment Station, Summary Version, Bulletin 788B, University of Illinois at Urbana-Champaign, June 1990, p. 20.

Hill, Lowell D., ed., Proceedings of "Uniformity by the Year 2000: An International Workshop on Maize and Soybean Quality," Department of Agricultural Economics, University of Illinois, Urbana-Champaign, 1991.

Hill, Lowell D., Marvin Paulsen, Aziz Bouzaher, Martin Patterson, Karen Bender, Allen Kirleis, "Economic Evaluation of Quality Characteristics in the Dry Milling of Corn," University of Illinois Agricultural Experiment Station, Bulletin 804, Champaign-Urbana, Nov. 1991.

Hill, Lowell D., Video, "Let's Meet the Competition," Department of Agricultural Economics, University of Illinois, Urbana-Champaign, 1991.

Ladd, George W., and Verophol Suvannunt, "A Model of Consumer Good Characteristics," *American Journal of Agricultural Economics*, Vol. 58, No. 1, February 1976.

Ladd, George W., and Marvin B. Martin, "Prices and Demands for Input Characteristics," *American Journal of Agricultural Economics*, Vol. 58, No. 1, February 1976.

Mehren, George W., "The Function of Grades in an Affluent Society," *Journal of Farm Economics*, Vol. 42, No. 5, December 1961.

Mittleider, John F., and Donald E. Anderson, *An Analysis of the Relationships Among Specific Quality Characteristics for Hard Red Spring and Durum Wheat*, Department of Agricultural Economics, North Dakota Agricultural Experiment Station, Bulletin No. 122, North Dakota State University, Fargo, August 1977.

Morsink, G. H., "Comparative Quality of U.S. Corn for Feed, Food, and Oil in the Netherlands," in *Corn Quality in World Markets*, Lowell D. Hill, ed., Interstate Printers and Publishers, Inc., Danville, IL, 1975.

Nichols, John P., Lowell D. Hill, and Kenneth Nelson, "Food and Agricultural Commodity Grading," in *Federal Marketing Programs in Agriculture—Issues and Options*, Walter J. Armbruster, Dennis R. Henderson, and Ronald D. Knutson, eds., Interstate Press, Danville, IL, 1983.

North American Export Grain Association, *Commitment to Quality: A Consensus Report of the Grain Quality Workshops*, June 1986, Washington, DC.

Paulsen, M. R., and L. D. Hill, *Corn Quality Factors Affecting Dry Milling Performances*, published for the British Society for Research in Agricultural Engineering, reprinted from *Agric. Engineering Research*, Vol. 31, pp. 255-263, 1984.

Peplinski, A. J., O. L. Brekke, E. L. Griffin, G. Hall, and Lowell D. Hill, "Corn Quality as Influenced by Harvest and Drying Conditions," *Cereal Foods World*, Vol. 20, No. 3, March 1975.

Petry, Timothy A., and Donald E. Anderson, *Comparative Analysis of United States and Canadian Wheat Grades*, Department of Agricultural Economics, Bulletin No. 99, North Dakota State University, Fargo, November 1974.

Phillips, C. Louise, *History of Grain Inspection in the United States, 1838-1936*, USDA, Washington, DC, 1936.

Rosen, Sherwin, "Hedonic Prices and Implicit Markets: Product Differentiation in Pure Competition," *Journal of Political Economy*, Vol. 82, No. 1, January 1974.

Watson, Stanley A., "Measurement and Maintenance of Quality," *Corn: Chemistry and Technology*, Stanley Watson and Paul Ramstad, eds., American Association of Cereal Chemists, St. Paul, MN, June, 1987, pp. 125-184.

Zusman, Pinhas, "A Theoretical Basis for Determination of Grading and Sorting Schemes," *Journal of Farm Economics*, Vol. 49, No. 1, Part 1, February 1967.

6

Pricing Grains

Richard G. Heifner and Bruce H. Wright

Paralleling the physical system for marketing grains is the pricing system, which provides for the interchange of bids, offers, and agreements to buy and sell among traders throughout the country and the world. The pricing system has two major functions to perform: (1) coordinating production and utilization decisions of farmers, merchants, processors, and consumers; and (2) helping to determine the distribution of income among these different groups.

The prices of grains, relative to each other and to other commodities, determine how much land, labor, machinery, and other inputs will be used in growing, storing, processing, and distributing grain and grain products. The price differences over time help determine in each period how much grain is consumed and how much is stored for future use. The price differences between locations provide incentives to transport grain to where the need is greatest, and the price differences between grades provide incentives to produce the kinds of grains that are in greatest demand.

For the grain farmer, the price and quantity of grain produced determines how much income is available for family living. For the consumer, grain prices are a determinant of the cost of food, affecting prices not only of cereal products but also products derived from livestock. And for the businesspeople engaged in assembling, storing, feeding, processing, and distributing grains and grain products, a few cents' difference in price per bushel can be the difference between a profit and a loss.

The Price Discovery Process

The process by which the price is found when quantities supplied by sellers are equated with quantities demanded by buyers is called *price discovery.* For grains, price discovery occurs through the many and varied contacts between buyers (or potential buyers) and sellers (or potential sellers). These contacts are dispersed geographically, except for the futures markets and a few spot markets where traders or their representatives gather on trading floors. Except in futures trading, most contacts are between one potential buyer and one potential seller at a time, such as a farmer talking to a country elevator manager, or an elevator manager talking to a processor-buyer by phone. Most buyers and sellers have contacts with more than one opposite party, and the overlapping contacts tie the entire market together.

For each trader, the price discovery process involves a search for the opposite party who offers the best available terms. The search typically begins with a review of market reports and then enters a period of negotiation where bids and offers are exchanged with one or more opposite parties. The search culminates when an oral or written contract or sales agreement is reached.

A *sales contract* is an agreement between two parties to exchange a commodity for money or for another commodity. The agreement must specify not only the price, or a formula for determining the price, but also the quantity, quality, and time and place of delivery as well as time and method of payment and provisions for guaranteeing performance. In negotiating contracts, some of the terms of trade other than price are frequently implicit, being set by custom or rule. In futures trading, the contracts traded are so highly standardized that only price and quantity need be specified during the negotiation process.

Not only futures contracts but most cash contracts are forward contracts in that they specify delivery at a later date. Some contracts call for *spot delivery*, which is interpreted as delivery within a few days.

Transactions Between Farmers and First Handlers

Sales of grains and oilseeds account for about one-fourth of U.S. gross farm income, so farm prices for these commodities are key determinants of income for agriculture as a whole. Farming operations vary greatly in size and type, ranging from part-time ones that may sell only a few hundred bushels per year to large operations that sell hundreds of thousands of bushels annually. The time and effort each farmer can advantageously devote to pricing grain depends on the amount of grain to be sold. Small farmers typically look to only one or a few buyers as outlets, but large producers may deal with a number of buyers including some at distant locations. Firms such as country elevators, subterminal and terminal elevators, processors, and feeders that buy directly from farmers are called *first handlers*. First handlers typically maintain daily contact with other buyers and sellers by telephone. Many buy and sell grain almost every day, particularly during the harvest season.

The Bid Pricing System

Under normal conditions, most first handlers determine and post spot bid prices each day for the types and grades of grain that they want to buy. Elevator operators determine their bid prices by subtracting operating margins and transportation costs from prices bid by the firms they sell to. Processors and feeders base their bid prices on current market quotations, adjusting their bids up and down to obtain the desired flow of grain into their plants. During periods of active price movements, first handler bid prices may be changed several times within the day. When prices are highly volatile, the bid price may not be posted at all; instead, it is quoted in response to the farmer's call, after checking the latest futures quotation. If the farmer has a large amount to sell, the buyer may call more than one outlet to confirm the price. Bid prices apply to a standard grade and quantity, such as No. 2 yellow corn, 15 percent moisture or lower. The elevator maintains a discount and premium schedule that is applied when a farmer delivers grain that differs in grade or quality from the standard.

Posted bid prices apply to any farmer's grain that is brought into the elevator and meets the grade requirements. Farmers usually sell at the posted price. Large farmers may occasionally be able to negotiate price or other terms that are more favorable than the elevator's posted bid.

How Sales Are Made

Nearly half of the wheat and soybeans and about a quarter of the corn in the United States are delivered directly from the field to the elevator; the remainder goes into on-farm storage. Grain brought to the elevator during harvest is weighed, graded, and sold or entered into storage in the farmer's account. Delivery of grain directly from the field to the local elevator at harvest eliminates extra handling. If the grain has not already been contracted, the farmer typically is allowed a few days to indicate whether the grain should be sold immediately or placed in commercial storage. Once storage is initiated, the farmer can sell to the elevator on any date thereafter, with storage charges being deducted from the returns on the grain. For grain delivered out of farm storage, price normally is set before delivery, subject to grade and moisture determinations.

Various types of contracts have evolved to allow the time of sale to differ from the time of physical delivery. The range of dates when sales can be made is illustrated in Figure 6.1. In choosing the date of sale, the farmer must consider price risk and make judgments as to possible price fluctuations.

Forward Pricing

Forward pricing involves setting the price for a sale or purchase before delivery. Crop producers can forward price their crops during the growing or storage season either by entering cash contracts or by selling futures contracts. By so doing, they protect themselves from losses due to price declines, but surrender the potential for gains from price increases. Because of yield uncertainty, the farmer's revenue risk is minimized by forward pricing only part of the expected crop before harvest. Alternatively, farmers can establish minimum prices for their crops by buying put options or by entering minimum-price cash forward contracts with buyers. The use of these two pricing tools remains limited but has been gradually increasing since the introduction of agricultural commodity options trading in 1984. (The use of options is discussed in Chapter 8.)

Pricing After Harvest

Grain prices typically rise after harvest as storage costs accumulate. Therefore, the farmer whose cost of storing grain is less than the expected seasonal price rise often can profit by holding the crop for sale some

FIGURE 6.1 Forward-selling opportunities during the planning, growing, and storage phases of annual crop production.

SOURCE: Authors.

months after harvest. Holding the crop, of course, postpones receipt of payment and entails continued exposure to price risks. Most of the seasonal price increase occurs during the first few months after harvest; carrying stocks more than six months is quite speculative.

Delayed Pricing

Many first handlers provide farmers another method to postpone pricing beyond harvest. In this method, called *delayed pricing*, the title to the grain is transferred from the farmer to the elevator without setting the price. The elevator sells and ships the grain to the next buyer, usually covering this short sale with a long hedge in the futures market. At a later date, chosen by the farmer, the farmer's price is set based on the elevator's then-current bid; the farmer pays intervening storage costs, the elevator lifts its hedge, and final settlement is made.

Delayed pricing allows the elevator to free its bin space to handle another farmer's grain while letting the farmer postpone pricing until a future date. With this arrangement, the farmer loses title to the grain before receiving final payment. Consequently, there is some chance of loss if the elevator should fail.

Transactions Among Grain Merchants and Processors

Participants in the grain trade beyond the level of the farmer and first handler include subterminal elevators, terminal elevators, processors, feeders, exporters, and foreign buyers. The size of transactions among these traders is much larger than in purchases from farmers. A common unit of trade is the hopper carload (about 3,500 to 4,000 bushels), but trainload- (up to 500,000 bushels) and shipload-size transactions are common for grain destined for export.

The commercial grain market is primarily a telephone market. The great bulk of transactions are negotiated by telephone between a buyer and a seller, both operating from their own offices. Agreements made by telephone are followed by written confirmation. A remarkable feature of this market is that verbal agreements, involving hundreds of thousands or millions of dollars, are used with a minimum of misunderstanding and disputes.

The Bid Pricing System

Price discovery occurs through several different kinds of trading processes. Grain futures are traded on open markets where buyers or sellers are equally likely to initiate transactions, and any number of traders may be involved. In contrast, most cash grain transactions are negotiated privately between individual buyers and sellers by telephone. In the cash grain market, the search for a mutually satisfactory price often starts with the buyer's bid, which is normally based on the latest futures price with adjustments for differences in grade, location, and time of delivery.

After the close of futures trading each day (about 1:15 p.m., Central Time), each of the major processors, exporters, and other merchants decides on a set of prices he or she will use to bid for grain until the futures market opens the next day. These bid prices are telephoned to the country elevators and other firms from which the merchants and processors regularly buy. The bid prices normally hold until futures trading opens again the next morning. Once the futures market opens, the bid prices may be adjusted to reflect changes in that day's futures price. To make these adjustments automatically during periods of fluctuating futures prices, bids are commonly made in terms of the *basis*.

Basis Pricing

The basis is the difference between a specific cash price and the price for a specific futures contract, normally the "nearby contract," the contract that is nearest to maturity. For example, an Illinois corn processor might bid "15 under" for 30-day corn. This bid means the firm is willing to pay 15 cents below the maturing future for corn deliveries to its plant during the next 30 days.

The potential seller can convert a basis bid to a cash price by simply applying the basis to the closing futures price. Sometimes sellers can compare basis bids by more than one buyer for the same location and delivery period, and thereby make their decisions about selling and about their own bids to farmers.

The advantage in quoting price in terms of the basis is that adjustments for changes in the general price level for the grain, as indicated by the futures price, are made automatically. For example, if a buyer bids "15 under" one afternoon, and the futures price opens up 10 cents the next morning, potential sellers know without being called that the bid is following the futures up. Thus, quoting price in terms of the basis provides an efficient means to exchange bids and offers. The practice is so common that traders use the term *flat price* to indicate when the full price is being quoted rather than the basis.

Sales agreements often specify price in terms of the basis. In the grain trade, this practice is commonly called *booking the basis*. Either the buyer or seller (by mutual agreement) is allowed a specified period of time to choose a date when the cash price for the transaction is determined by applying the agreed-on basis to the then-current futures quotation. Booking the basis sets the delivery terms and fixes price relative to a specific futures price, but it leaves both buyer and seller exposed to price level risk. Hence, the practice is frequently accompanied by hedging in the futures by one or both parties. Such a sales agreement is usually fulfilled by an exchange of futures, at the agreed-on futures price plus or minus the agreed-on basis. The futures exchanges have special rules to facilitate this type of trade, which is called an *exchange of futures for cash* or an *ex-pit transaction.*

Time of Delivery

Commercial grain sales agreements may call for "spot" delivery or "forward" delivery. *Spot delivery* normally means delivery within a few days. The bulk of the commercial cash grain trade involves contracts for forward delivery. More than two-thirds of the corn, wheat, and soybeans traded are priced for delivery more than 10 days in the future. Examples of common terms with regard to time of delivery are: "10 days," "30 days," "January," and "first half of March." By buying for forward delivery, processors and exporters are able to schedule uniform flows of products into their plants. Forward selling also enables country elevators to fix the price before they ship their grain and schedule their loading-out activities ahead of time.

Floor Trading on Organized Exchanges

An exception to the dominant practice of trading by telephone for forward delivery is floor trading in spot or cash grain at the Minneapolis Grain Exchange. In this market, cash grain trading shares the trading floor with futures trading. The grain bought and sold is in railroad cars, on a designated siding in the city, ready for delivery within a few days. Samples of the grains available for sale are displayed in pans on the seller's or broker's table on the trading floor. Potential buyers or their representatives circulate among the tables and negotiate with sellers on an individual basis. Transaction prices are posted around the trading floor.

Floor trading in spot grains reached its heyday in the era when country grain buyers commonly shipped grain on consignment to grain brokers at the terminal markets. The broker then took charge of the grain as it arrived on track in the terminal city and offered it for sale on the floor of the grain exchange. The consignment method of selling grain has largely been replaced by *to arrive* cash contracts between country elevators and terminal merchants in corn, soybeans, and most classes of wheat.

However, consignment sales by country elevators continue for malting barley and durum wheat at the Minneapolis Grain Exchange. These are grains for which the grade standards only partially reflect the quality factors important to buyers. Hence, buyers wish to see a sample before offering a price.

Pricing Export Grain

Grain pricing in the export market has much in common with domestic grain pricing. Overseas prices and domestic prices are closely related on a day-to-day basis. For many export sales, the price negotiation process is essentially the same as for domestic trades. Transactions are negotiated by telephone by traders operating from their offices and dealing with opposite parties they know and trust. But, because of the size and complexity of the transactions, direct personal contact between buyers and sellers or their representatives plays a larger role in the export trade. Some foreign buyers buy by calling for tenders or offers from all potential sellers at once, rather than by negotiating individually with sellers.

The export trade differs from the domestic grain trade in that most transactions are in shipload quantities or larger. Several different types of sales arrangements are used. The most common is an *f.o.b.* sale—f.o.b. signifying free on board ship at point of origin. Other types of sales arrangements where the buyer pays for transportation include *f.s.t.* meaning free on board, stowed, and trimmed, and *f.a.s.* meaning free alongside the ship (an ocean vessel designated to carry the commodity). When the seller pays for transportation, the sale can be *c. and f.*, meaning the seller pays cost and freight, or *c.i.f.*, meaning the seller pays cost, insurance, and freight. Some sales agreements call for the seller to deliver to an interior destination in the receiving country.

An export sales contract must contain the same basic provisions as a domestic contract, but much more detail is required. The contract must spell out what currency is to be used for payment. It must prescribe shipping arrangements in detail and explain what constitutes default and what recourse is available for the party defaulted against. Because of their complexity, export contracts commonly are written on standard forms developed by trade associations.

The need to charter ocean shipping and arrange for changing foreign currencies to U.S. dollars causes the export trade to differ from the domestic grain business. Because of the advantages of having representatives in grain-importing countries, and the special skills needed in managing ship chartering and foreign exchange transactions, the bulk of the export sales is handled by a few major firms who specialize in the export business.

Price Determination and Price Forecasting

What determines the price of grain? A quick answer is that price is determined by the interaction of supply and demand. A price rise, if correctly anticipated, will increase the amounts that farmers are willing to grow and sell while reducing the amount of grain products that consumers want to buy. Thus, for example, when carry-over from the previous year is low and anticipated exports high, there is some relatively high price that will just equalize the amount that domestic and foreign consumers are willing to buy with the amount produced. Similarly, in a year when yields are high, the extra supply will only be used at lower than normal prices.

In the United States, as in most of the world, grain prices are determined jointly by government programs and market forces. Through its price-support loan programs, the federal government can, in effect, set lower bounds for grain prices by accepting farmers' grains in lieu of loan repayment when price falls to the loan rate. The government may also limit price movements on the upper side by releasing stocks into the market when the market price exceeds a designated level. In addition, the government sometimes intervenes in the export market for grains and oilseeds through subsidies and embargoes.

Much of the time during the 1950s and 1960s, corn and wheat prices rested on support levels, leaving the government with substantial stocks. The initiation of grain sales to the former Soviet Union lifted prices above support levels where they remained during much of the 1970s. By the early 1980s, prices were again resting on supports. Price-support loan levels were lowered in the late 1980s to make U.S. grains more competitive on export markets. When prices are above the support level, and below the level where government stocks would be released, they respond relatively freely to market forces. Even when prices are resting on supports, the price differences associated with grade, location, and time within the year are primarily determined in the marketplace.

Market prices result from the actions of buyers and sellers at many locations, each seeking the most profitable outlet or source of supply. The price must be agreed on at each change in ownership as grain passes from the farmer through the processing and distribution sector to final consumption.

Grain price determination is best viewed as a dispersed, multidimensional process, with the futures markets serving as focal points. Futures trading brings out the time dimension in prices and provides central open marketplaces where the full range of price-making influences come to bear. Futures contracts call for delivery at elevators in the major cities—Chicago, Minneapolis, Kansas City, Toledo, and St. Louis, so futures markets price grains at the terminal elevator-processor stage of the marketing process. Futures contracts provide for only a few delivery locations and for only a limited range of quality specifications. The spatial and quality dimensions in grain pricing are provided mainly by the cash market.

To market grain successfully, the farmer or elevator manager must not only be familiar with trading arrangements, as just described, but also acquire up-to-date information about price levels for alternative outlets. Information about the current price is obtained from buyers and from other market information sources, as described in Chapter 9. Beyond this, however, farmers and grain merchants will more likely be successful if they understand the factors that determine price and possess some ability to predict prices.

Price forecasting calls for identifying and quantifying the factors that affect supply and demand. For example, if a drought occurs during the growing season, then yields are lowered and planted acreage cannot be changed; therefore, the supply will be less. This reduced supply will command a higher price in the market. Alternatively, a large un-anticipated export sale may occur after the crop is harvested. This unanticipated shift in demand, too, would result in a price increase as domestic and foreign buyers compete for the remaining quantity.

Seasonality in Grain Prices

Grain prices tend to be lowest at harvest and then gradually increase to a peak sometime prior to the next harvest. This seasonal price increase provides a return to those who store grain for use later in the year. Without such seasonality in price, individuals and firms would have no motivation to supply the needed storage services.

The monthly prices received by farmers for corn during three mar-keting years (1988-1989, 1989-1990, and 1990-1991) are graphed in Figure 6.2. The figure suggests a tendency for the price to reach a seasonal low at harvest and increase thereafter, but it also shows that the price movements over the season vary greatly from year to year, and the seasonality is frequently obscured by other factors. To obtain a better picture of the underlying seasonal pattern, we can average together a series of years so that the more or less random shifts in supply and demand cancel each other out. Figure 6.3 presents an index of the seasonal price pattern for corn based on data for marketing years 1981-1982 to 1990-1991.

For corn, an important characteristic of the seasonal price pattern is that price tends to rise rather rapidly and consistently during the first month or two after harvest, but with much less consistency during the

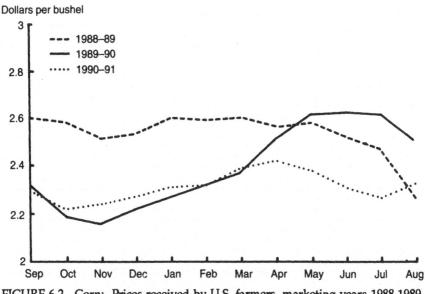

FIGURE 6.2   Corn: Prices received by U.S. farmers, marketing years 1988-1989 to 1990-1991.

SOURCE: U.S. Department of Agriculture, National Agricultural Statistics Service, *Agricultural Prices, Annual Summaries 1990 and 1991*, 1986 through 1992.

rest of the storage year. For wheat and other grains, the pattern is similar but generally less pronounced. This tapering off of the seasonal price rise is explainable in terms of the supply and demand for bin space. Owners of bin space are able to capture the largest return on storage immediately after harvest when the bins are full. As stocks are worked down, some bin space owners continue to hold stocks or make space available at lower expected rates of return rather than letting the space stand idle. Thus, the monthly returns from carrying stocks are normally highest immediately after harvest and decline thereafter.

Price Differences Over Space

Supply and demand also explain why grain prices differ between locations. Prices tend to be lowest in growing areas where production exceeds use and highest at processing centers and ports. These spatial price differences are needed to pay haulers for moving grain from

FIGURE 6.3   Corn: Index of monthly prices received by U.S. farmers, marketing
                 years 1981-1982 to 1990-1991 (average ratio of monthly prices to the
                 unweighted average price for each marketing year).

SOURCE: U.S. Department of Agriculture, National Agricultural Statistics
Service, *Agricultural Prices, Annual Summaries, 1985 through 1991*, 1986 through
1992.

growing areas to the points of utilization and export. The differences
tend to persist from month to month, but change as supplies and
utilization change at different locations. For example, a North Dakota
country elevator may ship wheat to Portland when the Japanese are
buying and to Minneapolis at other times.

State average prices received by farmers for the 1989 crop corn are
shown in Figure 6.4. The lowest prices tend to be in the Corn Belt states
with higher prices in areas of deficit corn production in the East, the
West, and the South. The possibilities for shipping from point to point
prevent price differences between any two locations from exceeding
transportation costs between them. But price differences between states
or regions may be less than transportation costs when there is no direct
movement from one to the other. Instead, relative prices at two different
points may be determined by the costs of shipping between each and a
third destination.

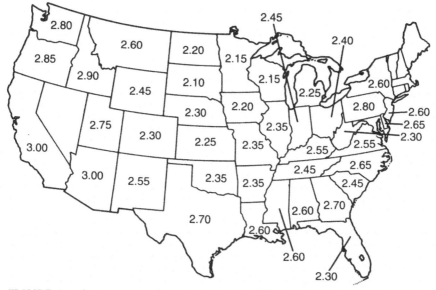

FIGURE 6.4   State average farm corn prices, 1989 crop (in dollars per bushel).

SOURCE: U.S. Department of Agriculture, National Agricultural Statistics Service, *Agricultural Prices, Annual Summary, 1989*, June 1990, page A-22.

Price Differences Due to Quality

Higher quality grains generally command higher prices. But just as spatial price differences change due to changes in supply and demand at different locations, so do quality differentials. To illustrate, Table 6.1 shows the differences in price paid during crop years 1980-1981 to 1989-1990 for No. 1 Dark Northern Spring wheat of different protein levels. Protein premiums were high in the middle 1980s due to a shortage of high-protein wheat. In contrast, the relative abundance of high-protein wheat in 1988-1989 and 1989-1990 resulted in low premiums for wheat with 14 percent or more protein.

The hard wheats often command a premium over the soft wheats because of their baking qualities and higher protein levels. But soft wheats are required for certain cakes and pastry products. Table 6.2 shows that hard wheats at Kansas City and Minneapolis commanded higher prices than soft wheat at Chicago for every crop year from 1980-1981 to 1989-1990.

TABLE 6.1   Protein Premiums for No. 1 Dark Northern Spring Wheat, Minneapolis, 1980-1981 Through 1989-1990 Crop Years

Crop Year	13 percent protein	14 percent protein		15 percent protein	
	Price	Price	Premium	Price	Premium
1980-1981	$4.57	$4.71	$ .14	$4.99	$ .26
1981-1982	4.26	4.29	.03	4.37	.08
1982-1983	4.05	4.09	.04	4.18	.09
1983-1984	4.26	4.30	.04	4.35	.05
1984-1985	3.90	4.08	.18	4.29	.21
1985-1986	3.64	3.94	.30	4.27	.33
1986-1987	2.85	3.07	.22	3.51	.44
1987-1988	2.98	3.15	.17	3.68	.53
1988-1989	4.32	4.36	.04	4.44	.08
1989-1990	4.15	4.16	.01	4.17	.01

SOURCE: U.S. Department of Agriculture, Economic Research Service, *Wheat Situation and Outlook Yearbook*, February 1991, (WS-292), pages 56 and 57.

TABLE 6.2   Price Premiums for Hard Wheats at Kansas City and Minneapolis Relative to No. 2 Soft Wheat at Chicago, 1980-1981 through 1989-1990 Crop Years

Crop Year	Chicago	Kansas City		Minneapolis	
	Price	Price	Premium	Price	Premium
1980-1981	$4.33	$4.50	$ .17	$4.57	$ .24
1981-1982	3.74	4.30	.56	4.26	.52
1982-1983	3.32	4.13	.81	4.05	.73
1983-1984	3.56	4.15	.59	4.26	.70
1984-1985	3.51	3.93	.42	3.90	.41
1985-1986	3.22	3.62	.40	3.64	.42
1986-1987	2.76	2.87	.11	2.85	.09
1987-1988	2.89	3.14	.25	2.98	.09
1988-1989	4.00	4.23	.23	4.32	.32
1989-1990	3.92	4.21	.29	4.15	.23

SOURCE: U.S. Department of Agriculture, Economic Research Service, *Wheat Situation and Outlook Yearbook*, February 1991, (WS-292), pages 52, 53, and 56.

Short-Term Patterns in Price Movements

Many of the price changes for grains, particularly those associated with weather, appear to be random and unpredictable. Indeed, the theory of efficient markets suggests that futures price changes will appear to be random if each futures price reflects all the currently available information about future demand and supply. In contrast, many traders believe that short-term futures price movements tend to follow certain patterns. Attempts to detect and predict such patterns using price charts or computers are called *technical analysis*. This method contrasts with *fundamental analysis*, which attempts to explain and predict price movements using the principles of supply and demand. Various types of technical analysis are extensively used by stock traders as well as commodity futures speculators, but the validity of these methods is widely disputed.

Pricing Strategies for Farmers

Let us now consider some of the strategies that a farmer might follow in pricing grains. Grain-pricing decisions do not begin at harvest. Instead, a farmer needs to form initial marketing plans at the same time that production plans are made. This decision calls for assembling information about prospective prices and costs during the production-planning stage so that alternative cropping and marketing strategies can be evaluated together.

Unlike decisions about crop acreage and fertilizer levels, decisions about when and where to market can be modified after planting time, as new information about supply and demand becomes available. For example, a farmer who has a good crop developing and expects an abundant harvest may forward contract more of the crop during the growing season than would be contracted otherwise. Thus, a crop producer's marketing strategy involves an initial plan and a series of revisions based on later developments.

In marketing grain, a farmer makes decisions about when to sell and to whom to sell. The following discussion will focus primarily on the first type of decision—the timing of sales.

Because grains are storable and active forward markets exist, farmers can price their grains at many different times, as was shown in Figure 6.1. Obviously, the farmer would like to sell when the net return is highest, but there is no infallible rule that guarantees the result. However, there are some principles that provide guidance. They involve taking advantage of seasonal price patterns, using government support programs, tax management, and risk management.

Taking Advantage of Seasonal Price Patterns

As discussed previously, grain prices tend to follow seasonal patterns that farmers need to consider in timing sales. This pattern does not mean that sales should be made during the same month every year. Instead, farmers need to consider their own grain storage costs, relative to the expected price increase, and continue to hold grain only so long as expected price appreciation exceeds their storage costs.

The costs of storage differ among farmers, and between farmers and elevators. They include interest, insurance, and taxes on the grain, charges for bin space, and quality losses (if any). Interest on the money invested in the grain normally is the largest single expense. In estimating the cost of bin space, it is essential to distinguish between short-term decisions (storing for an extra month, say) and long-term decisions (building new bins). For short-term storage decisions, only the variable or out-of-pocket costs for bin space need to be considered. The variable cost for bin space may be virtually zero if existing bin space would otherwise remain unused. Quality losses should be negligible if grain is stored at the proper moisture content and insect infestation is prevented.

A first and obvious principle is to fill storage bins at harvest to capitalize on the rapid price rise that normally occurs in the first month or two after harvest. Farmers storing in their own bins often can profitably hold grains later into the year than those renting storage space because their variable cost per month for bin space is virtually zero.

A comparison of futures prices with current cash prices shows what the market says about prospective returns from storage. If the futures price adjusted for basis exceeds the local cash price by more than storage costs, then storage is profitable. Otherwise, storage is not advisable unless the local basis is temporarily out of line.

Using Government Programs

Federal farm programs offer farmers price guarantees that are much like holding put options (see Chapter 8). By signing up for the program, a farmer can be assured of receiving at least the target price on eligible production and the loan rate on any excess production, while retaining the possibility of selling higher if prices rise. These price assurances are obtained without paying any option premiums or making any futures margin deposits, but the program participant must comply with acreage limitations.

Tax Consideration in Selling Grains

Farmers who pay income taxes on a cash basis can transfer tax liability from one tax year to the next by purchasing inputs in the current year and postponing crop sales to the next tax year. The taxes must, of course, be paid eventually, but in the meantime, the farmer has what amounts to an interest-free loan from the government equal to the amount of taxes postponed. For example, a farmer in the 28 percent federal income tax bracket, with 10,000 bushels of wheat worth $3.00 per bushel, might postpone 10,000 × $3.00 = $30,000 in taxable income and $8,400 in tax liabilities from one year to the next. This postponement amounts to an interest-free loan, which at a 10 percent interest rate, would be worth $840 a year or 2.8 percent of the value of the wheat involved. This example demonstrates that tax considerations are important in determining when to sell. Of course, to the extent that a large number of farmers, all using the same tax year, follow this practice, prices may be affected and the advantages for each reduced.

Some states assess property taxes on stored grain. This policy can provide a motive for farmers and others to sell grains ahead of the assessment date. This practice can result in price distortions if followed by enough sellers.

Risk Management

Farming is a risky business because of uncertain yields and prices. Before establishing a pricing strategy, farmers need to consider the amount of risk they are willing to accept or, equivalently, how much they are willing to pay for safety of return. Most farmers, like other businesspeople, are in business to make money, not necessarily to avoid risk, but most require some security and are unwilling to take unlimited risk for a small expected profit. Hence, most farmers neither disregard risk nor minimize it, but rather seek a satisfactory combination of risk and expected return.

Second, farmers need to appraise objectively their ability to forecast price changes. If they could accurately predict price changes, they could maximize profit by selling only when price is at or near its peak. Farmers have some advantages in price forecasting. They are among the first to know about local crop conditions that affect prices. But predicting price changes on a national or world market requires much broader information. If one were very good at making such predictions, one could probably make more money by full-time speculation than by farming. Most farmers or businesspeople are wise to follow a middle course, somewhere between pure speculation, which depends entirely on one's ability to forecast price movements, and pure risk avoidance.

Farmers can reduce their risks in crop production by buying crop insurance and by forward pricing with cash, futures, or options contracts. Neither buying crop insurance nor forward selling greatly changes average returns. On average, insurance indemnities approximately equal premiums paid. Similarly, most studies suggest that forward selling neither reduces nor increases average return over a period of years. Its main effect is to give producers a known price over the growing season, rather than a price that is unknown until the grain is marketed.

When yields are variable, the uncertainty in farmers' returns cannot be completely eliminated by forward selling before harvest. A poor crop may force the farmer who has sold too much forward to buy out forward sales contracts in lieu of making delivery. This situation can be costly if crop failure is widespread and results in a severe price run-up. For example, the 1988 drought in the Corn Belt raised many concerns about farmers' abilities to fulfill their forward sales commitments. Most studies suggest selling no more than a third to a half of the expected crop at planting time. More can be sold as output becomes assured.

The farmer who chooses to store a crop after harvest can reduce revenue uncertainty over the storage season by forward pricing with cash, futures, or options contracts. The entire amount in storage can be safely sold forward because output risk is not a major factor. Another possibility is to spread sales over time. The latter course leaves revenue undetermined until the last sale is made, but reduces the probability of selling the entire crop at a bad time.

Pricing Strategies for Country Elevators

In the country elevator business, returns depend on relatively narrow margins between prices received for grains and prices paid to farmers, plus fees charged for storage, drying, and other services rendered by the elevator, and profits from sideline activities such as selling feed and farm supplies. Three types of pricing strategies, ranging in order from the most passive to the most aggressive, are described here. First is ordinary margin pricing, where grains purchased from farmers are sold immediately to second handlers, and the elevator's bid price is set each day by subtracting a predetermined margin and appropriate transportation costs from the second handler's bid. Second is the use of hedging to give the firm flexibility in carrying inventories in its own account, while limiting its exposure to price risks. Third is pure speculation, carrying unhedged cash grain inventories or speculative positions in grain futures.

Margin Pricing with Minimal Net Positions

The basic pricing decisions made by all elevator managers are decisions about handling margins, fees charged for storage, and fees for other services. Elevators' margins are small relative to the price of the grain, normally only a few cents per bushel. By reducing margins and fees, the elevator manager may attract more business, but the profit per unit declines. Increasing margins and fees has the opposite effect. For the profit-seeking elevator manager, the basic pricing task is to adjust margins and fees to obtain the desired or maximum attainable level of total return over cost.

In conducting their businesses, elevator managers take title to large quantities of grain that often fluctuate markedly in value. A small percentage drop in the price of grain in inventory can easily wipe out the firm's margin and result in serious loss. One way that the country elevator can minimize its exposure to such price risks on grain inventories is to sell the grain as soon as it is bought from farmers. Many country elevator managers follow this practice, selling on the cash market each day approximately the same amount of grain that they buy, so that their net position remains small. Sometimes the grain is shipped as soon as it is sold; other times it is held until called for by the buyer. The advantages of this type of inventory management is that it is simple and relatively safe. Once the decision about the handling margin is made, the elevator manager need only subtract the margin and the appropriate transportation costs from the bids received from buyers to determine the bid price. This type of strategy can be conveniently spelled out by a board of directors and implemented by a manager who has limited time and ability for forecasting prices. A disadvantage of this approach is that use of the elevator's storage facilities is largely left up to other decision makers—farmers and second-level buyers. The elevator may find itself running grain through the plant and loading it out at a buyer's request, with no opportunity to earn storage and merchandising returns.

Hedging Owned Inventories

Elevator managers who wish to take a more active role in controlling their grain inventories and in seeking storage returns find the futures market a useful tool. They may, for example, want to accumulate inventories so they can offer potential buyers multicar shipments. A large spread between a futures price and the spot price offered by buyers may also make storage for the elevator's own account profitable. Hedging in the futures or options market enables elevator managers to take advantage of such opportunities by accumulating and controlling their own inventories without undue exposure to price risk.

Effective hedging requires knowledge of futures or options trading practices and understanding of basis relationships. Like other hedgers, country elevator managers must be able to predict the basis that will prevail when their hedges are to be lifted. Fortunately, the basis typically is much more stable and predictable than the price level itself. But unforeseen events, such as transportation bottlenecks and strikes, occasionally disrupt normal basis relationships, imposing losses on hedgers as well as others.

Hedgers must be prepared to meet margin calls when the price moves against their futures position. This situation generally calls for an arrangement with a lender to finance margin calls. It is in the lender's interest for borrowers who deal in commodities to hedge. When a borrower receives a margin call on a bona fide hedge, the lender who understands futures trading realizes that the value of the borrower's collateral has increased so that financing the margin call is sound business. Of course, the lender must be assured that the borrower maintains a valid hedge so that losses in the futures are balanced by gains in the value of the borrower's cash position.

Speculation

Like farmers and other grain merchants, country elevator managers can speculate in the futures and options market or hold unhedged inventories. Because they talk to knowledgeable people in the grain business and buy and sell grain daily, elevator managers may have some special insights about price prospects. But in speculating they must compete with other traders who may have wide access to information about prospective prices elsewhere throughout the country and the world. Some elevator managers who started futures trading as pure hedgers have become speculators, sometimes inadvertently, sometimes successfully, and sometimes to their subsequent regret.

Can the Grain-pricing System Be Improved?

Most indications are that the grain-pricing system is performing well, but it is nonetheless appropriate to ask if it could be made to work better. This question cannot be fully answered here, but some potential problems and possible solutions can be identified.

What distinguishes good performance from bad performance in a pricing system? If the system is working properly, each participant in the market—farmer, merchant, processor, or consumer—will receive or pay full value, but not more, for the grain or grain products sold or purchased. It would be impossible for any seller or buyer to obtain a more favorable price by switching to a different outlet or a different source of supply, once all the costs are taken into account. The market system tends toward this result, since each participant has a motive to search diligently for the most favorable price and source of supply or outlet.

But there are at least two potential impediments to effective functioning of the market that may prevent attainment of the best outcome. One is the tendency toward monopoly or monopsony-like behavior, when the numbers of firms on one side of the market is small enough that they can deliberately influence the market price. Second is randomness in prices and output coupled with incomplete information and expectations that cause decisionmakers to err. Such decision errors can lead to unwarranted price movements and wasted resources.

One place to look for pricing problems is where buyers or sellers are limited in number. For example, many farmers sell to only one elevator. If that elevator has no close competitors, it may, on the average, take a wider margin than necessary to cover cost and normal profit. Similarly, if the subterminal elevator, terminal elevator, or processor has no close competitors it, too, might take larger margins and larger than normal profits. Thus, whenever the number of buyers or sellers for a particular type of grain in a particular area is small, temptations to price noncompetitively arise. But countering this temptation is the threat of other firms entering the area should profits become too lucrative.

Another place to look for pricing failure is in excessive price fluctuations. The prices of grains and oilseeds change almost every day, and often markedly within a day. These changes reflect new information arising each day about supply and demand. But prices may sometimes fluctuate more violently than warranted by market conditions. Price manipulation is sometimes alleged when daily or weekly price changes are very large, as they were during 1972 to 1975. However, there is no proof that manipulation is a major factor contributing to price volatility. Traders may overreact or underreact to major news events, such as weather scares or export sales announcements. If there is an over- or underreaction, prices should eventually return to economically justified levels. In practice, even retrospectively, it is difficult to identify periods when the market overreacts to new developments. Many large price fluctuations are understandable when viewed in terms of the information available. There are no easy answers as to where and when the pricing system is defective.

Alternative Pricing Systems

The grain-pricing system now in existence has evolved one step at a time over many years. Continuing changes in communication technology, grain storage, transportation, processing technology, product uses, methods of measuring quality, and in the structure and organization of the grain industry call for further adjustments in the pricing system. Changes in pricing methods occur gradually because many people are involved and habits are hard to break. It is reasonable to ask whether the pricing system has kept pace with changes in the industry and in technology and what further changes are needed at this time. Could a better pricing system be developed and put into effect?

Grain pricing might be improved by developing a system using computers and computer terminals in merchants' and processors' offices to trade grains. Computers and electronic communication devices are well suited for processing the large volumes of information required in modern grain markets. They are already widely used for disseminating grain price information. But use of computers could be carried a step further, so that trades are actually executed and confirmed electronically. Each trader would be able to quickly scan all the existing bids and offers for the types of grain and locations of interest, and then enter bids and offers for wide exposure to other traders. The computer would determine when a bid and offer matched, notify each party that a trade had been completed, and produce written copies of the sales documents. Potential advantages of electronic trading would include more timely information for each trader about other traders' bids, offers, and completed transactions, and reduced time required in telephoning other traders one at a time. A possible disadvantage of electronic trading would be reduced personal voice contact with other traders that facilitates sharing of information such as local crop prospects and how the harvest is progressing.

In contrast to electronic trading and other schemes for making a dispersed pricing system operate more effectively, pricing decisions could be made more centralized. This centralization might involve installing a marketing board, a government or quasi-government agency with exclusive marketing privileges and powers to set prices and control quantities marketed, such as the Canadian Wheat Board. An advantage of the marketing board approach would be reduced price uncertainty for producers, merchants, and processors. Other possible gains might arise through concerted action in pricing grains on the world market. A major disadvantage would be the inefficiencies that would be engendered in attempting to set prices administratively. Canada and Australia have used marketing boards for grains with mixed success.

Summary

The grain-pricing system coordinates decisions of grain farmers, merchants, processors, and consumers and determines the allocation of income or buying power among these groups. The process by which these different groups interact with each other, exchange bids and offers, and agree on price and other terms of trade is called *price discovery*. The price discovery process for grains is keyed to futures trading and involves communicating bids for cash grains each day from buyers to sellers via telephone. The pricing process relies heavily on the system of grades and standards and the electronic communication network. Both of these factors enable buyers and sellers who are physically separated from each other and from the products involved to confidently trade with each other.

Grain prices are often quoted in terms of the basis—the difference between a cash price and a specific futures price. To make effective pricing decisions, farmers, merchants, and processors need to understand the factors that cause price to differ over time and space, and between different qualities of grain. In addition, knowledge of and ability to use futures markets is valuable. Indications are that the grain-pricing system functions well. Imperfect competition and erratic behavior by traders may sometimes distort prices, but such distortions appear to be small.

Selected References

Harwood, Joy L., Linwood A. Hoffman, and Mack N. Leath, "Marketing and Pricing Methods Used by Midwestern Corn Producers," *Feed Situation and Outlook Report*, FdS-303, pp. 33-39, USDA, Econ. Res. Ser., September 1987.

———, "Marketing Methods Used by Midwestern Soybean Producers," *Oil Crops Situation and Outlook Report*, OCS-15, pp. 26-31, USDA, Econ. Res. Ser., October 1987.

Heifner, Richard G., et al., *The U.S. Cash Grain Trade in 1974: Participants, Transactions, and Information Sources*, USDA, Econ. Res. Ser., Agric. Econ. Rept. No. 386, September 1977.

Hoffman, Linwood A., Joy L. Harwood, and Mack N. Leath, "Marketing and Pricing Methods Used by Selected U.S. Wheat Producers," *Wheat Situation and Outlook Report*, WS-281, pp. 19-24, USDA, Econ. Res. Ser., May 1988.

Paul, Allen B., Richard G. Heifner, and J. Douglas Gordon, *Farmers' Use of Cash Forward Contracts, Futures Contracts, and Commodity Options*, USDA, Econ. Res. Ser., Agric. Econ. Rept. No. 533, May 1985.

Tomek, William G., and Kenneth L. Robinson, *Agricultural Product Prices*, 2nd ed., Cornell University Press, Ithaca, NY, 1981.

Wright, Bruce H., et al., *Forward Contracting in the Corn and Spring Wheat Areas, 1988, Results of an Elevator Survey*, USDA, Econ Res. Ser., December 1988.

7

Futures: Markets, Prices, and Hedging

Reynold P. Dahl

Organized Commodity Markets

Organized commodity markets (grain exchanges) such as the Chicago Board of Trade, the Kansas City Board of Trade, and the Minneapolis Grain Exchange have played an important role in the development of an efficient grain-marketing system in the United States. These markets were organized in the latter half of the nineteenth century. They brought together buyers and sellers for grain trading in a central marketplace, facilitating open and competitive trading in grain. Market information and price discovery mechanisms were thereby improved, which increased competition and broadened the market for the farmers' grain.

In the early years, grain exchanges served as important markets for cash grain. But cash grain trade on the exchanges has declined substantially as buying and selling grain on a sample basis has been largely replaced by *to arrive* cash contracts between country elevators and terminal merchants. Single railroad cars of grain traded on the floors of grain exchanges have been reduced sharply as increased quantities of grain are shipped in unit trains (multiple-car units of 25, 50, or 100 cars) from country origins directly to ports for export or to domestic users. Trucks have also replaced single rail-car shipments over shorter distances. As the marketing of cash grain has become more decentralized, the role of grain exchanges in cash grain trade at major terminal rail markets such as Chicago, Kansas City, and Minneapolis has diminished. But the volume of futures trading in grain at the exchanges has increased substantially. Grain futures prices have become even more important under our decentralized grain-marketing system.

Futures trading through which futures prices for grain are derived takes place only on organized grain exchanges. By providing centralized futures trading under rules and regulations, organized commodity markets have helped make grain marketing a highly competitive business. These markets serve as clearinghouses of grain supply and demand information for efficient price discovery.

Evolution of Futures Trading in Grain

There is a widespread belief that futures markets are primarily speculative. However, the history of their development reveals that they developed out of an economic need for hedging. A study of the history of corn marketing in the 1850s indicates how futures trading developed out of that need.[1]

The opening of the Illinois-Michigan Canal, connecting the Illinois River with Lake Michigan, made possible the shipment of corn by water from central Illinois to Chicago. This route reduced transportation costs and facilitated the movement of corn into Chicago. Prior to the opening of the canal, transportation costs to Chicago were very high. There were no railroads going into Chicago, and most grain movements into the city were by horse and wagon. In 1884, the value of wheat in Chicago was about equal to the cost of hauling it 60 miles by horse and wagon. Corn prices were lower than those for wheat, so corn could be profitably hauled for even shorter distances than wheat.

After the canal opened in 1848, the amount of corn shipped to Chicago increased rapidly. Dealers invested in corncribs, which were constructed along the Illinois River, and purchased corn from farmers who hauled their corn to the river dealers by sleigh or wagon over the frozen roads in the winter. The dealers held the corn in cribs to permit drying, after which it was shelled and shipped in the spring or early summer.

Financing their accumulation of corn stocks strained the financial resources of these early dealers and limited their operations. Also, they assumed heavy price risks on the corn because they did not know what price they could sell the corn for the following spring in Chicago. Hence, they developed the practice of going to Chicago and finding a merchant who would buy corn for May delivery. In other words, they entered into time contracts, or "to arrive" contracts, with Chicago merchants. These contracts served two important economic needs. First, they enabled dealers to obtain financing from grain merchants who were then willing to advance them money for corn purchases. Second, they enabled dealers to shift price risks to someone else.

The first such contract was recorded on March 13, 1851, calling for delivery of 3,000 bushels of corn in June, at a price one cent lower than the quotation for corn that day. The use of these time contracts increased rapidly during the 1850s. These early time contracts were very informal, either verbal or simply evidenced by a memorandum. Only the quantity, price, and time of delivery were specified. Some had a proportion of the price paid in cash.

Merchants in Chicago began to trade these contracts, and speculators often bought them hoping to profit from correctly anticipating a price rise. As trading in the contracts increased, they were gradually standardized in written form with respect to number of bushels, delivery in specified calendar months, and the terms of payment.

The Chicago Board of Trade, which had been organized in 1848, established rules governing the trading of these time contracts in 1865. The transition to full-blown futures trading in corn took place sometime between 1870 and 1875 with the development of the standardized futures contract and the accounting of those contracts through a clearinghouse.

The clearinghouse took the role of intermediary between buyer and seller. After a futures trade was consummated, the buyer was deemed to have bought from the clearinghouse, and the seller was deemed to have sold to the clearinghouse. The clearinghouse calculated each day the gain and loss resulting for all outstanding contracts because of the price change that day. It then collected funds from the losing futures contracts and paid these funds to the futures having gains. Clearinghouse operations provided two important contributions that made a fully developed futures market possible. First, they reduced the risk of default on a futures contract to one day's market price change. Second, the clearinghouses made it easier for a trader to be released from obligations assumed under the original contract. Traders did not need to search out one another; instead, they merely needed to make an equal but opposite futures transaction.

If a trader had originally bought a futures contract, the obligation could be satisfied simply by selling a futures contract of the same month. Similarly, if a futures contract had been sold originally, the trader only needed to buy one of the same month. The clearinghouse could offset these two transactions, freeing the traders from their original obligation to either take or make delivery.

Trading in futures contracts and exchanging them through a clearinghouse, with its ability to offset one futures transaction with another, resulted in two economic effects. First, it opened futures trading to people without physical commodity-handling facilities. Thus, speculation in futures was facilitated. Second, it facilitated hedging by grain processors and warehouse managers. Grain merchants were able to transfer unwanted price risks to speculators without actual change in the ownership of grain.

In summary, the evolution of futures trading in grain clearly shows that futures trading developed out of an economic need for hedging. The early corn dealers in Illinois needed to shift price risks and to finance their inventories. The early time contracts that grew out of these needs ultimately evolved into fully developed futures trading, which made hedging more efficient. Our grain futures markets continue to serve this important economic need today.

Futures Markets and Commodities Traded

Futures trading in the United States had its origin in the marketing of grain, and it is in grain marketing where futures trading has achieved its highest degree of commercial use. But in recent years, futures trading has been extended to other commodities and areas such as financial instruments, which have grown rapidly. In fact, one of the most significant developments in the U.S. economy during the past two decades has been the explosive growth in futures trading. The total number of futures contracts traded on U.S. futures markets rose from 13.6 million in 1970 to 92.1 million in 1980 to 276.5 million in 1990. Interest rate futures, of which the U.S. Treasury Bond futures contract on the Chicago Board of Trade is the most important, have taken over first place in futures trading with 44.6 percent of total trade in 1990. Futures trading in agricultural commodities ranked second with 20.6 percent of total futures trade in 1990, down from 41 percent in 1983 (Table 7.1).

Futures trading in U.S. grains, soybeans, and products is conducted on four futures markets: the Chicago Board of Trade, the Kansas City Board of Trade, the Minneapolis Grain Exchange, and the Mid-America

TABLE 7.1 Total Futures Contracts Traded by Commodity Group, 1990

Commodity Group	Number	Percent
Interest Rate	123,410,201	44.6
Agricultural Commodities	57,088,348	20.6
Energy Products	35,441,219	12.8
Foreign Currency	28,880,844	10.4
Precious Metals	14,812,847	5.4
Equity Indices/Index	14,766,860	5.3
Nonprecious Metals	1,853,281	0.7
Other	272,217	0.1
Total	276,525,817	100.0

SOURCE: Futures Industry Association, Inc., Washington, D.C. 20006.

Commodity Exchange in Chicago. The Chicago Board of Trade, our nation's largest futures market, traded 34.6 million futures contracts, or over 90 percent of the total volume of trading in all four markets in 1990 (Table 7.2). Soybeans and corn had the largest 1990 trade volume, with over 10 million futures contracts traded in each on the Chicago Board of

TABLE 7.2 Futures Contracts Traded on U.S. Grain Futures Markets, by Commodity, Selected Years

Exchange and Commodity	Contract Unit	Thousand Contracts			
		1973	1980	1987	1990
Chicago Board of Trade					
Wheat	5,000 bu	1,567	5,428	1,929	2,876
Corn	5,000 bu	4,075	11,947	7,253	11,423
Oats	5,000 bu	183	321	291	434
Soybeans	5,000 bu	2,743	11,768	7,379	10,302
Soybean oil	60,000 lb	1,763	3,168	3,912	4,658
Soybean meal	100 tons	660	3,219	3,798	4,905
Total		10,991	35,851	24,562	34,598
Kansas City Bd. of Trade					
Wheat	5,000 bu	346	1,298	971	1,136
Grain Sorghum	5,000 bu	0	0	0	1
Total		346	1,298	971	1,137
Minneapolis Grain Ex.					
Spring wheat	5,000 bu	172	334	311	477
White wheat	5,000 bu	0	0	1	1
Oats		0	0	0	*
Total		172	334	312	478
Mid-America Commodity Ex.					
Wheat	1,000 bu	75	551	190	147
Corn	1,000 bu	103	441	312	455
Oats	1,000 bu	9	2	7	14
Soybeans	1,000 bu	56	1,053	418	1,566
Soybean meal	20 tons	0	0	3	5
Total		243	2,047	930	2,187
Total all markets		11,752	39,530	26,775	38,400

*Fewer than 1,000 contracts.

SOURCE: Futures Industry Association, Inc., Washington, D.C. 20006.

Trade. Soybean oil and soybean meal also had high trade volumes with nearly 5 million futures contracts traded in each of these soybean products. Wheat is the only grain that is traded on all four futures markets. The wheat futures contract on the Chicago Board of Trade had the largest trade volume in 1990 with 2.9 million. Four classes of wheat are deliverable on this futures contract, but it usually prices soft red winter wheat produced in the eastern corn belt. The Kansas City Board of Trade wheat futures contract calls for delivery of hard red winter wheat; hard red spring wheat is deliverable on the Minneapolis Grain Exchange wheat futures contract. The latter exchange also has a futures contract in white wheat that received a small trade volume in 1990. Wheat, corn, oats, soybeans, and soybean meal futures contracts are also traded on the Mid-America Commodity Exchange, but the contracts are only one-fifth the size of those on the other markets.

Changes in Volume of Futures Trading in Grain

The volume of futures trading in grain, soybeans, and products more than tripled from 11.8 million futures contracts in 1973 to 39.5 million in 1980. The latter year was an all-time record. This increase roughly paralleled the increase in grain exports, which also more than tripled during the decade of the 1970s. Grain production shortfalls, notably in the former Soviet Union but in other countries as well, increased the export demand for U.S. grain. Other economic factors contributing to the increase in exports were the weak U.S. dollar and rising incomes in many developing countries. The Commodity Credit Corporation (CCC), the price-supporting agency of the U.S. government, was able to dispose of its grain stocks that had been accumulated under past price support operations as prices rose. CCC stocks no longer served as a lid on market prices so grain prices became more variable as well. Increased grain exports in the 1970s along with greater price volatility and reduced levels of U.S. government-owned grain stocks combined to increase hedging needs, which pushed futures trading in grain to a record level of 39.5 million futures contracts in 1980.

These same economic variables operated in reverse during the early 1980s. World demand for grain declined as did world grain trade. U.S. grain exports bore the brunt of this painful adjustment. Aided by a strong U.S. dollar and the price umbrella provided by government price support programs, other grain-exporting countries increased their share of world grain exports. As U.S. grain prices fell to price support levels, grain began to accumulate rapidly under regular price support loan operations and the farmer-owned reserve programs. Government-owned stocks and price support loan operations again became dominant market factors reducing hedging needs and resulting in a decline in futures trading in grain to 26.7 million futures contracts in 1987. Volume increased in the next two years, reaching 38.4 million contracts in 1990 (Table 7.2). Experience in the past 15 years illustrates extremely well that futures trading in grain depends heavily upon commercial needs for hedging.

Cash and Futures Trading

Two types of trading are involved in grain marketing. First, cash or spot trading involves the sale and receipt of grain for immediate delivery, or forward delivery at some specified time and place. The cash market is highly decentralized with cash transactions widely dispersed geographically. This decentralization is facilitated by our excellent communications system. Buyers and sellers of grain often make cash transactions by telephone with follow-up written documentation of the trades. Second, futures trading occurs through the trading of standardized futures contracts. This trading is highly centralized and occurs only on an organized commodity market. Futures prices, as derived from the trading of futures contracts, are central to the entire grain-pricing and marketing mechanism.

Cash grain may be traded for immediate delivery or for delivery within some specified period, for example, within 30 days. The latter type of transaction is referred to as "to arrive" sales. In reality, they are forward cash contracts. A *to arrive* sale is a contract under which a buyer agrees to accept from a seller a certain quantity of grain to be shipped to an agreed-on destination within a specified time period at an agreed-on price for a base grade. The contract usually specifies price adjustments for such quality factors as moisture, test weight, and protein. Sometimes a maximum moisture, minimum protein, or other limits that the buyer will accept are also specified within the contract.

Characteristics of Futures Trading

The Futures Contract

Standardized contracts calling for the delivery of grain in some designated future month are traded in the futures market. Prices of futures contracts are determined by auction in a trading pit; the trading is conducted only at designated times and takes place under rules and regulations of the exchange. All terms of futures contracts, except the price, are standardized by the exchange. A futures contract calls for delivery of a specified quantity and quality of grain, at a specified place or places, in some designated month in the future. These contracts are binding, and their integrity is preserved under the rules and regulations of the exchange.

Example of a Futures Contract

Futures contracts and associated trading terminology are more readily understood through use of the following illustration:

June 25

Trader A
    Buys 5,000 bushels of December corn at $2.20.
Trader B
    Sells 5,000 bushels of December corn at $2.20.

Trader A has entered into a contract to take delivery of 5,000 bushels of corn in December and will pay $2.20 per bushel. Trader B has entered into a contract to deliver 5,000 bushels of corn in December and will receive $2.20 per bushel. If both traders are hedgers, each must deposit a minimum of $350 as margin money with the commission merchant, through whom the trade is made to guarantee contract performance. Also, after the trade, Trader A is deemed to have bought from the clearinghouse, and Trader B is deemed to have sold to the clearinghouse. This process eliminates the necessity for each trader to settle the contract by trading with each other either before or during the delivery month. The clearinghouse guarantees the performance on all futures contracts.

Assuming that neither of the above traders had a market position before the above trade was made, Trader A is a *new buyer* and is now "long"; Trader B is a *short seller* and is now "short."

If both traders leave their contracts open until the delivery month, the seller (Trader B) may deliver at any time during December up until the last few days, and it is up to the buyer (Trader A) to accept delivery and pay for the corn. If the seller decides to deliver on the contract, a warehouse receipt is delivered representing 5,000 bushels of U.S. No. 2 yellow corn stored in a public terminal elevator in Chicago, IL; Burns Harbor, IN; Toledo, OH; or St. Louis, MO, switching districts that have been declared regular by the Chicago Board of Trade. Deliveries at Toledo are made at a 3-cent discount, and deliveries at St. Louis are made at a 7-cent per bushel premium to the Chicago contract price. Alternate grades to No. 2 yellow are also deliverable at differentials stipulated in the contract.

If the seller delivers on the contract, a warehouse receipt must be presented to the clearinghouse, which in turn delivers it to the oldest long in the market. This may or may not be Trader B with whom the original contract was made.

Most futures contracts are not settled by delivery. They are usually settled by offsetting purchases or sales of futures contracts sometime before the delivery month, illustrated as follows:

July 23

Trader A
Sells 5,000 bushels of December corn at $2.25.
Trader B
Buys 5,000 bushels of December corn at $2.25.

Trader A gains 5 cents per bushel or $250 on the contract.
Trader B loses 5 cents per bushel or $250 on the contract.

In trade terminology, Trader A is *long liquidating*, and Trader B is *short covering*. Each of these traders has satisfied the obligation regarding delivery after offsetting trades are made. Traders A and B need not make offsetting trades by dealing with each other. Each may make an offsetting trade at any time after the original contract was made by dealing with another trader. This offsetting trade is possible because the clearinghouse serves as an intermediary in all futures contracts. The clearinghouse also marks all contracts as to the settlement price (the closing price) each day. In the preceding illustration, Trader A would have collected a net of 5 cents per bushel, or $250, from the clearinghouse, and Trader B would have paid a net of 5 cents per bushel, or $250 to the clearinghouse from June 25 to July 23.

Although most futures contracts are not settled by actual delivery, the privilege of making or taking delivery is important because it is the delivery mechanism that assures the convergence of the cash and futures prices in the delivery month. It is significant that futures contracts are not often used for the actual purchase or sale of grain. Instead, they facilitate hedging by serving as temporary substitutes for intended later transactions in cash grain.

Relationships of Futures Prices for Different Delivery Months

Grain production is highly seasonal, and the delivery months for grain futures contracts are related to the harvesting, marketing, and consumption of grain throughout the year. The delivery months for wheat, corn, and oats are March, May, July, September, and December. The delivery months for soybeans are January, March, May, July, August, September, and November.

Prices of future delivery months are usually different and change with market fundamentals. The price spreads between future delivery months also change. Price spreads between delivery months are related to storage costs and are often called *carrying changes*. They are some of the most important and interesting aspects of grain futures prices. Understanding the economics of these charges in futures price spreads is the key to the successful commercial use of futures markets.

Carrying Charges

Price spreads between delivery months can be calculated by subtracting the price of the near delivery month from the price of the next more distant delivery month. Sometimes, the spreads are positive, as was the case on April 19, 1988, for corn futures prices (Table 7.3). This positive spread is often referred to as a *carrying-charge market*, or as a market with

TABLE 7.3 Carrying-Charge Market, Closing Prices of Corn Futures, and Spreads Between Delivery Months, April 19, 1988

Delivery Month	Closing Price ($/bu.)	Spread[a] ($/bu)
May 1988	2.0550	
		.0775
July	2.1325	
		.0550
September	2.1875	
		.0825
December	2.2700	
		.0750
March 1989	2.3450	
		.0475
May	2.3925	
		.0225
July	2.4150	

[a]Spread is the near delivery month subtracted from the next distant delivery month.

SOURCE: Chicago Board of Trade futures prices.

positive returns to storage. On April 19, 1988, a trader could buy May corn for $2.055 per bushel and simultaneously sell July corn for $2.1325. He could then take delivery on the May contract, hold the warehouse receipt until July 1 and redeliver it in fulfillment of the July contract for a gross return of 7.75 cents per bushel. Such a return, however, is insufficient to cover his total storage costs or "carrying charges," which would include: (1) the commercial warehouse charge, (2) interest on capital invested, and (3) insurance. The commercial warehouse charge (which includes insurance) is 4.8 cents per bushel per month or 9.6 cents per bushel for two months' storage. If the prime interest rate is 9 percent, the capital charge would be 3 cents per bushel for two months, so the full carrying charge between May and July would be 12.6 cents per bushel.

Distant futures cannot exceed the price of near futures by more than full carrying charges because of the possibilities of the same type of arbitrage as explained above. So, the price of July corn could not exceed the price of May corn on April 16, 1988, by more than 12.6 cents per bushel, or the full carrying charge.

Price differentials between future delivery months rarely reflect full carrying charges for two reasons. First, grain merchants who own warehouses are willing to store grain for their own account at less than the commercial storage rate. Relevant storage costs to an elevator owner are the marginal costs of storing grain including interest on capital invested, shrinkage, and insurance. These costs may total significantly less than the commercial storage rate. Second, there is a convenience yield to the ownership of grain stocks because the possession of such stocks affords the grain merchant opportunities to merchandise these stocks at a profit at various times during the storage period. Also, grain processors such as flour millers can draw on the stocks for processing as needed. This convenience yield is positive and may be high enough to make the actual net costs of grain storage to a merchant quite low. Hence, competition among grain merchants to store grain forces futures prices to reflect something less than full carrying charges even when there is an abundance of grain available for storage. In reality, we can think of price spreads between adjacent future delivery months as representing a market-determined price of storage.

Futures markets with positive price spreads have positive prices of storage. They usually prevail in periods when "free" grain stocks are large and there is an abundance of grain to store. When grain can be bought at a lower price for current delivery than it can be sold for future delivery, there is an incentive to store and earn the carrying charge through hedging. So futures markets guide grain stocks into storage when storage is needed.

Inverse Carrying Charges

Sometimes distant futures sell for lower prices than near futures; hence, price spreads are negative. This negative spread is known as an *inverted market* or a market with inverse carrying charges.

Inverse carrying charges usually occur in years when grain supplies are short relative to the demand. Grain futures were inverted in 1973 when the export demand for grain was very strong relative to available supplies. In such periods, buyers of grain are willing to pay higher prices for current delivery than for future delivery. Hence, the market provides an incentive for owners of grain to sell for current delivery rather than store for future delivery. On June 13, 1973, for example, the corn futures market was inverted between all of the adjacent future delivery months (Table 7.4). If on this date a trader bought the July future at $2.32 and simultaneously sold the September future at $2.205, took delivery on the July future, and held the warehouse receipt for redelivery on the September future, the gross return would have been a negative 11.5 cents/bu (Table 7.4). Consequently, when storage prices are negative, merchants and processors will keep their inventories at a minimum. Nevertheless, some stocks will be stored because the convenience yield may be very high for a small level of stocks that is essential for continued business operations.

TABLE 7.4   Inverted Market, Closing Prices of Corn Futures, and Spreads Between Delivery Months, June 13, 1973

Delivery Month	Closing Price ($/bu.)	Spread[a] ($/bu)
July 1973	2.320	
		−.115
September	2.205	
		−.090
December	2.115	
		−.015
March 1974	2.100	
		−.010
May	2.090	

[a]Spread is the near delivery month subtracted from the next delivery month.

SOURCE: Chicago Board of Trade.

There is no automatic or theoretical limit on the possible premiums that near futures may go to distant futures. The amount of the premium depends entirely upon how much buyers are willing to pay for immediate delivery.

Price Relationships Between
Old-Crop and New-Crop Futures

Each harvest of grain brings a new marketing year with a new supply-and-demand situation. The wheat harvest begins in mid-May in the southern plains, so the U.S. Department of Agriculture's wheat marketing year is June 1 to May 31. The first future delivery month in the wheat marketing year is July, which is called the *new-crop* future. Successive delivery months are September, December, March, and May, the *old-crop* future.

Corn is harvested later in the year than wheat, so the USDA's corn-marketing year is September 1 to August 31. The first future delivery month in the corn-marketing year is September. However, because only limited amounts of corn are harvested and available for delivery in September, the September future is usually considered a transitional month between old-crop and new-crop corn. December is usually considered the "new-crop" future in corn because most of the new harvest will have been completed by December.

Sometimes old-crop futures prices differ markedly from new-crop futures. In a year when the supply of corn is tight relative to the demand, old-crop corn futures (July) typically sell at a substantial premium to new-crop futures (December) as illustrated by the closing corn futures prices reported on April 19, 1984 (Table 7.5).

The 51.75-cent negative price spread between the July and December future delivery months provides an incentive for merchants and others who own corn to sell the old-crop corn rather than carry it into the new-crop year. A customary explanation of the large discount of the December future to the July future was that the December future reflects the expectation of a large corn harvest between July and December, which will depress the price of the December future but will not affect the price of the July future. Research by Holbrook Working showed this explanation to be mistaken. It is only the supply of corn already in existence that explains the new-crop discount. It is the short supply of old-crop corn that results in the premium of the July future to the December future, not the expectations of a large harvest between July and December.[2]

The size of the premium of old-crop futures to new-crop futures can be very large in some years. In soybeans, for example, it has not been

TABLE 7.5   Small Carry-over, Closing Prices of Corn Futures, and Spreads Between Delivery Months, April 19, 1984

Delivery Month	Closing Price ($/bu.)	Spread[a] ($/bu)
May 1984	3.5725	
		.0075
July (old crop)	3.5650	
		−.3000
Sept. (transitional)	3.2650	
		−.2175
Dec. (new crop)	3.0475	
		.0900
March 1985	3.1375	
		.0300
May	3.1675	

[a]Spread is the near delivery month subtracted from the next delivery month.

SOURCE: Chicago Board of Trade.

uncommon in years of short soybean supplies for July (old-crop) soybean futures to sell at a premium of several dollars over November (new-crop) soybean futures in June.

On the other hand, in years when "free" supplies of corn are abundant and a sizable carry-over is in prospect, new-crop futures usually sell at a premium over the old crop, as illustrated by closing corn futures prices reported on May 2, 1988 (Table 7.6).

The 15.5-cent positive price spread between the July future (old crop at $2.13) and the December future (new crop at $2.285) reflects an abundance of "free" supplies of corn relative to the demand. Hence, warehouse managers have an incentive to carry more old-crop corn into the new 1988-1989 marketing year. In fact, the corn futures prices on May 2, 1988, showed positive storage prices all the way to the July 1989 future, the most distant delivery month for which prices were quoted on this date. The relationship among futures prices of the various delivery months is a highly useful guide to inventory policies of grain merchants and processors. The above analysis considers only futures prices. But we must also consider the relationship of cash and futures prices to understand hedging.

TABLE 7.6 Large Carry-over, Closing Prices of Corn Futures, and Spreads Between Delivery Months, May 2, 1988

Delivery Month	Closing Price ($/bu.)	Spread[a] ($/bu)
May 1988	2.0525	
		.0775
July (old crop)	2.1300	
		.0675
Sept. (transitional)	2.1975	
		.0875
Dec. (new crop)	2.2850	
		.0775
March 1989	2.3625	
		.0475
May	2.4100	
		.0200
July	2.4300	

[a]Spread is the near delivery month subtracted from the next delivery month.

SOURCE: Chicago Board of Trade.

Cash and Futures Price Relationships—The Basis

The difference between the cash price of a particular grain (corn, for example) at a specific location and the futures price is called the basis. The local basis for corn is calculated by subtracting the futures price (usually the near future) from the local cash price for corn prices prevailing in Minneapolis on June 2.

Cash − Futures (July) = Basis
#2 Yellow
$2.16 $2.34 −18¢

The Minneapolis corn basis is −18 cents, as the Minneapolis cash corn price is quoted at "18 cents under," or an 18 cents discount under the July futures price on the Chicago Board of Trade.

Sometimes cash grain will trade at a premium to the futures price for cash wheat in Minneapolis on June 2.

Cash − Futures (July) = Basis
No. 2 DNS Wheat
14% protein
$3.78 $3.43 +35¢

Hence, the Minneapolis basis for No. 2 DNS wheat 14% protein is +35 cents, as the Minneapolis cash wheat price is quoted at "35 cents over" or a 35-cent premium over the July futures on the Minneapolis Grain Exchange.

Cash grain prices are usually quoted, and cash grain is traded on the basis rather than as a "flat price." Basis trading is very common in grain marketing. To a grain merchant, a basis of 18 cents under for corn is more meaningful than a "flat price" of $2.16 in the above illustration; and 35 cents over for wheat is more significant than a "flat price" of $3.78 for several reasons. First, the basis is an indicator of how cheap or expensive cash grain is relative to the futures price. Second, most merchants practice hedging so they have positions in both cash and futures markets that are opposite each other. So it is a change in the basis that will affect revenue changes rather than a change in the cash price by itself. Finally, changes in the basis are also more predictable than changes in the cash price because cash and futures prices converge and become "equal" in the delivery month at the delivery point.

When cash grain prices are at wide discounts to futures prices, the basis is said to be weak. When the difference between cash and futures prices is small or cash prices are at a premium to futures prices, the basis is said to be strong. A movement in the basis from a weak to a strong one is called a strengthening of the basis, while the reverse movement is called a weakening of the basis. If the basis for corn changes from 18 cents under to 15 cents under, it has achieved a higher level and has strengthened. If the basis on wheat changes from 35 cents over to 27 cents over, it has achieved a lower level and has weakened.

At any particular time, the basis serves as a guide to merchant's inventory and hedging policy. When cash grain is at a sizable discount to futures prices (the basis is weak), a merchant will strive to be long cash grain and short futures. On the other hand, when the basis is strong and the merchant has to pay premiums for cash grain, the merchant will strive to be short cash grain and long futures.

In summary, we can think of the basis as the link between the general price level of a particular grain as represented by the futures price and the cash price of a specified quality at a specific location. Futures prices are the product of a considerable volume of trading and are very sensitive to new market information on supply and demand affecting the general level of market prices. Local cash grain prices reflect these changes in the futures price, but they also reflect local economic values such as transportation costs and availability; local supply-and-demand variations; and the availability of local storage.

How the Basis Changes Over Time

Changes in the basis over time are more predictable than changes in the flat price of cash grain. The change in the basis over time can be forecast because the cash and futures prices must be nearly equal in the delivery month at the delivery point. In other words, the cash and futures prices converge, and the basis approaches zero in the delivery month. This change occurs because of the privilege of making or taking delivery in the expiring futures delivery month.

The Basis at the Delivery Point

The convergence process and the behavior of the basis over time at the delivery point are graphically shown in Figure 7.1. When cash grain is at a discount to the futures price, such as in the bottom part of Figure 7.1, there is a positive price of storage. This is the time to store cash grain and hedge it through the sale of futures. As the delivery month approaches, cash prices strengthen relative to futures—the basis strengthens. The gross return earned on the storage is equal to the amount by which the basis strengthens. This strengthening in the basis usually does not occur evenly, but the trend will be for the basis to strengthen as the delivery month approaches. If, for example, a merchant purchases cash grain at 25 cents under the May future on October 1 and hedges by selling May futures, a gross return of 25 cents will be earned if the cash grain is sold and the hedge is lifted in May when the delivery market basis is zero.

When grain supplies are short relative to demand, the price of grain for current delivery often is at a premium to the price for future delivery. This behavior is illustrated in the upper part of Figure 7.1. If cash prices are at a premium to futures prices, the market reflects a negative price of storage. The reason is that if a merchant purchases cash grain at 25 cents over the May futures on October 1 and hedges by selling May futures, a 25-cent loss will occur if storage continues until May 1, when the hedge is lifted and the cash grain is sold. In a premium market, consequently, merchants have an incentive to keep their grain inventories at a minimum. In fact, they try to sell grain for forward delivery at high current prices and purchase futures as temporary substitutes for the cash grain until it is purchased for delivery.

The Basis at Local Points

The basis at local country points moves in a similar manner over time as the basis at the delivery point. However, the basis at a local point reflects the cash price of grain at a specified location in the country. If

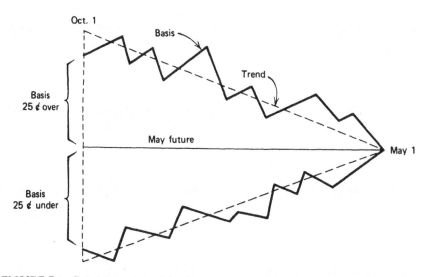

FIGURE 7.1   Behavior of the May basis over time in a discount and premium
market at a delivery.

grain normally moves from that location to the delivery point, the basis
will usually be lower at the country point by the cost of transportation
to the delivery point. For example, if on October 10 the basis for corn in
Chicago is 15 cents under the December futures, the basis for corn in a
country point in western Illinois may be 21 cents under. The 6-cent
difference reflects the transportation cost to Chicago. Cash corn prices in
Chicago would be higher than in western Illinois, reflecting the cost of
shipping the corn to Chicago.

On the other hand, if grain does not normally move from the local
point to Chicago, the basis at that location will reflect transportation costs
from Chicago. Hence, the basis at such a point will be higher than at
Chicago.

The basis at local country points is not only affected by the cost and
availability of transportation. Other factors affecting the local basis are
the supply and demand for corn at a local point and the availability of
storage. A small local supply of corn, for example, will strengthen the
local basis relative to Chicago. On the other hand, a large corn supply
at the country point will weaken the country basis relative to Chicago.

Figure 7.2 illustrates the movement in the basis over time at the
delivery point, Chicago, for corn futures, and the basis at some country
point where cash prices are typically lower than at Chicago because of

FIGURE 7.2 Behavior of the May basis over time at the delivery point and at a country point.

transportation costs. The difference in the basis between the two points reflects transportation costs and variations in the supply and demand for cash corn at the local point and at the delivery point. However, the basis tends to narrow as the delivery month approaches at the country point as well as at the delivery point.

Hedging

Hedging is an important marketing tool for grain merchants, processors, and producers. Hedging is best described as the purchase or sale of a futures contract on an organized commodity exchange as a temporary substitute for an intended later contract to buy or sell on other terms. There are many different types of hedging, but the common element in all of them is that futures contracts serve as temporary substitutes for intended later contracts.

Hedging is often oversimplified in textbooks on the principles of marketing, and even in more advanced economics books, because it is usually illustrated with an example of pure risk-avoidance hedging. Our discussion of hedging will begin with an illustration of pure risk-avoidance hedging as a convenient departure point from which to discuss other types of hedging that are more important and have more significant economic effects.

Pure Risk-Avoidance Hedging

Grain-merchandising operations of a country elevator are often used to explain this type of hedging. For example, on October 15 a country elevator may purchase corn from farmers at $2.20 per bushel. It will take time for the elevator to accumulate a sufficient quantity of corn for efficient transportation and sale. During the time the elevator holds the corn, it must assume the risk of a price decline. Consequently, to avoid this price risk, it hedges by selling an equivalent amount of December corn futures at $2.38. The price of cash corn then declines to $2.15 per bushel on November 15 when the cash corn is sold, resulting in a 5-cent loss. But this loss is offset by an equal 5-cent gain on the futures contract. Cash and futures prices move together, so the futures are purchased at 5 cents less when the hedge is lifted on November 15 than the price at which it was sold on October 15 when the hedge was placed (Table 7.7).

The effectiveness of pure risk-avoidance hedging is usually illustrated, as in the example, by a parallel movement in cash and futures prices. In other words, the basis is the same when the hedge is lifted as it was when the hedge was placed. It is true that price risks can be avoided through hedging, and over short time intervals hedgers may assume that the basis will not change. Over longer periods, however, the hedger can anticipate a change in the basis and profit from it. Hedgers can forecast this change in the basis because of the convergence principle —the tendency of the cash and futures prices to converge as the futures contract matures. Hence, hedgers often strive to make profits through hedging by correctly forecasting the change in the basis. Hedging, in essence, is arbitrage between cash and futures prices.

Returns to Hedging

Most hedging is undertaken in anticipation of a change in the basis, or that the buying basis will be different than the selling basis. The

TABLE 7.7 Pure Risk-Avoidance Hedging

	Cash		Futures		Basis
Oct. 15	Buy cash corn	$2.20	Sell December corn	$2.38	18¢ under
Nov. 15	Sell cash corn	$2.15	Buy December corn	$2.33	18¢ under
	Loss 5¢		Gain 5¢		

buying basis can be defined as the basis at which cash grain is bought and futures are simultaneously sold. The selling basis is the basis at which cash grain is sold and futures are purchased. Merchants and processors strive to buy grain on a weaker basis than they sell, and the difference should be large enough to cover costs along with a profit to the transaction.

In the example shown in Table 7.7, the country elevator fixes its buying basis at −18 cents on October 15 when it buys cash corn and sells futures, putting on what is commonly called a *short hedge*. The elevator does not establish its selling basis of −18 cents until November 15 when it sells the cash corn and buys the December futures to lift the hedge. The gross returns are as follows:

Selling Basis	−	Buying Basis	=	Gross Returns
Nov. 15		Oct. 15		
−18¢	−	(−18¢)	=	0

Had the elevator's selling basis on November 15 been −8 cents, the gross returns would have been +10 cents. So, a short hedger will gain when the basis strengthens, and the selling basis is stronger than the buying basis. This explanation shows why short hedges are usually most effective in a carrying-charge market when the basis is usually weak. As cash and futures prices converge with the approach of the delivery month, the basis strengthens, and the short hedge will yield a positive return.

The same principles apply, but in reverse, to the second generic type of hedging, namely, long hedging. A merchant who sells cash grain for forward delivery in advance of the purchase of cash grain and buys futures as a hedge becomes a *long hedger*. The long hedger fixes the selling basis first and has to estimate the buying basis, the basis at which cash grain can be bought and futures sold to lift the hedge. Long hedges are usually most effective in inverted markets when the basis is strong and cash grain often sells at a premium to futures prices. In such periods, merchants strive to sell grain for forward delivery at high current prices and purchase futures at lower prices as a hedge. The hedge will then show positive returns with the convergence of cash and futures prices as the futures contract matures. The long hedger, like the short hedger, strives to sell on a stronger basis than the basis at which cash grain is purchased. They differ in that a short hedger will first fix the buying basis and must estimate the selling basis. The long hedger, on the other hand, first fixes his selling basis and must estimate the buying basis or the basis at which cash grain can be bought at a later time.

Storage Hedging

Grain producers and commercial firms can earn returns on grain
storage through hedging. When the basis is weak (cash grain is cheap
relative to futures), grain can be stored and hedged through the sale of
futures. Returns on storage are earned through basis appreciation as cash
and futures prices converge with the maturity of the futures contract.

The key to successful storage hedging is understanding how the basis
behaves over the marketing year. Figure 7.3 is a graph of weekly cash
corn prices and July corn futures prices from September 2 to June 30 at
Clarkfield, Minnesota. It is evident that cash and July futures prices tend
to move together in the same direction. Also, these prices tend to
converge as the July futures approaches maturity. In other words, the
July basis strengthens from September to July. This strengthened basis
is shown even more clearly in Figure 7.4, which graphs the July basis.
In its construction, the July futures price each week is taken to be zero;

FIGURE 7.3   Clarkfield, Minnesota, cash and July futures prices, September 2,
1986, to June 30, 1987.

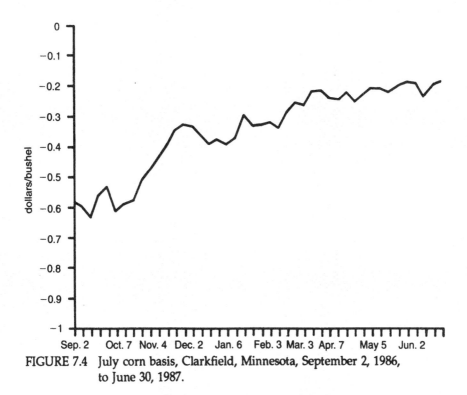

FIGURE 7.4 July corn basis, Clarkfield, Minnesota, September 2, 1986,
to June 30, 1987.

the basis (cash price – July futures price) is plotted as the number of
cents below the zero mark.

Seasonability in the July Basis. Figure 7.4 shows clearly that the July
basis is the weakest at harvest time in October. It then strengthens and
reaches its strongest level in May and June. On October 2, the Clarkfield
July basis was 58 cents under, and it was 21 cents under on May 5. So
the July basis strengthened 37 cents during this period.

How does the seasonal pattern in the July corn basis just described
compare with previous marketing years? Figure 7.5 shows the average
July basis patterns for the three marketing years 1983-1984 through 1985-
1986 and the three earlier years 1979-1980 through 1981-1982.   The
averages of the two periods reflect a similar seasonal pattern of weakness
in the fall and strength in late spring; that is, the basis strengthens
substantially with the passage of time. The reason for this typical basis
behavior is that corn supplies are usually large in the fall so elevator and
transportation facilities are often pressed to their limits.   Hence, the
market offers incentives to farmers and marketing firms to store corn to
relieve the glutted market. The basis is then weak, reflecting favorable
returns to storage. Later in the marketing year, as supplies diminish, the

FIGURE 7.5 Average July corn basis, Clarkfield, Minnesota, 1979/1980 to
1981/1982, and 1983/1984 to 1985/1986.

basis strengthens and encourages corn to be brought out of storage and
into marketing channels. However, there is variability in the July basis
pattern from year to year as market fundamentals change.

The difference between the average July basis of the two three-year
periods in Figure 7.5 is striking. In the earlier period, 1979-1980 to 1981-
1982, the July basis is much weaker on average throughout the period.
Also, there is considerably more basis appreciation from harvest to the
following spring. This was a period in which interest rates reached
record levels. High interest costs made corn storage expensive and
depressed cash prices relative to futures prices. Also, market prices for
cash corn exceeded price support loan rates so little corn was placed
under price support loan or in the farmer-owned reserve. In the later
period, 1983-1984 to 1985-1986, the average July corn basis was
considerably stronger. During these three years, government price
support programs were dominant market factors. Sizable amounts of
corn were placed under regular price support loan or in the farmer-
owned reserve. "Free" stocks of corn, or stocks outside of those held

under government programs, were often low. This tightness in stocks was reflected in the market through a stronger corn basis during this period. Also, interest rates were considerably lower, reducing corn storage costs.

Example of Storage Hedging. An example of storage hedging using actual prices is helpful in understanding storage hedging. The harvest time July corn basis was quite weak in the fall of 1986 relative to the three previous marketing years (Figures 7.4 and 7.5). For example, a farmer near Clarkfield harvested corn the first week of October. The Clarkfield elevator was bidding $1.22 per bushel on October 7, the July corn futures were selling for $1.83, and the basis was 61 cents under. The farmer recognized from past experience that this was a weak basis and that the basis might strengthen to about 20 cents under by April and May, if the average of the past three years was used as a guide. In this example, the farmer stored corn and hedged it through the sale of July futures on October 7, then sold the corn locally and lifted the hedge on April 7. The results are shown in Table 7.8.

The farmer's gross return to the storage hedge from October 7 to April 7 was 36 cents per bushel, or the amount by which the basis strengthened. Another way of looking at the results is through the price received by the farmer through storage hedging. If corn had been sold to the Clarkfield elevator on October 7, the price would have been $1.22 per bushel. By storing and hedging the farmer received $1.38 from a cash sale on April 7, plus 20 cents gained on the futures or $1.58 per bushel. This is 36 cents more, which equals the amount by which the basis strengthened.

Had corn been stored without hedging in anticipation of a seasonal price increase, the farmer would have received only $1.38, rather than

TABLE 7.8 Storage Hedging

	Cash	Futures	Basis
Oct. 7	Store 5,000 bu corn (local cash price is $1.22)	Sell July futures @ $1.83	$−.61
Apr. 7	Sell 5,000 bu corn $1.38	Buy July futures @ $1.63	$−.25
	Gain 16¢	Gain 20¢	
		Gross Return	$+.36

$1.58 through the storage hedge. Storage hedging will not always yield higher prices because in some years both cash and futures prices may rise, in which case the futures transaction will show a loss, and the price received through storage without hedging would have been higher. Storing without hedging incurs price risks but the hedge protects against a fall in corn prices.

Although the results of storage hedging may not compare as favorably with other marketing alternatives every year, storage hedging usually yields consistent returns to corn storage. Opportunities for storage hedging often appear in years of abundant corn supplies when the basis is usually very weak during the harvest period.

The above example shows only the gross return to storage hedging. To calculate the net returns, the farmer would have to subtract the variable costs of farm storage, which include the interest cost on capital invested in the corn, the cost of insurance on stored corn, and the physical loss of corn due to handling and shrinkage while in storage. These are the only costs that enter into short-run decisions on corn storage, if the farmer already has corn storage facilities on the farm. Many farmers have their own farm storage, which has expanded substantially in recent years.

Storage Hedges Can Be Lifted Sooner than Planned. In the above storage hedging example, the hedge was placed in the fall when the July basis is usually the weakest and lifted the following late spring when the July basis is usually the strongest. A grain merchant or farmer may place a storage hedge in the fall and anticipate lifting it in early May. However, if the basis strengthens unexpectedly before May, it may be a wise decision to lift the hedge and sell the grain sooner than anticipated. A recent corn-marketing year was a case in point. Figure 7.4 shows a substantial basis appreciation immediately after harvest that year. For example, the corn storage hedge placed on October 7 was lifted on January 20. The results are shown in Table 7.9. The gross return to the hedge if lifted on January 20 would have been 31 cents per bushel. The farmer may have decided to lift the hedge and take the 31-cent gross return for three months' storage rather than storing until late spring for the possibility of an additional return of 10 cents/bushel if the farmer anticipates the July basis to appreciate to its strongest level of 20 cents under in early May. An important advantage of storing cash grain and hedging it through the sale of futures when the basis is weak is the opportunity to sell the cash grain when the basis strengthens. Sometimes the basis will strengthen, at unexpected times, when there is an increase in the demand for cash grain for spot delivery.

The Best Delivery Month in Which to Hedge. A grain merchant or producer who had decided to place a storage hedge will hedge in the

TABLE 7.9   Storage Hedge Lifted Earlier than Planned

	Cash	Futures	Basis
Oct. 7	Store 5,000 bu corn (local cash price is $1.22)	Sell July futures @ $1.83	$−.61
Jan. 20	Sell 5,000 bu corn $1.39	Buy July futures @ $1.69	$−.30
	Gain 17¢	Gain 14¢	
			Gross Return $+.31

future delivery month that has the highest price relative to the others after considering the storage costs of carrying the grain.

For example, on October 2 these were the corn prices:

Clarkfield cash price	$2.67
December corn futures	3.43
March corn futures	3.55
May corn futures	3.61
July corn futures	3.62

December is the near futures on this date, so the near basis is 76 cents under. The carrying charge between December and July is 19 cents. The July basis is the sum of the two, or 95 cents under.

A storage hedger placing a hedge on October 2 and, feeling that the carrying charge of 19 cents between December and July is as wide as it will get, will hedge directly in the July future. But if the hedger thinks that this carrying charge will widen, the first hedge will be in the December futures and, before it expires, the hedge will be moved forward to the March, May, or July futures. Which of the latter three futures is selected depends on one's judgment of the carrying charge. The hedge is moved forward by buying back the December futures and selling the March, May, or July futures.

It should be emphasized that it is useful to think of the basis relative to the near futures plus a carrying charge between the near and distant futures. In the grain trade, the basis is usually quoted relative to the near futures.

Hedging Forward Cash Sales of Grain and Processed Products

Access to futures markets for hedging enables grain merchants to make forward cash sales at a fixed price even though they do not own the grain at the time the sale is made. Forward cash contracts are used extensively in the merchandising of grain and grain products such as flour, feed, soybean meal, and soybean oil. Merchants and processors can fix the price at the time the forward cash sale is made by using futures prices as a guide, with adjustments for their operating costs. After making such forward cash sales, they purchase futures contracts, which serve as temporary substitutes for the cash grain until it is purchased for delivery or processing. This is called *long hedging* because the merchant is short cash grain or products and long futures.

Long hedging is usually most effective in inverted markets when cash prices are at a premium over near futures, and near futures are higher priced than distant futures. Such price relationships often occur in years when supplies are short relative to demand. Negative prices of storage prevail, so merchandisers and processors have little incentive to carry inventories. In fact, since the market is paying higher prices for current delivery than for future delivery, they keep their inventories at minimal levels. They also look for opportunities to make forward cash sales of grain they do not own at high current prices and hedge those sales through the purchase of futures—in other words, engaging in long hedging. As the cash and futures prices converge as the delivery month approaches, the spot premium usually declines, yielding a profit on the hedge.

Long Hedging by Grain Exporters. Nearly all export grain sales are made on forward cash contracts calling for delivery up to a year in advance. If export contracts fix the price of grain, they are called flat-priced contracts. Exporters are able to quote forward prices even on grain not owned because futures markets are available for pricing and hedging.

For example, on June 1 a U.S. grain exporter sells 30,000 tons of soft red wheat f.o.b. the Gulf of Mexico to an importer for delivery in August at $4.68 per bushel. The exporter hedges the sale by purchasing an equivalent quantity of Chicago September wheat futures at $4.41. The futures contracts serve as temporary substitutes for the cash wheat until it can be purchased. At the time of the sale, the exporter estimates the buying basis for country cash wheat in Indiana and Illinois to be 33 cents under. The selling basis (+27 cents) minus the buying basis (–33 cents) equals an estimated (+60 cents) gross return to the transaction. This return must cover all of the exporter's cost of purchasing, transporting, and loading the wheat on the importer's vessel in August at the Gulf of Mexico along with a "profit" to the transaction.

Table 7.10 shows that wheat prices increased after the sale was made on June 1, but the exporter purchased cash wheat on July 15 at 33 cents under, hence, the gross return was 60 cents per bushel. Had the buying basis been stronger than 33 cents under, the gross return and "profit" margin would have been reduced. So, the success of this forward export sale depends on the accuracy with which the exporter can forecast the actual buying basis—the basis that will prevail when cash grain is purchased. This forecast must be made at the time of the forward cash sale. The grain merchant uses knowledge of historical relationships in the basis as well as personal judgment of market factors to estimate the basis at the time of purchase.

Basis Contracts and Exchange of Futures for Cash. Sometimes grain exporters make sales of grain to importers on basis contracts. A basis contract does not specify the flat price but only the basis relationship to a designated futures price. If the contract for the grain export sale just discussed was basis priced rather than flat priced, the designated futures price would be Chicago September wheat. For instance, the agreed basis is 27 cents over Chicago September wheat futures, as shown in Table 7.10.

This basis price sale does not initiate any flat-price risk for the exporter. It leaves the exporter open only to the risk that the basis will change against him. This risk is much lower than the risk of a flat-price change since cash and futures prices tend to move together.

When the exporter acquires the cash wheat for delivery to the importer in July, the exporter simultaneously sells September wheat futures to hedge the cash purchase. Then the exporter is protected against a price level drop prior to final flat-price determination.

TABLE 7.10 Long Hedge of a Flat-priced Export Sale of Wheat

	Cash	*Futures*	*Basis*
June 1	Sell wheat for delivery f.o.b. Gulf of Mexico in August $4.68	Buy Chicago Sept. Wheat Future $4.41	$+.27
July 15	Purchase wheat in the country $4.27	Sell Chicago Sept. Wheat $4.60	$−.33
	Gain 41¢	Gain 19¢	
		Gross Return $+.60	

For the importer, too, the 27-cent basis is already fixed. When the Chicago September wheat futures price is considered to be at a favorable level by the importer, he can lock in the flat purchase price by buying Chicago September wheat futures. At this point, the importer becomes exposed to flat-price risk.

When the importer makes the decision to fix the flat purchase price of the contract, the long futures position is turned over to the exporter, offsetting the exporter's short futures position. Hence, the futures hedges of both the exporter and importer are lifted and the exporter delivers the wheat to the importer at the agreed upon time and place. The flat price of the sale is arrived at by adding 27 cents to the "board price"—the price of the Chicago September wheat futures at the time of pricing the contract. The futures trade is carded by the pit brokers' X-pit at the futures price when pricing takes place. This transaction is done to lift the hedges of both the exporter and the importer at the same price, maintaining the integrity of the basis contract. Basis price contracts such as these are called exchange of futures for cash because the buyer, or importer, exchanges his long futures position for the spot commodity at the time of flat-price determination. Basis price contracts are a classic example of pricing flexibility afforded by futures markets.

Long Hedging by Grain Processors. Grain processors such as flour millers, feed manufacturers, and soybean crushers also make extensive use of futures to hedge the forward sale of processed products before they own the raw material and the processing takes place. For example, as explained by Robert Parrott, Central Soya sold soybean meal and oil products in January for delivery in October, November, and December.[3] They were able to make such forward sales because of their access to the futures market, even though the soybean crop would not be planted until May or June. November soybean futures were purchased when the product sales were made, and these futures served as a temporary substitute for the cash soybeans that they intended to purchase in the fall. When the cash soybeans are purchased, the soybean futures are then sold to lift the hedge. The soybeans are then processed, and the products—meal and oil—are delivered to the buyers. A hedge of this nature is a means of protecting against the price risk instead of charging larger margins to compensate for the unknown degree of risk.

Flour millers also frequently practice long hedging when they hedge the forward sales of flour. It is common practice for bakers to buy flour from millers on forward cash contracts. Even though there is no organized futures market in flour, millers are able to quote prices on flour for forward delivery because they can hedge those sales through the purchase of wheat futures. The wheat futures serve as temporary substitutes for the cash wheat for milling until it can be purchased. As the miller purchases cash wheat and mills it into flour, wheat futures are sold.

This hedging practice has been called operational hedging by Holbrook Working because it facilitates the operations of a grain merchandising or processing business. Flour millers typically do not store large quantities of wheat. They can, nevertheless, make large forward sales of flour because of their access to the futures market for hedging. Millers can purchase large quantities of futures immediately after forward flour sales are made. Later, the miller can purchase the cash wheat of the desired classes, grades, and protein needed to mill the flour. Such purchases can usually be made more advantageously over time.[4] As the cash wheat is purchased for milling, the wheat futures are sold, lifting the hedge.

Hedging Processing Margins

Soybean crushers have access to futures markets for both their raw material (soybeans) and finished products (soybean meal and soybean oil). This access allows considerable flexibility in hedging operations. In fact, some of the more sophisticated and complicated uses of futures markets are found in the hedging operations of soybean processors.

When soybeans are processed, meal and oil are obtained simultaneously. The difference between the value of the meal and oil obtained from a bushel of soybeans and the value of the soybeans is often called the *crushing margin*—a rough measure of the profitability of crushing soybeans.

Soybean crushers frequently utilize soybean, soybean meal, and soybean oil futures to establish or lock in their crushing margin. This is commonly known as *putting on the crush*. It is accomplished by buying soybean futures and selling equivalent amounts of soybean meal and soybean oil futures.

For example, on August 15, the following January futures prices are quoted for soybeans, soybean oil, and soybean meal.

January soybeans	$7.37 per bushel
January soybean oil	$0.27 per pound
January soybean meal	$210.00 per ton

These prices reflect a "board" crushing margin per bushel (60 pounds) in January futures as follows.

11 pounds of oil × $.27	=	$2.97
48 pounds of meal		
(0.022 tons × $210 per ton)	=	4.62

Total product value per bushel	=	$7.59
Less January soybeans per bushel	=	7.37
Gross crushing margin per bushel	=	$0.22

A processor who desires to lock in this 22-cent crushing margin on August 15 will purchase January soybeans and simultaneously sell equivalent amounts of January soybean meal and oil futures. In making this decision, the processor must forecast the buying basis for cash soybeans (the price relationship between January soybean futures and cash soybeans) that will prevail in November when soybeans will be purchased for crushing. Similarly, the processor must forecast the selling basis for soybean meal and oil (the price relationship between January soybean product futures and cash prices of the products) that will prevail in December when the products will be sold.

Basis forecasts are made using knowledge of historical basis patterns and personal judgments of market factors prevailing in the current year. It is easier to forecast the basis than the price level in price-volatile commodities such as soybeans, soybean oil, and soybean meal. Nevertheless, after putting on the crush, the processor is still subject to basis risk. If the actual buying basis for soybeans is stronger than anticipated and/or the selling basis for soybean oil and meal is weaker than anticipated, the 22-cent margin expected from the hedge may not be realized. As cash soybeans are purchased for crushing in November, the January soybean futures are sold. When the cash products (soybean meal and soybean oil) are sold in December or January, the January soybean meal and oil futures are purchased, lifting the hedge.

Soybean processors also have opportunities to sell meal and oil on forward cash contracts to users of these products. Having established board crushing margins, as previously described, a processor may find it advantageous to sell meal and oil on forward cash contracts, in which case the futures in the products would be repurchased. Then the crush hedge would consist of long January soybean futures versus short forward cash contracts of equivalent amounts of meal and oil.

Access to spot markets for current or forward delivery and futures markets in both soybeans and their products opens up a wide variety of trading alternatives for soybean processors.

Hedging by Agricultural Producers

Farmers can practice hedging in three principal ways. First, they can practice storage hedging by storing grain when their selling basis is weak and hedging it through the sale of futures. Because the basis normally strengthens as the delivery month approaches, the farmer earns a return on storage equal to the amount by which the basis strengthens. Second, farmers can use futures to price grain in advance of production. And third, futures can be used by livestock producers to price grain for livestock feed in advance of purchase. These latter two types of hedging are called *anticipatory* hedging. The purpose of anticipatory hedging is to take advantage of the current price. In both uses of anticipatory hedging, the futures contract serves as a temporary substitute for a merchandising contract that the producer intends to make at a later time.[5]

Pricing Grain in Advance of Production. The objective of pricing crops under production is to lock in a price in advance of harvest that will cover production costs and allow a reasonable return. To do this requires knowledge of the normal basis, the relationship between the cash price at the farmer's local market and the futures price in the delivery month closest to harvest. The farmer can localize the futures price by subtracting this normal basis from the futures price, which yields the "lock-in" price. The best method of determining the normal basis is to study the historical relationships between cash and futures prices.

For example, on May 5, a farmer near Clarkfield, Minnesota, has planted corn and is considering using the futures market to forward price the corn, which will be harvested during the first week in November. On May 5, the price of December corn futures in Chicago is $3.57 per bushel. The farmer also knows that over the past three years the cash corn at the Clarkfield elevator during the first week of November has averaged 30 cents under Chicago December corn futures (Figure 7.6). This is the normal basis, which, when subtracted from the December futures price, yields a lock-in price of $3.27 per bushel. December corn futures are then sold to set this price, assuming the farmer finds it satisfactory.

Prices for the remainder of the growing season then gradually decline; the farmer harvests corn on November 1, sells it to the Clarkfield elevator for $2.65 or 35 cents under the December corn future, and lifts the hedge by buying back December corn futures. The results of the hedge are shown in Table 7.11. The actual net price the farmer received was $3.22 per bushel, or 5 cents less than the lock-in price estimated on May 5. The actual selling basis was 35 cents under rather than 30 cents under, as anticipated. On a selling hedge, the realized price will be higher than the lock-in price when the actual selling basis is stronger than anticipated. Conversely, if the selling basis is weaker than anticipated, the actual price will be less than the estimated lock-in price.

If the futures price had risen between May 5 and November 1, there would have been a loss on the futures, but as the cash price would also have risen, the lock-in price would be achieved if the selling basis on November 1 was 30 cents under the December futures.

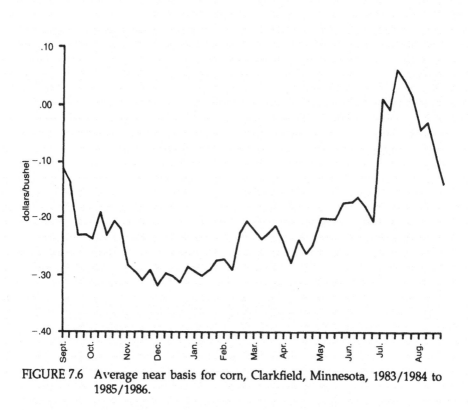

FIGURE 7.6 Average near basis for corn, Clarkfield, Minnesota, 1983/1984 to 1985/1986.

TABLE 7.11 Hedging to Price Corn in Advance of Production

	Cash	*Futures*	*Basis*
May 5	Wants to lock-in corn selling price at $3.27	Sells Dec. corn $3.57	−$.30
Nov. 1	Harvest Sells corn $2.65	Buys Dec. corn $3.00	−$.35
	Result:	Cash sale price $2.65 Plus futures gain +.57 Net selling price $3.22	

Farmers may also price grain in advance of production through a cash forward contract with their country elevator. Country elevators typically offer to purchase grain for forward delivery at a price agreed on at the time the contract is made. The country elevator then hedges to protect itself against adverse price changes.

More farmers are finding it is advantageous in marketing management to price at least some of their crops in advance of production. Each must decide which method of forward pricing is most suitable. Such forward pricing is made possible because of the futures market. The farmer can forward price grain through direct transactions in futures or through forward cash contracts, which, in effect, are indirect transactions in futures. In the latter case, such purchases are hedged in futures by country elevators or by a terminal merchant to whom the country elevator has sold grain for delivery.

Pricing Livestock Feed in Advance of Purchase. Pricing in advance of purchase is a third way in which farmers may use futures markets for hedging. For example, on November 1, a farmer is feeding cattle near Clarkfield, Minnesota, and knows that his available corn will last until sometime in May at present feeding rates, but he will have to buy more corn at that time. The feeder wants to lock in a favorable price for purchasing corn in May and avoid the risk of a price increase between November and May. The futures market offers this opportunity.

In May, the corn basis averages about 20 cents under (Figure 7.6). On November 1, the May corn futures price was $3.25. Subtracting the normal basis in May from this price gives an estimated local corn price of $3.05. If the farmer thinks this is a reasonable price, allowing a normal return on the cattle-feeding operation, then to lock in this price, May corn futures are purchased. Later, on May 1, cash corn is purchased locally, and the May corn futures are sold to lift the hedge. The results of the hedge are shown in Table 7.12.

The net price paid for the corn was $2.95, the local purchase price of $3.10 less the 15 cents gain on the futures. This is 10 cents less than the estimated lock-in price. The actual buying basis in May was 30 cents under rather than 20 cents under, as estimated in November. A gain is made on a long hedge when the basis weakens or does not strengthen as much as anticipated. This is the exact opposite of a short hedge when gains are made on strengthening of the basis, a distinction that is an important one to remember.

Cost of Hedging

Futures contracts are bought and sold through futures commission merchants (FCMs), sometimes called *brokers*, who are members of most exchanges. Their main business is to execute futures trades for others,

TABLE 7.12 Hedging to Price Corn in Advance of Purchase

	Cash	Futures	Basis
Nov. 1	Wants to lock-in corn purchase price at $3.05	Buys May corn $3.25	–$.20
May 1	Buys corn at $3.10	Sells May corn $3.40	–$.30

Result:	Cash purchase price	$3.10
	Less gain on futures	–.15
	Net purchase price	$2.95

such as hedgers or speculators, who are the real principals of the futures transactions. FCMs charge a commission rate for their services. The principals of futures transactions must also deposit margin money with the FCM to guarantee performance on their contracts. FCMs, in turn, deposit margin funds with the clearinghouse, which guarantees performance on the futures contracts. Any excess margin money held by the FCM must be in segregated accounts apart from other assets of the firm.

The cost of the futures transactions to the hedger consists of the futures trading commission rate, and the interest cost on the margin money on deposit with the FCM.

Table 7.13 shows the "round turn" commission rates (covering the purchase and sale of one futures contract) charged by one FCM as of December 1, 1987. The commission rate on one corn futures contract (5,000 bushels) is $65; the futures commission transaction cost to the corn hedger is 1.3 cents per bushel. Commission rates are negotiable, and it is likely that large traders would get lower rates than those shown in Table 7.13.

Margin requirements on hedging and speculative futures transactions are also shown in Table 7.13. A corn hedger would deposit $250 per contract with an FCM and must maintain the margin at this level. For example, if the price of corn futures advance 2 cents per bushel after a hedger has sold corn futures as a hedge, the hedger must deposit an additional $100 per contract with the FCM to make up for the loss of equity resulting from the price increase. Hedgers must be prepared to

TABLE 7.13   Margin Requirements and Commission Rates on Grain Futures
Contracts of a Futures Commission Merchant, December 1, 1987
(dollars per contract)

| Commodity | Hedging | | Speculative | | Round Turn |
	Initial	Maintenance	Initial	Maintenance	Commission[a]
Wheat	400	400	500	400	65
Corn	250	250	350	250	65
Oats	250	250	350	250	65
Soybeans	600	600	750	600	65
Soybean Meal	450	450	550	450	65
Soybean Oil	300	300	350	300	65

[a]Covers the purchase and sale of one futures contract.

SOURCE: Cargill Investor Services, Inc.

meet these "margin calls" promptly. Otherwise, their futures positions
will be liquidated by the FCM.

Initial margin requirements on speculative transactions are higher than
for hedging because speculation involves more risk. But the maintenance
margin on speculative accounts is lower than the initial margin and is the
same as the hedging maintenance margin.

Economic Benefits of Hedging

The best known and most widely accepted economic benefit of hedging
is that it reduces business risks.   This decrease, in turn, reduces
marketing margins, with the result that producer prices are higher and
consumer prices are lower than they would be in the absence of hedging.
Hedging reduces marketing margins by enabling business firms to
operate on lower margins when they can reduce the risks involved in
carrying price-volatile inventories. The latter also facilitates and reduces
the cost of financing. Commercial banks are willing to loan a high
percentage of the value of grain inventories at prime rates when they are
hedged. Huge amounts of bank credit are needed each year to finance
inventories of seasonally produced annual crops such as wheat, corn, and
soybeans.

A second economic benefit of hedging is that it reduces the variability
in cash or spot commodity prices. This effect stems from the fact that the

hedger is in a better position to judge what can be paid for the spot commodity because the futures price is a known reference point. The basis is much more meaningful to the grain merchant or processor than the flat price of the commodity; therefore, basis trading is dominant in all grains that are traded on futures markets.

A third economic benefit of hedging is derived from "promoting the stockpiling of commodities in private hands in times of surplus, inducing the economical storage of such stocks, and promoting their release for consumption at appropriate times."[6]

Hedging exerts this economic effect through the mechanism of spot-futures price relationships. In times of surplus, spot prices are at discounts to futures prices; storage prices are positive. This price difference encourages stockpiling, as merchants have an incentive to purchase spot grain for storage and hedge it "to earn the carrying charge." Conversely, spot-futures price relationships discourage stock-piling when supplies are short. In such periods, spot prices often go to premiums over futures prices. Storage prices are then negative and, hence, merchants are induced to sell rather than store. Spot-futures price relationships are a highly serviceable guide to the level of inventories that merchants and processors carry.

The impact that hedging has on stockpiling in times of surplus deserves reemphasis today since the need for government reserve stock programs is receiving considerable attention by policymakers. Some people who are strong advocates of government reserve stock programs overlook the fact that the futures market mechanism, and the temporal dimension it gives to prices, provides an economic inducement for private firms to carry stocks forward when they are in surplus. This inducement also applies to carrying surplus stocks from one crop year to the next.

The Importance of Hedging to Futures Trading

There is a widespread view that futures markets are primarily spec-ulative markets, but this is in error. In fact, as Gray argued, "the importance of hedging is best expressed in the categorical statement that futures trading depends upon hedging."[7] Considerable empirical evidence in grain futures trading supports this conclusion.

Open contracts in grain futures, for example, rise and fall over the marketing year with the commercial stocks of the commodity. In corn, open contracts tend to rise in the fall each year as the new corn crop is harvested and moves into marketing channels. Open contracts usually reach a peak when commercial stocks reach their maximum. Open contracts then decline as the marketing year progresses and corn stocks are used. This decline strongly suggests that the seasonal pattern of open contracts is dictated by hedging use.

Second, as previously discussed, futures trading in grain declined substantially from 1980 to 1987. During this period, government-owned stocks and price support operations became dominant market factors. Lower levels of "free" stocks and less price variability reduced hedging needs resulting in a decline in the volume of futures trading.

Finally, a study of the distribution of the open interest (open contracts) in grain futures shows the importance of hedging on both the long and short sides of the grain futures markets. As shown in Table 7.14, 50 percent of the open interest on the long side of the corn futures market was held by large hedgers (holders of a futures position of more than 500,000 bushels). These traders must report their positions daily to the Commodity Futures Trading Commission (CFTC) and classify them as either hedging or speculation. Small traders (traders with smaller positions) held 37 percent of the long open interest in corn. Because these traders do not report to the CFTC, we do not know the proportion that is hedging. If we assume that one-half of their long open interest (about 18 percent) was hedging, then 68 percent of the total long open interest in corn futures was held by hedgers. Although smaller percentages of the long open interest in soybeans and wheat were held by large long hedgers, the importance of long hedging in grain futures markets has

TABLE 7.14   Average Open Interest by Type of Trader for Corn, Soybeans, and Wheat, January-December 1987

| Grain | Open Interest | Large Traders | | | | Small Traders | |
| | | Speculative | | Hedging | | | |
		Long	Short	Long	Short	Long	Short
		----------- % of Open Interest[a] --------------					
Corn	648.9[b]	8.4	4.4	50.2	46.5	37.2	44.9
Soybeans	441.5	13.1	5.4	33.5	48.0	45.3	38.6
Wheat	163.9	14.9	5.9	25.2	40.1	54.0	48.1

[a]Long and short percentages each must equal 100. The omitted figures are spread positions held by large speculators. To obtain this number either for the long or short side, add the three columns and subtract from 100.
[b]In million bushels.

SOURCE: Commodity Futures Trading Commission, Washington, D.C.

increased over the years with increases in grain exports. Most export sales are made on forward cash contracts with the price fixed at the time of sale, so they are hedged by exporters through the purchase of futures contracts.

Short hedging, the sale of futures resulting mainly from the hedging of grain inventories, has been important in grain futures for many years. As shown in Table 7.14, 40 percent or more of the open interest in corn, soybeans, and wheat was held by large hedgers.

Futures trading developed in grain marketing over a century ago because of an economic need for hedging. Futures markets have achieved their highest degree of commercial use in grain marketing where the economic need for hedging continues to be the greatest.

Speculation Facilitates Hedging

Although grain futures markets depend on hedging business for their existence, futures markets that attract a large volume of speculation are better hedging markets. Speculation can be defined as the holding of a long or short position in futures for the sole purpose of profiting by correctly forecasting the movement of futures prices.

Speculators in futures perform the important economic function of facilitating hedging. They stand ready to absorb the price risks that hedgers do not wish to assume. Futures markets that attract a substantial volume of speculative trade are better hedging markets for two reasons. First, speculation adds to futures market liquidity, which enables hedges to be placed and lifted with a minimum price effect. In a small, illiquid market a large short hedger may find, for example, that the futures price may be depressed several cents by selling large quantities of futures to hedge a long position in cash grain. A liquid market, however, can absorb large trades with a much smaller price effect. The wheat market at the Chicago Board of Trade gets more speculative trade than the Kansas City or Minneapolis wheat futures markets because Chicago has a better balance between hedging and speculative trade and can absorb larger hedging transactions at lower costs than the other two markets.

Second, futures markets that depend mainly on hedging business are frequently characterized by "imbalance."[8] In other words, they are often lopsided and tend to favor the longs or the shorts. Such lopsidedness stems from the fact that hedging is rarely balanced. Only infrequently would short hedging be equally balanced by long hedging. In the grains, for example, the futures markets must absorb large quantities of short hedges immediately after harvest when large quantities are sold by farmers and move into marketing channels. Short hedging in this period may exceed long hedging by a considerable margin. The excess of short over long hedging must be absorbed by long speculation. If speculators are not there to absorb this excess, the market may display a price bias that favors the long side. Such a price bias may appear in small futures markets that depend almost entirely on hedging trade. Such price behavior does not appear in large futures markets such as Chicago corn, wheat, and soybeans that have a better balance between hedging and speculation.

So even though futures markets have often been criticized as being "too speculative," research evidence supports the conclusion that futures markets that receive a large volume of speculation are the best hedging markets. Speculation facilitates hedging and, therefore, serves an important economic purpose.

Regulation of Futures Trading

Futures trading is regulated by two principal means. First, there is self-regulation, as provided for under the rules and regulations of the exchanges or boards of trade on which the trading takes place. Second, futures trading is regulated by the federal government, under laws passed by the U.S. Congress.

The rules and regulations of each exchange constitute a code of ethics applicable to practically every aspect of futures trading. The government of the exchanges also provides a means of enforcing these rules and regulations. Enforcement begins with various committees of members who are appointed by the board of directors and charged with investigating charges and/or rumors of rule violations.

The most important committee charged with rule enforcement of futures trading is the business conduct committee, which is charged with the duty of preventing manipulation of prices. This committee also has general supervision and surveillance over all futures trading conducted by members, particularly insofar as such trading affects nonmember customers and the public.

It is in the exchanges' own best interest to have an effective mechanism of self-enforcement of their own rules and regulations. The members themselves are frequently in the best position to detect rule violations that may occur. If the exchanges do not do a good job of self-regulation, the federal government gets involved in prescribing various penalties and courses of action, as authorized by law.

Over the years, federal regulation of futures markets has broadened significantly. In 1922, Congress passed the Grain Futures Act to regulate trading in contracts for future delivery of grains and flaxseed. In 1936, Congress brought futures trading in cotton, rice, mill feeds, butter, eggs, and Irish potatoes under regulation and called the amended law the Commodity Exchange Act. This act was subsequently amended 13 times to bring additional commodities under regulation and to strengthen its provisions. The Commodity Exchange Act was designed to accomplish three principal objectives: (1) to assure that the exchanges make and enforce rules for the maintenance of competitive trading, (2) to prevent manipulation of prices, and (3) to protect the public from fraud resulting from misappropriation of funds. The latter was to be accomplished by the registration of all futures commission merchants and the requirement that all customer margin monies be kept in segregated accounts.

The Commodity Exchange Act was administered by the Commodity Exchange Authority, an agency of the USDA. There was good reason for such administrative organization because the act regulated futures trading in agricultural commodities only. Futures trading in imported foods (such as coffee, cocoa, and sugar), metals, and several other commodities was not regulated under the act.

From 1963 to 1973, the futures contracts traded on U.S. futures markets more than doubled, from 5.8 to 11.8 million contracts. A substantial share of this increase was attributable to many new futures contracts such as frozen pork bellies, live beef cattle, live hogs, lumber, plywood, silver, and other commodities that were successfully introduced. Many of these did not come under regulation by the Commodity Exchange Authority. Furthermore, in 1972 and 1973, there was substantial increase in the volume of futures trading in the grains. World shortages in these and other commodities caused prices to more than double. Price volatility also increased as markets allocated available tight supplies among many eager buyers. Futures markets again came under public attack for poor performance and facilitating "too much speculation" in commodities. It was argued that the Commodity Exchange Authority had insufficient resources and power to effectively regulate the expanded volume of futures trading.

After holding extensive hearings on various proposals to modify and expand federal regulation of futures trading, Congress passed the Commodity Futures Trading Commission Act in 1974. This act amended the Commodity Exchange Act of 1936. Although the objectives of federal regulation, as contained in the latter, remained intact, the new legislation was designed to strengthen the regulation through three principal means.

First, it created a new independent body, the CFTC, consisting of five commissioners to be appointed by the president of the United States. The CFTC was to be "independent" in that it was not to be a part of any existing government agency but was to report directly to the U.S. Congress. This is in contrast to its predecessor, the Commodity Exchange Authority, which was an agency of the USDA. The CFTC was also to have a larger budget, which would enable it to have a larger staff to expand regulation and surveillance of futures trading. The CFTC was designed to have the stature and regulatory power of futures trading that the Securities and Exchange Commission (SEC) has had for many years in regulating financial securities.

Second, the CFTC Act of 1974 brought all futures trading under federal regulation. Third, it broadened the authority of the new CFTC to regulate futures trading and the exchanges or boards of trade on which trading takes place. The CFTC has the authority under the 1974 act to go directly to the courts to obtain injunctions. It also has the authority to impose stiffer penalties or fines for violations.

The CFTC Act, subject to the four-year sunset provision, was reauthorized by Congress in 1978, 1982, 1986, 1990 (under a continuing resolution), and 1992 (under the Futures Trading Practices Act of 1992). The 1982 act lifted the long-standing ban on domestic agricultural options and futures options that had been imposed under the Commodity Exchange Act of 1936. It also clarified the jurisdiction between the CFTC and the SEC in the regulation of futures and options.

The 1982 Act also authorized the CFTC to create a quasi-public regulatory organization independent of the exchanges and funded through assessment fees on transactions in the markets. The National Futures Association (NFA) was created and charged with regulating sales practices, trader registration, the arbitration of disputes and the financial condition of FCMs that are not members of the exchanges. The NFA performs regulatory activities that the CFTC would otherwise have to perform. The NFA strengthened futures market regulation and freed the CFTC to concentrate on broader regulatory issues.[9]

Finally, the 1992 Act expanded the CFTC's regulatory authority over U.S. futures markets. It also strengthened intermarket coordination and enabled the CFTC to facilitate the development of new financial products.

Summary

Futures markets play an important role in the pricing and marketing of grain. They provide an efficient mechanism for price discovery and a means of shifting price risks to those willing to bear them. Futures trading in the United States had its origin in the marketing of grain; and it is in grain marketing that agricultural commodity futures trading has achieved its highest degree of commercial use.

Cash-futures price relationships are useful guides to successful grain marketing. Changes in the basis (the difference between cash and futures prices) are more predictable over time than changes in the level of cash prices alone. Understanding the economics of the basis and carrying charges between future delivery months is crucial to successful hedging.

The risk-avoidance motive in hedging has been overemphasized. There are many different types of hedging that are motivated by economic considerations in addition to the need to shift price risks. A common element in all hedging is that futures contracts serve as temporary substitutes for intended later contracts to buy or sell on other terms. Hedging results in several important economic benefits that contribute to the efficiency of the U.S. grain-marketing system. Grain-marketing costs would be considerably higher in the absence of hedging on futures markets.

Grain futures markets attract a substantial volume of commercial hedging business. In fact, these markets are highly dependent on the trade of hedgers. However, futures markets that attract a substantial volume of speculative trade are also the best hedging markets, providing greater liquidity, which reduces hedging costs.

Notes

1. Harold S. Irwin, *Evolution of Futures Trading*, Mimir Publishers, Inc., Madison, WI, 1954, pp. 69-83.

2. Holbrook Working, "The Theory of Price of Storage," *Selected Writings of Holbrook Working*, Chicago Board of Trade, Chicago, IL, 1977, pp. 25-31.

3. Robert B. Parrott, "A Professional's View of Trading," *Views from the Trade, Readings on Futures Markets*, Vol. 3, A. E. Peck, ed., Chicago Board of Trade, Chicago, IL, 1978, pp. 197-198.

4. Holbrook Working, "Hedging Reconsidered," op. cit., pp. 126-132.

5. Holbrook Working, "New Concepts Concerning Futures Markets and Prices," op. cit., pp. 251-252.

6. Holbrook Working, "Hedging Reconsidered," op. cit., pp. 132-136.

7. Roger W. Gray, "The Importance of Hedging to Futures Trading and the Effectiveness of Futures Trading for Hedging," *Views from the Trade, Reading on Futures Markets*, Vol. 3, A. E. Peck, ed., Chicago Board of Trade, Chicago, IL, 1978, pp. 223-225.

8. Roger W. Gray, "Why Does Futures Trading Succeed or Fail?: An Analysis of Selected Commodities," op. cit., 1978, pp. 235-242.

9. For an excellent discussion of these regulatory issues see G. D. Koppenhaver, "Futures Market Regulation," *Economic Perspectives*, Federal Reserve Bank of Chicago, Vol. 11, Issue 1, January/February 1987, pp. 3-15.

Selected References

Commodity Trading Manual, Chicago Board of Trade, Chicago, IL, 1985.

Peck, A. E., ed., *Selected Writings on Futures Markets*, Vol. II, Chicago Board of Trade, Chicago, IL, 1983.

————, *Views from the Trade, Readings on Futures Markets*, Vol. 3, Chicago Board of Trade, Chicago, IL, 1978.

Selected Writings of Holbrook Working, Chicago Board of Trade, Chicago, IL, 1977.

Williams, Jeffrey, *The Economic Function of Futures Markets*, Cambridge University Press, New York, 1986.

8

Commodity Options

William W. Wilson

History of Options on Grain Futures

Agricultural options provide a risk management alternative for most participants in the agricultural marketing chain. Grain options began trading on the Chicago Board of Trade as early as the Civil War. Shortly after their introduction, however, the Board of Trade tried to halt the trade in such options because of perceived abuses. The state of Illinois banned option trading both on and off exchanges, but trading continued despite this statutory ban. In 1887 the Board of Trade restricted option trading among its members. Agricultural options were periodically traded and banned between 1887 and 1936, at which time Congress completely banned the trading of agricultural options by enacting the Commodity Exchange Act of 1936.

A number of factors precipitated the 1936 ban, including the following: Options were not being used for traditional risk-shifting purposes; small traders "lured" into options markets to speculate usually lost money; large traders could use options to cause artificial price movements; terms and conditions of options were not standardized; and congestion near the close of the market could occur because many options were good for less than a day.[1] A specific case occurred in the early 1930s when options were blamed for excessive price movements in wheat, which led to the collapse of the Chicago Board of Trade's wheat market in 1933. Problems continued to occur in the late 1960s and early 1970s when gold options were being sold fraudulently.

The problems seemed to stem from the lack of one specific governing body. After the gold incident, the Commodity Futures Trading Commission (CFTC) was given full control of all commodity futures and options trading. The CFTC worked to develop regulations for exchange-traded options and, in September 1981, approved a three-year pilot program, including options trading in gold, treasury bonds, and sugar. Following this, the Futures Trading Act of 1982 lifted the 1936 ban on options on agricultural futures. The result was development of an experimental three-year program of organized trading in selected agricultural options, which was introduced in October 1984.

Definitions

An option is a contingent contract between two parties to buy or sell a particular *underlying* commodity under specific conditions. Basic definitions and relationships associated with agricultural options are as follows:

Call Option. A call option gives buyers the right, but not the obligation, to assume a long futures position in the underlying futures contract at the exercise price on or before the expiration date.

Put Option. A put option gives buyers the right, but not the obligation, to assume a short futures position in the underlying futures contract at the exercise price on or before the expiration date.

Premium. Option buyers pay a negotiated cash premium to option sellers in return for the right associated with holding the option. This premium or option price is determined competitively in the trading pit by open public outcry. The premium consists of two components: intrinsic value and extrinsic or time value.

Exercise or Strike Price. The exercise or strike price is the price at which a call (put) option holder may buy (sell) and a call (put) option writer is required to sell (buy) on the holder's demand. Exercise prices are established at prescribed intervals above and below the current futures price.

In-the-Money. A call (put) is in-the-money (ITM) if its exercise price is below (above) the current price of the underlying futures contract.

At-the-Money. An option (call or put) is at-the-money (ATM) if its strike price is equal to the current market price of the underlying futures contract.

Out-of-the-Money. A call (put) is out-of-the-money (OTM) if its exercise price is above (below) the current price of the underlying futures contract. The option, call or put, has no intrinsic value.

Intrinsic Value. Intrinsic value is a component of the option premium that is measured in a dollar amount (if any) by which the current market price of the underlying futures contract is above the exercise price for the call option and below the exercise price for the put option. Both at-the-money and out-of-the-money options have no intrinsic value.

Time or Extrinsic Value. Another component of an option premium, time or extrinsic value is measured by the amount the premium exceeds the intrinsic value of the option. Time Value equals Premium minus (or less) Intrinsic Value.

Expiration Date. This is the last day that a holder of a put or call option may "exercise" the right to buy or sell the underlying futures contract at the option exercise price. After the expiration date, the option contract becomes null and void. The expiration dates parallel futures contract months. For example, an option on a September futures contract expires at least ten days before the first September futures contract notice day.

Break-even Level. Call and put options have separate break-even levels. For call options, the break-even level equals the exercise price plus the premium. For put options, the break-even level equals the exercise price minus the premium.

Option Trading

Merchandisers and, to a lesser but growing extent, producers are familiar with the advantages of hedging with commodity futures. The development of options, in some cases, complements and, in others, substitutes for futures hedges, depending on the individual's risk aversion and potential return. The primary difference between options and futures for hedging purposes deals with the price that is established. A hedge in futures allows the hedger to "lock in" a specific price in the future, while use of options locks in a floor or ceiling price in the future. An uncovered cash position is characterized as being potentially risky. When a futures position is taken opposite to the cash position, the hedger offsets potential cash market losses but also negates any cash market gain.

With the use of options, gains in the option market offset losses in the cash markets but losses in the futures or options do not negate gains in the cash market. Option trading has four basic positions: buying puts, buying calls, selling puts, and selling calls. Puts do not offset calls and vice versa; each is a separate type of option. Each option position depends on the user's cash position along with other factors specific to each merchant. A long option position involves limited risk and potentially large returns. Conversely, a short option position entails potentially large risks and limited returns.

Buying Puts

A hedger with a long cash position can establish a floor price and can take advantage of a price rise by having a long put position. The cost involved is the premium the purchaser paid at the time the transaction was initiated. For example, if there is a long cash grain position, the appropriate position in the option market would be to buy puts, allowing the merchandiser to lock in a floor price equal to the strike price minus the premium. Thus, if prices decline further, the option can be exercised, assuming a short futures position at the strike price when their actual value is less than the strike price. The difference between the two prices minus the premium would be the net return. Alternatively, if the underlying futures price rises, the put would not be exercised; the effective price would be the higher cash price minus the premium.

Buying Calls

A long call position allows a hedger to establish a ceiling price yet benefit from a price decline. A short cash grain position can be protected by purchasing call options. Once again, the cost incurred is the premium. For example, a merchandiser who has sold cash grain for future delivery but does not have the physical commodity on hand is subject to the risk of a price increase before actually having bought or priced the grain. The merchant could hedge in the futures market but would not be able to benefit if prices declined.

Purchasing calls would lock in a ceiling price equal to the strike price; and, if prices increased, the merchant could realize the ceiling price by exercising the call option. This decision would entail assuming a long futures position at the strike price when the actual value is greater; the difference between the two prices less the premium would be the net return. Conversely, if prices fell, the calls would not be exercised; the effective purchase price would be the cash price plus the premium.

Selling Puts

Long option contracts are analogous to buying price insurance, with the cost of the insurance being the premium. Selling option contracts can be thought of as selling or writing price insurance. The option writer receives a premium in return for assuming the risk. Consequently, option sellers may be able to augment their income by the amount of the premium. However, sellers must also be prepared to face margin requirements and the possibility of a mandatory futures position if prices move against them.

A merchant with a short cash or forward position could sell puts with the intention of increasing income by the amount of the premium. This alternative is attractive only if price expectations are bullish or neutral, because, if prices decline, the futures position may be assumed at a loss. The merchandiser receives the premium right away but must also comply with margin requirements over the life of the contract. If prices go up sufficiently, the buyer does not exercise the puts, and the merchant gains the premium. If prices go down, the options may be exercised, and the merchant would be assigned a long futures position at the strike price.

Selling Calls

Selling calls is a strategy that a commercial merchant or producer with a long cash position might use. A short call position would be taken if the trader has bearish or neutral expectations about the particular grain market. Like selling puts, selling calls allows the merchant to increase current income or margins by the amount of the premium. When the calls are sold, the premium is earned. Once again, the trader must also face margin requirements, including possible margin calls if the options are exercised. If, however, prices fall or remain at approximately the same level over the life of the contract, the calls would expire. Consequently, the merchandiser would be successful in increasing income by the option premium less depreciation if the physical grain was being stored.

If prices appreciated significantly during the life of the option, the calls would be exercised. The merchant would be obligated to take a short futures position at the strike price when it actually is valued higher. The difference between these two prices plus the premium would result in the net loss. If the physical grain were being stored, the merchant would also realize an appreciation in the value of the inventory. Thus, a ceiling price would be locked in for the commodity.

Options Versus Futures

Options and futures are inherently related. This relationship is evident by the fact that, like futures, two basic option transactions can be initiated. One is to purchase options, and the other is to sell or write options. If the trader wants to lock in a floor or ceiling price, the appropriate action would be to buy options. The primary characteristic of this strategy is unlimited profit potential with the premium as the cost. Alternatively, if the goal is to increase income, options can be sold. In this case, the risk of an adverse price move would be unlimited, while the gain would be limited to the premium.

Table 8.1 summarizes the characteristics of options versus futures hedges for producers and merchants with long and short cash positions. Typically, a long cash position could be hedged in the futures market by selling futures contracts. This hedge requires margin money and, in addition, locks in the sale price so futures prevent hedgers from

TABLE 8.1   Cash Positions and Use of Futures and Options

Cash Positions	*Traditional Futures Positions*	*Alternative Option Positions*
Long Cash (Inventory or Forward Purchase)	Sell Futures - requires margin - protects against price decrease - cannot benefit from price increase	Buy Puts - locks in floor price - benefits from price increase - must pay premium - no margin Sell Calls - increases income by amount of premium - requires margin - risk of price decrease - possibility of having short futures if exercised
Short Cash	Buy Futures - requires margin - protects against price increase - cannot benefit from price decrease	Buy Calls - locks in ceiling price - benefits from price decrease - must pay premium - no margin Sell Puts - increases income by amount of premium - requires margin - risk of price increase - possibility of having long futures position if exercised

SOURCE: Author.

benefiting from a price increase. However, a long cash position could also be hedged in the options market, either by purchasing puts or selling calls. Buying puts allows the user to lock in a floor price equal to the strike price minus the premium. Selling calls increases income by the amount of the premium, although margin money is required and unlimited risk of a price decrease is realized. In addition, if the underlying futures price increases so that the calls are exercised, the seller would have to take a short futures position at the strike price when they are actually valued higher.

Conversely, a short cash position is traditionally hedged in the futures market by purchasing futures contracts. Once again, this purchase locks in a price so the hedger is not able to benefit from a price decrease, while also requiring margin money. Alternatively, this cash position could be hedged in the options market in one of two ways: buying calls or selling puts. Buying calls locks in a ceiling price. Selling puts increases income by the premium but requires margin money and exposes the hedger to risk of a price increase. However, if prices decline significantly and if the puts are exercised, the seller will be required to take a long futures position at the strike price when they are actually valued lower.

Terminating Option Positions

Traders with long option positions have three basic choices that must be made during the life of the contract: exercise the option, offset the option, or let the option expire. The direction of price movement over the contract's life affects this decision. The alternatives for terminating a long option position depend on price movements (Table 8.2).

TABLE 8.2 Alternatives for Terminating a Long Option Position

Long Option Contract	Prices Increase	Prices Decrease
Put	Let Expire or Offset	Exercise or Offset
Call	Exercise or Offset	Let Expire or Offset

SOURCE: Author.

The choice between letting the option expire and offsetting the option is relatively easy. If the option is left to expire, no gain will be realized directly. When it is sold to offset the position, the option might have some remaining value reflected in the premium, which can be earned by selling the option or by exercising it. However, offsetting would incur a commission cost. Thus, the choice between offsetting the option or letting it expire depends on commission charges and premium values.

The choice between exercising and offsetting the option is not as straightforward. Once again, offsetting would result in a possible premium gain and no subsequent futures position. Exercising would lead to a specific futures market position at the strike price. With this also come the responsibilities of a futures position, such as margins and commission costs. Either a futures position can be offset immediately and a gain realized, or it can be held until the cash position is liquidated. The primary decision criterion in this case is whether the trader wants a futures market position. If the trader needs a futures position (i.e., if there is still a cash position), the option should be exercised; if not, the option should be liquidated.

Application of Options by Producers and Merchants

Traditionally, elevators, domestic merchants, processors, and exporters have used the futures market extensively in hedging cash positions. Producers hold similar cash positions, and the same principles apply. They can use the options market as a substitute and/or supplement to the futures market in the hedging role. Each of these participants performs unique functions in the grain marketing chain. However, all are similar in the sense that they can have similar cash positions. This similarity allows for generalizations across producers and merchants about the effects of alternative option transactions used with alternative underlying cash positions.

In this section, a number of detailed examples illustrate the relationships between hedges placed in futures and options markets.[2] Each of the examples includes a number of assumptions:

- The purchase or sale of an option is made at an exercise price, which is at-the-money.
- Premiums are based on Black model calculations and are representative for example purposes. (The Black model will be explained in a subsequent section.)
- Options contracts are liquidated or offset in each case instead of exercising them or letting them expire.

- The basis does not change. A stable basis is assumed to demonstrate the mechanics and isolate the effects of changes in cash, futures, and option values.
- No commission charges.

In addition to these assumptions, numerous other strategies involve speculative positions and spreads among and between options. These strategies are not developed in this chapter, but detailed examples are contained in Schwager and various issues of *Futures*.

Long Cash Positions

For a merchant with a long cash position, the greatest risk is that associated with the price upon sale of the grain. A producer has similar risk when planting or storage decisions are made. The only difference between a producer's and a merchant's cash position is that before harvest the producer also has an element of risk associated with production (i.e., weather-related yield risk). Both participants want to protect the value of their cash market position. When the futures market is used, a future price is "locked in," thereby minimizing risk. However, they would be unable to benefit if futures prices increase after the futures position is taken. In this situation an option contract would be desirable. Merchants or producers with long cash positions can either buy puts or sell calls (Table 8.1). Examples of each are described in the next section.

Long Cash/Long Puts. A merchant or producer with a long cash position, who wants protection against a bearish price move, could lock in a floor price and also benefit from a potential price rise by purchasing put options. A long put position gives buyers the right but not the obligation to assume a short futures position at the exercise price. However, instead of either exercising the option or letting it expire, the buyer also may offset the position by selling the puts. The price movements of the underlying futures and corresponding options affect the buyer's decision. If the underlying futures price increases significantly over the life of the option, the buyer could either let it expire and realize the greater cash price or sell the puts, which would have fallen in value with increased futures.

If the opposite occurred, that is, the underlying futures price dropped, the buyer could again pick one of two possibilities: exercise the option and assume a short futures position at the strike price; or offset the option, receive a premium, and be out of the futures market completely. In either case, the risk is limited to the premium paid while possible returns are unlimited. The mechanics of this type of transaction are best

illustrated in Example 8.1. The premium earned on the liquidation of the
put in Part A is equal to zero because the option has no intrinsic or
extrinsic value left.

Conversely, when the underlying futures price declines, the intrinsic
value of the put increases and earns a greater premium from offsetting

EXAMPLE 8.1   Long Cash/Long Puts (in cents)

Setting: July 31

Buy 20,000 bushels	@ 300	
December futures	270	
Basis	30	
December 270 put	13	

A. Sell cash on November 1 at which time <u>December futures increased</u> to 310¢.
   (Basis is unchanged at +30¢.)

   1. Sell (offset) put or let expire
   2. Calculation of returns

Return from cash sale	340
Less purchase price	−300
Less premium paid	−13
Plus premium earned	0
Net result	27
Effective sales price	327

B. Sell cash on November 1 at which time <u>December futures decreased</u> to 240¢.
   (Basis is unchanged at +30¢.)

   1. Could exercise option or sell (offset) put
   2. Calculation of returns

Return from cash sale	270
Less purchase price	−300
Less premium paid	−13
Plus premium earned	30
Net result	−13
Effective sales price	287

the position. However, the net result is a much reduced but still negative figure because the intrinsic value would not increase enough to cover both the cash loss and the original premium. Thus, the returns have unlimited upside potential while the maximum loss is limited to the premium.

A traditional hedge against a long cash position would be to sell futures. In Example 8.1, because the basis is assumed unchanged, the net result would be zero, and the effective sales price would be 300 cents. Thus, if prices increase sufficiently, an option position is preferable to a futures hedge and vice versa for price decreases.

Long Cash/Short Calls. Another alternative for a merchant or producer with a long cash position is to sell call options. A short call position allows the seller to augment current income by the amount of the premium and is particularly attractive if prices are expected to remain relatively stable.

Example 8.2 illustrates the mechanics of this type of transaction when the options are offset. In this case, the seller receives the premium at the time of the sale. The return is limited to this premium. The choice is between holding the call position with the actions of the buyer determining the results or buying the calls before they are exercised or expire.

The return in Example 8.2 is limited to the premium earned from the call sale, which, in this case, is 13 cents. When the futures price declines and calls are offset, the premium paid equals zero because the option has no intrinsic or extrinsic value left. However, in Part A where the futures price increased, the options gained intrinsic value and, thus, resulted in the 40-cent premium paid for offsetting the calls.

A merchant or producer with a long cash position has two option alternatives: to buy puts for locking in a floor price or to sell calls to supplement current income. An important difference between the two choices is the risk level involved. Buying puts is preferable to traditional hedging if futures prices increase because losses in the futures do not offset the gain in cash prices. A long put position limits risk and provides unlimited profit potential, while a short call position limits the return to the premium.

Short Cash Positions

Another typical position of grain merchants, processors, and, to a lesser extent, producers is to be short cash grain. If a futures market is used as a hedge, the purchase price is locked in as long as the basis is unchanged. However, the merchant cannot take advantage of a decrease in prices after the contract is made. An options hedge, on the other hand, permits the buyer to benefit from a price decline.

EXAMPLE 8.2   Long Cash/Short Call (in cents)

Setting: July 31

                  Buy 20,000 bushels   @ 300
                  December futures   270
                  Basis   30
                  December 270 call   13

A. Sell cash on November 1 at which time <u>December futures increased</u> to 310¢.
    (Basis is unchanged at +30¢.)

    1. Buyer may exercise calls, or seller may offset (buy back) the call
    2. Calculation of returns

              Return from cash sale   340
              Less purchase price   -300
              Option premium earned   13
              Less premium paid   -40

                    Net result   +13

              Effective sales price   313

B. Sell cash on November 1 at which time <u>December futures decreased</u> to 240¢.
    (Basis is unchanged at +30¢.)

    1. Purchaser will not exercise option
    2. Calculation of returns

              Return from cash sale   270
              Less purchase price   -300
              Option premium earned   +13
              Less premium paid   -0

                    Net result   -17

              Effective sales price   283

   A merchandiser with a short cash position could establish a ceiling
price and take advantage of a fall in prices. A common scenario is for a
merchant or processor to sell grain for some future month but not have
the physical grain on hand. Purchasing call options would establish a
ceiling price for the eventual cash purchase. Merchants can either buy

calls or sell puts against a short cash position. The following examples illustrate the mechanics involved and possible outcomes.

Short Cash/Long Calls. A merchant or producer with a short cash position who wants to protect against a bullish price move and benefit from a potential price decline could lock in a ceiling price by purchasing call options. Example 8.3 illustrates the mechanics of this type of

EXAMPLE 8.3   Short Cash/Long Call (in cents)

Setting: July 31

Sell 20,000 bushels	@ 300
December futures	270
Basis	30
December 270 call	13

A. Buy cash on November 1 at which time December futures increased to 310¢. (Basis unchanged at +30¢)

    1. Option can be exercised or sold to offset
    2. Calculation of returns

Return from cash sale	300
Less purchase price	−340
Less premium paid	−13
Plus premium earned	40
Net result	−13
Effective purchase price	313

B. Buy cash on November 1 at which time December futures decreased to 240¢. (Basis unchanged at +30¢)

    1. Option could be left to expire or offset
    2. Calculation of returns

Return from cash sale	300
Less purchase price	−270
Less premium paid	−13
Plus premium earned	0
Net result	+17
Effective purchase price	283

transaction. This example shows that the loss is limited to the premium paid while the potential returns are virtually unlimited. When the underlying futures price increased, the net result was a loss of 13 cents—the premium. The premium earned from the liquidation of the calls just offset the larger purchase price because the intrinsic value of the calls had increased. In Part B, the premium earned equaled zero because no intrinsic or extrinsic value was left. The net result is greater in Part B because of the lower purchase price.

Buying calls is preferable to traditional hedging if futures prices decrease because losses in the futures do not offset the gain in the cash price. If futures prices increase, buying futures yields a greater return by the cost of the option premium.

Short Cash/Short Puts. Buying call options is not the only strategy for a merchant or producer with a short cash position. Put options also could be sold. The best outcome would occur if the market remains relatively stable so that the options would not be exercised, but yet would retain some intrinsic value. Selling puts would produce a limited return equal to the premium. In addition, the merchant is required to have margin money available.

Overall, a short put position has virtually the same fundamental characteristics as a short call position. Sellers can choose only between liquidating their position before the puts are exercised or waiting for the buyer's actions. The procedures and outcomes involved in a short cash/short put position when the futures price increases and decreases are illustrated in Example 8.4. The option position would be offset in both instances to insure consistency with previous examples.

Part B of Example 8.4 shows that a positive return occurs when the underlying futures price decreases, and this return is equal to the premium of 13 cents. The merchant paid 30 cents to buy the puts back because of the gain in intrinsic value as futures fell. In contrast, the cost of liquidating the short put position in Part A was risky because the option had no intrinsic or extrinsic value remaining. The net result was negative in Part A because of the greater cash purchase price. However, if options had not been used at all, the net result would have been –40 cents. Conversely, the outcome in Part B was positive because of the smaller cash price.

Forward Contracts with Floor and Ceiling Prices

The options market has the possibility of giving a new dimension to forward contracts. Typically, forward contracts are made by elevators to producers, by exporters to importers, and by merchants to processors. In the past, these types of contracts have been implemented via the

EXAMPLE 8.4   Short Cash/Short Puts (in cents)

Setting: July 31

Sell 20,000 bushels	@ 300
December futures	270
Basis	30
December 270 put	13

A. Buy cash on November 1 at which time December futures increased to 310¢.
   (Basis unchanged at +30¢)

   1. Seller could leave option to expire or offset
   2. Calculation of returns

Return from cash sale	300
Less purchase price	−340
Premium received	13
Less purchase of put	0
Net result	−27
Effective purchase price	327

B. Buy cash on November 1 at which time December futures decreased to 240¢.
   (Basis unchanged at +30¢)

   1. Buyer may exercise option, or seller may offset (buy back)
   2. Calculation of returns

Return from cash sale	300
Less purchase price	−270
Premium received	13
Less purchase of put	−30
Net result	13
Effective purchase price	287

---

futures market. The result has been a forward contract with a fixed price. The disadvantage of this is that a producer would not benefit from a rise in prices and an importer would be unable to take advantage of a fall in prices. Agricultural options stand to change this scenario by making forward contracts more attractive to producers and end users.

Merchants can offer a forward contract with a floor or ceiling price by incorporating the use of options. The logic behind this plan follows from previous examples. For instance, an elevator can lock in a floor price for its own grain sale by purchasing put options and then pass this floor price on to producers through a forward contract. The cost of the put options, or the premium, would be included in calculating the producer's floor price. Likewise, a merchant could lock in a ceiling price on purchases by buying call options, which could then be passed on to end users via a forward contract. Once again, the ceiling price would reflect the exporter's cost of buying the calls.

The mechanics of these transactions are demonstrated next. A number of advantages to both merchants and their customers can be realized by incorporating options into forward contracts. The biggest advantage to producers and end users is that they could lock in a floor or ceiling price and could benefit from a favorable cash price move; with traditional forward contracts, they would not. Furthermore, the new forward contracts would permit individuals to take advantage of options indirectly through merchants without being directly involved in the option market or needing to understand the mechanics themselves. Consequently, forward contracts based on options offer another merchandising alternative, which has potential applications throughout the grain exchange system.

Specific examples of forward contracts with floor and ceiling prices are discussed in the next section.

Forward Contracts with Floor Prices. Forward purchase contracts have typically been made at fixed prices. One of the most common transactions with this type of contract is between an elevator and a producer. Part A of Example 8.5 illustrates how an elevator manager calculates the traditional forward contract price and hedges in the futures market.

This type of contract allows the producer to lock in a sale price of $2.50. The disadvantage of this contract is that the producer cannot benefit from a rise in prices.

A forward contract with a floor price can be offered if the merchant incorporates put options into the scheme. Part B of Example 8.5 shows how this floor price is derived and the consequences of a change in the underlying futures price over time. In the following examples, the option positions are once again assumed to be offset.

In this example, the merchant purchases at-the-money put options at a premium of 13 cents. The floor price quoted to the farmer is less than the flat forward contract price by the amount of this premium. Thus, the put premium would affect the attractiveness of the price floor contract.

If the underlying futures price increased over the life of the contract, merchants either could let the option expire or liquidate their position.

EXAMPLE 8.5 Derivation of Forward Contracts for Purchase (i.e., to Producers) with a Price Floor (in cents)

Setting: July 31

December futures	@ 270
Basis (terminal market)	30
Trans. and handling	50
December 270 put	13

A. Derivation of conventional forward contract price: Hedge by selling December futures

Calculation:

December futures	270
Basis	+30
Trans. and handling	−50
Producer price	250

B. Derivation of forward contract prices with price floor: Buy December put options

Calculation:

Strike price	270
Basis	+30
Trans. and handling	−50
Less put premium	−13
Producer price	237

1. If December futures increase to 310¢ by November 1
   a) Option will not be exercised but will be offset
   b) Derivation of producer price:

December futures	310	
Basis	+30	
Trans. and handling	−50	
Less option premium	−13	
Plus option value (Nov. 1)	0	
Producer price		277

2. If December futures decrease to 240¢ by November 1
   a) Option can be exercised or offset
   b) Derivation of producer price:

December futures	240
Basis	+30
Trans. and handling	−50
Less option premium	−13
Plus option premium (Nov. 1)	+30
Producer price	237

The position is assumed to be liquidated; however, no intrinsic or extrinsic value is left in the option so the merchant does not receive a premium on the sale. The merchant does realize the greater futures price, which is then passed on to the producer. Consequently, the producer received $2.77 with the price floor contract rather than $2.50 under the conventional forward contract.

If the futures price decreased, the merchant could have exercised or offset the option. Once again, it is offset. The merchant does receive a premium of 30 cents on the sale in this case because the option has gained intrinsic value. However, this value is discounted by the futures price decline, and the producer would realize the floor price.

Forward Contracts with Ceiling Prices. Traditionally, domestic and export merchants offer forward sales contracts at fixed prices, and the futures market is an important component in determining the specific price. Part A of Example 8.6 demonstrates how the conventional forward price is derived and the appropriate futures position taken. This type of contract is potentially appealing to buyers such as importers. However, by doing so, they do not benefit from price reductions.

Options allow merchants to change the typical forward contract and offer forward sales contracts with a ceiling price, hedged by purchasing call options. The mechanics are identical to the earlier example of a merchant with a short cash position. By purchasing calls, merchants lock in a ceiling price and can pass this on to their customers via a forward contract. The only stipulation is that the cost of the calls is also passed on to the customer, built into the ceiling price. Importers, as well as other end users, should find this contract more appealing because they can still take advantage of a price decline and can benefit from options without actually trading them.

The calculation of the forward ceiling price is demonstrated in Part B of Example 8.6 along with the results of an increase and decrease in the underlying futures price. The merchant bought at-the-money call options for a premium of 13 cents. The ceiling price is then greater than the flat price by the amount of this premium.

If the underlying futures price increases, the merchant has two alternatives: exercise or offset the option. For the sake of consistency, the calls are again assumed to be liquidated. The increase in the underlying futures price results in a gain in the option's intrinsic value, which is reflected in the 40-cent premium earned on the call sale. This premium offsets the rise in the futures price and is passed on to the importer. Consequently, the price the importer realized is equal to the ceiling price that was originally quoted.

If the futures price decreases instead, the merchant again has two possibilities: to let the option expire or offset the position. The latter

EXAMPLE 8.6   Derivation of Forward Contracts for Sale (e.g., to Importers) with a Price Ceiling (in cents)

Setting: July 31

December futures	@ 270
Basis (terminal market)	30
Trans. and handling	50
December 270 call	13

A. Derivation of conventional forward contract price: Hedge by buying December futures
     Calculation:

December futures	270
Basis	+30
Trans. and handling	+50
Offer price	350

B. Derivation of forward contract prices with price ceiling: <u>Buy December call</u> options
     Calculation:

Strike price	270
Basis	30
Trans. and handling	+50
Plus call premium	+13
Offer price	363

   1. If December futures <u>increase</u> to 310¢ by November 1
     a) Option can be exercised or offset
     b) Derivation of importer price:

December futures	310
Basis	30
Trans. and handling	+50
Plus option premium	+13
Less option value	-40
Effective purchase price	363

   2. If December futures <u>decrease</u> to 240¢ by November 1
     a) Option will not be exercised but is offset
     b) Derivation of importer price:

December futures	240
Basis	30
Trans. and handling	+50
Plus option premium	+13
Less option premium	0
Effective purchase price	333

was chosen here. The option's value drops to zero because of the decline in the futures price so no premium was recovered on the liquidation. However, because of the flexibility of options, the merchant can pass the lower futures price on to the buyer. Thus, the buyer must decide if locking in a ceiling price that is greater than a flat forward price is worth the benefits received. The conventional forward contract is more beneficial if prices rise.

A forward contract with a floor or ceiling price is a new marketing concept, the success of which depends on the conceptual understanding by merchants of premium values and effective prices relative to farm program prices. Country merchants can offer producers an array of marketing alternatives, ranging from fixed price to various floor price contracts. Similarly, merchants could offer end users an array of marketing alternatives, ranging from fixed price to various ceiling price contracts. The primary advantage of incorporating options into forward contracts is that both the customer and the merchant can lock in a floor or ceiling price, which not only allows the customer to benefit from options without direct options market involvement but also can attract business for the merchant.

Other Applications

With the advent of options trading, additional applications exist for risk management in grain marketing, notwithstanding the increase in alternatives for speculation. Three additional grain-marketing applications using options are briefly discussed here.

Producers store grain after harvest for a number of reasons, one of which is the potential appreciation in postharvest prices. Storing grain for this purpose is essentially speculative but is a widespread practice. A problem, particularly during the early 1980s, was that grain storage often resulted in net losses due to the high opportunity cost of capital. However, selling at harvest precluded benefits of postharvest price appreciation. An alternative, using options, would be to purchase calls concurrent with the cash sale of grain.

If price appreciated, so would the call, while storage and interest costs would have been saved. A cash price decrease would result in a lower call value, but the loss would be limited to the initial premium. Analyzing this alternative requires a careful comparison of storage and interest cost to the cost of the premium. Savings in storage and interest may offset the cost of the call premium.

Another alternative that became attractive during the late 1980s was to use options to protect against deficiency payment reductions. During this time, grain prices were escalating, bringing the potential for deficiency payments to decrease. By purchasing call options, producers could increase their revenue so long as futures increased. This increase then could offset a lower deficiency payment associated with higher cash prices.[3]

Another viable use of options is as an alternative to hedge against quantity risk, which imposes problems for traditional hedging for various participants involved in grain marketing. Merchants are exposed to quantity risk after having tendered an offer to buy or sell grain. A long option position can be used to hedge against this risk by limiting the possible loss to the premium cost. If the futures market was used, the potential futures loss of an adverse price move is unlimited. In this sense, options are more appropriate than futures for relieving risk when quantity risk is present.

A merchant who has tendered an offer to sell grain but will not know for several days whether the offer will be accepted can buy call options to hedge this position. If the offer is accepted, the exporter would have been short cash/long calls. If the offer was rejected, the results again would depend on the futures price. The largest return in this scenario would be realized with an increase in the futures price. An elevator also may post a price to buy grain with the offer remaining valid for a specified number of days (e.g., overnight or on a weekend). In addition, the estimated purchase may only be partially fulfilled. Put options can be bought as a hedge against this type of transaction.

The maximum loss associated with a long put position is the premium cost. If the offer was accepted, the elevator would have a long cash/long put position. If the offer was rejected or partially fulfilled, the largest return would have occurred with a decrease in the futures price. The premium would not have been recovered if the futures price had increased. However, the loss would not be any greater than the premium. Thus, a long option strategy appears to be more advantageous than a futures hedge when a merchant has tendered an offer for either a purchase or a sale.

Producers likewise are exposed to quantity (yield) risk in preharvest hedging, and, as such, their alternatives are similar to the elevator example (i.e., long cash/long puts). In general, when quantity risk exists for producers, using options may be more appropriate than futures. Using options to reduce risk is generally cheaper than using futures when quantity risk exists. A quantitative analysis of this is fairly extensive and is not presented here.

In-the-Money Vs. Out-of-the-Money Options

An important consideration when purchasing options is whether to buy in-the-money (ITM) or out-of-the-money (OTM) options. To review some terminology, an *ITM put* (call) is one whose strike price is higher (lower) than the current futures market level, and an *OTM put* (call) is one whose strike price is lower (higher) than the current futures market level. ITM options have a higher premium than OTM options, and the exercise price-futures relationship depends on the level of acceptable risk exposure.

One way to consider this problem is to look at the risk exposure for different strike prices under long and short cash situations and compare possible net returns when the futures price increases or decreases. The examples demonstrate that ITM options provide greater protection from adverse price changes but generate less gains from favorable price moves. On the other hand, OTM options provide less protection from adverse price changes but generate greater profits from favorable price moves.

Long Cash/Long Puts. The decision for a trader or producer with a long cash/long put position is whether to buy ITM puts to insure a

EXAMPLE 8.7   Risk Exposure for Different Strike Prices with Long Cash/Long Puts (in cents)

Setting: July 31

	December futures	270
	Long Cash based on December futures @	270
	Buy December puts	

	Strike Price	Premium*	Net Minimum Sales Price for Futures	December Futures on Purchase	Maximum Loss Potential**
OTM	240	2.9	237.1	270	32.9
OTM	260	8.6	251.4	270	18.6
ATM	270	13.2	256.8	270	13.2
ITM	280	18.9	261.1	270	8.9
ITM	300	33.4	266.6	270	3.4

*Derived using values for December futures and results from the Black/Scholes Model.
**Derived as the difference between the December futures at which long cash (270¢) and minimum sales price were taken.

a higher selling price and pay a larger premium or to buy OTM puts, which lock in a lower selling price but at a smaller premium. Example 8.7 illustrates the risk exposure, or potential loss, at various strike prices.

Buying ITM puts results in a smaller exposure to risk than purchasing OTM puts. Further, the deeper ITM the put is, the smaller the risk exposure or potential loss. The calculations used to arrive at the risk exposure, or the maximum loss potential, are straightforward. Subtracting the premium from the strike price results in the net minimum sales price for futures or the effective floor price that is locked in. The risk exposure is then found by subtracting the floor price from the underlying (December) futures price. The buyer should make sure that the option strike price selected does not set a floor so low that too much of the original margin (return) is being risked. For example, an elevator operator likely would not use an option in which the potential loss exceeds the margin.

Example 8.8 demonstrates the calculation of net returns for ITM and OTM puts under conditions of rising and falling futures prices. Part A shows the effect on net returns when the futures price is increased from $2.70 to $2.90 at various strike prices. The options are assumed to be offset rather than left to expire; therefore, a premium received on the offset is included in calculating the net returns. The greatest return under this scenario occurred with an OTM put (11.44 cents). This return is due to the favorable cash price move coupled with the relatively low premium cost. However, the net return from the ITM put was still positive and obviously beneficial. Part B illustrates the results of a decrease in the underlying futures price from $2.70 to $2.50. In this case, the OTM puts led to the greatest net loss because the premium received from the sale did not increase proportionally with that of the ITM put.

The fundamental conclusion that can be drawn from this example is that purchasing OTM puts can result in both the largest potential return and greatest potential loss, depending on the change in futures prices. Thus, a risk-averse buyer wanting to reduce the variability in net returns would buy ITM puts. The potential gains will not be as great, but the possible loss will be much less relative to OTM puts.

Short Cash/Long Calls. The only difference in a short cash/long call position is that the merchant wants to "insure" a specific purchase price. First, the risk exposure or potential loss for various strike prices can be determined. Example 8.9 indicates that the lowest amount of risk occurs when ITM calls are bought, which is identical to the put scenario. Likewise, the deeper ITM the call is that coincides with lower strike price levels, the lower the risk exposure and return. Conversely, the deeper OTM the call is, the greater is the potential loss and return. In this

EXAMPLE 8.8   Purchase of In-The-Money Versus Out-Of-The-Money Puts (in cents)

Setting: July 31

Buy cash wheat based on December futures of 270¢
Buy December puts
November 1, puts liquidated, cash sold and no change in basis

A. December futures <u>increased</u> to 290¢ November 1
   1. Options will not be exercised but are offset
   2. Calculation of returns using different strike prices

Strike Price*	Sale Price	Option Cost	Sale Price of Option	Purchase Price	Net Returns
OTM 260:	290	– 8.6	0.04	–270	= 11.44
ATM 270:	290	–13.2	0.3	–270	= 7.1
ITM 280:	290	–18.9	1.7	–270	= 2.8

B. December futures <u>decreased</u> to 250¢ November 1
   1. Options could be exercised but are offset
   2. Calculation of returns using different strike prices

Strike Price**	Sale Price	Option Cost	Sale Price of Option	Purchase Price	Net Returns
OTM 260:	250	– 8.6	11.3	–270	= –17.3
ATM 270:	250	–13.2	20.2	–270	= –13.0
ITM 280:	250	–18.9	30.0	–270	= – 8.9

---

*Lower valued strike price gives greater return.
**Lower valued strike price gives greater loss.

---

situation, the maximum purchase price or guaranteed ceiling price is found by adding the call premium to the strike price. Subtracting the December futures price from this ceiling price results in the risk exposure or maximum loss.

Example 8.9 demonstrates that lower ceiling prices offset the higher premiums of ITM calls, whereas buying OTM calls results in a price ceiling that exposes the buyer to a relatively greater risk. In each case,

EXAMPLE 8.9 Risk Exposure for Different Strike Prices When Short Cash/Long Calls (in cents)

Setting: July 31

December futures		270¢
Short cash based on December futures @		270¢
Buy December calls		

	Strike Price	Premium*	Net Maximum Purchase Price for Futures	December Futures on Sales	Maximum Potential Loss**
ITM	240	32.9	272.9	270	2.9
ITM	260	18.5	278.5	270	8.5
ATM	270	13.2	283.2	270	13.2
OTM	280	9.1	289.1	270	19.1
OTM	300	3.9	303.9	270	33.9

*Derived using current values for December futures and Black/Scholes Model.
**Derived as the difference between the December futures at which short cash was taken (270 cents) and maximum purchase price.

the net return was negative, but the smallest loss resulted from deeper ITM options.

The net returns for ITM and OTM calls when futures prices increase are demonstrated in Example 8.10. Part A illustrates that a 20-cent rise in the futures price resulted in the smallest net loss, 8.5 cents, when ITM calls were purchased. The explanation for this is that the premium from the ITM call sale increased more in proportion to the OTM call sale. The greatest loss occurs with OTM calls.

The returns in Part B are exactly the opposite, where the largest gains, 10.9 cents, were realized with OTM calls. The futures price had declined from a $2.70 to $2.50, and the options were eventually liquidated, although only a small premium was earned from the sale of the OTM calls

EXAMPLE 8.10 Purchase of In-The-Money Versus Out-Of-The-Money Calls (in cents)

Setting: July 31

Sell cash wheat based on December futures @ 270¢
Buy December calls
November 1, calls liquidated, cash bought, and no change in basis

A. December futures increased to 290¢ November 1
   1. Options can be exercised or offset
   2. Calculation of returns using different strike prices

Components of Return

Strike Price*	Sale Price	Option Cost	Sale Price of Option	Purchase Price		Net Returns
ITM 260:	270	−18.5	30.0	−290	=	−8.5
ATM 270:	270	−13.2	20.3	−290	=	−12.9
OTM 280:	270	−9.1	11.7	−290	=	−17.4

B. December futures decreased to 250¢ on November 1
   1. Options cannot be exercised but will be offset
   2. Calculation of returns using different strike prices

Strike Price**	Sale Price	Option Cost	Sale Price of Option	Purchase Price		Net Returns
ITM 260:	270	−18.5	1.3	−250	=	2.8
ATM 270:	270	−13.2	0.2	−250	=	7.0
OTM 280:	270	−9.1	0.03	−250	=	10.93

---

*Lower valued strike price gives least loss.
**Lower valued strike price gives least return.

because the futures price move discounted the remaining option value. The reason for this large return was the relatively small option cost, 9.1 cents, compared to 18.5 cents for ITM calls. At the same time, the premium received from the ITM call sale was not large enough to bring the net cost closer to that of the OTM call.

Example 8.10 demonstrates a conclusion similar to that of the previous example. Buying OTM calls results in both the largest possible return and the greatest potential loss, depending on the move of the underlying futures price.

The thrust of this section has been to demonstrate the impacts of purchasing options with different strike prices as a hedge. The cheaper out-of-the-money options are preferable so long as prices do not move adversely. In-the-money option premiums are higher cost but offer greater protection in the case of an adverse price move. In general, use of in-the-money options results in the least loss potential and the least potential for gain, depending on subsequent movement in futures prices. Likewise, use of out-of-the-money options results in the greatest potential for gains and the greatest potential loss, again depending on subsequent movement in futures prices.

The use of ITM versus OTM options as a hedge depends on the merchant's or producer's capacity for absorbing risk. For example, a country elevator manager or a merchant should use deeper ITM options as a hedge so that the potential loss is less than the margins. Producers, on the other hand, have different capabilities for absorbing risk. Consequently, for some OTM options, being cheaper would be appropriate, and for others, deeper ITM options should be used.

Pricing of Commodity Options

A basic understanding of option premiums is important in making effective option hedging decisions. The purpose of this section is to briefly describe factors influencing option premiums; a more thorough description is available elsewhere (see Selected References, particularly Mayer; Barclay; Cox and Rubinstein; and all three Chicago Board of Trade references).

The option premiums used in these examples were derived using a theoretical model. In practice, option premiums are determined like futures, through open public outcry in trading pits. Review of the examples demonstrates that the relationship between call prices and the underlying futures is positive with an inverse relation between the price of puts and the underlying futures. This relationship to futures prices is intuitive since the value of the right to assume a long (short) futures position appreciates (depreciates) with increases in the futures prices, and vice versa.

Factors Influencing Option Values

An option premium has two components, normally referred to as the intrinsic and extrinsic values, which can be expressed as: Premium equals Intrinsic Value plus Extrinsic Value. Intrinsic value represents the amount that the option is in-the-money. For example, if the futures price for December wheat was 280 cents at Minneapolis and the call option's strike price was 250 cents, the intrinsic value of the option would be 30 cents. However, an option's premium typically exceeds the amount of intrinsic value, which the extrinsic value represents. Many factors affect the extrinsic value of an option, including: (1) the relationship between market and strike price, (2) interest ratio, (3) time to maturity of the option, and (4) the volatility of the underlying futures market.

Relationships Between Market and Strike Price. The relationship between market and strike prices involves several aspects. For call options, the premium decreases as the strike price increases relative to the

TABLE 8.3   Option Premiums for Variable Strike Prices

Futures Price (¢/bu)	Strike Price (¢/bu)	Call Premium (¢/bu)	Delta (Call) (%)	Put Premium (¢/bu)	Delta (Put) (%)
270	240	30.0	1.00	0.03	0.01
270	250	20.0	0.94	0.3	0.06
270	260	11.5	0.78	1.7	0.22
270	270	5.3	0.51	5.3	0.49
270	280	1.8	0.24	11.7	0.76
270	290	0.5	0.08	20.0	0.92
270	300	0.08	0.02	30.0	1.00

SOURCE: Author.

TABLE 8.4   Intrinsic and Extrinsic Values for Different Strike Prices of Call
             Option

Futures Price (¢/bu)	Strike Price (¢/bu)	Days to Maturity (days)	Market Volatility (%)	Interest Rate (%)	Call Premium (¢/bu)	Intrinsic Value (¢/bu)	Extrinsic Value (¢/bu)
270	250	90	10	5.5	20.06	20	0.06
270	260	90	10	5.5	11.53	10	1.53
270	270	90	10	5.5	5.27	0	5.27
270	280	90	10	5.5	1.82	0	1.82
270	290	90	10	5.5	0.46	0	0.46

SOURCE: Author.

futures price. The exact opposite occurs for put options. To illustrate, call and put premiums are calculated for options with different strike prices and an underlying futures price at 270 cents (Table 8.3). An option's extrinsic value is greatest when it is trading at-the-money, whether puts or calls as actual futures movement are most directly reflected in the option premiums. If the futures price is 270 cents, as the strike price changes from 280 cents to 260 cents, a call option is going from out-of-the-money to in-the-money, and the absolute level of the premium rises.

In Table 8.4, the intrinsic value of a call option with the futures price at 270 cents and strike at 260 cents is 10 cents per bushel. The extrinsic value can be found by subtracting the intrinsic value from the premium, which for a 270-cent strike is 1.5 cents per bushel. To compare this with the premium when the option is at-the-money, which reflects only the extrinsic value of 5.3 cents, one can see that the extrinsic value is higher when it is at-the-money. The same logic applies to put options.

Interest Rates. The levels of interest rates and option premiums are inversely related (Table 8.5). At higher interest rates, the return to interest-bearing investments increases, which means an increase in the opportunity cost of holding an option position. The competition between options and alternative investments will tend to decrease option premiums. The exact opposite occurs when interest rates are low.

TABLE 8.5   Option Premiums for Variable Interest Rate

Futures Price (¢/bu)	Strike Price (¢/bu)	Days to Maturity (days)	Market Volatility (%)	Interest Rate (%)	Call Premium (¢/bu)	Put Premium (¢/bu)
270	270	90	10	5	5.28	5.28
270	270	90	10	10	5.21	5.21
270	270	90	10	15	5.15	5.15
270	270	90	10	20	5.09	5.09
270	270	90	10	25	5.02	5.02

SOURCE: Author.

Time to Maturity. Everything else assumed equal, the more time an option has until expiration, the higher its premium. The underlying logic is that over a longer period, an unexpected event has a greater chance to develop, which would affect the futures price. Thus, buyers are willing to pay more for the longer term of protection. Likewise, sellers wish to receive a higher premium to compensate for the longer period of

TABLE 8.6   Option Premiums for Different Maturities

Futures Price (¢/bu)	Strike Price (¢/bu)	Days to Maturity (days)	Market Volatility (%)	Interest Rate (%)	Call Premium (¢/bu)	Put Premium (¢/bu)
270	270	30	10	5.5	3.07	3.07
270	270	60	10	5.5	4.32	4.32
270	270	90	10	5.5	5.27	5.27
270	270	120	10	5.5	6.06	6.06
270	270	159	10	5.5	6.93	6.93

SOURCE: Author.

protection offered. Table 8.6 shows that option premiums rise as their duration increases.[4] The premium is 4.3 cents (option premiums fluctuate in minimum fractions depending on the underlying commodity) for an option (either put or call) with 60 days left until expiration. But the premium for a 90-day option is 5.3 cents, an increase of 1.0 cents for the longer duration. Premiums for both puts and calls are the same when the strike prices are at-the-money. The intrinsic values for both are zero when the options are at-the-money, while the extrinsic values are the same because both options have the same market volatility, interest rate, and days to maturity.

Although options of greater durations have higher absolute costs, their average cost per day is substantially less. For example, the per-day cost of a 30-day option in Table 8.6 is 0.10 cents per bushel, but the per-day cost of a 90-day option decreases to 0.06 cents per bushel. However, the premium of an option does not depreciate uniformly over its duration. As shown in Figure 8.1, the curved line depicts how the

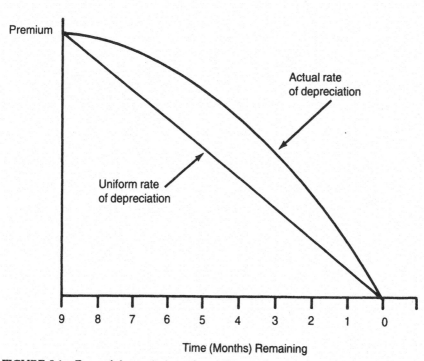

FIGURE 8.1 Rate of depreciation of option premiums over time.

premium depreciates with the passage of time. In the early days of trading an option, its premium depreciates slowly. As the option approaches expiration, the rate of depreciation in the extrinsic value accelerates.

Market Volatility. Market volatility in this context refers to the annualized standard deviation of returns. In other words, it is a measure of the fluctuation of percentage return on a particular commodity over the next year. If everything else remains equal, the premium would be higher if the volatility of the underlying futures market is greater. In times of high volatility, (1) more price protection is necessary, and buyers of options are willing to pay more; and (2) speculators will also pay more because of the higher potential return resulting from a volatile market.

On the other hand, the seller will expect to receive more (1) to justify the higher level of protection offered, and (2) to protect from any unfavorable price movements. As shown in Table 8.7, an increase of volatility from 5 percent to 10 percent increases premiums (both call and put) from 2.6 cents to 5.3 cents. The holders of options, whether puts or calls, have to pay higher premiums due to the higher volatilities in the underlying futures market.

In the case of grains, the most important factor influencing option premiums is that of market volatility. This is clear from the example

TABLE 8.7 Option Premiums for Different Levels of Market Volatility

Futures Price (¢/bu)	Strike Price (¢/bu)	Days to Maturity Days	Market Volatility (%)	Interest Rate (%)	Call Premium (¢/bu)	Put Premium (¢/bu)
270	270	90	5	5.5	2.6	2.6
270	270	90	10	5.5	5.3	5.3
270	270	90	15	5.5	7.9	7.9
270	270	90	20	5.5	10.5	10.5
270	270	90	25	5.5	13.2	13.2

SOURCE: Author.

presented in this section. Market volatility varies both within and between crop years. Within the crop year, market volatility increases when fundamental market conditions are uncertain.[5] Thus, during critical periods of uncertainty, such as soybean flowering, corn pollination, or announcements of export subsidies, volatility normally would be greater. Figures 8.2 through 8.5 demonstrate the volatility of the underlying

FIGURE 8.2 Standard deviation of Minneapolis Grain Exchange wheat, July futures contract, September to June.

futures price (using a 20-day moving average) at four futures exchanges for July contracts.

The important points are that the volatility behaved somewhat differently in each market, and that volatility changed substantially within

FIGURE 8.3   Standard deviation of Kansas City Board of Trade wheat, July futures contract, August to June.

the contract period. In Figure 8.5, soybean volatility decreased from September through November after a critical period in the U.S. crop and then increased in December through January, concurrent with the critical period of the South American crop. Option premiums also vary

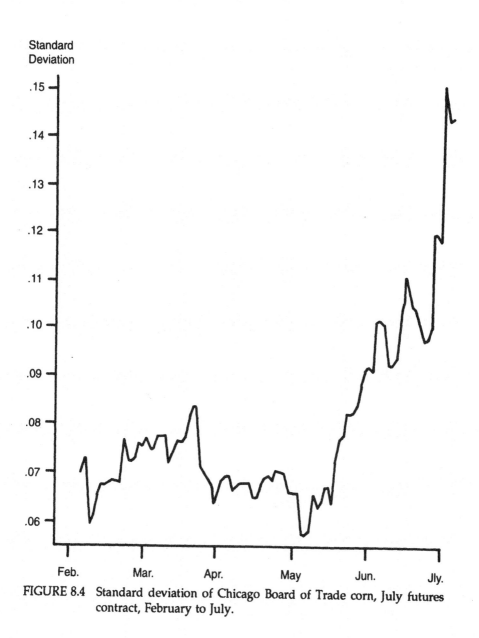

FIGURE 8.4   Standard deviation of Chicago Board of Trade corn, July futures contract, February to July.

substantially among years. Table 8.8 shows the average volatility of
wheat futures at three exchanges. Of particular importance was that
volatility decreased substantially in the 1980s mainly because of the
increase in U.S. and world stocks and the increase in the role of U.S.
farm programs during this period.[6]

FIGURE 8.5 Standard deviation of Chicago Board of Trade soybeans, July
futures contract, September to June.

TABLE 8.8 Average Volatility of Wheat Futures Prices for the July Contract (in percentages)

Contract Year	Futures Exchange		
	Minneapolis Grain Exchange	Kansas City Board of Trade	MidAmerica Commodity Exchange
1974	54.6	56.7	58.3
1975	30.8	33.3	37.4
1976	27.4	29.5	33.4
1977	16.7	20.7	23.7
1978	19.4	20.4	23.6
1979	17.7	19.7	22.2
1980	23.1	25.5	30.8
1981	16.2	21.6	25.7
1982	15.7	14.9	18.3
1983	13.9	12.6	19.1
1984	10.3	14.7	17.9
1985	7.7	8.8	11.9

The Black Pricing Model

Participants in the options markets use one of a number of different pricing models to evaluate market-determined premiums. These models incorporate the factors described above (i.e., time to maturity, interest, volatility, and the difference between the strike and futures price) in a probabilistic type of model to derive an estimate of the "fair market value" of the option. The most popular formula applied to commodities is that of Black, which is an adaptation of the Black-Scholes stock option model. The Black model for calls may be stated as

$$C = e^{-rt} [UN(d_1) - EN(d_2)]$$

where

$$d_1 = [\ln (U/E) + (sd^2 t) 2] / sd \sqrt{t}$$
$$d_2 = [\ln (U/E) - (sd^2 t) / 2] sd \sqrt{t}$$

C = call premium
U = underlying futures price
E = exercise price
r = short-term interest rate
t = term to option expiration (in days)
sd = market volatility (standard deviation of market returns on annualized basis)
N = normal cumulative probability distribution[7]
e = 2.7183 (base of the natural logarithm)
ln = the natural log of the term

Puts are priced similarly as:

$$P = -e^{-rt} [UN(-d_1) - EN(-d_2)]$$

The Black model makes the following five assumptions: that (1) the short-term interest rate is known and is constant through time; (2) the futures price follows a random walk in continuous time with a variance rate proportional to the square of the futures price. Thus, the distribution of possible futures prices at the end of any finite interval is log-normal. The variance rate of the returns on the futures contract is constant; (3) the option is not "European"; that is, it can be exercised any time before expiration; (4) transaction costs are negligible; and (5) there is no penalty to short selling.[8]

The premiums used in the examples in previous sections were derived using the Black model. Black premiums for wheat options are shown in Table 8.9 for various strike prices assuming time to maturity equals 90, volatility equals 10 percent, and interest equals 5.5 percent.

The call premium decreases as the strike price increases (as the call goes from deep in-the-money to deep out-of-the-money), while the put premium increases as the strike price increases (as the put goes from deep out-of-the-money to deep in-the-money). When an option is deep in-the-money, the premium equals the intrinsic value, and the extrinsic value is zero. This happens because a deep in-the-money option has so much protection against adverse price changes that the impacts of extrinsic variables are insignificant. The option premiums for both puts and calls have relationships and characteristics as expected.

Delta is an important and useful variable, which can be derived from the Black model. It is represented as the term $N(d_1)$ for a call and $N(-d_2)$ for a put in the formula. When the underlying futures price changes, option premiums also fluctuate to reflect this movement. The rate at which the price of an option changes in relation to the price change of the underlying futures price is referred to as delta. To be more specific, delta

TABLE 8.9 Option Premiums and Delta Computed from Black Model

Futures Price (¢/bu)	Strike Price (¢/bu)	Call Premium (¢/bu)	Delta (call) (%)	Put Premium (¢/bu)	Delta Put (%)
260	230	30.0	1.00	0.0	0.01
260	240	20.0	0.95	0.3	0.05
260	250	11.4	0.79	1.5	0.21
260	260	5.1	0.51	5.1	0.49
260	270	1.7	0.23	11.5	0.77
260	280	0.4	0.07	20.1	0.93
260	290	0.1	0.01	30.0	1.00
270	240	30.0	1.00	0.0	0.01
270	250	20.1	0.94	0.3	0.06
270	260	11.5	0.78	1.7	0.22
270	270	5.3	0.51	5.3	0.51
270	280	1.8	0.24	11.7	0.76
270	290	0.5	0.08	20.2	0.92
270	300	0.9	0.02	30.0	1.00
280	250	30.0	1.00	0.1	0.01
280	260	20.1	0.94	0.4	0.06
280	270	11.7	0.78	1.8	0.22
280	280	5.5	0.51	5.5	0.51
280	290	2.0	0.25	11.8	0.75
280	300	0.5	0.09	20.3	0.91
280	310	0.1	0.02	30.0	1.00

SOURCE: Author.

represents the percentage change of an option's premium if the underlying futures price changes one percent. As shown in Table 8.8, when an option is deep out-of-the-money, the delta factor would be close to zero, indicating the premium is not very responsive to minor changes in the underlying futures price. On the other hand, when an option is deep-in-the-money, delta is high and approaches one. When attempting to hedge a futures position with options, traders frequently refer to the delta factors. For example, a delta factor of 0.50 implies that two option positions should be established against every cash or futures position.[9]

Black premiums are an estimate of the fair market value of the option and are not necessarily equal to market premiums. Wilson, Fung, and Ricks compared the behavior of Black call premiums to those of the market for selected grains and oilseeds. The behavior of actual and

Cents/bu.

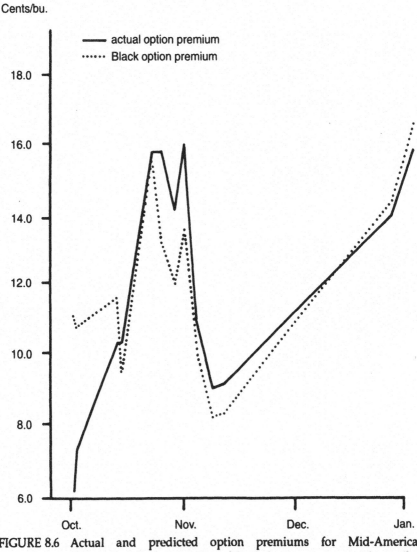

FIGURE 8.6 Actual and predicted option premiums for Mid-America Commodity Exchange wheat, March futures contract with a strike price of $3.30, October to January.

predicted option premiums are shown in Figures 8.6 through 8.8. In general, the actual and Black premiums are highly correlated; but, in some cases, persistent deviations exist. Descriptive statistics of option pricing error (OPE), which is defined as the actual less the Black premium, are

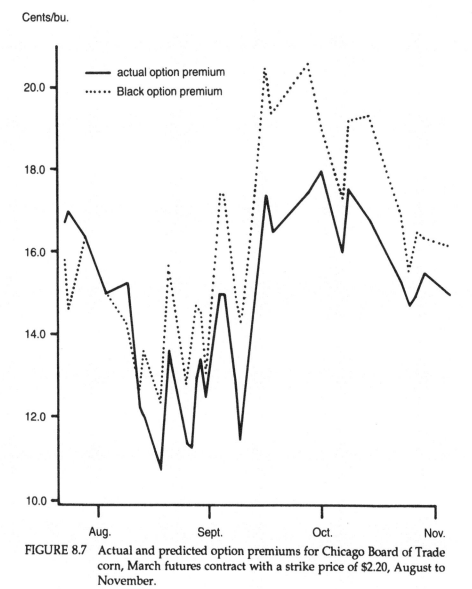

FIGURE 8.7 Actual and predicted option premiums for Chicago Board of Trade corn, March futures contract with a strike price of $2.20, August to November.

shown in Table 8.10. The results indicate that OPE on average for Chicago Board of Trade (CBT) soybeans did not differ significantly from zero. Those for the other exchanges all differed significantly from zero. Wilson, Fung, and Ricks provided a detailed explanation of factors influencing the relationship between actual and Black premiums.

FIGURE 8.8   Actual and predicted option premiums for Chicago Board of Trade soybeans, March futures contract with a strike price of $5.25, August to November.

TABLE 8.10 Summary Statistics for Option Pricing Error (in cents per bushel)

Contract	n^a	Mean	Standard Deviation	$t=0$
CBT Soybeans	1,346	−0.25	5.97	−1.50
CBT Corn	1,265	0.24	2.10	4.08*
KCBT Wheat	525	−0.68	3.29	−4.73*
MGE Wheat	378	−1.96	3.46	−10.98*
MACE Wheat	473	.26	2.65	2.14*

[a]n denotes number of observations; t, t-statistics.
*Differs significantly from zero at the 5 percent level.
SOURCE: W. W. Wilson, H. G. Fung, and M. Ricks, "Option Price Behavior in Grain Futures Markets," *Journal of Futures Markets*, Vol. 8, No. 1, February 1988, pp. 47-65.

Summary

Options on agricultural commodity futures began trading in October 1984 on an experimental basis and appear to be gaining in popularity. Introduction of options can be viewed as an innovation or as a new technology for marketing. Merchants or producers can use options in a multitude of ways—some as a complement and others as a substitute for traditional hedging. Their primary advantage for hedging is the ability to lock in a floor or ceiling price and at the same time allow the merchant/producer to take advantage of favorable price moves in the underlying commodity. The examples in this chapter demonstrate uses of options for long and short cash positions, forward contracting, and the implications of using in-the-money versus out-of-the-money options. In addition, the components of option premiums were discussed and followed by the development of a pricing model.

Notes

1. David L. Horner, and Eugene Moriority, "The CFTC Options Pilot Program: A Progress Report," *Education Quarterly*, Vol. 35, No. 3, pp. 9-14, 1983; Eugene Moriority, Susan Phillips, and Paula Tosini, "A Comparison of Options and

Futures in the Management of Portfolio Risk," The Commodity Futures Trading Commission, *Education Quarterly*, Vol. 2, No. 1, 1983, pp. 5-11.

2. The examples presented here are limited to those involving an underlying cash position, and comparisons are made to hedges in futures.

3. Chicago Board of Trade, "Protecting Deficiency Payments Using CBOT Ag Options," *The Commodity Futures Professional*, November 1987.

4. The option pricing model developed by Black is used to compute option premiums discussed in this section. See Fisher Black, "The Pricing of Commodity Contracts," *Journal of Financial Economics*, Vol. 3, 1976, pp. 167-179.

5. R. W. Anderson, "Some Determinants of the Volatility of Futures Prices," *Journal of Futures Markets*, Vol. 5, No. 3, 1985, pp. 331-348.

6. Michael Ricks, "Analysis of Option Premiums and Price Variance," unpublished M.S. thesis, Department of Agricultural Economics, North Dakota State University, Fargo, ND, 1986.

7. Value of normal cumulative probability distribution can be identified by using a table of normal probability distribution. For example, if the value of d_1 in $N(d_1)$ is positive, adding .5 to the table value will give the value of normal cumulative probability distribution. If d_1 is negative, subtract the table value from .5.

8. Fisher Black, and Myron Scholes, "The Pricing of Options and Corporate Liabilities," *Journal of Political Economy*, Vol. 81, May/June 1973.

9. For a more detailed discussion, see Terry S. Mayer, *Commodity Options: A User's Guide to Speculation and Hedging*, New York Institute of Finance, New York, 1983.

Selected References

"Agricultural Commodity Options," *Doane's Agricultural Options 46*, No. 47-5, 1983, pp. 15-16.

Anderson, R. W., "Some Determinants of the Volatility of Futures Prices," *Journal of Futures Markets*, Vol. 5, No. 3, 1985, pp. 331-348.

Barclay, William, *How Much Will Option Premium Be Theory and Reality*, Economic Research, MidAmerica Commodity Exchange, Chicago, IL, 1983.

Belongia, Michael T., "Commodity Options: A New Risk Management Tool for Agricultural Markets," Federal Reserve Bank of St. Louis, St. Louis, MO, 1983.

Black, Fisher, "The Pricing of Commodity Contracts," *Journal of Financial Economics*, Vol. 3, 1976, pp. 167-179.

Black, Fisher, and Myron Scholes, "The Pricing of Options and Corporate Liabilities," *Journal of Political Economy*, Vol. 81, May/June 1973.

Bowe, James, "Cutting Risk with Commodity Options," *Commodities*, December 1982, pp. 64-65.

Cargill Inventory Services, Inc., *Insight*, Chicago, IL, March 11, 1985.

Cargill Inventory Services, Inc., "Option Strategies—Straddles and Strangles," *Insight*, Chicago, IL, June 11, 1984.

Chicago Board of Trade, *Options on Soybean Futures—Contracts Fundamentals, Pricing, and Applications*, Chicago, IL, revised February 9, 1984.

————, *Options on Soybean Futures—Fundamentals, Pricing, and Applications*, Chicago, IL, June 1984.

————, "Protecting Deficiency Payments Using CBOT Ag Options," *The Commodity Futures Professional*, November 1987.

"Cotton Options—A New Risk Management Tool for the Producer, Merchant, and Mill," *CFTC AgReport*, Vol. 4, No. 4, 1983, pp. 8-10.

Cox, John C., and Mark Rubinstein, *Options Markets*, Prentice-Hall, Englewood Cliffs, NJ, 1985.

Dalton, James F., and Fred Bailey, *A Guide To: Options Strategies for the Farm Business*, J.F. Dalton Associates, Chicago, IL, 1984.

Figlewski, Stephen, and M. Desmond Fitzgerald, "The Price Behavior of London Commodity Options," *Review of Research and Futures Markets*, No. 1, May 1981, pp. 90-104.

Hauser, Robert J., and Dane K. Andersen, "Modifying Traditional Option Pricing Formulae for Options on Soybean Futures," paper presented at the 1984 AAEA Meeting, Cornell University, NY, August 5-8, 1984.

Horner, David L., and Eugene Moriority, "The CFTC Options Pilot Program: A Progress Report," *Education Quarterly*, No. 3, pp. 9-14, 1983.

Kenyon, David E., *Farmer's Guide to Trading Agricultural Commodity Options*, Agricultural Information Bulletin No. 463, USDA, Economic Research Service, Washington, DC, 1984.

————, *The Use of Futures Versus Put Options in Pricing Corn Production in Virginia*, Department of Agricultural Economics, Virginia Polytechnic Institute and State University, Blacksburg, VA, August 1984.

Klemme, Diana, "Ag Options, Missing Link in the Marketing Alternatives Chain?" *Grain Storage and Handling*, February 1984, pp. 26-33.

————, "Ag Options 2, A Merchandising Tool for the Country Elevator," *Grain Storage and Handling*, August 1984, pp. 29-45.

Labuszewski, John, "How to Produce Your Own Tables of Option Premiums, Deltas," *Futures*, Vol. 12, No. 10, October 1983, pp. 106-108.

————, "Volatility Key to Finding Fair Option's Premium," *Futures*, Vol. 12, No. 9, September 1983, p. 84.

Labuszewski, John W., and James F. Meisner, "Diagonal Options Offer Risk-Reward Alternatives," *Futures*, Vol. 13, No. 6, June 1984, pp. 87-92.

Mayer, Terry S., *Commodity Options: A User's Guide to Speculation and Hedging*, New York Institute of Finance, New York, 1983.

Meisner, James F., and John W. Labuszewski, "How `Worthless´ Options Can Wind Up `In-the-Money´," *Futures*, Vol. 12, No. 2, November 1983, pp. 76-77.

Moriorty, Eugene, Susan Phillips, and Paula Tosini, "A Comparison of Options and Futures in the Management of Portfolio Risk," The Commodity Futures Trading Commission, *Education Quarterly*, Vol. 2, No. 1, 1983, pp. 5-11.

"Options One Year Later: Index Contracts Big Winners," *Futures*, Vol. 12, No. 10, October 1983, pp. 99-101.

Ricks, Michael, "Analysis of Option Premiums and Price Variance," unpublished M.S. thesis, Department of Agricultural Economics, North Dakota State University, Fargo, 1986.

Rowlan, Tedi H., "Ag Options Update," *Com-Line*, Vol. 4, No. 4, 1984, pp. 6-7.

Schwager, J., "A Complete Guide to the Futures Market," John Wiley and Sons, New York, 1984.

Wilson, W. W., H. G. Fung, and M. Ricks, "Option Price Behavior in Grain Futures Markets," *Journal of Futures Markets*, Vol. 8, No. 1, February 1988, pp. 47-65.

9

Market Information

Dean Linsenmeyer and Dennis M. Conley

A Cleveland homemaker studies the grocery advertisement in the morning newspaper. An Illinois farmer briefly stops harvesting to telephone local grain elevators and checks their cash bid price for corn. A Nebraska broker scans the video screen for news on Japanese trade negotiations before taking a position in live-cattle futures. A USDA scientist develops a sophisticated computer model, using satellite photographs, to estimate winterkill in wheat in the Volga Valley of the former Soviet Union. A radio announcer in Iowa prepares hourly updates on slaughter hog prices from terminal markets and packing houses in the area.

Although the degree of sophistication varies, all of these individuals and millions of others like them are involved in the "information" business. The production of information, its dissemination to ultimate consumers, and its final use in making decisions are essential functions of marketing systems in all societies—be they private enterprise or centrally planned, developed or less developed.

In subsistence sectors of a developing country where producer, processor, and consumer are integrated into one individual, the need for information is similar. Where are the best sources of supply? What quantity will the household need in the future? The communication of such information may present no problem to such individals, but their very existence depends on the accuracy of the information they gather.

In the same manner, the decisionmaking bureau of a centrally planned economy needs information on current and anticipated supply-and-demand factors to devise its optimum national plan. The producer on the collective farm needs timely and accurate information describing that central plan as a guide to production efforts. In a market economy, each operator in the market searches out pertinent information on supply-and-demand factors, government policy, transportation conditions, current prices, and so on, to devise production and marketing strategies.

Buyers and sellers at all stages of the marketing channel need *technical information* on new innovations, storage practices, and other factors that may indirectly affect their marketing plans. To an even greater extent, they need *market information*. Buyers need an assessment of the quantity of a commodity being supplied to the market and the level of storage stocks at the time they plan to place their bids. They also need market intelligence on the expected quantity to be offered on the market in the near future to determine if it is more economical to buy now and hold for future use. In addition to supply information, buyers monitor the demands of other buyers in the marketplace. Buyers, other than final consumers, need market information on their own anticipated sales situation as well as that of their competitors. Will the final demand for their product justify current purchases of inputs? Will an aggressive sales policy by a competitor effect a change in their own product design or market share?

In the same fashion, sellers need to be informed of the quantity being demanded, and the product specifications desired. Sellers also gather information on the sales activities of other sellers to evaluate their competition. Government policymakers, who work toward a more efficient marketing system while they protect the rights of their producing and consuming citizens, are demanding more market information. They look toward information on carry-over stocks, expected production and consumption, the market share, and the returns to individual marketing firms to guide development of public programs necessary to improve the performance of the market.

An Essential Ingredient for Competitive Markets

Information on the production and consumption of a product provides the basis for establishing a price. The market agent's awareness and understanding of this information guide decisions concerning the appropriate type and amount of the product to buy or sell and the best bid to accept as well as the best time and place to enter the market. By being aware of competing bid prices from many independent millers, exporters, feeders, and others, the grain merchandiser can be assured of the true market value of the commodity.

It is the knowledge of alternative resource suppliers, or alternative bidders for a resource, that enables inputs to flow into their most valued use. Similarly, in the product market, it is only when buyers and sellers are equally informed of the market situation that goods and services can be distributed so as to maximize consumer satisfaction.

When buyers and sellers are not adequately informed of the relative supply-and-demand conditions between different marketing dates or different market centers, prices may not reflect the real cost of supplying that product. To the extent that price differences are not due to differences in the product (different time, place, or form characteristics), these price differences are an indication of the lack of information or uncertainty on the part of decisionmakers in the market.

When traders in market X are informed of the prices in market Y, they can determine if the price differential between X and Y is greater than the cost of transferring the good between the two markets. When such is the case, the potential for profits will motivate traders to buy a commodity in one market and transfer it for sale in the other market. This practice is known as *arbitrage*. Many market agents derive successful careers from arbitrage. Transferring between markets may involve transportation (locationally distinct markets), storage (different markets over time), or processing (markets for different products). For example, a soybean trader with knowledge of crushing costs and the estimated oil content of beans will carefully monitor the price spreads among beans, oil, and meal to know when arbitrage between the three markets is profitable.

Arbitrage across markets attempts to correct unjustified price dispersion, thereby improving the market's efficiency. Arbitrage breaks down, however, if the cost of obtaining information on alternative markets is prohibitively high.

The competitiveness of the market is conditional on the many buyers and sellers being aware of the others' market activity. When access to market information becomes the possession of a select few, the potential is there for widespread misuse of the market system. Price relationships easily become distorted from their true measure of either end-use value or relative resource scarcity. Protected information concerning factors that directly affect market prices is a powerful tool for reaping windfall gains from the marketplace. Consequently, the private market agent who possesses particularly valuable information has economic incentives to suppress its release to the general public. Misinformation or unbiased rumors can be purposely introduced to mislead or confuse competitors. The cost of verifying such rumors or critically analyzing misleading information must be considered as part of the total cost of information gathering.

For market information to be an effective guide in planning and coordinating the market channel, it must be *accurate, adequate, understandable* for the market user, and *equally available* to all traders.

The accuracy of market information depends on the data base from which the information is derived. Average daily price quotations may be based on a large number of transactions randomly selected from an active day's trading and thus accurately represent the true mean price. However, if the particular central market was relatively inactive or handled an insignificant share of the total commodity traded, its daily price report may provide a distorted picture of actual market activity. As actual cash markets are bypassed, with more grain being shipped directly from producers to terminals, the sample base and accuracy of reports from central cash markets decrease.

A more adequate source of market information may be one that either provides more detail in its information or more timely dissemination of its information. Adequacy is determined in relation to need. Therefore, for many market agents, information on the volume of U.S. No. 2 hard red winter (HRW) wheat transacted and its average sales price for the day may be adequate. However, for other buyers and sellers such as flour millers, this information is inadequate, because they need detailed information on the gluten content, protein level, and baking characteristics to evaluate the price for their purposes. Farmers frequently want more detailed information on quoted prices at local elevators. Although the price quoted by two elevators may be identical, the actual transaction may involve a difference in the premium and discount factors applied against the quoted price. These extra details determine the information's adequacy as a guide in making market choices.

Adequacy is also related to the timeliness of informaion. Is the information relayed to its potential user while relevant decisions are still open? The awareness of a profitable sales option is of little benefit after the commodity has been sold. Timeliness has two facets: frequency and promptness. Improved timeliness may require that information be disseminated more frequently. It may also require more rapid dissemination so that the information conveyed is a current description of the market, rather than past history. Published mailings of information are still an important source, but the telephone, teletype, radio, and television are now common sources because of their greater speed in disseminating timely information. Over 200 telephone-tape recording outlets provide market agents with instant information 24 hours per day. In addition, some 1,500 radio stations and 160 television stations periodically broadcast market information each day.[1] However, with the advances in technology, even these more common media are being displaced by electronic screens connected to computer and telecommunications networks. The cost of such access has significantly declined over the past decade; local retail agribusinesses and producers now have the same, nearly instant access to events and data as do larger participants in the grain markets.

The third major requirement of market information is that it must be understandable by the market user to be effective. It must describe the structure and activity of the market in a format that the user can quickly comprehend and relate to an immediate situation. The language of the market—its jargon and terms (*bullish, bearish, basis, to-arrive, on-track,* and so forth)—may render the information useless to people not directly involved in the marketplace yet dependent on market information to guide their own enterprises. The intelligibility of market information also depends on how data and relevant events have been composed to present the information. Data, which are usually a collection of numbers, cannot by itself yield much information. Yet with the increasing volume and faster delivery of data, the transformation of data into information becomes more difficult. In some cases, a stream of data on an electronic screen defies interpretation. For example, on learning that the former Soviet Union has experienced a 10 mmt shortfall in coarse grain production, most U.S. farmers are unable to incorporate that information into their production alternatives or marketing strategy. However, if a frame of reference accompanies the data, explaining that the 10 mmt shortfall represents, say, 30 percent of last year's production, and such a shortfall has not occurred in over five years, then the user can assimilate the information and make a decision. Only when a piece or collection of data is presented in a recognizable format and connected to a frame of reference does it become transformed into intelligible information.

To effectively enhance the competitiveness of the market, information must be equally accessible to all agents for whom it may have value. The more likely a bit of information is to directly impact a particular market trend, the greater the need that all agents have an equal opportunity to benefit by it. From the private trader's point of view, the more relevant the information, the greater is the incentive to tightly control its release. In this way, the individual marketer can capture a greater benefit from that person's elite position. When relevant information is not equally available, it results in commodities being underpriced or overpriced in relation to their true resource value to society. Market agents can each respond in an optimal manner when they all have equal access to the needed information; no one agent can obtain monopoly power by controlling the information flow to other agents.

It can be concluded that although the purely competitive market model assumes "perfect" knowledge by all market agents, market information in reality strives toward "relevant" knowledge equally accessible to all buyers and sellers. In fact, the perfect knowledge assumption could potentially be satisfied only in a single, completely centralized market with a relatively small number of agents. Here, any one agent potentially could know the prices quoted by all other agents at a particular time. However, for this potential to be realized, one additional assumption must be made. Perfect knowledge implies that additional information would have no economic benefits to the user. For it to be economically rational for any one agent to use information when its economic benefits are nil, it must be assumed that the cost of gathering that data is zero. Only if information is a free good, with unlimited supply, will the agent choose to use the maximum amount of this input in a decision.

Economics of Information

Market information is one of many inputs that are combined by a firm in the production process. Therefore, it may be helpful to treat market information as any other physical input in examining its production and use by market agents.

The Information Supply Industry

Scarce resources must be invested in the production of market information as in the production of physical inputs. Porta has estimated that the production, processing, and distribution of informational goods and services account for over a quarter of the U.S. gross national product.[2]

Market research and data collection may take various forms depending on the type of information needed and the size of the firm involved as well as the diversity and complexity of the commodity in question. In past experiences the individual buyer or seller may have invested time and resources heavily (i.e., trying various products or making valuable contacts with other market agents). Knowledge gained from past actions will be used by each individual and close associates of that person to determine which products to buy, the best time to sell, and who will offer the most favorable price.

Market intelligence gained by experience demands a high initial fixed investment in time and resources. Therefore, any one agent may only deal in a local market, where experience can be gained at a lower cost to him or her.

Larger firms commanding a greater share of the market also undertake a high fixed investment to produce market information but through different methods. Commodity market research departments, private meteorologists, and elaborate computer facilities are all indications of the fixed resource costs incurred by major grain firms in their efforts to obtain up-to-the-minute assessments of market conditions. Not only do the larger firms command sufficient resources to acquire such data-producing technologies, they also handle sufficient volumes to spread these high fixed costs over many units. In this way, they can reduce their per-bushel cost components of sophisticated and expensive data-gathering systems.

Once the initial investment is made and the data are gathered, the variable costs of disseminating and incorporating this information into production or marketing decisions is relatively minor, compared to the initial investment. The combination of high fixed costs and relatively low variable costs would indicate that the information-producing industry faces economies of scale over a wide range of output. Given such economies of scale, low-volume marketing firms would operate with higher average information costs per unit of output and would be at a competitive disadvantage. With a declining average total cost curve, the information-producing industry cannot be expected to remain competitive in the long run.[3] The alternatives are either a publicly subsidized monopoly responsible for producing market information, which would then be available to all market agents, or a few large marketing firms producing large volumes of market information for their private use.

A second critical characteristic of the information supply industry is that of risk bearing. The production function for most physical inputs can be estimated with relative certainty. However, the production function for information cannot be predicted accurately. Any investment in market research cannot guarantee either a particular outcome or that the outcome obtained will be useful in the decision process. No insurance policy can be purchased to cover the risk of unproductive research, so private firms investing in information gathering must carry the risk themselves. Larger firms are more capable of absorbing potential losses and, therefore, are more willing to undertake specialized market research.

Another aspect of the risk involved in the production of market information is that information can be a highly perishable product. Grain supply-and-demand conditions change constantly. Consequently, the information gained in the latest monitoring of those conditions may be obsolete, or even counterproductive, before it can be disseminated to decisionmakers and used.

In summarizing the economics of the information supply industry for grain markets, one could expect extremely high costs per unit of information if the industry consists of small, privately operated firms competing independently for market intelligence. These high costs are frequently hidden in the total transaction cost. The risk inherent in the information production process as well as the risk of rapid obsolescence would further contribute to a suboptimal investment by society in market information.

The Economics of Information Use

Market information is not a free input but is obtained with an expenditure of valuable resources, so economic rationality dictates that there is an optimal amount of information purchased by the market agent for a decision with less than perfect knowledge.

Classical economic theory indicates that the optimal amount of any input is determined where the increased value of the product derived from additional inputs equals the additional costs incurred in acquiring those inputs. Market information is used in the production of better marketing decisions. The market agent must decide between alternative actions in the marketplace—alternatives that are mutually exclusive events. Either one buys or does not buy, sells or does not sell. Before choosing one alternative, the agent attempts to evaluate the probable consequence of each alternative. What is the *expected* payoff if one buys now instead of buying later or sells at this location rather than another? Because the agent is faced with uncertain events, information is purchased to reduce uncertainty and to assist in creating the best marketing strategy. The more uncertain the outcome of a decision, the higher the value that can be attached to more information and, consequently, the higher the demand for information.

In the 1950s and 1960s, large grain surpluses and relatively stable administered prices reduced price uncertainty and, therefore, reduced the need for grain market information. In the early 1970s, the demand for more and better market information increased because of increased price variability in most agricultural commodities.[4]

The consequence of one market decision depends on the uncertain occurrence of other events. Today's sale of grain may be judged as a "good" or "bad" decision depending on how much the price rises or falls tomorrow. What are the chances that, when one has bought from a given source, a lower offer price could have been found for the same good from a different source? Market information is frequently used to determine the probability of each result for any single market act. The sum of the various results, each weighed by its probability, is the expected payoff of that marketing alternative. Therefore, the optimal course of action for the market agent is the one that yields the highest expected value. The marketer is willing to pay the cost of additional information if it will improve the predictability of market events. Information changes uncertainty to greater certainty and, therefore, assists in more accurately identifying the expected returns from a market action. As more information becomes available, the market agent revises the probabilities attached to the different events affecting that person's decision outcome. For example, a wheat farmer whose anticipated sale price is $3.00 per bushel revises the expected price as that individual begins to collect price quotations from a sample of local elevators. As bids over $3.00 are found, sales price expectations are adjusted upward. The value of additional information equals the expected profits with the increased certainty of events minus the expected profits under the previously more uncertain conditions. If the cost of additional information is greater than the expected increase in profits derived from that information, the market agent should not invest in it.

Although economic theory enables one to better understand rational decisionmaking, it does so within a set of simplifying assumptions. Particular characteristics of market information complicate its allocation process. Production functions using fertilizer inputs in corn production can be modeled with relative certainty to determine the marginal value product per unit of fertilizer. In contrast, the marginal value product of additional information increases with uncertainty. The exact value of the information is not definable until after the information has been incorporated into the uncertain setting and the user has taken a particular course of action in a specific decision.

Information has economic value to individual agents only if misinformation is present in the market. *Misinformation* may be current information inaccurately determined, and therefore misleading, or accurate information that is obsolete and no longer appropriate to the situation. When the information is exposed in the market, and the public can correct its expectations, then the information loses its value to the individual agent. If everyone knows that production in the former Soviet Union is greater or less than previously expected, that information loses its market power for the individual trader.

Each market agent acquires better information in anticipation of being able to identify the price dispersion and obtain a higher than average price. However, if all agents possessed the same information, price dispersion in a given market would collapse to a single average market price. Consequently, each agent would receive this average price less the cost of the information, making them in total no better off than before.[5] In other words, "It is only because prices do not accurately represent the true worth of the commodity . . . that the informed are able to earn a return to compensate them for the cost associated with the acquisition of the information."[6]

The quandary of market information, which distinguishes it from other private goods, is that the supplier of information cannot fully appropriate the returns from all users. Once the information is acted on by one agent in the market, it is revealed for other agents to benefit. Consumption by one market agent does not preclude consumption of the same information by other agents. This characteristic encourages individual firms to maintain the secrecy of market information so that they may capitalize on the market advantage that it gives them.

A review of both the economic structure of the supply and demand of information and the importance of open access to information in an efficient, competitive market shows that several qualities of market information are particularly important.

High fixed costs and economies of scale in information production have indicated that efficiencies could be gained when such market information is produced by a relatively few large-volume firms and/or public agencies. It was noted that investment in information gathering is a high-risk venture, with little certainty of the product gained. This uncertainty results in a comparative advantage for large firms who can absorb the risk. The comparative advantage in access to information translates directly into a competitive advantage in the marketplace and greater inequality in market power between small and large firms in the long run.

Because the value of information to the industrial user increases with uncertainty in the marketplace and cannot be fully recovered by the supplier of information,[7] society would benefit by more market information being produced and used than the quantity that private interests would choose to produce. Market information has many characteristics that make it a collective (or public) commodity, with advantages to be gained by publicly supplying this good.

The Component Part

Although it is true that no part of the world's economy operates in total isolation from the rest, some parts have a more direct impact on grain markets and therefore merit closer monitoring. Market prices are the common denominator or the grand indicator of all the world forces affecting a particular commodity. By identifying and understanding the impact of those forces on market prices, the market agent can select the components of informational input needed to improve the market decision.

Five categories of forces that market information must address are the availability of supply, demand for the commodity, ownership positions, physical market constraints, and those government programs and regulations that modify the framework within which supply and demand interact. For purposes of clarity, it is convenient to categorize these forces; in reality, they are continuously interrelated and dynamic.

Major Supply Factors

Information on the supply of grain capable of coming to the market in the current planning period is extremely important to the market agent. Because grain production in the temperate climates is a very seasonal enterprise, information on the quantity produced is crucial. This information facilitates the market's effective rationing of that supply among users until the next harvest becomes available.

The assessment of grain production begins five to six months before the crop is actually planted. Estimates of producer planting intentions provide an early indication of expected acreage of the particular crop. This indication allows the competitive forces of buyers and sellers to evaluate more accurately the existing supply in line with longer-term supply projections. Planting intentions are periodically revised until the figures for actual acreage planted become available shortly after planting is completed. But the length of growing season may become a constraint on the quantity or quality of grain produced, so the actual timing of the seeding is also important information. If the planting of regular-season varieties is delayed, the probability increases that fall crops will be frosted prematurely. Lighter test weight and lower protein or oil content may result as well from heat damage or drought during the period when new kernels are filling out.

Throughout the season, information on crop-growing conditions helps buyers and sellers revise their estimates of future supply. Unfavorable patterns of temperature, relative humidity, wind, and precipitation provide early warning signals to the market that particular crops may be under stress and, therefore, raise the possibility of reduced yields. Information on weather conditions is also important in projecting harvests of the current season as well as the succeeding season. Subsoil moisture levels are a measure of the water reserve held in the soil that next season's plants may tap. Other adverse growing factors such as fungi, insects, blights, diseases, or winterkill are directly affected by weather patterns. Wide distribution of such information allows all buyers and sellers to revise their market decisions and act accordingly. After the plants are mature, weather has an important impact on the actual bushels harvested. Crop losses increase as wind and precipitation hamper harvesting efforts. Weather also affects the amount of artificial drying necessary to maintain the crop in good condition during storage. Major grain producers and merchandisers need information on weather and other production variables in planning their market strategy.

Available supply is composed of current-season output plus the carry-over stocks of grain in storage from earlier periods. The ability to place grain in storage and remove it in later periods of relative scarcity facilitates more uniform consumption. Accurate data on the level and location of grain stocks inform traders of the size and accessibility of these additional supplies that may be tapped if needed. So long as the anticipated increase in price more than offsets the cost of storage, stocks are withheld from the marketplace.

Major Demand Factors

Information concerning the demand for grains must also be analyzed in relation to the total available supply. The demand for grain depends on its value as food for humans and as a livestock feed, its value in industrial products, and its productive value as seed. These sources of demand may be domestic or international. In the latter case, they would be reflected in our export demand. Information reflecting increases or decreases in swine farrowing, poultry numbers, cattle on feed, or breeding stock is important to grain marketers as these factors represent changes in the intermediate and long-run demand for grain. Crucial to the use of such figures is the marketer's ability to estimate their impact on price changes for grain. Low livestock numbers and subsequent high meat prices relative to the cost of grain inputs are reflected in wider feeding margins. The wider the feeding margin, the greater the economic incentive for feedlots to feed cattle to heavier weights before slaughter, thereby increasing the demand for grain. Equally important in the feed demand for grains are changes in consumers' tastes for lower-grade, leaner meat cuts. Information on such changes indicates long-term adjustments in the demand for grain-fed meats, which directly affect the demand for grain.

Grains are also used as a raw material in producing a wide array of industrial products as diverse as dynamite, cosmetics, and gasohol. Technical information on innovations in the manufacturing process carries an economic impact in grain markets. Successful technological changes may increase the marginal value product of grains, resulting in a new optimal level of grains used. Early access to accurate technical information improves the market position of the grain marketer.

Ownership Positions

Most of the data and information available on the supply and demand for a commodity relate to its physical position in the marketing channel. Yet, during the marketing process, a significant amount of grain is owned on paper at positions in the channel other than the corresponding physical position. For example, corn still growing in a Nebraska field during August may have already been sold in June to a local elevator by using a forward contract. It is possible the elevator manager made a forward contract sale to an export shipper, who in turn sold it to a feed mill in Europe, with the corn to be delivered in December. Title and ownership of the Nebraska corn is still held by the producer, but promises to deliver were passed along the marketing channel all the way to the European feed mill. All the forward sales are likely hedged with offsetting futures or options contracts, and the absolute price or price formula is determined at the farm, local elevator, exporter, and foreign feed mill well in advance of the physical bushels leaving the field.

In the government reports issued during the summer and fall on crop development and production estimates, the corn still growing in the Nebraska field is counted as part of the potential corn supply that will be physically available in Nebraska. Technically, because of the advance sales and movement of ownership positions, the estimate overstates the physical supply available to the *open marketplace* at harvest. An example helps clarify the point. For example, perhaps published sources estimate U.S. corn production to be a record 8.9 billion bushels, but unknown is the fact that 3.9 billion are already committed to advanced sales, leaving only 5.0 billion bushels available to the open market. The reporting of record production and the impression of ample supplies of physical bushels at harvest may lead a seller to conclude prices will not increase during the year and will likely fall. On the other side of a potential transaction, a buyer may conclude ample supplies will exist, and the purchasing strategy should be to buy only for immediate needs while taking advantage of declining prices during the marketing year. Without corresponding information on the advance sales and ownership positions, the buyers and sellers receive the wrong impression about available supplies.

Another example describes what happens when a shortfall in production reduces supplies of physical bushels. Forward contract purchases by grain handlers and processors in advance of harvest months, and in some cases before crop development is assured, allow these grain buyers to secure their physical needs during harvest and beyond. When a production shortfall occurs, or even when one is anticipated prior to harvest, producers are directly aware of the new crop prospects, and they anticipate higher prices. Buyers, on the other hand, have already secured their needed supplies with advance purchases and are not active in the market during harvest. Prices are not bid up as expected based on physical evidence, and sellers of grain see an apparent contradiction in the open marketplace, questioning the quality of competitive forces at work.

A classic example happened in the summer of 1972 when U.S. exports of grain led to a boom in the agricultural economy.[8] In early April of 1972, a trade delegation of U.S. agricultural officials departed for Moscow to explore the possible sale of grain and credit terms for such a sale to the Soviet Union. A subsequent meeting took place in Washington, attended by representatives from the Soviet Union and the U.S. assistant secretary of agriculture, to again discuss credit terms, barter agreements, and the possibility of buying direct from government-held stocks. A government-to-government deal of direct purchases was rejected by U.S. agricultural officials, and the representatives from the Soviet Union were told they would have to buy from private grain companies. A trading team arrived June 30 and over the next five weeks bought one-fourth of the U.S. wheat crop through a number of private companies. Later investigations into the advance sales raised a number of contentious issues about the release of market information and the role of government officials and grain company executives. However, it was clear only a few knew at that time that the advance sales were taking place. Market information was relatively slow in coming, showing the ownership positions held by the Soviet Union. But, when these positions were eventually revealed, the price of wheat shot up, regardless of the physical position of wheat and the previous illusory comfort of ample supplies.

The reporting of market information by both public and private sources focuses on the physical aspects of supply, demand, and positions in the marketing channel, with little to no information available on advance sales and ownership positions. An exception is when the government has required grain-exporting firms to report advance sales to unknown destinations and eventually to provide more specific information on those sales to the countries of destination. Developing comprehensive market information on ownership positions is difficult for many reasons, including the need to reconcile the chain of advance sales in the marketing channel, or the problem of "double counting"; the rapid movement and geographic range of these positions on paper, which make their measurement difficult to record; and the intrusion into proprietary business information that is part of the everyday operations of a grain firm. Yet, ownership positions are a force the marketing agent needs to consider along with the physical aspects of supply and demand.

Physical Marketing Constraints

In addition to being knowledgeable of changes in supply-and-demand conditions, the market agent must keep abreast of physical marketing constraints. When portions of the Mississippi, Ohio, or Missouri rivers freeze over in the fall, grain destined by barge for lower gulf ports must be sent by a different mode. Such bottlenecks in transportation depress cash bid prices in producing regions and raise sales prices to the final user or exporter. Other market agents who do not face a particular transportation bottleneck need information to locate and benefit from such premium markets.

Quantities of a particular grain supplied and demanded are seldom equal at a given location, and transportation must, therefore, be arranged to move the commodity. An accurate and adequate information source can assist the market agent in choosing the most economical mode of transportation. Grain merchandisers place a high priority on being informed of the availability and location of hopper cars, the freight structure of alternative modes to alternative destinations, and the limitations on load-carrying capacities over rail branch lines. Because grain movement is absolutely vital to being able to take advantage of any particular supply-or-demand source, information on the physical transportation constraints is central to any marketing strategy.

The market agent must also accurately assess other physical constraints such as the availability of storage, processing, and handling facilities. More on-farm and commercial elevator storage facilities are being built annually. Such facilities differ in their handling, drying/aeration, and storage costs. Information on which facilities have uncommitted space and what total costs are involved in utilizing each facility provides the basis for wise marketing decisions. Knowing the cash bid prices of a local milling company, soybean crushing plant, or livestock feeder facilitates greater flexibility among potential grain-marketing outlets. Because such localized demands may utilize a large share of local grains, basic patterns may change frequently. Alertness to changes in local markets distinguishes the aggressive marketer from all others.

Information regarding the physical constraints at port or terminal elevators may have an equally important impact on local prices and marketing choices. As port elevators become congested or embargoed, grains must be redirected to other ports, possibly using different modes of transportation. Knowledge of these adjustments and their relative marketing costs will soon be reflected in local cash bid prices to producers.

Government Programs and Regulations

Commodity markets always operate within the political framework of the country in which they are located and are affected by the political and economic situations in other countries. Some public regulations, such as antitrust laws, affect the structure of the market; others, such as price support legislation, affect market prices more directly. Regardless of their direct impact, these regulations ultimately result in an alteration of supply or demand and a new market equilibrium.

As soon as it is reasonably clear that impending legislation will be passed, market agents search for information that will clarify its probable impact on their operations. Wheat and feed grain producers contact their local Agricultural Stabilization and Conservation Service (ASCS) office to better understand and analyze all the provisions of the set-aside or price support program. These programs restructure the marketing alternatives that they must adjust to. Each operator must assess the impact of this new framework, be it a higher import quota for red meats or a three-year grain reserve program, on the supply of and demand for grains. Knowledge of such regulations and the ability to accurately assess their impact translate into profitable market positions for those firms that can respond quickly.

Being informed early of foreign nations' policy changes is difficult for most domestic grain merchandisers. Brazil's policy to subsidize fertilizer and other structural inputs in soybean production translates directly into more market competition for U.S. soybeans. A decision by the former Soviet Union to expand wheat acreage or livestock numbers has a direct economic impact on total world supply and demand for grains. The agent who gains information about such changes and can interpret their probable impact in the grain market ahead of other merchandisers can reap considerable benefits by reacting early. Such information is extremely valuable and when gathered by private operators is a carefully protected resource.

Public regulations also affect the demand for grains directly. With the passage of the Staggers Act, railroads had new flexibility to change rail rate structures or initiate unit train rates over new routes. Consequently, the purchasing pattern for grains in the affected areas was changed. Some firms benefited from such changes, but others were placed at a comparative disadvantage. Being aware of current and potential changes is important to people who need to make better investments in marketing facilities.

Changes in the demand for U.S. grains and consequent price adjustments are directly affected by the presence of import restrictions into foreign countries. All major firms dealing with the export of U.S. grains closely monitor changes in the grain import levies imposed by the EC. The firms must evaluate the changes and respond quickly to maintain their market position.

Government programs, regulations, and policy changes take a considerable amount of time to develop and implement with ample lead times to form expectations on market effects. However, the effects of political events are less predictable. Since the early 1970s, the international trade of grain has significantly increased, making markets more sensitive to unanticipated political events in certain countries. For example, when the Soviet Union invaded Afghanistan, the U.S. response, which came quickly and was perhaps anticipated by only a few, was to immediately embargo all grain sales to the invader. Other grain-supplying countries were asked to comply with the embargo, and some did. Regardless of the source of impact, governments and political events ultimately do alter supply, demand, and trade in grain and have an effect on the market. Because such information is so crucial to the life of a firm, vast amounts of private funds are expended annually to obtain accurate details before they are generally available. The more such information is readily accessible to all agents, the quicker agents can respond accordingly in their buying and selling practices, thus improving the physical and pricing efficiency of the entire system.

Sources of Market Information

The market agent seldom relies on any one source for market information. As the agent's needs vary in terms of subject matter, depth, and timing, those sources of information are selected that are most suited to that person's needs. Part of the information will come in published form such as newspapers, government circulars, radio or TV broadcasts, or commodity newsletters. Information derived from published sources usually covers topics that are either quite general and appeal to a large audience or topics that may be detailed but that do not become outdated quickly. These sources are useful to an agent who may not be operating daily in the market but who still needs to keep abreast of underlying market trends.

Other information will be obtained from direct-contact sources. These may be face-to-face discussions or some form of telecommunication interchange with other market agents. The advances in telecommunications and computer technology have created additional devices for rapid communications that, in some firms, reduce reliance on the telephone. These devices include computer networks with electronic mail capabilities, direct computer-to-computer transfer of information, facsimile machines, voice mail, cellular phones, and satellite networks that can globally and instantly transmit data and voice and video communications. Direct-contact sources have an important role in disseminating information that is quite detailed or that becomes obsolete very rapidly. One grain producer may visit directly with other producers or with several grain merchandisers to gain specific market information for a particular location. Here, an up-to-the-minute picture of a specific market can be received. Brokers, exporters, or other large-volume grain merchants who operate daily in the marketplace rely heavily on direct telecommunication interchanges to anticipate market changes before they occur.

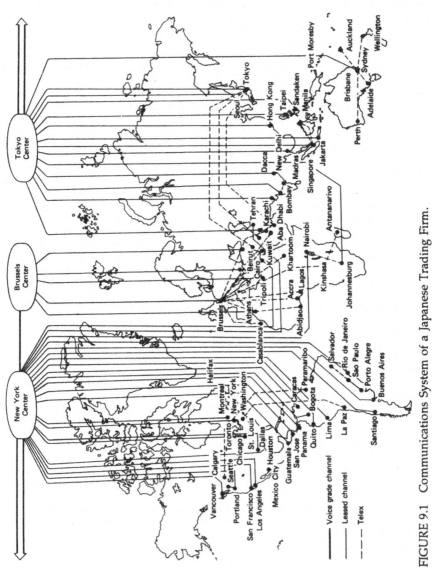

FIGURE 9.1 Communications System of a Japanese Trading Firm.

SOURCE: John Freivalds, "Trading Firms' Role in Japanese Economy," *Feedstuffs*, Vol. 50, No. 11, March 13, 1986, pp. 23-24.

The direct communication lines for one international grain-trading firm are illustrated in Figure 9.1. Direct-contact monitoring of conditions in production centers, political capitals, and transportation centers allows the firm to keep abreast of changes in the main factors influencing its worldwide operations. For example, this trading company has direct telex channels to Halifax, Nova Scotia. Although Halifax is relatively small and distant from major grain-producing regions in Canada, it is one of Canada's major winter ports handling its international grain exports.

Throughout any given day, a grain merchandiser may call several producers in different major producing regions for a snapshot of field conditions; several elevator managers to monitor prices, storage conditions, and transportation constraints; several brokers to evaluate the rumors and activity of the central markets; several major processors to analyze the strength of the demand for their products. Representatives at the port or on the trading floor will also be contacted for in-house information. Within a matter of minutes, the major factors affecting the market at that time can be pulled together from this network. The agent then proceeds with market operations and rechecks several of the more volatile factors throughout the day's trading. The accuracy of this information may be questionable at times, but the important issue is that *regardless of whether it is fact or fiction, the information has an immediate and direct impact on the market*. The challenge of obtaining timely information is to anticipate early what the market will do with that information when it becomes generally known.

The marketing agent must evaluate the reliability of information sources. Does the agent perceive the information as fact or fiction? Equally important is the question, "Will the market in total accept the information as fact or fiction?" The individual buyer or seller may judge the information to be false and, believing that the rest of the market will accept it as false, choose not to make a position adjustment in the marketplace. However, if the individual market agent perceives the information as false, but the rest of the market sees it as true, this individual may choose to enter the marketplace to take advantage of the market's inaccurate perception. The actual accuracy of the information may be immaterial so long as it is perceived in the market as being accurate and reliable.

In actuality, grain market agents use both direct-contact sources (to fine-tune each day's strategy) and published sources (to cross-check and revise their close-up estimates in light of underlying aggregate changes). Either direct-contact or published sources may be publicly or privately funded.

Public Sources

Sources supported by public funds have frequently been seen as a means of equaling access to accurate unbiased information for all interested market agents. This role is important to equalize market power among agents regardless of size and, in the process, to improve the market's efficiency.

The U.S. Department of Labor releases statistics on employment and wage levels by sector. These statistics provide some indication of changes in purchasing power. The department also computes and publishes the Consumers' Price Index, which measures the changes in price levels for major consumer goods, such as food. Other indexes of sales and industrial production are released by the Department of Commerce. Every five years it also publishes the Census of Agriculture. Such publications provide an evaluation of general demand levels in the economy.

Some of the agencies of the USDA that publish information more specific to grain marketing are the Agricultural Marketing Service, ASCS, the Economic Research Service (ERS), National Agricultural Statistics Service (NASS), the World Agricultural Outlook Board, and the Foreign Agricultural Service.

Timeliness and depth in reporting market information have created two basic types of information for market agents. An example of the first type is the *Market News Service* from the USDA's Agricultural Marketing Service. It provides daily cash prices that can be used in making immediate marketing decisions. The mass media and electronic vendors report these daily prices, but because of time and space restrictions, detailed background information and analysis are limited.[9]

An example of the second type of information is available from ERS and NASS. It comes in two forms: one is print, and the other is electronic. Table 9.1 illustrates selected periodicals and issues per year that are available to the public in print form. Table 9.2, listing printed form reports, shows the month for release of figures by NASS. Many of the printed reports listed in Tables 9.1 and 9.2 are also available electronically for immediate downloading from a host computer when the reports are released. Martin Marietta Information Systems is the private vendor for USDA's Computerized Information Delivery System (CIDS) with the vendor providing assembly and telecommunication services to users. Details on the specific date and hour of release are available from the source given in the footnote to Table 9.2. The purpose in scheduling specific release times is to allow all market agents equal access to the information at exactly the same time.

In addition to the print form, a number of electronic products are now being produced by ERS and made available on microcomputer disks or magnetic tapes. The scope of the data is similar to that contained in the printed reports listed in Tables 9.1 and 9.2. An example of an electronic product is *Feed Grain Data by States, 1949-86*, which gives data on acreage planted and harvested, yield, production, quantity used on farms and sold, season average price, value of production, and value of sales for corn, grain sorghum, oats, and barley.[10] As the title suggests, the feed grain data are available at the state level on an annual basis for the years 1949-1986. This product gives a historical perspective with much depth, but it is not timely for short-run decisions. These data are stored on four 5.25-inch disks and can be accessed using Lotus 1-2-3 software.

TABLE 9.1   Selected Periodicals from the Economic Research Service and World Agricultural Outlook Board

Periodical	Issues per Year
Agricultural Outlook	11
Food Review	4
Economic Indicators of the Farm Sector	5
Foreign Agricultural Trade of the United States	6
Rural Conditions and Trends	4
Situation and Outlook Reports:	
Agricultural Income and Finance	4
Agricultural Resources	5
Agriculture and Trade	5
Dairy	5
Feed	4
Livestock and Poultry	6
Livestock and Poultry Update	12
Oil Crops	4
Outlook for U.S. Agricultural Exports	4
Rice	3
U.S. Agricultural Trade Update	12
Wheat	4
World Agriculture	4
World Agricultural Supply and Demand Estimates	12

SOURCE: *Reports*, ERS-NASS, USDA, Washington, D.C., Summer 1991.

TABLE 9.2 1991 Schedule of Releases from National Agricultural Statistical Service for Selected Commodities

Report	Jan	Feb	Mar	Apr	May	Jun	Jul	Aug	Sep	Oct	Nov	Dec
Field Crops												
Acreage						X						
Crop Production	X	X	X	X	X	X	X	X	X	X	X	X
Crop Production-Annual	X											
Crop Values	X											
Grain Stocks	X		X			X			X			
Prospective Plantings			X									
Rice Stocks			X					X		X		X
Winter Wheat and Rye Seedings	X											
Livestock and Products												
Cattle		X					X					
Cattle on Feed	X	X	X	X	X	X	X	X	X	X	X	X
Cold Storage	X	X	X	X	X	X	X	X	X	X	X	X
Hogs and Pigs			X			X			X			X
Livestock Slaughter	X	X	X	X	X	X	X	X	X	X	X	X
Livestock Slaughter-Annual			X									
Meat Animals-Production, Disposition and Income				X								
Sheep and Goats		X										
Wool and Mohair			X									

Milk and Dairy Products												
Dairy Products	X	X	X	X	X	X	X	X	X	X	X	X
Dairy Products-Annual	X	X		X	X	X	X	X	X	X	X	X
Milk Production	X			X	X	X	X	X	X	X	X	X
Milk Production, Disposition and Income				X								
Poultry and Eggs												
Eggs, Chickens and Turkeys	X	X	X	X	X	X	X	X	X	X	X	X
Egg Products	X	X	X	X	X	X	X	X	X	X	X	X
Hatchery-Annual		X										
Layers and Egg Production-Annual	X											
Poultry Production and Value			X									
Poultry Slaughter	X	X	X	X	X	X	X	X	X	X	X	X
Turkeys	X											
Turkey Hatchery	X	X	X	X	X	X	X	X	X	X	X	X
Prices												
Agricultural Prices	X	X	X	X	X	X	X	X	X	X	X	X
Agricultural Prices-Annual	X			X								X

SOURCE: *1991 Calendar of Reports*, ERS-NASS, USDA, Rockville, MD, 1991.

The reports listed in Tables 9.1 and 9.2 and the electronic products provide current and historical supply-and-demand information as well as some predictions about future supply, demand, and price variables. This information assists the market agent in understanding past relationships of the variables and in forming expectations about the future based on past behavior. Thus, the first type of information, which is very timely but contains little depth or relationship to past conditions, is useful in making immediate decisions. The second type of information is not so timely but is comprehensive and relates to past conditions. It helps the agent in making medium- and long-term decisions.

A long-standing public source of information and analysis is that provided by the extension marketing specialists located at many of the land grant universities. Normally, they are faculty members in the Department of Agricultural Economics, College of Agriculture. In the major grain- and livestock- producing states, it is common to have at least one person specializing in each commodity area. The extension specialists continually assess events that affect the market and incorporate new facts and information into their analysis. Their output is distributed in many forms including situation and outlook reports, verbal commentaries over radio and television, telephone networks linking county extension offices, and various newspapers and magazines.

Private Sources

Private agencies also provide market information. In many respects, their operations lie between the extremes of direct contact and governmental publications. For an annual or monthly subscription fee, an individual may receive a variety of informational aids. Wire services have largely been replaced by electronic screens as a source of news and market information about the major grain markets. In addition, information is transmitted on weather conditions, government programs, and other national and international forces that will affect the cash and futures markets.[11]   A major advantage of instantaneous electronic services is the speed with which market information is collected and transmitted to the potential user.   Speed in delivery is extremely important because market news is a perishable, quickly obsolete product. Commodity News Service, Grain Instant News, Reuters' Economic Service, and the Data Transmission Network (DTN) are just a few of the private firms that provide electronic information services.

Subscribers to daily or weekly technical market reports receive an analysis of the forces that determine the major trading patterns for that day, week, or month. Although such reports are not as frequent as electronic reports, they have the advantage of providing the user with data on factual market happenings; they also explain and assist in analyzing the impact of various market factors. Examples of this type of private information service are Sparks Commodities and Grain Services Corporation.

Other private sources of market information are provided by brokerage and investment firms. They primarily issue market information on the futures prices of select commodities. Usually these reports provide a summary of major market trends and a brief analysis of factors likely to affect the market in the near future. *Futures Research In-Depth Report* on a specific commodity and *Futures Research Weekly Report* on selected futures markets, both from Shearson Lehman Brothers, are examples of such market information reports issued by brokerage firms.

The *Kansas City Grain Market Review* is an example of a daily publication of market information provided by a grain exchange, in this case by the Kansas City Board of Trade. The *Minneapolis Grain Prices and Receipts* and the *Weekly Commodity Review* are similar publications made available by the Minneapolis Grain Exchange and the Chicago Board of Trade, respectively. Such information sources frequently cover cash or futures trading prices at the close of the market, a statement of the volumes transacted, and a record of the receipts, shipments, and stocks at other primary markets.

One additional source of market information is the commercial market advisory service. This source provides the user with more specialized and detailed assistance in analyzing market forces than other private sources. In addition to a technical interpretation of daily price charts for commodities at central markets, the subscriber to this service may receive specific charting services and daily access to a toll-free telephone giving instant market reports for grains. *Helming Group Reports* and *Doane Market Watch* are just two examples of commercial advisory services.

Summary

One of the keys to the control and allocation of physical resources and the reaping of pure economic profit is the control of information. Patent laws protect the rights to *technical information* for a limited period so that the inventor may receive a return on the initial investment. However, the rights of the investor in *market information* are unprotected. The investor must limit dissemination of the product to obtain a return on a relatively high fixed investment.

In many respects, market information possesses qualities similar to other public or collective goods. As with most collective goods, the individual optimizer cannot take into account many of the positive externalities gained from market information. Consequently, the competitive market system will not allocate adequate resources to the production of information as the broader collective interests of society would prefer. For this reason, publicly supported information-generating agencies are engaged in the producing and disseminating information at subsidized costs to buyer and seller alike.

Public agencies provide market information that is released to all interested market agents at the same time. This practice helps to equalize the access to some market information and, therefore, to equalize market power.

As grain markets have become more widespread and specialized for particular grading characteristics, the need for more timely and detailed information has grown. Traditional cash market reports have become more obsolete as an ever-larger share of the transactions are made through direct sales arrangements or forward contracts. Such transactions are more widely dispersed and more individually negotiated. Therefore, the burden to provide adequate, accurate, and timely market information to a broad spectrum of market agents has become more difficult.

Notes

1. W. J. Manley, "Adapting to Change," paper presented to the 1977 National Marketing Service Workshop, Kansas City, MO, March 1977.

2. D. M. Porta, *The Information Economy*, Institute for Communication Research, Stanford University, Stanford, CA, 1975.

3. Steven Salop, "Information and Monopolistic Competition," *American Economic Review*, Vol. 66, No. 2, May 1976.

4. C. H. Reimenschnieder, "Economic Structure, Price Discovering Mechanisms and the Informational Content and Nature of USDA Prices," Agricultural Economics Staff Paper No. 77-19, Michigan State University, East Lansing, 1977.

5. Robert Wilson, "Information Economies of Scale," *The Bell Journal of Economics*, Vol. 6, No. 1, Spring 1975.

6. Stanford Grossman, and Joseph Stiglitz, "Information and Competitive Price Systems," *American Economic Review*, Vol. 66, No. 2, May 1976.

7. The "free rider" problem associated with other collective goods is also evident in the market information industry.

8. William Robbins, "Golden Grain: The Russian Wheat Deals," *The American Food Scandal*, William Morrow & Company, Inc., New York, 1974.

9. Allen Wellman, *Sources of Grain Market Information*, Nebraska Guides G82-622, Cooperative Extension Service, University of Nebraska-Lincoln, January 1989.

10. *ERS Electronic Products*, ERS-NASS, USDA, Rockville, MD, January 1991.

11. Ian T. Steward, *Information in the Cereals Market*, Hutchinson and Co., Ltd., London, 1970.

Selected References

Alchian, Armen A., "Information Costs, Pricing and Resource Employment," *Western Economic Journal,* Vol. 7, No. 2, June 1969.

Grossman, Sanford, and Joespeh Stiglitz, "Information and Competitive Price Systems," *American Economic Review,* Vol. 66, No. 2, May 1976.

Hayame, Yujwo, and Willis Peterson, "Social Returns to Public Information Services: Statistical Reporting of U.S. Farm Commodities," *American Economic Review,* Vol. 61, No. 1, March 1972.

Heid, W., Jr., F. Niernberger, and L. D. Schnake, *An Overview of the U.S. Wheat Industry with Emphasis on Decision Making and Information Sources,* USDA, Commodity Economics Division, Economic Research Service, Washington, DC, May 1977.

Manley, W. J., "Adapting to Change," paper presented to the 1977 National Marketing Service Workshop, Kansas City, MO., March 1977.

Porta, D. M., *The Information Economy,* Institute for Communication Research, Stanford University, Stanford, CA, 1975.

Reimenschneider, C. H., "Economic Structure, Pricing Discovering Mechanisms and the Informational Content and Nature of USDA Prices," Agricultural Economics Staff Paper No. 77-19, Michigan State University, East Lansing, 1977.

Rothschild, Michael, "Models of Market Organization with Imperfect Information: A Survey," *Journal of Political Economy,* Vol. 81, No. 6, December 1973.

Salop, Steven, "Information and Monopolistic Competition," *American Economic Review,* Vol. 66, No. 2, May 1976.

Steward, Ian M. T., *Information in the Cereals Market.* Hutchinson and Co., Ltd., London, 1970.

Theil, Henri, *Economics and Information Theory.* Amsterdam: North Holland Publishing Company, 1967.

Wilson, Robert, "Informational Economies of Scale," *The Bell Journal of Economics,* Vol. 6, No. 1, Spring 1975.

10

World Grain Trade

Gail L. Cramer

International trade is vital to U.S. agriculture. Currently, the United States exports the production from one out of every three and one-half acres of cropland. These exports generate about 25 percent of U.S. farm receipts. The U.S. proportion of world agricultural exports has varied from 12 percent in the early 1950s to 17 percent in the early 1980s to 14 percent in the late 1980s. Much of this increase and later decrease in the U.S. share of the world market has occurred in grains. U.S. exports of all agricultural products increased from $7 billion in 1970 to $44 billion in 1981 and then shifted back to $40 billion in 1990.

Grains, feeds, oilseeds, and oilseed products account for 54 percent of the value of U.S. agricultural exports (Table 10.1). Exports of grain

TABLE 10.1   Value of U.S. Agricultural Exports, by Principal Commodity Groups (Million dollars)

Commodity Group	1970[a]	1975[a]	1980[a]	1981[a]	1986[a]	1990[b]
Grains and feeds	2,531	11,561	17,168	20,310	9,476	15,697
Oilseeds and products	1,885	4,753	9,811	9,305	6,266	6,099
Livestock and products	829	1,666	3,771	4,115	4,367	6,587
Cotton (including linters)	328	1,055	3,016	2,230	692	2,719
Fruits, nuts, and vegetables	632	1,373	3,464	4,445	2,914	5,196
Tobacco (unmanufactured)	537	897	1,349	1,339	1,318	1,359
Other	215	549	1,902	2,044	1,279	2,564
Total exports	6,957	21,854	40,481	43,788	26,312	40,203

[a]Years beginning October 1.
[b]Preliminary.

SOURCE: *Agricultural Statistics*, 1990, U.S. Department of Agriculture.

and feeds increased considerably during the world food crises from 1972 to 1975 and from 1979 to 1981.

Some U.S. agricultural industries depend more heavily on the export market than do others. In 1990, the United States exported 44 percent of domestic soybean production excluding meal and oil, 46 percent of the rice crop, 39 percent of the wheat, including flour, 38 percent of the grain sorghum, and 22 percent of the corn produced plus most of the corn gluten by-product. From these figures, one can see how important the international market is to the grain and oilseed producers in the United States.

The Benefits from Trade

Basically, nations trade for the same reason that individuals trade—because of the benefits that can be obtained from so doing. Gains to nations include a larger total output of goods and services that is possible from specialization and division of labor, which increases total national productivity. A larger output, however, is beneficial only if some of this increased output can be traded for other producers' goods and services that are more highly valued than the additional goods produced domestically. Through the process of mutual and voluntary trade, individuals and nations increase their level of living. Voluntary trade will occur only if one or more parties to the trade are made better off without making another party worse off. In general, trade allows an individual to obtain more goods and services for a given money income than could be obtained without trade.

The direct benefits from international trade are evident from an analysis of the actions of importers and exporters. Importers, to maximize their profits, purchase goods from other countries if they can be obtained at a lower price than comparable domestically produced goods. Exporters, on the other hand, sell goods to other countries if they can obtain a higher price than they could get in the domestic market. Therefore, the *relative* prices of goods among countries determine what goods are imported and exported.

To more clearly demonstrate the benefits of trade, a simple example is used. The example assumes two countries, the United States and the former Soviet Union, each with a set of resources capable of producing two commodities—wheat and crude oil, are in a competitive environment. Although this illustration involves only two nations and two commodities, the results of this analysis readily extend to all countries and all commodities.

The amount of wheat and oil that each country can produce is shown by its "production possibilities" (or product transformation) curve,

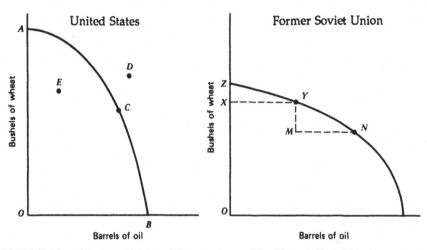

FIGURE 10.1 Production possibilities curves of the United States and the former Soviet Union.

Figure 10.1. The production possibilities curve for the United States (labeled *AB*) shows that, given all the resources available in the United States and its current technology, it can either produce *OA* bushels of wheat and no oil, *OB* barrels of oil and no wheat, or any combination of wheat or oil on or within the boundary of the production possibilities curve, such as point *C* in the diagram. The United States would not willingly produce at some point inside its production possibilities curve, such as at point *E*, because that represents a production combination with unemployed resources; neither can it produce at a point *D* because that combination of products lies outside of the capabilities of its resources. The former Soviet Union faces similar resource and technological limits to its wheat and oil production choices.

It may be noted that the production possibilities curve is not a straight line but a curve that is concave to the origin. A concave production possibilities curve illustrates diminishing marginal resource productivity (i.e., increasing opportunity costs). Resources cannot be shifted from wheat production to oil at a constant rate, only at increasing costs. The cost of an additional barrel of oil, in terms of wheat sacrificed, increases as additional oil is produced; more and more wheat has to be given up to get additional barrels of oil. Some resources are better suited to producing wheat than oil, and vice versa. As more and more resources are shifted from wheat to oil production (or from oil to wheat production) their productivity declines, thus the increasing cost shape of the production possibilities curve.

The diagram for the former Soviet Union in Figure 10.1 shows that it must sacrifice ZX bushels of wheat for XY barrels of oil if it has been producing OZ of wheat and no oil and wishes to increase oil output by XY. Further, if it is now producing OX wheat and XY oil (at point Y), an addition to its oil output of MN will require a sacrifice of YM wheat. The shift from point Y on its production possibilities curve to point N requires a much greater sacrifice of wheat than would the shift from point Z to point Y (given that MN = XY). The slope of the production possibilities curve demonstrates the *marginal rate of transformation* (MRT = ΔW/ΔO). The MRT shows how much the output of wheat must be decreased to increase oil output by one barrel.

Equilibrium Before Trade

A competitive economy that is not engaged in international trade will be in equilibrium when it produces a combination of products such that the marginal rate of transformation (ΔW/ΔO) equals the domestic price ratio for oil and wheat (P_o/P_w), shown by the line BT at point A in Figure 10.2. At point A, this country is producing the equilibrium quantities OW bushels of wheat and OC barrels of oil, because the cost of producing another unit of each good is just equal to the values of those goods to the

FIGURE 10.2 Equilibrium in a closed economy.

consumers. If the market price of crude oil increases relative to the price of wheat, as shown by the line *B'T'*, the optimal output of oil would increase to *OC'* and the optimal production of wheat would fall to *OW'*.

Before trade opens between the United States and the former Soviet Union, the United States is in equilibrium at point *A*, the former Soviet Union is in equilibrium at point *A'* (Figure 10.3). The domestic exchange ratio in the United States is two bushels of wheat for one barrel of oil, whereas the domestic exchange rate in the former Soviet Union is one bushel of wheat to two barrels of oil. In the United States, the opportunity cost of one bushel of wheat is one-half barrel of oil; in the former Soviet Union, wheat and oil exchange at one bushel for two barrels or a 1:2 ratio. Thus, the real cost of wheat is lower in the United States than in the former Soviet Union because the amount of oil sacrificed for a bushel of wheat is less in the United States than it is in the former Soviet Union. Hence, the United States has a "comparative advantage" (its greatest relative advantage) in wheat production. On the other hand, the former Soviet Union has a lower opportunity cost in oil. Its opportunity cost of one barrel of oil is only one-half bushel of wheat, as compared with two bushels of wheat in the United States; thus, the former Soviet Union has a comparative advantage in the production of

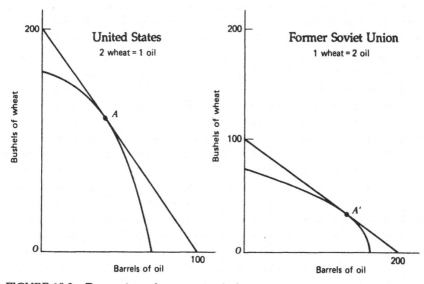

FIGURE 10.3 Domestic exchange ratios before trade between nations.

oil. The benefits of these advantages cannot be realized, however, until trade is permitted between these two nations.

Equilibrium with Trade

It is the price (or exchange) ratio that determines the comparative advantage for nations, and the goods that countries will produce and trade. Given free trade, the United States will emphasize the production and export of wheat to the former Soviet Union; the latter will produce and export its oil to the United States. Thus, it is *comparative* advantage, not *absolute* advantage, that determines what products enter international trade.

From their equilibrium positions before trade (Figure 10.3), at points *A* and *A'*, respectively, it may now be assumed that the United States and the former Soviet Union permit trade of wheat and oil between these two nations at market-determined prices and quantities. When these countries open trade, they each will specialize at the newly determined price ratio for wheat and oil, which is the same in both countries (the slopes of *AT* and *AT'* are the same; see Figure 10.4). The new equilibrium output in the United States will be at point *B*, producing more wheat and less oil than would be the case without trade. The former Soviet Union's new equilibrium will be at point *B'*, with a greater oil output and a reduced level of wheat production.

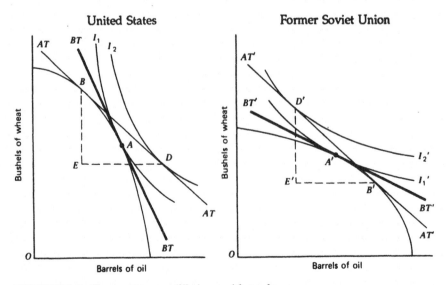

FIGURE 10.4 Two-nation equilibrium with trade.

Note that after trade opens, a country can consume anywhere along its price ratio line. The equilibrium point of consuming the two commodities depends on each country's intensity of demand and its ability to exchange specific quantities of the commodities at the AT ($=AT'$) ratio of prices. Equilibrium points of consumption, with trade, are D and D'; the United States produces combination B and now is able to consume combination D, while the former Soviet Union will produce at combination B' and consume at D'. The United States reaches a higher level of satisfaction (indifference curve I_2) through trade by exporting BE bushels of wheat to the former Soviet Union and importing ED barrels of oil from there. The former Soviet Union also reaches a higher indifference curve (I_2) by importing $E'D'$ ($=BE$) bushels of wheat from the United States and exporting $B'E'$ ($=ED$) barrels of oil to the United States. Consumers in both nations are better off with trade because they are able to consume wheat and oil in quantities that are beyond their individual resource base capabilities on a higher indifference curve than is possible without specialization and trade.

One can observe many examples of comparative advantage. For instance, a medical doctor may be able to type much more rapidly and accurately than the receptionist. The doctor's opportunity cost of typing is high, however, because of the greater medical practice earnings that would be sacrificed while typing. In spite of the fact that the doctor has an absolute advantage both in practicing medicine and typing, he or she has a comparative advantage only in practicing medicine; the doctor is unable to compete with the receptionist's comparative (or relative) advantage in typing.

When countries promote and engage in free trade, and each country also specializes in producing those goods and services in which it has a comparative advantage, the world's resources are used more efficiently and the world's consumers are able to share in a larger total world output, increasing their well-being. Any restraint on the volume of international trade will reduce the world's output of goods and services and cause an inefficient use of resources.

Trade Barriers

Many countries establish barriers to trade in the form of tariffs, quotas, or nontariff barriers. *Tariffs* are taxes levied on a commodity when it crosses a nation's boundary. Tariffs may be fixed per unit imported or *ad valorem* (percent of value imported). *Quotas* restrict the absolute quantity of a good that may be imported. *Nontariff barriers* are government regulations that reduce the free flow of goods in international trade. Examples of nontariff barriers are sanitary and phytosanitary regulations on imported foods. The economic effect of all such actions is to reduce the volume of trade and to increase the price of the product to domestic consumers.

As an example, a small country is assumed to be importing U.S. corn, a situation depicted in Figure 10.5. That nation, with domestic demand and supply curves labeled D_d and S_d, is in equilibrium when it produces and consumes quantity q_0 at market price P_0. With trade, this country is able to purchase all the corn it wants at the international price of P_1. At price P_1, domestic corn producers will provide q_2 bushels of corn, and q_1q_2 corn will be imported from the U.S. Total corn consumption in that country is q_1.

Now it is assumed that a tariff of x dollars per bushel of corn is imposed by that nation. The domestic price of corn will rise to $P_1 + x$. Domestic production increases from q_2 to q_4 but domestic consumption decreases from q_1 to q_3. Imports of U.S. corn drop from (q_1 to q_2) to (q_3 to q_4). With tariffs, consumers pay higher prices for their goods and services and consume less of those goods than if free trade is permitted.

As shown in the diagram, domestic producers gain from a tariff because with a tariff they are able to produce more at a higher market price than they could without the tariff. It is producer groups that pressure for tariffs because of the benefits they see for themselves, but it is the consumers that lose when tariffs are applied. Other compensatory methods, such as direct payments, are more efficient in supporting farm incomes, if that is their goal.

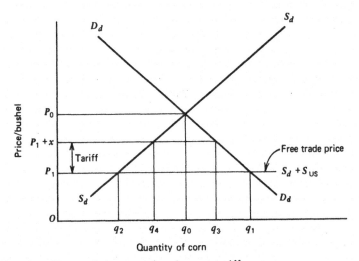

FIGURE 10.5   Domestic impact of an import tariff.

Most countries implement trade barriers with the intent to protect the incomes of their producers, but their use reduces worldwide production efficiency. Export policies, such as the variable levy system in the EC, and the quotas and tariffs of Japan, reduce U.S. exports of agricultural commodities and depress their prices in the United States.

Policy Issues

GATT and Trade Liberalization

As shown in the theoretical portion of this chapter, there are substantial benefits to free trade. If individuals were allowed to follow their comparative advantage, their welfare would be increased by exporting goods they can produce relatively cheaply and importing those they can produce at relatively higher costs. On the other hand, economic distortions such as government farm programs can pervert prices and resource allocations, thus reducing a nation's income and consumer welfare.[1]

The Uruguay Round of the General Agreement on Tariffs and Trade (GATT) was initiated in September 1986 to improve the world markets for agricultural products. The emphasis on trade liberalization was a result of expanding production caused by government assistance programs and technological advances in developed countries. Growth in production was faster than the increase in demand for farm products, primarily due to the world recession in the early 1980s. As a result, exporting countries such as the United States and the EC were left with large commodity stocks and rising costs of government farm programs. Tensions developed between the United States and the EC as the EC reverted from a net grain importer to a net grain exporter using high internal price supports and liberal export subsidies. The United States countered the EC-subsidized grain exports with the Export Enhancement Program. This program targeted export subsidies to specific countries to compete with EC subsidies. This basic subsidy conflict, which has been fueled by other trade barriers, forced the entire agricultural trade issue to the forefront of multilateral trade negotiations.

At GATT the United States and the Cairns Group (14 agricultural exporters) have proposed reforms that would eliminate domestic farm policies and trade policies that distort international agricultural markets. The U.S. proposals have included tariffication to replace nontariff barriers with tariffs to be phased out over a 10-year period, elimination of export subsidies over a 5-year period, and reduction of internal policy support. These free-trade proposals, however, have been unacceptable to the European Community, Japan, and Nordic countries (Norway, Sweden, Iceland, and Finland). The EC has proposed lower farm-support levels with the continuation of the Common Agricultural Policy. The EC has also supported a reduction in trade barriers on a reciprocal basis. In contrast, Japan's emphasis is on food security and the need to provide a certain level of food security. Finally, the Nordic group wants a gradual change in the level and form of border protection. Differences in positions on free trade are large, so there is no wonder that trade liberalization will be very slow in coming. Such liberalization may require several years rather than months of negotiations.

The USDA has estimated the impact of the Uruguay Round Agreement on agricultural trade liberalization, assuming a 1986-1987 base and only the industrialized market economies liberalized. World agricultural prices would increase about 10 to 20 percent, and real output would increase about 10 percent or $35 billion. The major beneficiaries would be the EC with $14 billion in annual gains, Japan with $6 billion, and the United States with about $9 billion. The EC and Japan benefits are derived primarily from lower consumer prices, while the U.S. gains are from reduced government support payments.

U.S. farm output would drop because of reduced government support prices although market prices would increase. Compared to current target prices, producer prices would decrease 44 percent for wheat, 33 percent for coarse grains, 59 percent for rice, and 7 percent for oilseeds. Wheat production could drop 6 percent, coarse grains 4 percent, and rice 11 percent. Oilseed production would increase 2 percent. U.S. wheat trade volume would decrease 3 mmt, coarse grains 8.3 mmt, and rice 0.4 mmt; soybeans would increase 1.7 mmt. Rising world prices from trade liberalization would increase net export trade values for U.S. wheat by $7 billion, coarse grains by $3 billion, oilseed by $4 billion, and rice by zero. These trade value increases are the result of world wheat prices rising 37 percent, coarse grains 26 percent, rice 26 percent, and oilseeds 6 percent.

The United States would improve its agricultural trade balance by about $3 billion. This improvement would result from decreases in beef import costs and increases in grain export revenues. The EC and Japanese trade balances would worsen by $9 billion and $7 billion, respectively.

The welfare implications for the world would be positive. The producers in the United States, the EC, other Western European countries, and Japan would be the big losers. Consumers would gain in the EC and Japan, while consumers would lose in centrally planned economies and the developing importing countries. The taxpayer gains would be large in the United States, the EC, other Western European countries, and Japan. However, if trade liberalization occurs, it is highly unlikely taxpayers would gain because producer losses would probably be compensated for by decoupled payments that would presumably not distort production and trade. Many of the U.S. farm programs currently do not distort production (because of set-asides) and international trade, and therefore would continue to exist under GATT proposals.

Bilateral Agreements

For many years, the USDA has used a variety of domestic agricultural policies in attempting to stabilize the production and prices of grains. With unstable foreign demand causing domestic and world prices to fluctuate, especially since 1972, the United States and importing countries have intensified their efforts to moderate the effects of variations in world production and demand.

In the past the United States has signed long-term, bilateral agreements with several countries, including the former Soviet Union, the People's Republic of China, Poland, and Mexico. By virtue of their minimum sales guarantees, such agreements temper somewhat the effects of variations in foreign demand, and aid the USDA in achieving domestic policy objectives. Importers are also assured of a more stable source of supply to meet their grain needs. At present, these bilateral agreements account for only a small proportion of total production in grain-exporting countries. In normal circumstances, bilateral agreements have a relatively small impact on U.S. producers. Some economists are concerned, however, about their effects when U.S. supplies are tight or free reserves are low. During such periods, prices could be quite unstable, causing even more price variability than without those agreements.

The United States signed its first long-term grain supply agreement with the government of the Soviet Union in October 1975. The reason for the agreement was to stabilize the price effects of the Soviet Union's unexpectedly large purchases of U.S. grain and to encourage them to hold larger grain reserves. The first agreement was for a five-year period, and required the Soviet Union to import 6 million metric tons (mmt) of wheat and corn per year. Up to 8 mmt of grain could be purchased under that agreement, but quantities greater than 8 mmt were dependent upon U.S. grain supplies and required prior consultation with the U.S. government as well.

Despite the perceived advantages of long-term agreements, the contract with the Soviet Union has been caught up in broader U.S. foreign-policy changes. Upon expiration of the original agreement, the content and length of agreements have depended upon the political relationships among the countries. Now that the Soviet Union has divided itself into a group of independent states with relatively open markets, the market system should be able to handle their needs in the world market system.

Cartels

Wheat producers in Canada and the United States have suggested the formation of a wheat cartel composed of the United States, Canada, Australia, and Argentina. Each country in the cartel would remain as a separate business organization, but these countries would jointly make decisions regarding marketings and prices of their grains.

The cartel would be composed of independent nations, so each would have its own policies and goals, and internal conflicts between cartel members could be expected. Presently, the marketing strategies of these countries differ widely. Private grain traders in the United States adopt policies to maximize their profits. The Canadian and Australian Wheat Boards, as government and producer marketing boards, respectively, seek to maximize consumer and producer gains. Australia, however, follows a strategy of selling most of its grain crop each year because of its limited storage capacity. Argentina's objective is to maximize foreign exchange earnings.[2]

Besides using different marketing approaches, each country has a different marginal cost of producing and marketing wheat. Consequently, at some point, one or more of the cartel members may find it advantageous to market individually rather than staying with the cartel. In addition, if the cartel succeeds in increasing the world price of wheat, it must control supply within each member country to maintain that price. At the new, higher price, domestic and international consumption would fall, countries not in the cartel would increase their production of wheat, and other countries would intensify their efforts to develop substitute products. All such actions and efforts would weaken the cartel and threaten its effectiveness.

The fundamentals of forming a cartel were tried formerly in several International Wheat Agreements and finally in an International Grains Arrangement. Member countries failed to comply with the intent of these agreements, and further cordination agreements have not been attempted since the late 1960s.

Embargoes

Most nations, at one time or another, and for a variety of reasons, have embargoed the movement of goods and services to or from another country. On numerous past occasions, the United States has also utilized the embargo to achieve a specific foreign-policy objective.

As a result of the massive U.S. grain and oilseed sales in 1972, the need for a better monitoring and export-reporting system was recognized in order that domestic supplies of these commodities might be better maintained. On June 27, 1973, an embargo was placed on new sales of soybeans, soybean meal, and cottonseed, and previously committed sales of these commodities were reduced on a pro rata basis. On October 4, 1974, the administration temporarily suspended the sale of 3.4 mmt of grain to the Soviet Union.

Another partial grain embargo was placed on the Soviet Union on January 4, 1980, on all sales in excess of 8 mmt. This embargo was in response to the invasion of Afghanistan by the Soviet Union. This suspension resulted from broader U.S. foreign policy, rather than from inadequate supplies at home. The embargo lasted 18 months, but its impact on the Soviet Union was small; it forced the Soviet Union to reduce its grain purchases only slightly. The Soviet Union paid somewhat higher prices for grain and it reduced domestic meat supplies a bit, but it also developed new grain-trading partners for these import supplies as a result of U.S. actions. The effect of the embargo within the United States was reduced corn prices through the following winter. The impact was partially offset by CCC grain purchases and changes in the farmer-held grain reserve program.

Exporters operating in the United States are required by law to report large export sales on designated commodities to the Foreign Agricultural Service (FAS), a division of the USDA, on a daily and weekly basis. The commodities reported on a daily basis include wheat (by class), barley, corn, grain sorghum, oats, soybeans, soybean cake and meal, and soybean oil. Large sales are defined as 100,000 metric tons or more of one commodity in one day to a single destination or 200,000 tons or more of one commodity during the weekly reporting period. (Lower limits for reporting soybean oil exports are 20,000 tons and 40,000 tons, respectively.) The daily reports are released the same day they are received by the FAS.

Weekly reports are required, regardless of transaction size, for all of the previously mentioned commodities as well as wheat products, rye, flaxseed, linseed oil, cottonseed, cottonseed cake and meal, cottonseed oil, rice (by class), and other nongrain agricultural items. Reports include information regarding sale quantity, type and class of commodity, marketing year of shipment, and the destination.

Reporting requirements were mandated in 1973 by the U.S. Congress due to the large, unexpected purchases made by the Soviet Union in 1972, which created an unexpected increase in U.S. food prices and depleted U.S. reserve stocks. The reporting provides up-to-date export information to the general public and to the U.S. government, enabling all players in the grain market to assess changes and make adjustments accordingly, thus creating a more efficient and stable system. Exporters are also required to file declarations on exports with U.S. Customs at the time of shipment.

Buffer Stocks

Buffer stocks of grain have been proposed to reduce the large fluctuations in grain prices and to provide emergency food aid. On an international basis, there is little agreement on which countries should contribute to the stocks and hold reserves. Under present conditions of plentiful supplies, importers do not want to pay the added cost of storage. Although some of the food-deficit countries have increased their stocks of food grain, producers are not advocating worldwide buffer stocks because producer prices would most likely be depressed by the existence of such reserves. The U.S. grain reserve program is designed to act as a buffer stock, increase price stability, and provide emergency food aid.

Balance of Payments

The U.S. balance of payments is a statement that shows all international transactions between residents of the United States and all other nations. This record includes the transactions of both private and government units for a given period, usually one year. This accounting record is useful in determining changes over time in the economic strength of the United States compared to other countries, and often is used in determining specific trade policies.

Any international trade transaction can be divided into either a *plus item* (credit) or a *negative item* (debit). Those transactions where U.S. consumers receive foreign currency (dollar inpayments) are *credit* items, whereas those transactions where foreigners receive U.S. dollars (dollar outpayments) are *debit* items. The United States has a deficit in its balance of payments when debits are greater than credits; a surplus exists when credits are greater than debits.

Under freely fluctuating exchange rates, the balance of payments should come close to balancing. If the United States runs a deficit in its balance of payments, its currency depreciates, which makes U.S. goods cheaper to foreigners and foreign goods more expensive to U.S. consumers. Therefore, U.S. exports should increase, and U.S. imports should decrease, with a resulting improvement in the U.S. balance of payments. The reverse would occur for a country that is running a surplus in its balance of payments.

Agricultural exports since 1971 have helped strengthen the dollar, even with large expenditures for oil from the Organization of the Petroleum Exporting Countries (OPEC). In 1990, agricultural exports exceeded imports of agricultural products by $15.6-billion. This surplus helped offset a $156.0 billion nonagricultural deficit. Even so, the United States continues to run large current account deficits, which have continued to put downward pressure on the dollar relative to other currencies. Therefore, it is important for those involved in public policy to be aware of the effects of restricting agricultural trade among nations. In fact, much more effort is needed to encourage freer international movement of grains and other agricultural products.

Trade in Food Grains and Feed Grains

Grains may be used for either food or feed. Food grains normally are considered to be wheat and rice. Feed grains (or coarse grains) include corn, barley, sorghum, oats, and rye. In the developed countries, feed grains are used primarily for livestock feeds, with lesser quantities utilized directly for human consumption, or indirectly in brewery or starch products. In many developing countries, however, coarse grains are used as a food grain, with only small amounts used in livestock feeds or for industrial purposes.

The amount of grain exported from a nation depends on many factors other than price. Some of the major factors are national agricultural and food policies and red meat production. Countries will export only that output in excess of domestic requirements plus carry-over. Exports are used to reduce domestic supplies to maintain prices and incomes for domestic producers.

Agricultural policies are particularly important in world trade, since almost all countries protect their domestic agriculture from open market forces. In fact, many countries explicitly plan the amounts of grains to be produced and to enter world trade.

Supply management is a major policy goal of many grain-producing countries. Supply management is used to expand domestic production to achieve self-sufficiency or to reduce reliance on foreign sources of supply. It may also be used to control domestic production so that supply is balanced with demand at prices that provide adequate incomes to producers.

Governments are sensitive to consumer prices of food grains, and some countries, therefore, control the wholesale and retail prices of grains. This control has been especially common in those centrally planned countries that subsidize grain production to keep consumer prices at a fairly stable level.

We take for granted the predominant role of the United States in world grain trade, but this nation has not always been in that position. As

TABLE 10.2   World Net Imports and Exports of Grain, Selected Periods (million
metric tons)

Region	1934 -1938	1960 -1963	1969 -1972	1972 -1973	1975 -1976	1980 -1981	1990 -1991
North America	+5	+43	+55	+91	+100	+137	+111
Latin America	+9	+1	+3	0	+1	-8	-2
Western Europe	-23	-26	-22	-18	-18	-7	+22
Eastern Europe and the former Soviet Union	+4	0	-3	-27	-33	-47	-36
Africa and the Middle East	+1	-4	-9	-9	-14	-27	-46
Asia	+2	-16	-28	-35	-36	-41	-62
Oceania (Australia and New Zealand)	+3	+7	+11	+6	+12	+14	+14

Net imports indicated by (–) and net exports by (+).

SOURCE: Economic Research Service, *World Agricultural Outlook and Situation*,
USDA, Washington, D.C., December 1990.

shown by Table 10.2, North America was not the leading exporter of
grain until after the 1930s. Most countries were exporting to Europe at
that time, with Latin America the leading exporter. More recently, the
United States, Canada, the EC, and Australia have dominated food and
feed grain exports.

The developing countries of the world have become large importers of
grains. Cereal grain imbalances occur occasionally because of drought or
other natural disasters. Imbalances in food grains are due primarily to
increases in population and incomes in Asia, Africa, the Middle East, and
Latin America. In Eastern Europe and the former Soviet Union, the
imbalances are in feed grains.

World Wheat Trade

Total world wheat trade amounted to 49.8 mmt in 1968-1969, increased
to 73.7 mmt in 1975-1976 and totaled 94.0 mmt in 1990-1991. The U.S.
share of total wheat trade was 30 percent (14.8 mmt) in 1968-1969, 43
percent (31.9 mmt) in 1975-1976, and 30 percent (28.5 mmt) in 1990-1991.
About 20 percent of total world wheat production enters international
trade.

The world's leading wheat exporters are the United States, Canada, EC-
12, Australia, and Argentina, accounting for 91 percent of all the wheat

that is exported. From 1988-1991, Canada exported 71 percent of its total wheat production, Australia—76 percent, the U.S.—56 percent, Argentina—47 percent, and the EC-12—25 percent. Over this period, the United States accounted for 35 percent of the world's wheat trade, Canada—18 percent, Australia—12 percent, Argentina—5 percent, and EC-12—22 percent.

Most of the U.S. wheat exports go to Brazil, Israel, Pakistan, Tunisia, China, Japan, South Korea, Philippines, Algeria, Italy, Morocco, and the former Soviet Union.

Canada's wheat markets are Belgium, Luxembourg, Italy, United Kingdom, the former Soviet Union, China, Japan, Algeria, Bangladesh, Indonesia, and Egypt. Australia exports grain to China, Indonesia, Japan, and Egypt. Argentina exports wheat to Iran, the former Soviet Union, Indonesia, and China. Most Western European countries are trading wheat among themselves, with some wheat being exported to African nations, Asia, and South America.

World Feed-Grain Trade

The major coarse-grain-producing countries are the United States, the former Soviet Union, China, France, Canada, India, Brazil, and Germany. World coarse grain production totaled 834 mmt in 1990-1991. It is estimated that about 67 percent of this output was used as livestock feed. About 85 percent of world coarse grain exports is for use as livestock feed. Historically, coarse grains were traded among developed countries, but recently many middle-income countries have increased their use of grain for feed.

Corn is the most important feed grain traded, accounting for 68 percent of all coarse grains exported in 1990-1991. Barley was second with 21 percent, then grain sorghum at 9 percent; small amounts of oats and rye are exported.

Wheat is classified as a food grain rather than a feed grain, but significant quantities of wheat are used for feed in the United States, the former Soviet Union, the EC, and Eastern European countries. Most of the wheat that is diverted to feed use in these countries is either denatured or of low quality.

World coarse grain exports increased more than two and one-half times from 1968-1969 to 1989-1990, increasing from 37 mmt to 100 mmt per year, but decreasing to 84.8 mmt in 1990-1991. Corn, barley, sorghum, and rye exports have almost tripled while oats trade has remained relatively stable.

The increased exports of feed grains are a result of rising world demand for meat and increasing per capita incomes around the world.

The growth of feed grain exports was also a result of poor weather conditions that caused poor harvests in much of the world, especially in the former Soviet Union. Other factors include the devaluation of the dollar, and better political relationships with the former Soviet Union and the People's Republic of China. The major exporters of coarse grains are the United States, Argentina, Canada, South Africa, Western Europe, Australia, Thailand, and China. The United States, Argentina, South Africa, Thailand, and China are the main exporters of corn. The large barley exporters are Canada, EC-12, the United States, and Australia. The United States and Argentina together supplied more than 90 percent of the world's grain sorghum exports in 1990-1991.

The largest importers of corn are Japan, the former Soviet Union, Mexico, EC-12, Eastern Europe, Taiwan, and South Korea. The former Soviet Union, Eastern Europe, Saudi Arabia, and Japan import most of the barley; Japan, Mexico, and Israel import most of the grain sorghum. The main exporters of oats are Argentina, Australia, Canada, Finland, and Sweden. The main importers of oats are the United States, EC-12, Switzerland, Japan, and the former Soviet Union. In fact, the United States has gone from being the largest producer of oats in the 1960s to being a net importer of oats in 1982-1983 and to being the largest import market for oats since 1984-1985. The main exporters of rye are Canada and EC-12, while the main importers are Japan, the former Soviet Union, and Western Europe.

Overall, the United States is the largest feed grain exporter, with over 60 percent of the world market in 1990-1991; other exporters have relatively small individual market shares. The major importers (Japan, EC-12, the former Soviet Union, and Eastern Europe) buy about 50 percent of all feed grain imports.

World Rice Trade

The world's major rice producers are China, India, Bangladesh, Indonesia, Thailand, Burma, and Japan, accounting for more than 80 percent of total world production. Most of these countries produce mainly for their own consumption, exporting less than 1 percent of their production.

Of the 13 mmt of rice exported in 1990-1991, 34 percent came from Thailand, 18 percent from the United States, 10 percent from Pakistan, 8 percent from Vietnam, and 8 percent from EC-12.

The major rice importers, with no one nation buying more than 1.1 mmt, are EC-12, Indonesia, Iran, Saudi Arabia, former Soviet Union, Brazil, South Africa, Senegal, Peru, Malaysia, and the Ivory Coast.

The United States shipped rice to about 110 nations in 1990, with the largest amounts going to Peru, Saudi Arabia, Mexico, Indonesia, Iraq, and Canada. Thailand's major markets are Indonesia, Malaysia, Singapore, Saudi Arabia, Senegal, and Hong Kong.

World Soybean Trade

Soybeans are the leading oilseed in world trade and are very important in the world oil and meal markets.[3] Soybean oil and meal are joint products from processing operations. Soybean oil is an edible oil used mainly in margarine, shortening, and cooking and salad oils; soybean meal is used primarily as a high-protein feed supplement.

The ten major oilseeds or oilseed substitutes are soybeans, cottonseed, sunflowerseed, peanuts, palm kernel, rapeseed (canola), sesame, copra, linseed, and castorseed. As of 1989-1990, 75 percent of oilseed exports were soybeans compared to 81 percent in 1979-1980. Sunflowerseed and rapeseed combined increased their share from 12 percent to 16 percent during the same period. The United States exported 49 percent of production from all 10 oilseeds in 1989-1990, a decrease from 74 percent in 1979-1980.

Ninety-six percent of the world's 1990 exports of soybeans, a total of 26 mmt, were obtained as follows: the United States (59 percent of all exports), Argentina (12 percent), Brazil (15 percent), Paraguay (5 percent), and China (5 percent). The major importers are EC-12 (51 percent), Japan (17 percent), Taiwan (9 percent), South Korea (4 percent), and Mexico (4 percent). The United States exports mainly to EC-12 (41 percent), Japan (23 percent), and Taiwan (13 percent).

The world's 17 sources of edible oils and fats are soybean, cottonseed, sunflowerseed, peanut, palm, palm kernel, olive, rapeseed, sesame, corn, coconut, linseed, castor, and fish oil, as well as butter, lard, and tallow and grease. Soybean oil exports totaled over 3.5 mmt in 1990. The major exporters are Argentina (27 percent), EC-12 (31 percent), Brazil (21 percent), and the United States (14 percent). The major importers are China (13 percent), Iran (13 percent), Pakistan (9 percent), and Bangladesh (8 percent).

Palm oil, which comes from the fruit of palm trees, is the dominant export oil. Palm and palm kernel oil combined have increased from 23 percent of exports in 1979-1980 to 36 percent in 1989-1990. The main exporters are Malaysia (70 percent), Indonesia (14 percent), and Singapore (8 percent). Rapeseed oil, which is better known by consumers as canola oil, has also increased its market share. Canola is a relatively new hybrid version of rapeseed (rapeseed oil has a high erucic acid content rendering it usable for industrial uses only by U.S. standards), which has proven to have a very well-balanced healthful oil that is low in saturated fat and cholesterol.[4] Exported sunflowerseed and rapeseed oil combined increased from 10 percent in 1979-1980 to 15 percent in 1989-1990. Animal oils and fats experienced the largest decrease, from 30 percent to 19 percent over the same 10-year period. Soybean oil's share of the export market has slightly decreased from 19 percent in 1979-1980 to 17 percent in 1989-1990. The U.S. exports of the 17 oils and fats decreased from 20 percent in 1979-1980 to 10 percent in 1989-1990. Malaysia had the greatest gain from 14 percent to 26 percent for the same period. The United States exported over 500,000 tons of soybean oil in 1990. Over half of its soybean oil is exported to Pakistan. The Pakistan market, however, is very dependent on PL-480 funding.

The major oil meals include soybean, cottonseed, peanut, sunflowerseed, rapeseed, linseed, copra, palm kernel, and fish meal. These meals differ in quality and quantity of protein, so they are not perfect substitutes for one another.[5] Between 1980 and 1990, soybean meal has remained stable at 58 percent of the export meal market while sunflowerseed and rapeseed meal combined have increased from 5 to 9 percent of the export market. The U.S. market share of the 12 oilmeals has decreased from 35 percent in 1979-1980 to 22 percent in 1989-1990. Argentina and Brazil combined experienced the greatest increase in market share, from 20 to 38 percent for the same period.

Soybean meal exports totaled over 25 mmt in 1990. The major exporters are Brazil (34 percent), Argentina (32 percent), the United States (18 percent), China (7 percent), and India (4 percent). The major importers are the EC-12 (52 percent), the former Soviet Union (11 percent), Eastern Europe (9 percent), Japan (3 percent), Philippines (2 percent), and Western Europe (excluding the EC-12) (2 percent). The United States exports mainly to the former Soviet Union (37 percent), Canada (12 percent), Algeria (8 percent), Venezuela (7 percent), and the EC-12 (6 percent).

The United States is very influential in the soybean market despite the market share declines it has been experiencing. It produces nearly 50 percent and consumes approximately 30 percent of total world soybean production in addition to providing a large share of export trade.

The Role of Food Aid

Under the Agricultural Trade Development and Assistance Act of 1954, as amended, also known as PL- 480 or the Food for Peace Program, the U.S. government administers concessional sales[6] and donations of agricultural commodities. The policy objectives of the 1990 revision of the act emphasize using the abundant U.S. agricultural productivity to promote U.S. foreign policy by enhancing the food security of the developing world.

The USDA states that the plans for achieving this objective are to:

- combat world hunger and malnutrition and their causes;
- promote broad-based, equitable, and sustainable development (including agriculture);
- expand international trade;
- develop and expand export markets for U.S. agricultural commodities;
- foster and encourage the development of private enterprise and democratic participation in developing countries.[7]

There currently are three titles to PL-480, as amended. Title I covers concessional sales and a market development program; Title II, donations and relief; and Title III, a government-to-government grant program.

Title I, administered by the USDA, provides for concessional sales made to foreign governments. The credit sales can be dollar or local currencies, with the maximum repayment term of 30 years with a 7-year grace period at low interest rates. Countries eligible are developing countries with a shortage of foreign exchange earnings and countries unable to meet all food needs through commercial channels.

Title II, administered by the Agency for International Development (AID), provides grants of U.S. agricultural commodities to governments, private voluntary organizations, cooperatives, and the World Food Program, mainly for humanitarian and economic development purposes.

Under Title III, also administered by AID, the United States provides government-to-government grants of food assistance. The recipient country may sell the food on the domestic market, but then must use that money for economic development purposes.

Seventy-seven nations benefited from a total of $1.24 billion in aid from U.S. government-sponsored aid programs in 1981. The value of U.S. farm products exported under government-financed programs amounted to slightly less than $1.5 billion in fiscal 1990. Almost 50 percent of the aid in 1990 was received by Egypt, Bangladesh, Pakistan, Poland, and India.

Structure of the Grain Export Market

The conditional requirements of a purely competitive market cannot be met if either governmental activities or regulations of the market prescribe decisions for individuals or firms, or if the market is dominated by a relatively few firms able to exert power in the market. The worldwide grain export system is an industry where there are many small producers but their products for export are handled by a few large exporters (private trading firms and national marketing boards) and by importers (private traders and state trading agencies) who also are few in number and might exert some market power. However, the variety of programs and policies of governments throughout the world probably have more influence on the process of pricing agricultural commodities than any of the firms and agencies in the industry.

When a government agency, influenced (or directed solely) by domestic policy objectives, is buying or selling grain in the international market, or when a very large multinational firm is negotiating with a buyer or seller (either an agency of government, or a private firm), chances are slight that the forces of supply and demand will be the only meaningful influence in price determination.[8]

As previously discussed, almost all countries enact internal agricultural or food policies to protect their farmers and consumers from world price fluctuations. Domestic farm policies in the United States, such as the set-aside, acreage reduction, and farmer-owned reserve programs reduce international grain supplies relative to demand and enhance prices in the short run.

EC Farm Policy

The EC has a high commodity price-support program that encourages production in the EC, discourages consumption, and reduces its imports from other nations. Its policy to subsidize exports of surplus commodities increases the supply in world trade, which reduces prices. The net impact of the EC variable levy system is an increase in the supplies of commodities produced within the EC, which exerts downward pressures on world prices outside the EC even though EC prices are maintained at high levels.

On the importing side of the international market, the EC variable levies and Japanese domestic pricing policies have reduced grain imports below what they would have been in those nations. Thus, the international market is one of imperfect competition, and it influences prices received by U.S. producers.[9]

Grain-trading Firms

As the United States is the world's largest exporter of feed grains, wheat, and oilseeds, millions of tons move through the marketing channels each year from U.S. farms to foreign buyers around the world. These commodities are exported through the private sector by trading firms.

Several large multinational trading firms, including U.S.- and foreign-owned corporations, account for the majority of U.S. exports of grain and oilseeds. The remainder is handled by a number of smaller firms and several grain cooperatives. These multinational firms are highly diversified businesses that maintain sales and procurement offices in a number of countries. They have access to or control of the many functions required in procuring, handling, selling, financing, and delivering commodities to buyers around the world. They compete in a world market—buying and selling commodities without considering country of origin.

Several large cooperatives handle grain and oilseeds for export. A large proportion of their commodities is delivered to a U.S. port, then sold to one of the international trading firms to be sold to foreign buyers. There are some cooperatives, however, that maintain commodity control from the point of origin to the port in the importing country.

The internal marketing channels of importing countries differ from country to country. The large international trading firms may sell U.S. grains to private trading firms, such as those in the EC, or to government agencies such as the Japanese Food Agency or the Grain Trading Agencies in various countries.

There are two principal ways of making an export sale. One is by private negotiation. Sales to the former Soviet Union and China as well as to most buyers in the EC are of this type. The other type of sale is by *tender*. Under a tender system, the potential buyer issues specifications as to kind of grain, class, grade, time of shipment, destination, and so forth. Normally, all potential sellers must submit sealed offers by a specific time at a specified place.

Once accepted, an export transaction creates a short cash position for the exporting firm. This firm, however, may be in a net long or net short overall position, depending on the management's assessment of future grain prices and its willingness and ability to assume uncertain price risks.

Much of the price risk in such forward selling is covered by purchasing call options or futures contracts. In hedging this grain, the firm is interested in the basis between its forward cash selling price and the price of futures. This basis includes the firm's margin as well as other cost factors. The call option gives the firm the right to buy the grain in the future at a specified price as in a long futures position.

Grain firms must be relatively secretive about their large grain sales as any informational leaks can drive up futures prices before they can purchase futures contracts to hedge the forward sale. An information leak could reduce the firm's basis and, hence, its margin on the sale. In closing this long hedge, the firm desires secrecy because any discount in cash purchasing prices versus futures prices will increase its margin and the profitability of the transaction.

A study conducted for the secretary of agriculture analyzed 29 large grain and soybean export sales during the years 1975-1978.[10] Exporters were found to be net long in futures, and in overall positions before and after export sales were reported, but there was little buying of futures during the week of the reported sale. Cash market purchases during the week of the sale exceeded futures purchases.

The futures market provides a reference price or index of value for grain, and it can be used to shift the price risk of grain ownership to speculators, if that is deemed appropriate. Such times would be when the firm's management thinks future cash prices may rise, or when there is considerable uncertainty over the direction of future price movements. Hedges are used for both short and long positions in cash grain.

U.S. futures markets are quite competitive,[11] given the theoretical norm for competitive markets. It is not necessary for some foreign buyers to contract with an exporter to trade in the futures market. It is possible for international buyers to purchase futures contracts, hold them for delivery, and then have the grain shipped to a foreign port. Or the buyers could purchase the cash grain at a port or inland terminal and have it shipped to the foreign destination. These actions on the part of buyers are rarely taken, but they restrict the potential for margin-widening actions by multinationals or state trading agencies and boards. Sellers, likewise, are not wholly at the mercy of international buyers—multinationals or state trading agencies—because some could simply sell futures contracts and deliver on that contract at its maturity. However, very few futures contracts are settled by making or taking commodity delivery.

Summary

International trade is important both to U.S. agriculture and the general U.S. economy as, with specialization and trade, more goods and services are obtained at a lower real cost than could be obtained without trade. Exports pay for imports, and make available many products that would otherwise be available only at a high cost, if at all.

The United States is most clearly at a comparative advantage in producing several agricultural commodities and is especially efficient in the production of grain and soybeans. These products account for more than one-half of the value of all U.S. agricultural exports.

Since the beginning of World War II, the United States has become the world's leading exporter of feed and food grains, and oilseeds and oilseed preparations, supplying more than 40 percent of the total of these commodities traded in the international markets. The United States provides 35 percent of all world wheat exports, over 60 percent of the coarse grain, 18 percent of the rice, and 59 percent of the soybean market. The developed countries rely on the United States primarily for feed grains, whereas the developing countries depend on the United States for much of their food grain supply.

Barriers to free trade include tariffs, quotas, and nontariff barriers. These devices are used to reduce a nation's volume of imports, for domestic, protective purposes, but they also reduce the gains from specialization and trade and thus reduce the real wealth of the world community.

Imperfect competition in markets is of concern to many people because of the market power that might be exercised by the participants. The international grain market is imperfectly competitive, with a marketing system so structured that the bulk of agricultural commodities is handled by state trading agencies and large multinational firms. With often huge quantities of grain involved in a single transaction, the market prices of those commodities may be influenced by the actions and decisions of the negotiating parties, rather than only by the forces of supply and demand. In spite of this market's structure, there is competitive behavior among private grain-exporting firms. Government-imposed trade barriers, export subsidies, and taxes as well as a variety of domestic agricultural programs more effectively restrain competition in the grain export industry than do presently existing market imperfections.

Notes

1. Alex McCalla and Andrew Schmitz, "The World Market for Wheat," in *How Prices Are Determined for Montana Wheat*, KEEP Public Affairs Forum at Great Falls, MT, September 1977, Montana State University, Bozeman, 1977.

2. This section on world trade in soybeans is based on Thomas Mielke, ed., *Oil World Statistics Update*, Hamburg: ISTA Mielke GmbH, 1991, and Thomas Mielke, ed., *Oil World Annual 1990 Advance Issue*, Hamburg: ISTA Mielke GmbH, 1990.

3. Fereidoo Shahidi, ed., *Canola and Rapeseed Production, Chemistry, Nutrition, and Processing Technology*, Van Nostrand Reinhold, New York, 1990.

4. James P. Houck, Mary E. Ryan, and Abraham Subotnik, *Soybeans and Their Products*, University of Minnesota Press, Minneapolis, 1972, p. 21.

5. Concessional sales are those sales in which the United States grants the buying country more favorable terms (in currency accepted, length of time for repayment, or interest rate) than the buyer could obtain when trading on the open market.

6. Mark E. Smith, Karen Z. Acherman, Ann Fleming, and Nydia R. Suarez, *Title XV-Agricultural Trade*, Agriculture Information Bulletin No. 624, ERS, USDA, June 1991.

7. Some argue, on the other hand, that trading firms are concerned only with their margins and not with the absolute prices of the commodities that are being traded. See, for instance, Richard E. Caves, "Organization, Scale, and Performance

of the Grain Trade," *Food Research Institute Studies*, Vol. 16, No. 3, Stanford University, Stanford, CA, 1977-1978.

8. Caves, ibid., concludes that the export grain industry "exhibits competitive behavior."

9. Richard Heifner, Kandice Kohl, and Larry Deaton, "A Study of Relationships Between Large Export Sales and Futures Trading," USDA, U.S. Government Printing Office, Washington, DC, June 8, 1979.

10. In a recent survey, it was found that four major grain exporting firms held 24 percent of the short futures contracts in corn, wheat, and soybeans, and 6 percent of the open long contracts at the major grain exchanges.

11. This section is based on Vernon O. Roningen and Praveen M. Dixit, "Economic Implications of Agricultural Policy Reforms in Industrial Market Economies," Economic Research Service, USDA, Washington, DC, August 1987, Staff Report No. AGES89-36. It is also based on Andrew Schmitz, "GATT and Agriculture: The Role of Special Interest Groups," *American Journal of Agricultural Economics*, December 1988, pp. 994-1005.

Selected References

American Enterprise Institute, *Food and Agricultural Policy*, Washington, DC, 1977.

Canadian Wheat Board, *Annual Report*, 1960-1961 to 1978-1979, Winnipeg, Manitoba.

Caves, Richard E., "Organization, Scale, and Performance in the Grain Trade," *Food Research Institute Studies*, Vol. 16, No. 3, 1977-1978.

Caves, Richard E., and Ronald W. Jones, *World Trade and Payments*, Little, Brown, and Company, Boston, MA, 1977.

Economics, Statistics, and Cooperatives Service, *Alternative Futures for World Food in 1985*, USDA, Foreign Agricultural Economic Report No. 146, Washington, DC, April 1978.

Ellsworth, Paul T., *The International Economy*, 4th ed., The Macmillan Company, New York, 1969.

Heller, H. Robert, *International Trade: Theory and Empirical Evidence*, 2nd ed., Prentice-Hall, Inc., Englewood Cliffs, N.J., 1973.

Helmuth, John W., *Grain Pricing*, Economic Bulletin No. 1, Commodity Futures Trading Commission, Washington, DC, September 1977.

Ingram, James C., *International Economic Problems*, 3rd ed., John Wiley & Sons, Inc., New York, 1978.

Kindleberger, Charles P., *International Economics*, 4th ed., R. D. Irwin, Inc., Homewood, IL, 1968.

Kreinin, Mordechai E., *International Economics: A Policy Approach*, 2nd ed., Harcourt Brace Jovanovich, Inc., New York, 1975.

McCalla, Alex F., "A Duopoly Model of World Wheat Pricing," *Journal of Farm Economics*, Vol. 48, No. 3, Part 1, August 1966.

McCalla, Alex F., and Timothy E. Josling, eds., *Imperfect Markets in Agricultural Trade*, Allanheld, Osmun and Company, Inc., Montclair, NJ, 1981.

McCalla, Alex F., and Andrew Schmitz, "Grain Marketing Systems: The Case of the United States Versus Canada," *American Journal of Agricultural Economics*, Vol. 61, No. 2, May 1979.

Schienbein, Allen, *Cost of Storing and Handling Grain in Commercial Elevators, Projections for 1974-75*, USDA Commodity Economics Division, Economic Research Service, Washington, DC, February 1977.

Schmitz, Andrew, Alex F. McCalla, Donald O. Mitchell, and Colin A. Carter, *Grain Export Cartels*, Ballinger Publishing Company, Cambridge, MA, 1981.

Shepherd, A. Ross, *International Economics: A Micro-Macro Approach*, Charles E. Merrill Publishing Company, Columbus, OH, 1978.

Snider, Delbert A., *Introduction to International Economics*, R. D. Irwin, Inc., Homewood, IL, 1975.

Speaking of Trade: Its Effect on Agriculture, Special Report No. 72, University of Minnesota, MN, 1978.

Taplin, J. H., "Demand in the World Wheat Market and the Export Policies of the United States, Canada and Australia," unpublished Ph.D. dissertation, Cornell University, Ithaca, NY, 1969.

Thompson, Sarahelen R., and Reynold P. Dahl, *The Economic Performance of the U.S. Grain Export Industry*, Agricultural Experiment Station Technical Bulletin No. 325, University of Minnesota, St. Paul, 1979.

United States Senate, hearings before the Subcommittee on Multinational Corporations of the Committee on Foreign Relations, Part 16, U.S. Government Printing Office, Washington, DC, 1977.

Wexler, Imanuel, *Fundamentals of International Economics*, Random House, Inc., New York, 1968.

Wheeler, R. O., Gail L. Cramer, Kenneth B. Young, and Enrique Ospina, *The World Livestock Product, Feedstuff, and Food Grain System*, Winrock International, Morrilton, AR, 1981.

11

Government Policy

Bob F. Jones

The U.S. grain-marketing system can be characterized as one in which private individuals and private firms perform the major functions and bear the consequences of their actions. The system is largely a free-enterprise, private system in contrast to systems in many other countries. However, the U.S. system is not free of government involvement, particularly at the federal level; neither did it evolve without government efforts to shape the system into what it is today.

In addition to affecting the marketing system, the federal government affects grain prices through various programs. The government became involved in the grain-marketing system in 1929, when it began programs to raise farm incomes from depressed levels. It has used a wide variety of programs to influence prices and production and has varied the types and amount of involvement in accordance with changing economic conditions.

More direct government involvement in agriculture was a part of the general trend in the early 1930s as the government became more involved in private economic activity. This involvement occurred in response to depressed economic conditions that included the failure of the general economic system to provide adequate employment opportunities to many of its citizens and adequate income to the nation's farmers.

Public Decisionmaking

Policy Objectives

Public policy is a special kind of group action designed to achieve certain aspirations held by members of society.[1] It is distinct from private firm policy in that it involves group action by individuals who frequently have conflicting or diverse aspirations or objectives. However, an adopted policy requires at least minimal agreement on what is to be achieved. For a policy to be adopted, expected achievement must be consistent with or bear some correspondence to the values held by society.

Architects of U.S. farm policy, particularly since the late 1920s, have been guided by their attitudes toward political and social stability, economic stability, a particular type of economic organization, economic growth, equality of opportunity, and a desire to share U.S. agricultural abundance with other less fortunate people. These preferences are manifestations of values held by members of the society and play an important role in determining what it is that society wants to attain, both for the agricultural sector and the general economy.

Values correspond to society's subjective perceptions of what "ought to be." Beliefs and facts pertain to the situation as it currently is. In this context, a belief pertains to a person's or group's perception of what the situation is. When "what is" does not correspond with what "ought to be," a policy problem may exist. When a policy problem exists, a decision must be made as to what to do about it. Society must decide whether to attempt to solve the problem, ignore it, or learn to live with it. Beliefs, even when factually incorrect, have an important role in policy formulation because they influence problem definition and choice among policy alternatives. Therefore, one task of policy formulation is to sort fact from belief, with emphasis on verification of the "true" situation.

President Lyndon B. Johnson's "Message on Agriculture" in 1964[2] contained representative statements of the objectives of farm policy, as perceived by the executive branch of government. That message stated that farm program objectives were "to maintain and improve farm income, strengthening the family farm in particular," and "to use our food abundance to raise the standard of living both at home and around the world."[3] The preamble of the Agriculture and Food Act of 1981 stated the objectives to be "to provide price and income protection for farmers, assure consumers an abundance of food and fiber at reasonable prices, and continue food assistance to low income households, and for other purposes."[4]

Policy Formulation

Grain production and marketing policy is established at the federal level by the House of Representatives and the Senate and is subject to approval by the president. Policy is formulated in each legislative body after study and recommendation by the Senate Committee on Agriculture and Forestry and the House Agriculture Committee. The executive branch of government, working through the USDA, contributes input to the legislative committees, often taking the lead in designing programs and serves in an advisory capacity to the committees. Farmers contribute to the process, individually and through the general farm organizations and commodity groups; other business groups, consumer groups, private citizens, and other governmental agencies also provide input to the legislative committees.

Approaches to Public Policy for Agriculture

Public policy for grain production and marketing has followed two approaches. Policies are designed to make the private production and the market system work more efficiently. Where it is believed noncompetitive behavior is either present or conditions exist that would allow its emergence, policies may be designed to enforce competition. The second approach involves direct government intervention into the production and marketing processes.

Policies to Promote Competition. Examples of this approach include establishment of grades and standards, collection and dissemination of market information, and provision of funds for agricultural research and extension activities.

In this approach, public funds are entitled to make the marketing system work more efficiently and thereby lower marketing costs. Lower marketing costs can result in lower costs to consumers and/or higher prices to producers. In a competitive economy, all consumers of agricultural products benefit by being able to buy food products at lower cost and stand to gain from public expenditures for these types of activities.

Consumers receive greater benefits when the system is more competitive and the benefits of lower costs are passed through to them. The regulation of railroads that began in 1887 when rail shipment was the primary mode of shipment of grain was visualized as necessary to assure competition and equitable freight rates for shipment of grains. This type of regulation was discussed in Chapter 4. Regulation of commodity markets was believed to be necessary to reduce fraud and increase competition in commodity exchanges. These forms of government involvement, which affect the infrastructure within which grain marketing takes place, have evolved over time, continue in various forms today, and are subject to continual review and modification.

More Direct Government Involvement. Depressed conditions in agriculture in the 1920s led to passage of the McNary-Haugen bills. This legislation authorized establishment of a government-sponsored board that would determine fair prices for agricultural products to be sold in the domestic market and would sell surpluses of wheat and seven other basic commodities abroad. The bills passed Congress twice but were vetoed by President Calvin Coolidge. They were considered to be price fixing and against the economic principles of free enterprise.

As agriculture continued to suffer from depressed economic conditions, other approaches were tried. The Federal Farm Board was established by the Agricultural Marketing Act of 1929. It was set up with a revolving fund of $500 million, had authority to make loans to cooperative associations, and could buy up surplus grains. However, the board had no control over production or acreage and soon exhausted its revolving fund as wheat prices on the Chicago market dropped from $1.20 in 1929 to $0.39 a bushel in 1931.

As the general economy was severely depressed in 1932, additional approaches for agriculture were tried in the form of the Agricultural Adjustment Act of 1933. With passage of this act, the federal government began programs to influence grain prices and farm incomes through acreage allotments, direct payments to farmers, and storage programs. These programs focused on the objective of raising farm prices by restricting production. The approach assumed demand for agricultural products was inelastic and that, by reducing production, prices would rise sufficiently to increase farm income.

The Commodity Credit Corporation (CCC) was created in 1933 to carry out loan and storage operations as a means of supporting prices above the level that would have prevailed in a free market. At that time, export demand had nearly disappeared as a result of depressed world market conditions and a progressively more restricted export market as tariff walls were raised worldwide. U.S. consumers' incomes had dropped, causing domestic demand for farm products to shrink further still. At the same time, the quantity of grains available for marketing off the farm was increasing as draft horses, which ate home-grown grain, were being replaced by tractors.

Restoration of purchasing power for agriculture was a key feature of the Agricultural Adjustment Act of 1933. This goal became known as "parity for agriculture." The original legislation designated wheat, corn, cotton, hogs, rice, tobacco, and milk as basic commodities that would be supported. It was assumed that support for a few key commodities could raise farm incomes.

The advent of World War II relieved the crisis of overproduction and low farm prices. Prices rose as stocks were used up. Concern over surpluses turned to concern over producing enough food with which to fight the war.

In part to stimulate production during the war, and also to relieve farmer concerns over low farm prices that were expected to return following the end of the war, the Steagall Amendments were passed in

1941. They authorized that farm prices be supported at relatively high levels for two years following the declared cessation of hostilities. The amendments expired on December 31, 1948.

Agriculture's Changing
Technical and Economic Environment

World War II to 1972

Farm prosperity during and after the war enabled farmers to invest heavily in new production technology. Output expanded rapidly as new chemical, biological, and mechanical technologies were acquired in agriculture. As wartime demand declined and output expanded, grain stocks again accumulated in CCC storage. Public debate arose over whether fixed price supports were more appropriate for simultaneously supporting farm income and facilitating resource adjustment to changing technological and market conditions, or whether flexible supports would be more appropriate.

Grain policy was considered the central element in price and income support, with the CCC loan rate a key component of the policy. Grains were considered the central element because of their storability and their role in livestock production. It was believed that farm incomes could be maintained through use of the loan rate and government storage of grain.

Justification for continuing price and income programs for agriculture following World War II was based on the following lines of reasoning. Domestic demand for food had recovered as a result of general economic recovery associated with the war. Export demand had expanded during and immediately following the war but had declined following recovery in Europe. The U.S. physical plant for producing farm commodities had expanded as a result of wartime demands and was capable of producing more than could be sold at acceptable prices.

Although farm prices and incomes were significantly above the levels of the 1930s, per capita incomes in agriculture continued to lag behind per capita incomes in the nonfarm sector. Given the low income elasticity of demand for food, and the tendency for farm output to continue to grow as a result of the adoption of new technology, farm incomes were likely to continue to lag behind nonfarm incomes. Proponents of continued government support for agriculture argued that farmers needed help in making the necessary resource adjustments as new technology was being acquired. The surplus commodities tended to be looked on as temporary surpluses that would likely be needed as weather and crop conditions changed. Storage programs would enable producers to store commodities past the low-price harvest period, and if prices stayed low during the year, the CCC would acquire grain that could be disposed of in subsequent years. Thus, storage programs were considered necessary as a means of both stabilizing farm prices and raising farm incomes.

Dramatic resource adjustments were occurring in agriculture. The number of farms declined from 6.3 million in 1930 to 4.8 million by 1954. Yet the adjustments were not occurring rapidly enough to bring farm incomes up to the level of nonfarm incomes, and governmental help in the form of price supports was considered necessary.

The continued concern over equitable treatment of the farm sector, and the desire to maintain political, social, and economic stability, maintained political support for a continuation of governmental involvement in supporting farm prices. Emphasis shifted toward disposing of government-owned stocks in the export market with passage of the Agricultural Trade Development and Assistance Act, better known as Public Law 480. Finally, the magnitude of excess capacity in agriculture was becoming more apparent as grain stocks accumulated and storage costs escalated.

Although greater emphasis was placed on surplus disposal with passage of PL-480 in 1954, the loan rate and storage programs continued to play a key role in price and income policy. The Feed Grain Program for 1961 included direct payments to farmers who would idle a portion of their feed grain acreage as a means of supply control. This provision represented a slight shift from dependence on price supports (the loan rate) toward direct payments to producers as a means of income support. This trend grew in importance and eventually evolved into the target price/direct payment approach for supporting farm income. Every major Food and Agricultural Act since 1973 has continued the target price/deficiency payment approach that became the centerpiece of price and income support in that act.

1972 to 1980

A series of events in 1972 to 1974 changed the economic and technical environment in which U.S. agriculture operates. These events dramatically affected the grain production and marketing system as well as its participants. Three events had key roles in changing the environment—one was the large increase in exports of wheat to the Soviet Union starting in 1972, the second was the quadrupling of crude oil prices between 1972 and January 1, 1974, and the third was the shift from a fixed rate of exchange for the U.S. dollar to market-determined exchange rates.

As a result of adverse weather conditions in the Soviet Union in 1972, the country's wheat crop was expected to be sharply reduced. Prior to 1972, planners in the Soviet Union had embarked on a five-year plan to increase food production, especially livestock products. The plan was to import grain to maintain livestock herds if domestic production was reduced. In earlier periods of crop shortfall, the Soviet Union had reduced livestock numbers to accommodate the lower level of grain production. With the sharply reduced crop in 1972, they implemented the new plan and sought grain from the world market. Because the United States held large stocks of grain in government storage and was eager—in retrospect, probably too eager—to dispose of its surplus stocks, a bargain was struck. The United States would supply wheat to the Soviet Union at the prevailing world trading price of about $1.65 per bushel. U.S. wheat support prices were above this level, so export subsidies were required to move the wheat. During mid-1972, grain buyers in the Soviet Union agreed to purchase over 400 million bushels of wheat from the United States. A purchase of this magnitude enabled the CCC to unload its stocks of grain so that by the end of 1974 the CCC held no wheat under its control.

The second event consisted of a dramatic change in world oil prices. In a series of moves starting in 1973, OPEC raised the price of crude oil from $3.01 to $11.65 per barrel.[5] The Arab-Israeli War and the Arab oil embargo dramatized to the world the dependence of many countries on imported oil. Importing countries continued to buy oil at these prices and were either unable or unwilling to get OPEC to lower its prices. The era of cheap energy came to an abrupt end.

The third event involved shifting from fixed exchange rates to market-determined exchange rates that significantly increased the demand for U.S. grain. These changes are described in a subsequent section entitled General Economic Policy.

A combination of the surge in exports in 1972-1973, sharply reduced U.S. grain stocks, and a relatively poor corn crop in the United States in 1974 led to much higher grain prices in the United States and in world markets. Thus, supply conditions and worldwide inflation, fueled by expansionary monetary and fiscal policy in the United States, contributed to higher grain prices.

As a result of the increase in exports and the depletion of government grain stocks, no cropland was held out of production under set-aside programs in 1974 through 1977. Most of the 50 to 60 million acres of cropland that had been held out of production under various programs during the late 1960s and early 1970s came back into production.

Agriculture in the mid-1970s appeared to be more nearly in equilibrium than at any time in the previous 40 years. Growth in demand was primarily from the export market. U.S. agricultural exports expanded from $4.5 billion in the early 1960s to over $43.8 billion in 1981. During this period, agricultural exports as a percentage of sales of farm products increased from 14 percent to 30 percent. Clearly, agriculture had become more dependent on the foreign market as a source of its receipts. Moreover, agriculture became more dependent on the export market than any other major sector of the economy.

Conditions for the expansion of U.S. agricultural export products had been developing throughout the 1960s. The steady long-term growth was a result of rising real per capita incomes worldwide and growing world population. Poor crops in the Soviet Union, Australia, and Brazil in 1972 contributed to the sudden growth in demand for U.S. products. Changes in the world monetary system that resulted in devaluation of the dollar relative to the currencies of Japan and several Western European countries also contributed to expansion of U.S. agricultural exports.

Labor continued to move out of agriculture during the 1960s, and the number of farms continued to decline. Most of the excess labor had been removed from agriculture. After the return of set-aside land to production, additional production was possible through bringing new land into production, and through increasing productivity of existing resources. Some evidence suggested that the long-term growth in productivity of agriculture was declining.

With expanding exports and U.S. agriculture more nearly in an equilibrium situation in the mid-1970s, the set of questions concerning consumers and policymakers changed; it became more concerned over high and unstable farm prices and the adequacy of grain stocks. The situation of higher prices and no government stocks brought into being the use of the target price/direct payment approach contained in the Agricultural Act of 1973 as the principal means of income support for agriculture.

Ever since storage programs came into existence in the 1930s, they have been combined with various types of programs to restrict and manage the supply of selected farm products to maintain prices. The amount of restriction and control over supplies has depended on the changing economic environment for agriculture. Control programs have taken the form of acreage allotments, marketing quotas, and land diversion under various names. They all represent attempts to tailor production to demand in a manner similar to that used by industrial producers.

During the period between passage of the 1973 and 1977 Agricultural Acts, government price supports had a very minor role in supporting farm prices and income. Market prices were above minimum support levels. With growing agricultural export markets, grains and oilseeds moved abroad rather than into government storage bins.

With market prices for wheat and feed grains much higher than they were prior to 1972 and production costs on an upward trend, loan rates were increased by authority of the 1977 act. That act set minimum target prices for grains and required the secretary of agriculture to annually adjust them upward to reflect changes in the cost of production. The increase in loan rates set the stage for renewed acquisition of grain by the CCC through loan forfeiture when market prices failed to exceed loan rates plus carrying costs.

The Food and Agriculture Act of 1977 was up for renewal or change in 1981. During the four years prior to 1981, both the volume and value of agricultural exports increased, grain production continued its upward trend, and production costs continued to rise. High rates of inflation after 1973 and intermittent concern over world food shortages accentuated by a U.S. drought in 1980 influenced the level of price supports that were authorized by the Agricultural Act of 1981.

After 1980

The Agriculture and Food Act of 1981 continued the system of price and income supports contained in the 1977 Act. Loan rates continued to be the key component of the income support system. The minimum support price for corn was established at $2.55 per bushel for the 1982-1985 crops. Target prices for the 1982-1985 corn crops were stipulated to be not less than $2.70, $2.86, $3.03, and $3.18 per bushel, respectively.

The minimum wheat loan rate was established at $3.55 for the four-year period. Target prices for the 1982 through 1985 wheat crops were established at levels to be not less than $4.05, $4.30, $4.45, and $4.65, respectively.

The ink for the 1981 act was barely dry when the economic environment for agriculture changed dramatically. The U.S. inflation rate, measured by changes in the consumer price index, reached a peak of 13.5 percent in 1980, then declined to 1.9 percent by 1986 as a result of actions taken by the Federal Reserve Board starting in late 1979. Inflation rates outside the United States dropped also. The prime interest rate charged by banks increased to the 18.9 percent level in 1981 in contrast to the 6.8 to 12.7 percent rates during the 1970s. The exchange value of the dollar relative to other major currencies increased by over 50 percent in a period of five years.

The export market for U.S. agricultural products declined from its peak of $43.8 billion in 1981 to a low of $26.3 billion in 1986. Once again, large quantities of grain were acquired by the CCC as a result of loan takeovers. Grain also accumulated in the farmer-owned reserve and was held there as a result of the large difference between market prices and the price level prescribed by release rules. As a result of the increased loan rates authorized by the 1981 Act, the rising value of the dollar, and declining export market conditions, the United States was pricing itself out of the export market.

In the early 1980s, many farmers were encountering severe financial stress as a result of the collapsing export market, the high interest rates they were paying, and the overexpansion of credit that occurred in the 1970s. In the major grain-producing areas, land values dropped by half in a period of about four years. With declining asset values and large debts outstanding, farm bankruptcy rates rose sharply.

Due to above-average grain crops in 1981 and 1982 and the sharp decline in exports, grain stocks grew so large that drastic action was taken to reduce acreage planted in wheat and feed grains for 1983. In addition to the acreage reduction required to receive program benefits, producers were offered an option of idling additional base acreage (up to the total base acreage) for pay. Payment for the additional acreage was in the form of payment-in-kind (PIK) rather than cash. The 1983 PIK program, together with the acreage reduction program and paid land diversion, diverted nearly 32 million acres of cropland from corn production.[6] The reduction in acreage planted in corn in 1983 and a severe drought over much of the country resulted in a 1983 corn crop of only 4.175 billion bushels in contrast to 8.235 billion bushels in 1982.

Participating farmers in both the corn and wheat programs were issued PIK certificates for a part of their program payment. To participate in the 1983 corn program, a producer was required to idle 20 percent of his or her base acreage. He or she received no payment for the first 10 percent and a cash payment for the second 10 percent. The producer had an option of idling an additional 10 to 30 percent of the base. Inkind payment was at the level of 80 bushels of corn for each 100 bushels of ASCS-established yield base. An additional option enabled a producer to reduce corn acres (the final 50 percent) to zero acres. This program was on a bid basis with payment made in PIK certificates.

Producers were issued certificates that were denominated in bushels. They could exchange these certificates for No. 2 corn from CCC stocks or for corn which they, themselves, had under either the regular nonrecourse loan or for corn they held in the farmer-owned reserve. The stored corn could be from any location where the CCC controlled grain.

Although the release price and sales rules established by the act of 1981 remained in effect, issuance of these certificates enabled corn and wheat to be withdrawn from government stocks and in effect removed the floor under prices as established by the loan rate. Nevertheless, market prices for corn remained above the loan rate until 1985. Market prices for wheat dropped below the loan rate in 1983 and 1984. The use of PIK certificates enabled corn stocks to be reduced from 3.5 billion bushels for the 1982 crop year to 1.0 billion for the 1983 crop.

In 1984 and 1985, most of the United States experienced more normal growing conditions for corn and wheat, with yields back on trend. (Wheat, a more drought-tolerant crop, did not experience the yield decline in 1983 that corn did. Also, the drought came later in the wheat production cycle.) With a small acreage set-aside in 1984 and 1985 compared to the 1983 so-called PIK year, production again exceeded annual use, and stocks accumulated rapidly.

Discussions about replacing the 1981 act started a year ahead of the more normal schedule for passage of legislation. The pessimistic price and stock outlook focused attention on changes that needed to be made in the approach for supporting price and income for grain farmers. The export market continued to decline, and grain continued to be acquired by the CCC as a result of loan takeover. Corn stocks at the end of the 1985 crop year exceeded 4 billion bushels. Wheat stocks approached the 2-billion-bushel level. The financial condition of many farmers, especially cash grain farmers, remained precarious. Either lower prices were needed to move grain into markets rather than into government bins, or more stringent production controls were needed. Otherwise, government stocks would continue to grow. Given the financial condition of many farmers as well as input suppliers and farm credit providers, price reduction would further weaken farmer income positions.

The Food Security Act of 1985 was passed in the closing days of the 1985 congressional session. Although the act continued the approach used in the 1981 act, it put more emphasis on supporting farm income and less emphasis on supporting grain prices. The target price for corn was set at $3.03, $3.03, $2.97, $2.88, and $2.75 for the 1986-1990 crops, respectively. The loan rate for 1986 corn was $2.40 per bushel with a provision for downward adjustment if the secretary of agriculture determined that was necessary to maintain U.S. corn's competitive position in world markets.

Target prices for wheat were set at $4.38 in 1986 and 1987. Rates of $4.29, $4.16, and $4.00 were set for 1988-1990. The loan rate was set at $3.00, with downward adjustment provisions similar to those for corn. Acreage limitation programs continued to be authorized.

The 1985 act authorized a generic certificate program as a tool for moving grain and other stored commodities out of CCC warehouses and as a way of reducing government cash costs of various commodity programs. The 1985 act authorized the USDA to make 17 different program payments on a noncash (certificate) basis.[7] These authorizations allowed commodities to be moved out of government ownership or storage programs while maintaining the cash release price rules.

Generic PIK certificates became an important component of the income support program during the 1986-1990 period, and an important marketing tool for farmers and the grain-marketing industry. The introduction of generic marketing certificates added a new complexity to grain marketing, which required participants to become acquainted with how to use them.

To understand the impact that generic certificates had on market prices relative to CCC release prices and loan rates, one needs some understanding of how the certificate program worked. The USDA made payments of dollar-denominated generic certificates directly to the producer, often referred to as the initial holder. The initial holder could choose one of three options:

1. Sell or transfer the certificate;
2. Use the certificate to redeem regular commodity, reserve, or special producer storage loans; or
3. Exchange the certificate for cash at the issuing county office no sooner than five months following issuance.

A subsequent holder could choose one of the following courses of action:

1. Sell or transfer the certificate;
2. Use the certificate to redeem CCC loans; or
3. Exchange the certificate for commodities from CCC inventory.

Subsequent holders could not exchange their certificates for cash from the Agricultural Stabilization and Conservation Service.

Certificates were dollar denominated. When certificates were redeemed for commodities, the face value of the certificate was converted to a bushel equivalent based on the posted county price (PCP) of the commodity. PCPs were determined on a daily basis and were market prices at the terminal and were adjusted for location.

A market quickly developed for generic certificates. When PCPs were below the loan rate incurred by a producer, it was often to his or her advantage to acquire certificates and pay off the loan at the lower rate (price), especially if the producer was in the market for livestock feed. Furthermore, because PCPs were determined on a daily basis and were valid for a 24-hour period, opportunity arose for gain from daily price changes.

Certificates often sold at a premium to face value. These premiums varied considerably since they were first issued in May 1986. They have ranged from 30 percent above face value to 1 to 3 percent below.[8]

Variations in premiums can be attributed to one or more of the following:

1. The difference between redemption values and the actual market price;
2. The elimination of carrying charges consisting of interest and storage charges; and/or
3. Access to CCC inventory.

Avoidance of carrying charges became an important use of PIK certificates. Certificates allowed producers to receive the benefit of the loan rate without storing the crop for nine months. This practice came to be called *PIK and roll*. Without certificates, producers were required to hold their crops for the life of the loan—usually nine months—before they could forfeit the commodity to the CCC. The producer would incur storage costs for nine months if the grain were forfeited plus interest on the loan principal if the loan were repaid.

Certificates gave the original loan holder an additional way to repay the loan prior to maturity. He or she could take out the loan and immediately repay it with PIK certificates, saving the storage costs.[9]

From 1981 to 1985, as agricultural exports declined, the United States lost market share as a result of U.S. pricing policy, increased strength of the U.S. dollar, and increased competition from expanding grain production in the EC. During the 17-year period from 1970 to 1986, the EC shifted from being a net importer of 22.3 mmt of grain to being a net exporter of 19.5 mmt. A large part of EC grain exports was made possible only by use of export subsidies. In an effort to regain export markets lost to the EC, the 1985 Food Security Act provided for use of export subsidies under the Export Enhancement Program (EEP). Initially, it was a program where U.S. exporters would be provided with a sufficient subsidy payment to make U.S. grain competitive in markets that had been lost to EC suppliers. This program made payments to U.S. exporters in the form of PIK certificates. Exporters acquired PIK certificates directly from the CCC as a result of competitive bids or from initial holders (producers). Exporters used these certificates to acquire grain from CCC inventories.

The 1986-1990 period can be characterized as one in which grain prices declined sharply, in part as a result of the CCC dumping of stocks via USDA issuances of over $20 billion of generic certificates, increased use of deficiency payments as the means for supporting farm income (not prices), and much higher rates of participation in commodity programs.[10] The federal outlay for farm programs increased from $2.8 billion in fiscal 1980 to $25.8 billion in 1986. A combination of lower prices and the use of export subsidies contributed to an upturn in U.S.

exports starting in 1986, with exports approaching the $40 billion level in 1989-1990. These developments set the stage for the debate over the 1990 Agricultural Act.

The Food, Agriculture, Conservation, and Trade Act of 1990 continued the types of programs contained in the 1985 act. The act provided less budget support for agriculture than the previous bill. Target prices for feed grains and wheat were frozen at the 1990 level for the next five years. By introducing a flex-acres concept, less than 100 percent of base acres were eligible for payment.[11] Additional budget savings were to be obtained by changing the way that deficiency payments would be calculated in 1994 and 1995. For those years, these payments would be calculated using a 12-month average market price rather than a 5-month average market price as was the case for 1991-1993.

The 1990 act continued authorization for making specific program payments with generic certificates. Acreage reduction programs were reauthorized. Provision for continuing the EEP was made with authorized expenditure to be a function of progress in trade negotiations under the General Agreement on Tariffs and Trade (GATT).

Analysis of Price and Income Support Programs in a Closed Economy

Analytical Approach

The general impacts of government commodity programs can be analyzed using conventional supply-demand concepts, as briefly discussed in Chapter 2.[12] Programs can be categorized into those that affect demand and those that work through supply. Various forms of demand expansion have been used since the 1930s. They include commodity distribution programs (which have made surplus commodities available to persons on welfare), child nutrition programs (such as the school milk program), and donations of surplus commodities to school lunch programs. Food stamps are supplemental income for consumers and represent another form of demand expansion. Surplus distribution in the export market under authorization of PL-480 and the Mutual Security Administration has taken various forms that represent demand expansion.

The supply of farm commodities that can be marketed has been affected through marketing quotas, domestic allotments that limit the acreage planted, and soil bank or acreage set-aside programs that reduce either a specific crop or the total acreage available for the production of grain crops.

Loan and Storage Programs. Line *D* in Figure 11.1 represents the demand curve for wheat that domestic consumers would buy at various possible prices in a given period, for example, one year.[13] Line *S* represents the quantity of wheat that producers would supply at various prices, and there is no carry-over. In the absence of government intervention, the price would be P_0 and the quantity produced and sold would be Q_0. If the government determines that P_0 is too low to provide equitable income to producers, and further determines that the price should be at P_1 and makes loans available to producers at that rate, prices will tend to rise to P_1. With the price at P_1, consumers will purchase only Q_1, and the government will need to store Q_1Q_2 of wheat as a result of loan takeover.[14]

If the storage program is a voluntary one and has requirements such as a reduction in acreage planted that some producers may not wish to meet, not all grain producers will be eligible for loans; so, market prices may drop somewhat below the loan rate. If the government places no restrictions on the amount produced or sold, except through limiting acreage planted, producers will gain from the loan programs so long as the government is willing to continue to store grain. If technological progress is occurring that causes the supply curve to shift to the right over time, the government will be faced with the problem of acquiring larger quantities of grain each year. This problem tended to occur in many of the years from 1948 through the 1960s.

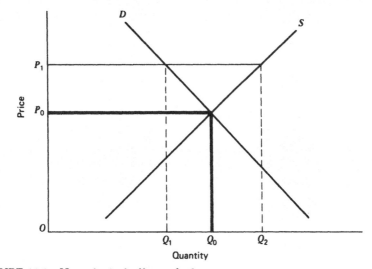

FIGURE 11.1 Hypothetical effects of a loan rate.

The Agricultural Act of 1949 provided that the CCC should not sell any storable commodity in the domestic market at less than 5 percent above the support price plus reasonable carrying charges. The secretary of agriculture may (and has) set a higher minimum level. For feed grains and wheat, the minimum selling price was raised 15 percent above the support price plus carrying charges for the 1974-1977 crops.[15] If the grain was in danger of going out of condition with continued storage, the CCC was allowed to sell grain at less than the established release price.[16]

The release policies of the CCC kept market prices near the loan rate

dollars per bushel

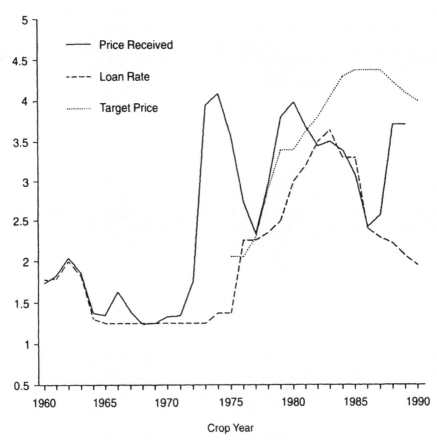

FIGURE 11.2   Wheat support levels and farm prices.

SOURCE: Joy L. Harwood, C. Edwin Young, *Wheat, Background for 1990 Farm Legislation*, USDA, ERS, CED, Oct. 1989.

dollars per bushel

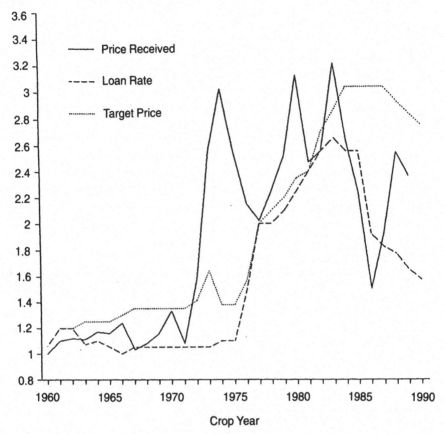

FIGURE 11.3 Corn support levels and farm prices.

SOURCE: Stephanie Mercier, *Corn, Background for 1990 Farm Legislation*, USDA, ERS, CED, Sept. 1989.

when they had large quantities of grain in storage, in effect, putting a ceiling on how much grain prices would increase. As a result of large stocks and the operation of the resale rules, grain prices tended to be relatively stable from the early 1950s through 1965, at which time the drought in southeast Asia caused world grain prices to rise (see Figures 11.2 and 11.3).

When the CCC supports the price of grain above the market price, as in Figure 11.1, producers offer a larger quantity to the market. Furthermore, when technological progress is causing the supply curve to shift to the right at a more rapid rate than demand shifts (due to low-income elasticity of demand and slow population growth), the quantity that farmers wish to put under loan continues to grow over time. If the government chooses to restrict its expenditures on commodity programs it may do so with several types of restrictive devices. If it were to choose marketing quotas, specifying that a given quantity of wheat could be marketed by producers, the situation would be as illustrated in Figure 11.4. If the marketing quota is set at Q_1, the price would be maintained at P_1. Producer incomes would increase if the demand for wheat is inelastic, as it was believed to be. On the other hand, if the demand for wheat is price elastic (greater than one in absolute value), producers would gain from selling more wheat at lower prices.

Marketing quotas were used for wheat during the 1950s and early 1960s. A producer's marketing quota was the amount of wheat grown on the allotted acres. Acreage allotments were tied to the acres of wheat planted on the farm in previous years. An allotment of this type represents a less precise quantity than implied in Figure 11.4, because total production on a fixed acreage allotment would vary as yield per acre increases or decreases.

Whether demand is elastic or inelastic is crucial in determining whether programs that restrict supply will benefit producers. Research evidence used in formulating policies in the 1950s and 1960s clearly indicated that demand for grains, especially wheat, was indeed inelastic

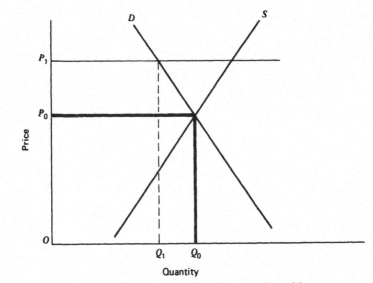

FIGURE 11.4 Example of quota restriction on quantity sold.

in the domestic market. Evidence of demand elasticity on the foreign market was less certain, but it, too, was presumed to be inelastic.

When supply restrictions are imposed on one input, the supply curve tends to shift upward and to the left (e.g., S_2 in Figure 11.5). However, other inputs tend to be substituted for the input being limited. Therefore, the supply curve shifts less than might be expected. When programs have required that production of a given crop be limited to an acreage allotment, farmers have supplied more fertilizer and other variable inputs to the planted acres and have been able to increase average yield on allowable acres in the program. Slippage in the program also occurs from another source. Since farmers tend to idle their less productive acres, the average yield on planted acres increases. Research has shown that the productivity of idled acres in the Eastern Corn Belt is about 85 percent of the cropped acres on the farm. After all adjustments, the reduction is equivalent to a leftward shift in the supply curve to S_1, as illustrated in Figure 11.5.

When grain prices are supported above the market equilibrium price, market prices tend to be higher, producers have an incentive to increase production, consumption is less, and stocks accumulate in government storage. The existence of government-owned stocks and the operation of release rules tend to keep prices within a narrow range. The loan rate becomes the floor price, and the ceiling price is only slightly above it.

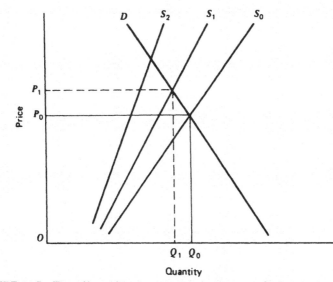

FIGURE 11.5 The effect of acreage restriction on supply.

The risk of carrying stocks is largely transferred from farmers and the grain-marketing system to the government. When stocks become too large, greater emphasis tends to be placed on methods of expanding domestic demand and surplus disposal, even thought it is at a loss to the government.

Deficiency Payments Plus a Loan Program. The Agriculture and Consumer Protection Act of 1973 incorporated a new concept in the commodity price support system. Target prices were established for wheat and feed grains. Grain prices were allowed to be determined by market forces. Direct payments per unit of the commodity were made to eligible producers if the market price received by farmers for the first five months of the marketing year was below the target price as established by the legislation. The deficiency payment could be no larger than the difference between the target price and the loan rate. To be eligible for program benefits, participants were required to limit production by setting aside a percentage of their planted acres on which no grain or forage crop could be grown for harvest. The secretary of agriculture could require set-aside acres if supply-demand conditions warranted it.

The effects of this method of price support depend on the level of the loan rate and the target price relative to the market equilibrium price. With the loan rate below the market equilibrium price, as in Figure 11.6, grain would not be acquired by CCC through loan forfeitures. Some farmers would place grain under loan, using the program as a source of credit rather than as a price-supporting mechanism, although the loan would keep prices from dropping below that level. Because not all farmers choose to participate in voluntary programs, prices could drop below the loan rate, as has occurred occasionally. Prices shown in Figure 11.6 indicate a deficiency payment of the difference between P_0 and P_T would be paid to eligible participants. Program participants would receive the market price for their grain plus a deficiency payment, so they would tend to expand their output from Q_0 to Q_1. If a set-aside requirement were in effect, production would expand less than Q_0 to Q_1. Conceptually, the supply curve would rotate to the left as land was removed from production (as illustrated in Figure 11.6).

During the period from 1973 to 1976, market prices tended to be above both the target price and the loan rate for wheat and feed grains (Figures 11.3 and 11.6). In these cases, no deficiency payments were made, and the CCC did not accumulate grain as a result of loan takeover. In later years, when the market price dropped below the loan rate (both target prices and loan rates were increased from 1976 to 1985), some grain was acquired by the CCC as a result of loan takeover.

Market prices for corn and wheat seldom dropped below the loan rate and then by only a small amount prior to implementation of the Food

FIGURE 11.6   Example of target price and loan rate.

Security Act of 1985 (Figures 11.2 and 11.3). Producers could put their grain under loan where it remained until market prices exceeded prices established by legislated release rules. With implementation of the 1985 act, grain was released from government storage through the issuance of generic PIK certificates. This release of grain, although appearing to keep release sales rules intact, allowed market prices to fall below the loan rate. Farm income was maintained by increasing the deficiency payment. Only program participants were eligible for deficiency payments, so producers had additional incentive to participate in these programs. Consequently, program participation rates increased significantly.

The use of the target price/loan rate approach allows market prices to vary with changing supply-demand conditions. It encourages greater consumption and export of grain as supplies increase rather than causing an accumulation of grain in government storage. This approach allows exports without export subsidies when the loan rate is equivalent to or below the world trading price for the grain.

Inventory Management Policy. Producers who participate in loan programs may store grain on their farms or in approved commercial facilities. In either case, the producer is responsible for storage costs for the initial period of the loan (typically 9 to 11 months). From a national perspective, most of the wheat stored under government loan is stored off the farm; corn usually is stored on the farm. If a producer decides to forfeit the grain at the end of the loan period, it must be delivered to an approved storage facility designated by the CCC. The CCC has tended to use commercial facilities close to areas of production when available and encouraged construction of commercial storage facilities, prior to 1981, through use guarantees.

When the CCC held large stocks of grain during the 1950s and 1960s, and commercial facilities were inadequate, CCC-owned bins were erected in major grain-producing areas to receive grain from loan takeovers. Storage capacity on farms was increased through operation of the storage facility loan program. Credit was extended to farmers under favorable terms, and rapid write-off as an expense for calculating federal income tax liability was allowed and encouraged.

Storage programs from the 1950s through the mid-1970s required farmers to pay the costs of storing grain under loan whether stored on the farm or in commercial facilities. In some years, a reseal program was offered after the initial storage period. During the reseal period, farmers were paid a storage allowance. When grain was acquired by CCC, and not under reseal, storage costs were paid by the government. The CCC owned large stocks of grain from mid-1950 through the early 1970s; thus, the government assumed a large share of the cost of carrying the nation's grain inventory.

During the period from 1974 to 1977, when the government held almost no stocks of grain, the private sector was carrying the grain inventory. Changes in inventory management following passage of the Agricultural Act of 1977 are discussed in the section on price stabilization and food security.

Evolution of Trade Policies

Foreign markets have been an important but variable source of income to U.S. farmers for almost two centuries. The abundance of rich agricultural land and a relatively small population have enabled U.S. farmers to produce more farm products than could be consumed domestically.

The quantity of agricultural products that enters into foreign markets is a function of demand conditions in other countries, the competitive position of U.S. suppliers relative to other supplying nations and importing countries, and the policies that countries follow toward regulating the flow of products in international markets. Importers can significantly influence the volume and terms of trade through both their domestic and trade policies. Similarly, the United States can influence the amount it sells through its domestic and trade policies. Having a historical perspective on U.S. trade policy as it has evolved over time, especially since the turn of the century, makes it easier to understand changes in the volume of trade.

Pre-World War I to the Great Depression

From the end of the Civil War to World War I, the United States became a major producer and exporter of grains to Europe. However, conflicts between liberal and protectionist advocates in the United States kept trade policies and tariffs fluctuating, affecting the volume of exports and imports. Tariffs reached their highest level with the Dingly Tariff of 1897.

Just prior to World War I, some reduction in U.S. tariff barriers was accomplished with passage of the Underwood-Simmons Tariff of 1913. Trade in agricultural products expanded rapidly in response to mobilization needs and to meet needs created by war-time disruptions in European grain production. Grain production increased in the United States, Canada, and Australia to meet the goal of winning the war and to supply immediate needs, not in response to long-time economic needs. Following the armistice and recovery of European agriculture, much of the export market for grain collapsed. Farmers around the world petitioned their governments for restrictions on imports. Although U.S. farmers were exporters of agricultural products and not importers, they saw their interests as coinciding with the interests of industry and called for higher tariffs against foreign imports. Congress responded in 1930 by passing the Hawley-Smoot Tariff, one of the highest in U.S. history and a watershed in U.S. tariff levels.

In retrospect, it is difficult to understand trade policies from World War I to the beginning of the Great Depression, and why farmers took the position they did. After being a debtor nation since its founding, the United States had emerged from the war as a creditor nation. It had entered into agreements requiring that the losers in the war pay reparations to the winners. Yet, the United States and other nations kept raising their tariff walls, thereby making it more difficult if not impossible for reparations to be paid because the shipment of goods is the principal way by which one country repays its debts to another. The impossible situation that developed was a factor in the Great Depression, in the conditions leading to World War II, and in repudiation of war debts to the United States.

Reciprocal Trade Agreements, 1934

The Reciprocal Trade Agreements of 1934 represent a turning point in U.S. trade policy. The trade agreements act emphasized the need to expand markets for U.S. products and authorized the president to enter into mutually advantageous bilateral trade agreements. The act contained a "most favored nations" provision that allowed the benefits of bilateral negotiations to have a multilateral effect. Agreements were concluded with over 30 countries, which moved the United States toward freer trade in industrial products, a necessary condition for expansion of agricultural trade.

While U.S. tariffs were being lowered for industrial products, agricultural trade was being curtailed. The introduction of direct market intervention in domestic agriculture had international trade repercussions. Government support of domestic agricultural prices above world trading price levels required restrictions against the entry of competing products and the use of export subsidies to sell products abroad.

The Charter for the International Trade Organization

Passage of the Reciprocal Trade Agreements Act was a first step in a long line of activities by the U.S. government to promote freer world trade. The United States has been a participant in international organizations and multilateral negotiations that had the intent of expanding international trade, including trade in agricultural products.

Following World War II, an attempt was made to formulate a set of principles for carrying out international trade. The charter for the International Trade Organization (ITO), drawn up in 1948, specified principles for the reduction of tariffs, the elimination of quotas, and the creation of conditions for expansion of multilateral trade. It set up rules for international commodity agreements. The charter recognized the need for national governments to relate foreign trade policies to domestic programs. President Harry Truman submitted the agreement to Congress for approval in April 1949. Hearings were held by the House Foreign Affairs Committee, but the charter did not go further.[17]

The General Agreement on Tariffs and Trade

Although the ITO failed, many of the basic provisions survived on a more modest scale in the General Agreement on Tariffs and Trade (GATT) in which the United States has been a participant since its inception. This agreement contains a code of principles and rules and provides a continuing forum with a permanent staff and headquarters for periodic negotiating conferences to be called by member nations. More than 100 governments, including the United States, participate fully; about 30 others have associate or observer status.

The GATT has five basic principles, each of which has some impact on the growth and conduct of international trade in grain. The basic principles are as follows:

1. Trade must be nondiscriminatory. All contracting countries receive equal treatment regarding import and export duties and charges.

The original agreement allowed exceptions for less developed countries and regional trading groups.

2. Domestic industries receive protection mainly by tariffs except that agriculture was granted special treatment. For example, Article 11 permits import quotas on agricultural products if domestic production restrictions are in effect. (This authorization was critical if the United States were to retain Section 22 authority.)

3. Agreed-on tariff levels bind each country. If tariff levels are raised, compensation must be paid to injured countries.

4. Consultations are provided to settle disputes.

5. When warranted by economic or trade circumstances, principally balance of payments difficulties, GATT procedures may be waived (or escape provisions allowed) if other members agree and compensation is made.[18]

From its inception through 1993, eight major "rounds" of trade negotiations have been held under GATT auspices. Negotiations led to significant reductions of industrial tariffs. Much less has been accomplished for trade in agricultural products. The escape clause that allows exceptions when domestic agricultural programs are in effect to restrict production has permitted the retention of import quotas.

Despite the difficulties encountered when domestic and trade policies conflict, progress has been made in facilitating trade in agricultural products. The willingness of countries to negotiate and resolve differences has been a positive factor. For commodities that had no duties, an agreement was reached to continue to allow duty-free entry. This agreement is an important factor for U.S. exports of soybeans, soybean meal, and certain by-product feeds to the EC. An agreement also has been reached on the expansion of quotas on imports.

A thorough study of the GATT and trade negotiations is beyond the scope of this chapter. However, one needs some understanding that an international institution exists whose objective is to expand trade among nations. Trade expansion is no simple task when one recognizes the problems associated with reconciling policies of over 100 sovereign nations, each with its own internal institutions, monetary systems, economic conditions, and national objectives.

Analysis of Price and Income Policies in an Open Economy

Analytical Approach

The preceding section in this chapter considered only the domestic market for agricultural products. This section explicitly introduces the foreign market as a component of total demand for agricultural products. The modification in approach is presented to: (1) explicitly recognize the interdependence between markets in two countries when international trade occurs, and (2) develop analytical devices for determining the effects of domestic policies on international trade. The approach shows that quantities exported and imported are a function of price and not just a result of a *surplus* in the exporting country or of a *shortage* in the importing country. The model permits determination of the surplus and/or shortage. The graphical approach represents a partial equilibrium model, given the usual assumptions of a perfectly competitive model.

For example, one can assume two countries, each of which produces and consumes an agricultural product, such as wheat. Production in each country is represented by the conventional supply curve which slopes upward to the right. Similarly, each country's demand can be represented by the conventional demand curve, which slopes downward to the right. These conditions can be illustrated graphically, as in Figure 11.7.

For now, one assumes that these two countries are isolated from each other and that no trade between them occurs. Given these assumptions, the U.S. equilibrium price would be P_1, and consumers would demand Q_1 of wheat. Only at this price-quantity combination would both producers and consumers be satisfied. In the Netherlands, consumers would demand Q_2 of wheat and would pay P_2 per unit for it. In this example, the price of wheat in the Netherlands would be significantly higher than in the United States.[19]

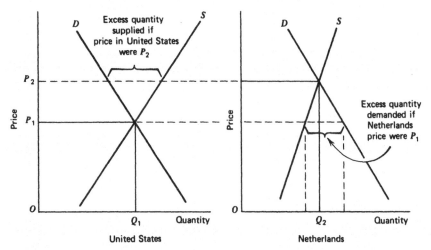

FIGURE 11.7 Two-country supply-and-demand model without trade.

If the two economies are closed to trade, the price can be significantly higher in one country than in the other. When resources are less well suited for production of a particular product, per unit cost tends to be higher. The number of people in the country and their per capita incomes tend to cause differences in demand. When these conditions cause the equilibrium price to be different in two countries, opportunities for trade exist.

If the price were P_2 in the United States, an excess supply would exist (i.e., producers would be willing to supply more than U.S. consumers would purchase at that price). At price P_1, no excess supply would exist in the United States. For prices above P_1, the horizontal difference between the supply and demand curves in the United States would represent the excess supply of wheat in the United States.

If price P_1 were to exist in the Netherlands, consumers would demand more wheat than producers would be willing to supply at that price. At price P_2, no excess demand would exist in the Netherlands. Therefore, at prices below P_2, the horizontal difference between the supply-and-demand curves in the Netherlands would represent the excess demand curve for wheat in the Netherlands.

Now, one may assume that trade is possible between the two countries. In this case, U.S. producers would like to sell at the Netherlands' price, P_2. Netherlands' consumers would like to buy at the U.S. price, P_1. An equilibrium trading price would lie somewhere between these two extremes. It can be determined by transferring the information in Figure 11.7 to a new set of excess supply and excess demand curves (Figure 11.8).

In Figure 11.8, the left panel represents supply and demand conditions in the United States. The right panel represents supply and demand conditions in the Netherlands. The center panel supplies the connecting

FIGURE 11.8 Two-country supply-and-demand model with trade.

link. The excess supply curve, or export supply curve, is derived from the U.S. market. The excess demand curve, or import demand curve, is derived from the Netherlands market. The intersection of the excess supply and excess demand curves in the center panel determines the world trading price for wheat, P_0. In equilibrium, the price of wheat is the same in both countries; it has increased from P_1 to P_0 in the United States.[20] The amount of wheat exported from the United States, OQ_0, is determined in the center panel. It is equal to the amount exported from the United States, Q_1Q_3, as well as the amount imported in the Netherlands, Q_5Q_6. An important lesson from this example is that internal prices in the United States are determined by both domestic market conditions and world market conditions. The exportable surplus is a function of the world trading price. Similar statements pertaining to price and the shortage can be made about the Netherlands. The model can be used to trace through the effects of technical, economic, or policy changes in the United States or the Netherlands. The important point is to start with shifting the appropriate curve. For example, unusually good weather in the United States would cause the US_S curve to shift to the right. This shift would cause the export supply curve to shift to the right. All other curves would be assumed to remain fixed. New prices and quantities could be determined. In a similar manner, an increase in income in the Netherlands would cause a rightward shift in the N_D curve, which generates a rightward shift in the import demand curve. Using this approach, one change at a time can be analyzed.

The model can be expanded to handle more than two countries by making the simplifying assumption that demand-and-supply curves can be aggregated for all importing countries into one set of curves for the importers and labeling the region the *rest of the world* (ROW). The world market, as in the center panel of Figure 11.8, becomes an analytical link between the United States and the ROW. The model then becomes a useful tool for analyzing the effects of policy changes in the United States. Both domestic policy changes and trade policy changes can be analyzed.[21]

U.S. Deficiency Payments in an Open Economy

The Agricultural Acts of 1977, 1981, 1985, and 1990 retained the concept of a target price-deficiency payment as a means of supporting income of U.S. farmers. The effects of this program were discussed and analyzed in a previous section but did not include selling grain in the export market. Because the export market for U.S. grain has become such an important component of total demand for U.S. grains, it is necessary to add that dimension to the analysis.

When the government guarantees farmers a price above the market price for their grain, larger production is stimulated. However, with the deficiency payment approach, the government does not buy grain and store it. Instead, all grain enters the market, and the government makes up the difference between the market price and the target price.

In Figure 11.9, P_0 is assumed to be the equilibrium price in the United States and the ROW before any government intervention. The price is determined by the intersection of the export supply curve and the import demand curve in the world market. Exports are OQ_0.

A deficiency payment does not cause the supply curve to shift. Instead, production moves along the supply curve. Farmers produce and sell Q_4, which is determined by price P_1, the target price. In this case, farmers receive P_2 from the market, plus P_1P_2 as deficiency payment from the government. In effect, the U.S. supply curve below price P_1 is irrelevant and can be considered to be vertical below P_1. The effect of this modification is to cause the excess supply curve to become kinked at price P_1. The new excess supply curve is $ABES_{US}$. The intersection of the kinked supply curve with ED_{ROW} determines the price P_2, which exists in ROW, the world market, and in the United States. U.S. farmers get the deficiency payment in addition to the market price.

As a result of the target price being above the market price that would have prevailed, U.S. output expands, U.S. exports are larger, and the price in the world market is lower. U.S. consumers buy a larger quantity, Q_1 to Q_2. Producers in the ROW receive a lower price for their production. Production in ROW is reduced, but consumers buy more because of the lower price that occurs with increased imports.

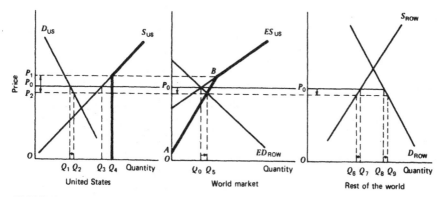

FIGURE 11.9 Effects in the United States and the rest of the world of a U.S. deficiency payment policy.

To minimize government cost of the deficiency payment program and to minimize complaints from ROW producers, the United States may impose production restrictions in the form of an acreage set aside. In this case, the quantity produced could be restricted so that the quantity exported would remain at the level that existed before the deficiency payment program went into effect.

The Impact of EC Variable Levies on U.S. Exports

Other countries frequently have domestic programs that lead to restrictions on exports of U.S. products. The Common Agricultural Policy (CAP) of the European Community is a well-known example that affects a large market for U.S. grain and oilseeds. The CAP uses a system of target prices for grain (somewhat similar to the system used in the United States since 1973, although the mechanisms for implementing the price support are different) that guarantees a specific price level to producers. Market prices are maintained near that level by imposing a variable levy on imports. When the import price drops below the target price, the levy increases to offset the difference. Figure 11.10 is a graphic analysis of the effects of the variable levy on the United States as well as in the EC.

It is assumed that the initial equilibrium price and quantity exported are P_0 and Q_0 as indicated in the center panel. When the EC sets its target price at P_1, the rest of the world would attempt to increase production and exports to Q_1. To avoid this flood of imports, the EC would charge a variable levy of P_1P_2 on each bushel of imports. The price P_2 would be determined by the intersection of the new kinked ED_{EC}

FIGURE 11.10 Effects in the rest of the world and the EC of a price support policy including a variable levy in the EC.

import demand curve and the ES_{ROW} curve. This price would prevail in the ROW. Regardless of how low the price fell in the ROW, if, for example, the S_{ROW} shifted right, imports would not increase.

The consequences of this policy are higher prices in the EC, larger domestic production, less domestic consumption, and reduced imports. In the ROW, there would be lower producer and consumer prices, reduced production, and increased consumption. How large these effects will be depends on the size of the EC market relative to its suppliers, the amount by which the target price exceeds the market price, and the elasticities of demand and supply in both the EC and the ROW. Because the United States was the world's largest supplier of both feed grains and wheat to the EC when the CAP was implemented, a major part of the impact on ROW suppliers initially fell on the United States.

As EC producers expanded their production in response to the supported prices, U.S. producers of wheat and feed grains were virtually squeezed out of the EC market. As production of wheat continued to expand in the EC, the community became a large net exporter of wheat. Exports were possible only through the use of export subsidies. As the United States lost its export market in the EC and in markets that received subsidized wheat from the EC, the United States responded by using export subsidies in an effort to regain export market shares. U.S. and EC competition for markets in third world countries generated complaints from other traditional suppliers of wheat such as Argentina, Australia, and Canada that they were the victims of the subsidy wars engaged in by the United States and EC. These developments have contributed to strongly held positions by members of the GATT in the Uruguay Round of trade negotiations.

Government Programs to
Expand Foreign Demand for U.S. Grain

The U.S. government has participated in two types of programs to expand foreign demand for agricultural products. Public Law 480 facilitates the shipment of about $1 to $1.5 billion of agricultural products per year. Since 1954, the dollar amount shipped under PL-480 has remained relatively stable. In the mid-1950s, PL-480 shipments varied from 25 to 40 percent of all U.S. exports of agricultural products. By 1980, commercial exports had grown to such an extent that PL-480 shipments were only about 3 to 4 percent of total shipments. About one-third of all PL-480 shipments are donations of food to charitable organizations that distribute food to very low-income people. The other two-thirds of PL-480 shipments are to countries that buy the grain using long-term credit arrangements with the United States. Credit may be extended for as many as 40 years, with interest rates of 2 or 3 percent per year.

Market Development Activities under PL-480

The government also participates in market development activities under PL-480 authorization. Market development typically consists of activities jointly sponsored by the government and U.S. producer groups. The United States provides market development services through its system of agricultural attachés that is a part of the U.S. foreign service stationed abroad. Examples of market development activities include introducing new products in foreign markets, providing technical services to potential users of products, training foreign persons in the use of U.S. products, and participating in school lunch programs abroad.

CCC Credit for Export Sales

In addition to PL-480 market development activities, the United States provides a limited amount of credit for purchase of U.S. commodities. Credit is extended through the CCC under terms that are similar to those extended by commercial lenders. Money is loaned at a rate slightly above the U.S. prime rate.

The 1990 act reauthorizes the Short-Term Export Credit Guarantee Program (GSM-102) and the Intermediate-Term Export Credit Guarantee Program (GSM-103). The GSM-102 program guarantees repayment of credit of up to 3 years to finance export sales of privately owned stock. The GSM-103 program guarantees repayment of loans of between 3 and 10 years that will directly benefit U.S. agricultural producers. Minimum funding levels for these two programs are $5 billion and $500 million per year, respectively.[22] If these amounts of credit are truly a net addition to credit available for financing grain sales abroad, they would represent some expansion in demand for U.S. grain exports. It is an attempt to meet the competition for U.S. grain sales when U.S. competitors, primarily Canada and Australia, provide credit to facilitate their grain sales.

Price Stabilization and Food Security

Grain prices tend to fluctuate from day to day and over time for several basic reasons. Demand for grain tends to be both price and income inelastic. Supply tends to be inelastic in the short run. The quantity produced in any production period may be significantly changed because of weather conditions. Markets are characterized as having a large number of producers (sellers), each with no control over price.

Price Stabilization Prior to 1972

Since the early 1930s, the federal government has tried to remove some of the instability from the grain market. Usually, it relied on loan programs and government purchase of grain as means of placing a floor under prices. Emphasis was on keeping prices up rather than holding them down. Although specific price stability objectives were not stated in legislation or by administrators of the legislation, the availability of grain in CCC storage served as a stabilizing factor in the market until 1972. The existence of large government stocks and resale policies adopted by CCC kept prices in a relatively narrow range (see Figures 11.2 and 11.3).

Large grain exports from the 1972 and 1973 crops, and the rapid draw-down of government stocks followed by a short U.S. corn crop in 1974, led to much greater price variability than had been encountered in the two previous decades. Policy efforts were directed toward reducing the amount of variability in U.S. grain prices. Much of the instability originated from the foreign markets, so managing grain exports became of major interest.

Embargoes and Trade Suspensions

As the U.S. produces much more grain and soybeans than are consumed domestically, internal price variability could be reduced through control of exports. The United States tried this strategy three times during the 1970s. Embargoes were placed on additional shipments of grain and/or oilseeds and were kept on for short periods until supply conditions became known with greater certainty. The embargoes were applied at times when grain prices had risen sharply and caused short-run prices to fall.

The use of embargoes generated much unfavorable reaction from farmers and the grain trade. It was argued that the immediate repercussions of the embargoes had fallen on farmers who had been urged to expand production. Furthermore, the long-run effect would be to stifle export growth. The use of embargoes became a sensitive political issue.

Antiembargo support led to a provision in the Agricultural Act of 1977 that required that loan rates be raised to 90 percent of parity for the commodity if, based on determination of short supply, the president or any member of the executive branch suspended the commercial export sale of any named commodity. Wheat, corn, grain sorghum, soybeans, oats, rye, barley, rice, flaxseed, and cotton were included.

The January 4, 1980, suspension of sales of grain to the Soviet Union in response to its invasion of Afghanistan was not covered by the above provision. Specifically, sales were not suspended because of "determination of short supply" but as a means of attaining foreign-policy objectives.

U.S.-USSR Grain Agreement

Because a major source of price instability in U.S. grain markets was believed to be a result of unpredictable sales of grains to the Soviet Union, the United States sought to stabilize its purchases. The instability arises principally from variable import needs that are caused by variations in the Soviet Union's grain production. Historically, the Soviet Union has entered world grain markets when its production was reduced below expected needs. When production recovered, imports were reduced or stopped.

The U.S.-USSR five-year agreement, signed in 1975, called for the Soviet Union to purchase at least 6 mmt of grain per year from the United States. Up to 8 mmt could be purchased without further U.S. approval. Minimum purchases were to be divided between corn and wheat and were to be spread evenly over the year. The government of the Soviet Union could request to purchase larger quantities, and the United States agreed to offer more if production conditions warranted.

For the first four years under the agreement, purchases were as follows: 1976-1977, 6.1 mmt; 1977-1978, 14.6 mmt; 1978-1979, 14.8 mmt; and 1979-1980, 7.94 mmt. Prior to the suspension of sales in 1980, the Soviet Union had contracted to buy about 21.7 mmt. The United States had given authority for purchase of 25 mmt. As a result of the suspension, sales of 13.7 mmt were canceled, and approval of the additional sales was withdrawn.

Research to determine the impact of the agreement on U.S. grain sales used a simulation model that took into account the Soviet Union's yield variability and variable import needs.[23] The research showed that the combined purchases of the Soviet Union and Eastern Europe would increase by 2.8 mmt of grain per year with the agreement. The research also showed that the agreement would probably have no effect unless the purchase agreements between the United States and Poland remained in force. Without the latter agreement, the Soviet Union tended to buy U.S. grain and sell it to Poland in years when the Soviet Union had crops sufficient to meet domestic needs. In those cases, Polish purchases from the United States were decreased.

The United States and the former Soviet Union have continued the purchase agreement concept with the most recent agreement, which was signed in June 1990. It covers the five-year period through December 31,

1995. The agreement provides for a minimum purchase of 4 mmt of wheat and 4 mmt of feed grains. The Soviet Union agreed to purchase an additional 2 mmt of wheat and/or feed grains and/or soybeans and/or soybean meal provided that one ton of soybeans and/or one ton of soybean meal could be counted for two tons of wheat and/or feed grains. This long-term grain agreement became ineffective with the breakup of the Soviet Union. Current grain exports to the former Soviet Union are made under various food assistance and credit programs.

The Farmer-owned Grain Reserve

The Food and Agriculture Act of 1977 authorized a farmer-owned grain reserve. The act provided for the accumulation of grain in the reserve when prices were relatively low. Ownership of the grain was to be retained by farmers who would store grain on their farm or in commercial facilities for periods of not less than three nor more than five years. Participants would receive a nonrecourse loan on the commodity from the CCC. They could enter the grain in the reserve at the end of the initial storage period or sooner if authorized by the secretary of agriculture. Grain would be released from the reserve as prices rose above specified levels. Farmers would receive storage payments for grain stored in the reserve. The secretary of agriculture was authorized to waive interest charges on grain in the reserve.

The legislation authorized a wheat reserve of not less than 300 million bushels nor more than 700 million bushels. The secretary of agriculture was given authority to develop a reserve of feed grains, but no quantity was specified. In administering the law in 1978, he established an initial target of 17 to 19 mmt of feed grains.

Grain was to enter the reserve at the existing loan rate. Grain could be released from the reserve when the market price reached a specified percent of the loan level. Loans were to be called when market prices rose to a higher specified percent of the loan rate. The release level for wheat as specified in the act was to be not less than 140 percent of the loan level nor more than 160 percent of the then current level of price support. Release and call levels for corn were set at 125 and 140 percent of the loan level.

Penalties were specified for withdrawal of grain from the reserve before the price rose to the release level. A penalty of one-fourth the loan rate plus repayment of storage payments was specified. The CCC was restricted from selling any grain that it owned at less than 150 percent of the loan rate.

In 1979 and 1980, the United States was gaining experience from operation of the reserve. During mid-1979, grain prices rose above release levels for wheat and feed grains. When the size of the 1979 corn harvest became evident, corn prices declined below the release level. The authorization for release was withdrawn. Corn began to reenter the reserve, especially after the secretary of agriculture authorized additional incentives to get corn in the reserve. Following the 1980 suspension of sales to the Soviet Union, additional corn entered the reserve. Later, when it became apparent that the 1980 corn crop would likely be reduced because of drought and heat damage, prices again exceeded the release level and approached the call level.

Although the farmer-owned reserve appeared to be working as intended—accumulating grain during periods of low prices and feeding it back into the market during periods of higher prices—the reserve did not have specific price-stabilization objectives. In 1978, 1979, and 1980, it probably kept prices from falling as low as they would have without the reserves. The reserve does not contain sufficient grain stocks to keep prices from rising significantly if a major shortfall were to occur in world grain production. Because farmers retain control over the grain, subject to prescribed rules, grain may not be fed back into the market as prices exceed the release level. Farmers may choose to hold for further price rises. Some analysts argue that the government should have more control to require farmers to put grain on the market when prices reach the release level rather than let farmers continue to speculate. This argument assumes that the government is more capable of assessing future production and consumption needs than the larger number of participants in the market.

Food Security

In response to heightened world food concerns, a World Food Conference was held in Rome in 1974, under the auspices of the United Nations. Since that conference, the United States has played a more active role in efforts to achieve greater world food security. PL-480 has made it possible for the United States to pledge grain to meet the goal of 10 mmt from all nations. In 1979, the United States raised its pledge to 4.7 mmt of cereal grains as our minimal annual commitment to food aid while negotiations for a new Food Aid Convention were in progress. Although this was a multilateral commitment on the part of the United States, it was intended to encourage other donor countries to implement pledges they had made earlier.

Creation of a food security reserve represents some increase in demand for wheat, at least during the time the reserve quantity is being acquired. More importantly, it provides a type of food insurance at a time when food supplies might be short. Grain could be purchased from the market, but experience shows purchases for humanitarian needs get scaled down after prices have risen sharply.

General Economic Policy

General economic policies of the nation affect both the prices that farmers receive for their grain and the incomes of grain-marketing firms. Monetary and fiscal policies affect interest rates and foreign-exchange rates, which, in turn, affect the prices that farmers receive for grain.

Changes in interest rates affect prices mainly through changes in the cost of holding grain in storage. In early 1980, the prime interest rate rose from about 12 to 20 percent, as the Federal Reserve Bank sought to slow inflation by curtailing the rate of growth in the money supply. An interest-rate increase of 8 percentage points would raise the interest cost for holding a bushel of soybeans (valued at $6) by 24 cents for a six-month period. This increase would cause more soybeans to be placed on the cash market and would lower prices accordingly. A similar change in interest rates would raise the six-month cost of holding a bushel of corn valued at $2.50 by 10 cents. During periods of tight credit, some farmers would probably be unable to get operating credit and would be forced to sell grain to meet expenses. This situation would put further downward pressure on prices.

From the Bretton Woods Conference in 1944, the United States and its major trading partners operated until 1973 with fixed exchange rates for each currency. Under this system, the U.S. dollar had an important role as a major reserve currency. As a supplier of a major reserve currency, the United States ran a persistent deficit in its balance of payments.

This system worked reasonably well until inflation accelerated in the United States in the late 1960s. Gradually, the U.S. dollar became overvalued at fixed exchange rates relative to other currencies. Foreigners became less willing to hold U.S. dollars at established prices.

In a 13-month period in 1972 and 1973, the United States moved to a market-determined exchange rate. With persistent high rates of inflation in the United States during the 1970s, the dollar declined in value relative to other major world currencies. Rates fluctuated relative to the German mark and Japanese yen during the 1970s, but the general trend was downward.

A declining exchange rate contributes to higher U.S. grain prices. A devaluation of the dollar means a revaluation of the other currency, so holders of the other currency (e.g., marks) have to put up fewer units of them to buy a unit of U.S. grain. At lower prices, foreigners tend to purchase more U.S. grain. This increased buying leads to higher prices in dollars, especially if the supply of grain in the United States is relatively inelastic.

Monetary and fiscal policy has other diverse effects on grain prices. During inflationary periods, people seek to protect their wealth by acquiring assets that increase in value along with the rate of inflation rather than hold their wealth in money. They may buy precious metals, commodities, or land in an attempt to protect their wealth. Thus, money may move in and out of the commodity market affecting commodity prices as a result of changing monetary and fiscal policies.

Inflation affects commodity prices in other less direct ways. Although commodity prices do not necessarily advance with the rate of inflation in the short run, they tend to advance in the long run because of changes in the cost structure for producing grains. Farmers purchase over 80 percent of their inputs from the industrial sector, which operates on a cost-plus basis and is able to pass on its costs due to its form of industrial organization. As higher costs are passed on to farmers, they may be in a cost-price squeeze in the short run since they cannot pass on higher costs. In the longer run, however, producers adjust production and/or revalue their production assets to cover their full costs of production. Farmers receive help from government in doing this when target prices are adjusted upward based on changes in the cost of production.

The introduction of this chapter noted that the federal government attempts to affect performance of the economy by trying to make it more competitive. Grain marketing, especially in export markets, is concentrated in the hands of a small number of firms. It has been estimated that over 50 percent of all U.S. grain exports are handled by the six largest international firms. Cooperatives handle only about 7 percent of U.S. grain exports.[24]

The federal government does not have any explicit policies that deal with market structure in terms of the number, size, and location of plants and firms in the grain-marketing industry.[25] Instead, it tends to be concerned over mergers and their potential impacts. In recent years, the federal government has attempted to promote competition in export grain marketing by encouraging cooperatives to increase their share of the market. These actions have been undertaken through the USDA's Farmer Cooperative Service.

Policies for Alternative Marketing Systems

As noted in the introduction, the U.S. grain marketing system is essentially a privately owned system, rather than a government-owned or operated system. The U.S. government influences the system through the many policies discussed in this chapter.

The large increase in grain exports during the 1970s and the large price variations that occurred have raised questions about the conduct and performance of the system. These questions arise, in part, because of the way the world grain-marketing system is organized. The concentration of market power, and the differences among countries in the organization of their markets, raise questions about equitability of treatment of the many participants in the system, especially U.S. farmers.

Grain marketing in Canada and Australia is conducted by government-sanctioned marketing boards. The boards are jointly controlled by representatives of farmers, consumers, and governments and are given monopoly control over most domestic sales to flour mills and the export marketing of grain. Farmers pool their wheat and, in effect, turn over the marketing functions to the board. They share in proceeds from the pool after all wheat is sold. Therefore, farmers receive an average-for-the-year price for their wheat.

Many countries, such as the former Soviet Union, China, and India, have state purchasing agencies that have responsibility for purchase of all imported grain. In countries such as the former Soviet Union, the purchasing agency has at least one advantage when dealing with firms in a market economy. They have a high degree of control over information about domestic crop conditions, internal feed and food needs, and government policy decisions about imports. This control of information may enable them to manipulate world markets and exert a degree of monopsony power in purchasing grain.

It has been estimated that as much as 90 percent of all wheat that enters world trade involves state traders or marketing boards on at least one side of the transaction. Given this large variation from the competitive model, various proposals have been made to change the U.S. system to one that would more effectively deal with state traders. These proposals include establishment of a marketing board for the United States, or turning all export marketing over to the CCC. Other proposals include joining with Canada, Australia, the EC, and Argentina to form a cartel for marketing wheat.

These and other proposals will likely continue to receive discussion and study. However, interest in these approaches may diminish as a result of the breakdown of communism and central planning in Eastern Europe and the former Soviet Union. The current state of knowledge is too inadequate to determine whether U.S. producers would benefit from any of the proposals, or whether any of the proposals would be feasible in the long run.

Summary

Grain marketing in the United States is conducted by a privately owned grain-marketing system. Although the system is private, it has been molded by various government policies and programs. These policies affect both the system and the prices that farmers received for their grain.

Policies have been designed to attain economic, political, and social stability, economic growth, and equality of opportunity for U.S. citizens. The United States has also been willing to share its agricultural abundance with other countries. The process by which the U.S. system formulates policy is a slow-moving, complex one, which tries to reconcile the diverse interests of many conflicting groups.

Agricultural policies have followed two general trends. On one hand, policies were designed to make the private production and marketing system work more efficiently. These types of policies predominated prior to the 1930s and continue today, but are often overshadowed by other policies, which have their roots in the 1930s. Since the 1930s, the government has intervened directly in markets to affect both the marketing system and the prices that producers receive for their grain.

Policies have changed over time and will continue to do so in response to changing technical and economic conditions in the United States and in the world. Trying to understand government policies for grain production and marketing without study of the technical and economic environment is like trying to understand the behavior of a boat in the river without observation and study of the water and the currents in which it is operating.

A graphic approach was presented for analysis of commodity programs in a closed economy. The partial equilibrium approach enables the analyst to determine the primary effects of a program on price, production, consumption, storage, and government costs. Inferences can be drawn about the effects of the program on the marketing system.

A brief history of trade policies in the United States points out variations in U.S. trade policies, from restricting trade to freeing up trade, and back again to restricting trade. Since the mid-1930s, efforts have been made to promote freer trade even though protectionist tendencies always lurk near the surface.

Recognizing the dramatic growth of trade in agricultural products that occurred during the 1970s, the graphic approach was expanded to analyze programs in an open economy. This approach makes it possible to show how U.S. grain prices are determined by the interaction of supply and demand in the United States with supply and demand in importing countries. The tendency for many countries to develop domestic programs to influence farm prices and incomes within their own country has important repercussions for trade in agricultural products. Domestic objectives generally take precedence over trade objectives, so trade policies tend to be modified accordingly.

Because the export market has become such a large component of total demand for U.S. grain, U.S. prices have become more variable than they were during the 1950s and 1960s. The government has used embargoes, an export-reporting system, and long-term agreements with several nations to try to stabilize world and U.S. grain prices. A farmer-owned grain reserve, which accumulates grain during periods of abundance and feeds it back into the system during periods of below-trend production, has been developed in the United States. The United States continues to contribute to food security programs for less developed countries.

General economic policy, particularly monetary and fiscal policy, affects grain marketing. Prices are influenced through changes in interest rates and foreign exchange rates.

Six multinational grain firms dominate private world trade in grain. Their ability to serve U.S. farmers' needs when facing a world characterized by national marketing boards and state trading agencies continues to be questioned and evaluated.

Notes

1. Dale E. Hathaway, *Government and Agriculture*, The Macmillan Co., New York, 1963, p. 3.

2. No President since Lyndon Johnson has sent a special message on agriculture to the Congress.

3. "President Johnson's Message on Agriculture," transmitted to the 88th Congress, 2nd Session, January 31, 1964, printed in the *Congressional Quarterly Almanac* (Congressional Quarterly Service, Inc., Washington, DC, 20: 887, 1964).

4. Public Law 97-98, 85 Stat. 1213, 97th Congress, December 22, 1981.

5. Ronald Cotterill, *Cartel and Embargoes as Instruments of American Foreign Policy*, Agricultural Economics Report No. 373, Michigan State University, East Lansing, April 1980, p. 39.

6. Stephanie Mercier, *Corn, Background for 1990 Farm Legislation*, USDA, ERS, CED, Washington, DC, September 1989, p. 33.

7. William C. Bailey and James Langley, "The Role of Generic Certificates in U.S. Commodity Programs," *Agricultural Food Policy Review*, USDA, ERS, AER No. 620, November 1989.

8. Bailey and Langley, ibid., p. 109.

9. See Bailey and Langley, ibid., for an example, p. 110.

10. In 1982 the participation rate in the corn program was 29 percent. By 1987 the rate exceeded 90 percent of base acreage. For wheat, participation rates grew from 48 percent to 87 percent over the same time period.

11. For example, in 1990 a corn producer could receive price support payments on 85 percent of base acres after meeting the 7.5 percent set-aside requirement. The so-called flex acres could be planted to alternative nonprogram crops such as soybeans and still retain base history for future program payments.

12. Initially, the analytical approach includes only domestic consumers. It

assumes all trade is domestic or that the international market is unimportant to the analysis. The international market is explicitly introduced in subsequent sections.

13. For simplicity it is assumed all consumers demand unprocessed wheat. Introducing marketing margins and derived demands would only complicate the analysis and add little to the analysis at this point.

14. An important feature of CCC commodity loans has been that they are nonrecourse loans to eligible producers. The producer is loaned a specified amount of money per unit of the commodity (the loan rate). At the end of the loan period, usually one year or less, the borrower must pay off the loan plus interest or forfeit the grain to the CCC. If market price is less than the loan rate, the CCC considers the loan as fully paid with no further recourse to the borrower. The bulk of the grain acquired by CCC has been acquired in this manner.

15. Agricultural Stabilization and Conservation Service, *Acquisition and Disposal of Farm Commodities by CCC*, USDA, ASCS Background Information, B.I. No. 3, Washington, DC, October 1975, p. 5.

16. This provision sometimes generated controversy over administration of the program. Because determination of "going out of condition" (deteriorating in quality) requires a degree of subjective judgment, managers of the inventories were sometimes accused of using this provision to encourage participation in voluntary programs.

17. Robert L. Tontz, "U.S. Trade Policy: Background and Historical Trends," *U.S. Trade Policy and Agricultural Exports*, Elizabeth S. Ferguson, ed., Iowa State University, Ames, 1973, pp. 17-48.

18. Mary E. Ryan and Robert L. Tontz, "A Historical Review of World Trade Policies and Institutions," *Speaking of Trade: Its Effect on Agriculture*, Gail McClure, ed., University of Minnesota, Minneapolis, Special Report No. 72, 1978, pp. 5-19.

19. An additional assumption is important for simplifying the analysis. Although each country has its own form of money—dollars and guilders—we assume the two currencies and all prices are quoted in dollars.

20. For simplicity, transportation costs have been assumed to be zero. When transportation costs are introduced, prices tend to be different in the two countries by the amount of the transportation cost per unit. Price would be lower than P_0 in the United States and higher than P_0 in the Netherlands.

21. Only two cases are illustrated here. A larger number of policies are analyzed in "Interrelationships of Domestic Agricultural Policies and Trade Policies," by Bob F. Jones and Robert L. Thompson in *Speaking of Trade: Its Effect on Agriculture*, Gail McClure, ed., Special Report No. 72, University of Minnesota, Minneapolis, 1978, pp. 37-68.

22. Susan L. Pollack and Lori Lynch, eds., *Provisions of the Food, Agriculture, Conservation and Trade Act of 1990*, Agricultural and Trade Analysis Branch, ERS, USDA, Agric. Info. Bull. No. 624, 1991, p. 70.

23. James H. Hilker and Bob F. Jones, *A Stochastic Simulation Analysis of the U.S.-U.S.S.R. Grain Purchase Agreement*, Department of Agricultural Economics, Agricultural Experiment Station Bulletin No. 356, Purdue University, West Lafayette, December 1981.

24. Coops are involved in a large share of U.S. grain exports but not through all stages of the exporting process. The five large companies sell grain both c.i.f. and f.o.b., leaving the final stages to the private firms.

25. Dale C. Dahl, "Public Policy Changes Needed to Cope with Changing Structure," *American Journal of Agricultural Economics*, Vol. 57, No. 2, May 1975, pp. 206-213.

Selected References

Agricultural and Trade Analysis Division, *Agricultural and Food Policy Review: U.S. Agricultural Policies in a Changing World*, USDA, Economic Research Service, Ag. Ec. Report No. 620, November 1989.

Agricultural Stabilization and Conservation Service, *Acquisition and Disposal of Farm Commodities by CCC*, USDA, ASCS Background Information B.I. No. 3, Washington, DC, October 1975.

Cotterill, Ronald, *Cartel and Embargoes as Instruments of American Foreign Policy*, Agricultural Economics Report No. 373, Michigan State University, East Lansing, April 1980.

Dahl, Dale C., "Public Policy Changes Needed to Cope with Changing Structure," *American Journal of Agricultural Economics*, Vol. 57, No. 2, May 1975.

Export Grain Sales, hearing before the Subcommittee on SBA and SBIC Authority and General Small Business Problems of the Committee on Small Business, House of Representatives, 96th Congress, First Session, June 11, 1979, U.S. Government Printing Office, Washington, DC, 1979.

Hathaway, Dale E., *Government and Agriculture*, The Macmillan Company, New York, 1963.

Hilker, James H., and Bob F. Jones, *A Stochastic Simulation Analysis of the U.S.-U.S.S.R. Grain Purchase Agreement*, Department of Agricultural Economics, Agricultural Experiment Station Bulletin No. 356, Purdue University, West Lafayette, IN, December 1981.

Jones, Bob F., and Robert L. Thompson, "Interrelationship of Domestic Agricultural Policies and Trade Policies," in *Speaking of Trade: Its Effect on Agriculture*, University of Minnesota, Special Report No. 72, Minneapolis, MN, 1978.

Mercier, Stephanie, *Corn Background for 1990 Farm Legislation*, USDA, ERS, CED, Washington, DC, September 1989.

Rasmussen, Wayne D., and Gladys L. Baker, *Price Support and Adjustment Programs from 1933 through 1978: A Short History*, USDA, Economics, Statistics, and Cooperative Service, AIB-424, Washington, DC, February 1979.

Ryan, Mary E., and Robert L. Tontz, "A Historical Review of World Trade Policies and Institutions," in *Speaking of Trade: Its Effect on Agriculture*, Gail McClure, ed., University of Minnesota, Special Report No. 72, Minneapolis, MN, 1987, pp. 5-19.

Talbot, Ross B., and Don F. Hadwiger, *The Policy Process in American Agriculture*, Chandler Publishing Company, San Francisco, 1968.

Thurston, Stanley K., Michael J. Phillips, James E. Haskell, and David Holkin, *Improving the Export Capability of Grain Cooperatives*, USDA, Farmer Cooperative Service, FCS Research Report No. 34, Washington, DC, June 1976.

Tontz, Robert L., "U.S. Trade Policy: Background and Historical Trends," in *U.S. Trade Policy and Agricultural Exports*, Elizabeth S. Ferguson, ed., Iowa State University Press, Ames, IA, 1973, pp. 17-48.

12

Canada's Grain-marketing System

Colin A. Carter

Almost 90 percent of Canada's grain is produced in three western prairie provinces: Alberta, Saskatchewan, and Manitoba. Of this production, more than 50 percent is exported. Grain was the major contributor to the economic development of western Canada, and the grain industry today remains as one of Canada's most important sectors. It ranks second, behind the lumber industry, as an earner of foreign exchange.

The development of the early system of marketing grain in Canada closely paralleled that of the United States. Because the establishment of the grain economy in Canada lagged behind that of the United States by about 40 years, it was inviting for Canadians to adopt the U.S. open market grain-marketing system, one that had many of its inefficiencies already worked out.

The first export shipment of wheat from the Canadian prairies took place in 1876. From the time of this shipment until World War I, the marketing was handled by private companies, as it was in the United States. Circumstances surrounding the two world wars, and the Great Depression between them, resulted in Canada taking a much different approach than the United States to the marketing of its major export commodity, grain.

The following discussion traces the origins of the present grain-marketing policies and institutions in Canada, and it outlines the workings of the Canadian grain-marketing system. There is an intricate mixture of government, cooperative, and private enterprise in the Canadian grain markets, and for this reason, it is a complicated system. Only its major characteristics and institutions are described in this chapter.

Grains are marketed in Canada through one of three channels: (1) the Canadian Wheat Board (CWB), (2) the dual CWB-open market system, or (3) the open market. For some types of grain, only one channel is available as a market, and for others, the producers have a choice.

As an aid to the following discussion, Table 12.1 presents production and export figures for the major Canadian grains. Normally, about 50 percent of Canada's seeded acreage is planted to wheat every year, making it by far the most important crop. About 90 percent of the 26 mmt of wheat produced in an average year is hard red spring, and the remainder is mostly durum, which is used for pasta products.

Canada normally exports more than 75 percent of its wheat, and it is the second largest exporter in the world, behind the United States. At one time Canada was the largest wheat exporter; however, during the 1960s and 1970s, its market share fell behind that of the United States. Most of the wheat produced in Canada is marketed through the CWB.

In world barley production, Canada ranks second behind the former Soviet Union. Of the 12.5 mmt produced, about 4.8 million are exported, on average, making Canada the world's largest barley exporter, with France a close second. Less than one-half of the barley crop is marketed through the CWB. Canada is the world's largest producer of rapeseed (canola), the second largest producer of flaxseed (behind Argentina), and a major exporter of both of these oilseeds. These oilseeds are handled in the open market. The majority of the oats produced in Canada is marketed for feed usage outside of the elevator system through farm-to-farm and farm-to-feedlot sales.

Approximately 6.5 mmt of corn are produced annually, primarily in eastern Canada. This production is outside of the designated CWB area, and its marketing is mainly through local direct sales for feed purposes. On an individual basis, the remaining crops of rye, soybeans, sunflowers, fababeans, lentils, and forage seeds are produced on a smaller scale.

Historical Development

Prior to the turn of the twentieth century, the main marketing problems in Canada were related to grading standards and the preservation of competition at local delivery points, where railways were granting line elevator companies monopoly loading rights. The passing of the General Inspection Act (1886), the Manitoba Grain Act (1900), and the Canada Grain Act (1912) by the federal government, cumulatively established stringent statutory grades and provisions for the supervision of cleaning and shipping grain. This legislation corrected the early marketing problems, encouraged competition in the industry, and established a reputation for Canada as a producer of quality grain. This valuable reputation for high quality standards has been maintained throughout the years.

TABLE 12.1 Canadian Grain Production and Exports (thousands of metric tons)

Grain Crop	Crop Year (August 1 through July 31)					
	1986/87	1987/88	1988/89	1989/90	1990/91	Average (1986-1991)
Wheat						
Production	31,378	25,992	15,996	24,575	32,709	26,130
Exports	20,783	23,519	12,419	17,425	22,104	19,250
Oats						
Production	3,251	2,995	2,993	3,546	2,851	3,127
Exports	257	286	728	737	383	478
Barley						
Production	14,569	13,957	10,216	11,674	13,925	12,868
Exports	6,719	4,594	2,879	4,506	4,635	4,667
Rye						
Production	609	493	268	873	713	591
Exports	201	221	115	295	342	235
Flaxseed						
Production	1,026	729	373	497	936	712
Exports	690	624	455	498	494	552
Rapeseed (Canola)						
Production	3,787	3,846	4,288	3,096	3,281	3,660
Exports	2,126	1,750	1,949	2,048	1,888	1,952
Corn						
Production	5,911	7,015	5,369	6,379	7,157	6,366
Exports[a]	(499)	188	(946)	(534)	(381)	(434)
Soybeans						
Production	960	1,270	1,153	1,219	1,292	1,179
Exports	(70)	35	135	(94)	24	6

[a]Numbers in parentheses are imports.

SOURCE: Canada Grains Council, *Statistical Handbook '91*, Canada Grains Council, Winnipeg.

Price formation in the early years took place on the Winnipeg Grain and Produce Exchange, which was established in 1887. This exchange was originally a cash grain market but, following the model of the Chicago markets, it commenced trading in wheat futures in 1904. The marketing of all Canadian grain was handled by the open market until World War I broke out in 1914. The war had a major impact on the structure of the grain markets, primarily because the British government took control of grain export licenses from Canada during this period. The import purchasing agency of the British government cornered the wheat futures market in the spring of 1917; as a result, the futures market was closed. For the first time the Canadian government became directly involved in the marketing of grain through the appointment of the Board of Grain Supervisors, who were charged with selling the 1917 and 1918 wheat crops.

For the entire 1919 crop the government temporarily appointed a Canadian Wheat Board. It borrowed the Australian system of pooling sales, which gave producers an initial cash payment at the time of delivery and then a subsequent final payment. The final payment was based on net returns to the Board during the crop year.

This Canadian Wheat Board was never intended as a permanent agency; consequently, wheat futures trading resumed in Winnipeg in 1920. During the 1920s it became the most active wheat futures market in the world. At that time the price of a membership on the Winnipeg exchange was as high as $35,000, while memberships on the Chicago Board of Trade were approximately $2,000. Today the Chicago memberships sell for over $400,000, compared to less than $5,000 in Winnipeg.

The farmer cooperative movement began sweeping western Canada in the early 1920s, as farmers placed some of the blame for low grain prices on the private grain merchants. Producers were requested to join voluntary marketing pools for a minimum of five years. The campaign was very successful, and the participation rate was high. By 1930 the Alberta, Saskatchewan, and Manitoba Pools had over 50 percent of farmers' marketings. At this time they also had control of approximately 50 percent of the total country elevator capacity. This control was a tremendous feat to achieve in a short period of about seven years. Disaster struck the pools at the pinnacle of their success however. On behalf of the three prairie pools, their Central Selling Agency (CSA) marketed their grain. It advanced an initial payment to farmers at the beginning of the year and pooled returns much as the CWB had done in 1919. However, the CSA's failure to forward sell, or hedge, the 1929 crop resulted in massive losses as the grain prices fell drastically that year. This incident marked the end of pure farmer cooperative grain marketing on a large scale in Canada.

The federal government intervened at this point to guarantee the bank loans on the debt amassed by the pools, and it became directly involved in the grain markets, controlling the operations of the CSA for the next five years. The government supported wheat futures prices by purchasing wheat futures contracts, and in 1935 it reappointed the CWB. It was set up as a voluntary marketing board with a government guaranteed floor price. During World War II, the demand for grain raised prices, and in 1943 the CWB was made a compulsory marketing board to control prices in line with the government's anti-inflationary policy. The Winnipeg wheat futures market was simultaneously closed in September 1943. The compulsory board was intended to last two years at the most, but instead it has had sole authority over the marketing of all subsequent wheat crops. The reason the CWB was retained after the war was that most of Canada's wheat was exported to Britain under bilateral agreements, and the CWB made the administration of the agreement much simpler. In 1949 the CWB was also given the sole marketing rights for oats and barley as well as wheat, and in 1967 the passage of the Canadian Wheat Board Act made it a permanent board.

During the 1940s and 1950s, the open market with futures trading continued to operate for the merchandising of flaxseed, rye, and intraprovincial feed grain sales. The CWB participated in this market to a limited extent. In the early 1960s, rapeseed (canola) production increased rapidly in western Canada, and the Winnipeg Commodity Exchange initiated open market trading in this edible oilseed crop.

The monopoly power the CWB had over domestic feed grain sales was relaxed in 1974 when the government allowed for off-board sales of feed grains interprovincially. However, the CWB retained its sole control over feed grain exports. CWB authority over oats was discontinued with the end of the 1988-1989 crop year. This latest change means the CWB is now responsible for only two grains: barley and wheat.

The Canadian Wheat Board

The CWB is an agency of the government of Canada and is primarily a sales agency. It owns no physical facilities for the handling of grain. However, it is the largest grain merchant in the world. As set out in the Wheat Board Act of 1967, its major objectives are the following:[1]

- Market wheat, oats, and barley delivered to it to maximize producer returns;
- Provide producers with initial payments established and guaranteed by the federal government;

- Pool selling prices for the same grain so that all producers get the same basic return for the same grain and grade delivered;
- Equalize deliveries through quotas so that each producer gets a fair share of available markets.

The CWB is composed of between three and five commissioners, who are appointed by the government and one of whom is the chief commissioner. The total staff of the CWB numbers approximately 430. The commissioners seek advice from an advisory committee elected by producers, but they are not responsible to them. Unlike the Australian Wheat Board, the CWB is a government rather than a producer board. The CWB employs the services of both private and cooperative elevator companies to carry out the logistics of physically handling the grain that it buys and sells.

When selling to the CWB, producers' marketing costs are deducted in two stages. Freight costs and primary elevator handling costs are deducted from the initial payment at the time of delivery. The other costs (which include interest, insurance, storage, terminal elevator handling charges, and the board's operating costs) are charged later against the pool.

For the two grains it currently handles, the CWB has monopoly rights over both their exports and domestic sales for human consumption. However, a large percentage of the barley production is marketed locally and does not even enter the elevator system. Approximately 95 percent of the wheat marketings enter the elevator system, in contrast to 60 percent of the barley. An average of about 97 percent of the wheat that enters the elevator system is delivered to the CWB.

At the beginning of each crop year, which runs from August 1 through July 31, the government of Canada establishes initial producer prices. All CWB sales within a given crop year are pooled. Producers thus receive the initial payment at the time of delivery, in some years an adjustment payment during the crop year, and then a final payment once the pool is closed and the CWB deducts its administrative expenses from the pool. Each producer receives the same price (before freight deductions) no matter when the grain is sold to the CWB during a particular crop year. The CWB has four pools: one each for wheat, amber durum wheat, barley, and malting barley. Table 12.2 presents initial, adjustment, and final payments for the two major pools between 1986-1987 and 1990-1991.

Prior to August 1989, domestic sales of wheat by the CWB to millers took place at prices that were insulated from world price levels. This policy was referred to as the two-price wheat policy. It was in effect from 1967 to 1989.

During most of the 1970s the Canadian government fixed the domestic

TABLE 12.2   Canadian Wheat Board Prices Paid to Producers, In-store, Thunder Bay, Ontario (Canadian dollars per bushel)

Crop and Payment	Crop Year				
	1986/87	*1987/88*	*1988/89*	*1989/90*	*1990/91*
Wheat[a]					
Initial	3.54	3.26	4.63	4.49	3.67
Interim	0.00	0.00	0.41	0.19	0.00
Final	0.00	0.38	0.33	0.00	0.00
Total for Year	3.54	3.64	5.37	4.68	3.67
Barley[b]					
Initial	1.74	1.41	2.61	2.18	1.96
Interim	0.00	0.00	0.00	0.26	0.00
Final	0.00	0.20	0.09	0.27	0.00
Total for Year	1.74	1.61	2.70	2.71	1.96

[a]No. 1 Canadian Western Red Spring.
[b]No. 1 Feed.
Note: Farmers receive this Thunder Bay price less freight and handling charges.

SOURCE: Canadian Wheat Board, *Annual Report* (various issues), Winnipeg, Manitoba.

price to mills at relatively low levels and thus subsidized consumers when prices were above these levels.   From 1979 to April 1986, the domestic price was allowed to vary, according to changes in world prices, within a band of $5 to $7 per bushel (for No. 1 Canadian Western Red Spring [CWRS] in-store, Thunder Bay, Ontario), and the government quit subsidizing either producers or consumers if the world price fell outside of this range.   The two-price system was a controversial and complicated pricing policy, and the Canadian government abolished the policy for two major reasons.

The first reason was that the price discrimination policy was encouraging wheat production outside of the CWB designated area—the prairie provinces.   Wheat production in Ontario (outside of the designated area) was growing rapidly, and this development threatened the CWB's control over the domestic market.   The program was unsuccessful as a mechanism to transfer income to farmers.   The two-price system's cost to taxpayers was about $500 million.   Producers gained marginally from the program (in its last few years), and consumers received benefits of close to $500 million.   Most of this "consumer" benefit was probably captured by middlemen, such as the flour millers. During 1986 and 1987 there were significant transfers from consumers to producers because world wheat prices fell dramatically.

The second reason the two-price policy was abandoned was due to the signing of a Canadian-U.S. trade agreement. The agreement goes by the acronym CUSTA, and it took effect in 1989. CUSTA contained provisions for the removal of Canadian grain import licenses. Prior to (but in anticipation of) the removal of import licenses the Canadian government eliminated the two-price wheat policy. Domestic milling wheat prices are now set according to Chicago and Minneapolis futures prices. The CWB guarantees millers a fixed basis vis-à-vis the futures price.

Export sales of wheat and barley by the CWB involve mainly direct sales to national trading agencies representing importing countries. These direct sales account for between 70 and 80 percent of the CWB's exports. This policy is in sharp contrast to the practice of the CWB selling to the private grain trade in the 1950s and 1960s. The rising importance of sales to centrally planned economies and declining importance of sales to western Europe account for this major shift away from sales to intermediaries and toward direct sales.

Less than one-fourth of Canadian wheat exports are currently shipped under long-term agreements (LTAs) with importers. These agreements are signed for various lengths of time and normally cover a minimum volume of trade to be carried out each year during the agreement. Separate sales contracts are negotiated under each LTA, and pricing is often done on a semiannual or quarterly basis. When the LTA is initially signed, specific details such as grades or prices are not usually agreed upon. Before the grain starts moving under the agreement, these are worked out, and normally a flat price applies to shipments for short time periods, after which it is renegotiated. At the present time the CWB has LTAs with different countries including Japan and the former Soviet Union. China failed to renew its LTA with Canada.

The Dual Feed-Grain Market

The CWB had monopoly selling privileges for all interprovincial feed grain movements from the late 1940s until 1974. The Canadian government changed the feed grain policy in 1974 and created a dual marketing system for feed grains. This change in policy allowed for domestic sales of feed grain either through private (e.g., Cargill, Continental, N. M. Patterson, Pioneer) and cooperative (e.g., United Grain Growers, Alberta Wheat Pool, Saskatchewan Wheat Pool, Manitoba Pool) grain companies or through the CWB. To facilitate the "open" or "non-board" market, futures trading in feed barley, feed wheat, and feed oats was started on the Winnipeg Commodity Exchange. The private and cooperative grain companies trade actively on the market, with the cooperatives being the major participants.

The 1974 change in market structure was radical, and it was brought about to correct regional feed-grain price differences that greatly exceeded transportation costs. Livestock feeders outside of the prairie provinces did not have free access to low-priced western Canadian feed grains, with the CWB acting as the sole supplier. The CWB's participation in the domestic feed-grain market was reduced significantly with this policy change. The open market is now the major supplier of prairie feed grains to eastern Canadian and British Columbian livestock feeders. There is also a significant amount of feed grain sales that does not enter the elevator system. They are particularly common in Alberta where 50 percent of the feed grains produced is used locally. Over the years, eastern Canada has also increased its corn production to the extent that it has become less and less dependent on western Canadian feed grains. Export and local markets have thus become the most important outlets for the western feed-grain producer.

The Open Market

As previously mentioned, the open market is made up of both farmer-owned cooperative and privately owned grain companies. These companies operate much the same as the grain companies in the United States, primarily as middlemen in the market. However, in Canada they have far more regulations and constraints to cope with.

The open-market sales of feed barley and feed wheat are for Canadian domestic markets only. Rapeseed, flaxseed, rye, and oats are also traded through the open market and, because these crops are not handled by the CWB, they are marketed in both domestic and international channels by the grain companies. A producer plebiscite in the early 1970s rejected the marketing of rapeseed through the CWB.

Additional open-market grains in Canada are classified as specialty crops. These include soybeans, corn, fababeans, sunflowers, lentils, canary seed, and forage seed. Producers of specialty crops generally market through grain companies, and most often these crops, except for corn and soybeans, are exported.

The majority of grain production outside of the prairies takes place in Ontario. The total nonprairie production is approximately 8 mmt, of which 80 percent is in Ontario and 12 percent is in Quebec. Wheat marketing in Ontario is controlled by the Ontario Wheat Producers' Marketing Board. Feed grain sales, which consist mainly of corn, are made on the open market.

Another important component of open market sales is the amount of feed grain marketed outside of the elevator system. This component accounts for about 30 percent of prairie grain production and represents on-farm usage, farm-to-farm, and farm-to-feedlot sales. Generally, the Winnipeg Commodity Exchange price is used as a reference point for pricing in this market.

Market Regulations

The Canadian grain market is heavily regulated. Some of these regulations have developed to provide equity to producers, some to facilitate ease of marketing, and some to appease special interest groups. Major regulations include grain freight rates, producer delivery quotas, rail-car allocation, and grain licensing and grading.

Figure 12.1 shows the major grain transportation routes in Canada, from primary elevators to export ports. The bulk of the grain is moved by rail rather than truck or barge. Historically, rail rates for grain transportation have been at very low levels. These levels were essentially established in 1897 when the Canadian Pacific Railway (CPR) and the

FIGURE 12.1   Routes and approximate volumes (in thousand tons) of Canadian grain moved by rail to export position in the 1985-1986 crop year.

SOURCE: Canadian Grain Commission: Annual Report (1986).

Canadian federal government signed the Crow's Nest Pass Agreement. As the major part of the agreement, the CPR lowered its freight rates on grain and promised to thereafter maintain the lower level. This promise was in return for a $3.4 million subsidy that enabled the CPR to build a rail line through the Crow's Nest Pass from southern Alberta to southern British Columbia.

These grain freight rates became the statutory rates in 1925, and they applied to grain and certain grain products moved by rail from western Canada to an export position. The rates, as fixed, bore a relationship to distance from the prairies to the ports (i.e., Thunder Bay, Vancouver) and averaged about 20 cents per hundredweight.

An important development in Canadian grain policy was the passage of the Western Grain Transportation Act (WGTA) in November 1983. This act removed the historically low, grain freight rates in Canada and facilitated the transportation and handling of western grains destined for exportation. It provides for an annual federal government contribution of C$658.6 million and higher freight payments by farmers to railways. In the late 1980s, farmers were paying about 22 percent of the actual freight rate for the movement of grain, and the remaining 78 percent was effectively paid by the government. The government's share will decline over time. The WGTA is aimed to ensure that Canada will be more efficient in the transportation and handling of grain in future years.

Producer delivery quotas are another important set of regulations. The CWB regulates producer deliveries to primary elevators through quotas on CWB grains only—wheat and barley. Prior to 1922, the CWB also set delivery quotas on canola, flaxseed, rye, and oats. The federal government determined these quotas were unduly restrictive for farmers, and thus, the quotas were discontinued in 1992.

Delivery quotas have been in place since 1940, and their original purpose was to provide equitable access to markets by all producers. In the 1950s, they were also used to provide producers with equitable access to an elevator system that did not have sufficient capacity to handle the grain as fast as producers wished to deliver. In the 1960s and 1970s, the sales strategy of the CWB and the capacity of the transportation system began to influence quota levels and justify the continuation of the quota system. However, the quota system was far from equitable. Grain producers were each assigned a base acreage whether the land was seeded or not. Beginning with the 1982-1983 crop year, producers who did not summerfallow started receiving bonus base acres.

Producers allocate their base acreage according to seeded acres, and quotas are announced by the CWB as a fixed number of bushels per assigned acre for each grain. No allowance is made for varying yields across the prairies, and there has been further concern that the delivery quotas give the CWB undue regulatory power to determine which grains are to be marketed. Quotas were particularly restrictive in the 1968-1971 and the 1977-1979 periods, and farmers were forced to carry burdensome on-farm inventories.

The quota system came under review during the 1987-1988 crop year, and a major change in the quota policy was recommended by the quota review committee in 1989. The committee suggested that the CWB move toward a delivery system based on grain available for sale rather than based on assignable acres. This recommendation has not yet been implemented. The CWB has implemented this type of system on a trial basis with special contracting with farmers. Special delivery contracts have been available for both wheat and barley, and the use of contracting has been expanded by the CWB.

As mentioned above, the Crow's Nest grain freight rates have contributed to an inadequate grain-transportation system in Canada. Subsequent government regulation was required to provide farmers equitable access to the limited transportation service. The Grain Transportation Authority (GTA) was formed in 1980 with the responsibility to monitor the performance of Canada's transportation system and to allocate rail cars for the movement of grain.

The government regulatory agency that is responsible for the quality control of Canadian grain and for the supervision of its handling is the Canadian Grain Commission (CGC). Among the most important of its functions are the inspection and the grading of grain. The official inspection of grain is done on a visual basis, and any new variety licensed for sale must be visually distinguishable from any existing variety. In addition, new wheat varieties must have high milling and baking characteristics.

Canadian regulators have historically stressed the quality aspect of Canadian wheat, and this policy has brought Canada the reputation of supplying highly uniform, good-quality wheat, but it has come at a cost in terms of lower yields. Many of the semidwarf spring wheats that are grown in the United States are higher yielding than Canadian varieties, but because most of them are not visually distinguishable from existing Canadian varieties, they are not licensed.

Some farmers "smuggled" seed into Canada in the early 1980s and started growing those wheats and selling them as "unprescribed" varieties. This label meant they were sold for feed prices. Most of these varieties were not visually distinguishable from CWRS varieties, and there was a fear of possible mixing into CWRS grades. However, the CGC found the contamination of CWRS grades with unlicensed varieties to be a problem of minimal proportion (Report of the Committee on Unprescribed Varieties). It has been estimated that the economic costs of this licensing regulation are high, and the costs lie between 5 and 17 percent of annual net farm income in Canada.[2]

By 1985 there were approximately 500,000 acres of wheat seeded to unprescribed varieties.[3] The census figures for 1986 indicate there were close to 600,000 acres in that year. In response to farmers' desire to produce semidwarf wheats, the Canadian government followed the advice of the Committee on Unprescribed Varieties and licensed Oslo wheat in 1987. Oslo is visually distinguishable from Neepawa, and it became eligible for the newly established "Prairie Spring" grade.

International Trade

Canada has historically been the second largest exporter of wheat, with a market share ranging between 18 and 26 percent over the past 15 years. The other major exporters are the United States, the European Community, Australia, and Argentina. The United States and Canada account for over 60 percent of the wheat trade. Several different categories of wheat are traded internationally, but Canada specializes in high-quality wheat and durum wheat.

Canada's ability to compete in the international market is enhanced by the fact that it offers a high-quality uniform product. High-quality wheats (No. 1 and No. 2 CWRS) represent over one-half of Canadian exports, whereas medium-quality wheats account for almost two-thirds of U.S. exports. No. 1 CWRS is by far the dominant grade exported as it accounts for 45 percent of exports on average. The major importers of these high-quality wheats from Canada have been, in order of importance, Japan, the United Kingdom, the former Soviet Union, China, Cuba, and Brazil. Table 12.3 reports the destinations (in percentage terms) for Canadian wheat shipments from 1970 to 1990. The largest customers are the former Soviet Union, China, Japan, Brazil, and the United Kingdom.

The high-quality wheat market has been growing very slowly compared to that for medium-quality wheats. Improvements in baking techniques worldwide permit flour of lower protein content to be used without sacrificing bread quality, which, in turn, reduces the need for high-quality Canadian wheat in blends.

Global trade in wheat increased from 54 mmt in 1970 to almost 105 mmt in 1990. Large gains were made in the 1970s when the grain trade grew approximately twice as fast as world production. Canada's wheat exports grew by about 30 percent. There was also an important distributional shift in the pattern of the world wheat trade. Wheat imports by the developed countries stagnated, and the centrally planned economies (CPEs) increased their purchases dramatically. The Canadian

TABLE 12.3 Share of Canadian Wheat Exports by Country of Distribution (in percentages)

Crop Year	Brazil	China	Egypt	India	Italy	Japan	United Kingdom	USSR	West Germany	Others
1970-1971	2.9	20.3	4.1	6.0	2.8	8.6	13.4	4.2	3.4	34.3
1971-1972	2.6	20.8	0.6	4.5	3.7	10.1	10.3	19.4	1.7	26.3
1972-1973	2.3	28.0	0.2	2.8	2.3	8.7	8.0	26.7	1.6	19.4
1973-1974	6.8	11.8	—	3.0	5.9	14.4	10.2	13.6	2.8	31.5
1974-1975	8.8	21.2	—	4.5	5.2	10.6	14.4	2.8	0.8	31.7
1975-1976	4.3	9.9	—	3.9	5.6	13.2	10.0	26.0	1.0	26.1
1976-1977	7.6	14.9	1.7	1.1	3.9	10.2	10.2	9.3	2.6	38.5
1977-1978	5.4	20.9	3.4	—	5.9	8.5	10.1	10.6	0.4	34.8
1978-1979	7.8	23.5	1.2	—	3.0	9.1	9.8	14.0	0.1	30.7
1979-1980	6.9	17.6	0.2	—	4.6	8.6	9.2	13.7	—	39.2

1980-1981	8.3	16.9	0.1	0.2	5.1	8.5	7.8	25.5	—	27.7
1981-1982	7.3	17.3	1.8	0.5	2.9	7.6	7.6	28.0	0.1	27.1
1982-1983	7.2	21.1	0.1	—	3.0	6.4	5.3	33.2	—	23.7
1983-1984	6.4	16.1	3.1	2.4	3.5	6.2	4.5	31.8	0.1	25.9
1984-1985	6.7	16.3	2.6	—	1.3	7.7	3.7	35.3	—	26.4
1985-1986	5.7	14.8	2.7	—	2.1	7.3	4.0	30.1	—	33.3
1986-1987	3.9	20.1	1.0	—	0.4	6.7	2.5	26.7	0.1	38.7
1987-1988	1.9	32.7	0.4	0.2	—	6.4	1.8	19.4	—	37.1
1988-1989	0.1	23.2	—	—	—	11.2	3.4	21.9	0.1	40.2
1989-1990a	1.3	26.6	—	0.1	0.1	8.5	1.6	20.3	—	41.7

a1989-1990 data are preliminary.

SOURCES: Canada Grains Council, *Wheat Grades for Canada*, p. 77 (for 1970-1971 to 1980-1981). Canadian Wheat Board, *Annual Report*, 1989-1990, Table 11, pp. 11-12 (for 1981-1982 to 1990-1991).

Wheat Board established a firm position in this market as it now exports more than 50 percent of its wheat to the former and current Communist bloc countries.

During the 1970s, Canada sold about 20 percent of its wheat to Western Europe, and this declined to 10 percent in the early 1980s, largely as a result of decreased sales to the United Kingdom. Sales to Japan as a percentage of total Canadian exports have also become less important. The markets in Eastern Europe, the former Soviet Union, and Latin America have increased in importance. Sales strategies of the CWB have also changed over the last 15 years. As mentioned above, most of the Canadian wheat sales in the 1960s were made to multinational grain companies who, in turn, sold to the importer. In the 1970s, the CWB moved away from this technique and began dealing directly with the importer. This, of course, was facilitated by the growing importance of the CPEs in the market and their use of state trading agencies to import grains. The multinational companies now play a smaller role in marketing Canadian wheat as the CWB deals directly with the customer in many cases.

Although the international trade in feed wheat is relatively small, Canada is a major exporter along with the European countries and Australia. The former Soviet Union is the largest feeder of wheat in the world, and in the mid-1980s about 25 percent of its wheat imports (approximately 4 mmt) was of feed quality. This wheat was supplied primarily by Canada and the Europe.

Canada is the third largest contributor of world food aid. Of the wheat and flour exported each year from Canada, about 4 percent (over 700,000 tons) is exported under a food aid program. Approximately 50 percent is donated under bilateral programs, and the remainder is distributed through the World Food Program. In the early 1970s, Canadian exports of flour were about 5 percent of wheat exports, but this amount has fallen to less than 2.5 percent. Commercial markets for Canadian flour (e.g., the United Kingdom) have disappeared, and almost all of Canada's flour exports are now in the form of food aid shipments.

As mentioned above, the CUSTA contained provisions for lower bilateral trade barriers in grains. The CUSTA has led to greater trade flows and increased competition in the Canadian-U.S. wheat (and wheat product) market. Prior to the CUSTA, Canadian import barriers on grains were high while those in the United States were relatively small. Of importance were Canadian import licenses on wheat (and wheat products) and the U.S. tariff. As a result of the CUSTA, a formula was developed to allow for removal of the Canadian import licenses, and in return the United States agreed to eventually eliminate its tariff of 21 cents per bushel on wheat.

Under the CUSTA, the trigger for the removal of Canadian import licenses was equalization of wheat producer subsidy equivalents (PSEs). PSEs measure government support as a percentage of the value of production plus direct payments.[4] In compliance with article 705 of the CUSTA, import licenses were eliminated in May 1991, based on the average of 1988-1989 and 1989-1990 PSEs. Canadian farmers received an average PSE of 31 percent, compared to 27 percent in the United States over that two-year period. The same calculation showed barley PSEs in the United States remained higher than in Canada. Therefore, Canadian import permits for barley and barley products will remain in place until the PSEs are equalized.

An interesting aspect of the increased integration of the North American market is the growing amount of southbound wheat sales. There has been a sharp rise in U.S. imports of Canadian durum wheat beginning in the mid-1980s, from zero to almost 400,000 metric tons in 1990-1991. Almost 25 percent of the U.S. durum consumption is now supplied by Canada. The Canadian Wheat Board had never been precluded from selling into the U.S. market but the CUSTA provides a means of legitimizing such sales. With the CUSTA there is less threat of imposition of trade barriers under Section 22 (of the Agricultural Adjustment Act of 1933), which allows the U.S. secretary of agriculture to impose quotas on imports if it is determined that such imports are threatening U.S. price support programs.

Of perhaps greater importance is the role of U.S. export subsidies on wheat (i.e., the U.S. Export Enhancement Program—known as EEP). The export subsidies raised the price on the domestic U.S. market above that on the international market. Natural arbitrage pressures will result in more Canadian wheat flowing into the United States.[5] The CUSTA also eliminated Canadian subsidized freight rates on grains exported to the United States through the west coast of Canada. However, export subsidies were retained for exports to the United States through the Great Lakes. This issue has become litigious. Durum wheat producers in North Dakota viewed these export subsidies as a violation of the CUSTA, and they raised the issue in 1989. Under article 701.3 of the CUSTA, public entities cannot export agricultural goods to the other country at less than their acquisition price. The United States Trade Representative (USTR) at that time determined that Canada had not violated this article because the freight subsidy applied to all shipments to Thunder Bay, whether destined for export or domestic use. However, the U.S. Congress did instruct the United States International Trade Commission (USITC) to examine the "conditions of competition" between the U.S. and Canadian durum industries. The USITC report concluded that the drought of 1987-1989 was the main reason for increased durum imports from Canada. Price differences were not found to be a factor.

The issue of Canadian durum sales into the United States arose again as a point of contention. The USTR and the Canadian Department of External Affairs held discussions on the topic in late 1991. The United States maintained that the CWB was violating section 701.3 of the CUSTA. The claim is that U.S. customs import data on unit values show that the CWB is selling below acquisition cost (including storage, handling, and freight). The Canadian government disagrees. But the two governments are only looking at the durum issue from the narrow bilateral trade standpoint. If one takes a more global view, it could be shown that Canadian durum sales into the United States are not necessarily harmful to the United States. The Canadian sales to the United States cannot be treated in isolation of the EEP.

The export subsidy program drives a wedge between U.S. domestic and world prices. The CWB has a strong incentive to arbitrage this price wedge by removing sales to third markets (such as Algeria) and increasing sales to the United States. The removal of Canadian third-market sales increases U.S. sales abroad and tends to offset the impact of the loss of domestic sales.

In late 1992, the bi-national panel established under CUSTA announced a ruling on durum wheat and that ruling was favorable to Canada. The panel found that the CWB is not violating the CUSTA agreement. The U.S. wheat industry argued that the CWB was selling into the United States below the CWB's "acquistion" price (i.e., the price it pays farmers). U.S. durum producers argued that the acquistion price should be the sum total of the initial, adjustment, and final payments. However, the bi-national panel ruled that the acquisition price is the initial payment only and does not include the adjustment and final payments.

Government Support

The largest amount of government financial support in the Canadian grain industry has been in the form of transportation subsidies. The federal government has funded part of the "crow gap" that the railways have suffered as a result of moving grain at rates below cost. The feed-grain freight assistance program is also financed by the government.

Unlike the U.S. government's involvement in the grain markets, Canadian policy typically does not involve direct income or price supports. Notable exceptions were the Lower Inventories for Tomorrow program in 1970 (whereby farmers were paid to set aside acreage), the 1986 and 1987 special grains program (SGP), and the 1991 Gross Revenue

Insurance Program (GRIP). The CWB's initial payment is guaranteed by the federal government, but it has seldom acted as a support price. However, in 1985-1986, in 1986-1987, and in 1990-1991 the CWB experienced deficits in the barley pool. In the 1990-1991 crop year there was a large deficit ($673 million) in the spring wheat pool and a deficit of $70 million in the durum wheat pool.

The Western Grain Stabilization Act (WGSA) was a voluntary income stabilization program that was jointly funded by farmers themselves and the federal government. It was enacted in 1976 and ended in July 1991. There were seven eligible crops: wheat, barley, oats, canola, flaxseed, rye, and mustard seed. The basic objective of the WGSA was to provide protection against variations in net cash flow from grain sales. It was designed so that the aggregate net cash flow in the prairie region would not fall below the previous five-year average. A participating producer received stabilization payments in proportion to levies paid into the program. There were payouts in 1978 ($115 mil.), 1979 ($253 mil.), 1984 ($223 mil.), 1985 ($522 mil.), 1986 ($850 mil.), 1987 ($1,395 mil.), 1988 ($693 mil.), and 1989 ($176 mil.). The WGSA was widely recognized for its effectiveness in stabilizing farm income. In fact, a special study of the U.S. Congress in 1984 recommended a cash flow stabilization program similar to the WGSA be considered as an option for U.S. farm policy. The WGSA ran into deficit problems in the late 1980s and, as a result, received large government payments. The program was discontinued because it was too costly.

A one-billion-dollar ($C) SGP was announced by the Canadian government in late 1986. Individual crop producers received a maximum payment of $25,000 under the program. The payment for each crop was inversely related to the relative price decline attributable to the trade war. A second payment of approximately $1.1 billion was made one year later. The WGSA program was replaced by the Net Income Stabilization Account (NISA) and GRIP. The federal and provincial governments reached agreement on both the NISA and GRIP programs early in 1991. GRIP was designed as an income stabilization program. Under the program, individual farmers can "insure" a target revenue per acre for any major crop. The target level is based on historical yields and historical prices. Farmers pay 33 percent of the program's costs, the federal government pays 42 percent, and the provincial government pays 25 percent. A payout is made whenever actual revenue falls below the "insured" target revenue for an individual farmer.

NISA is more of an income "smoothing" program, and it was designed to complement GRIP. Farmers participating in NISA are permitted to contribute 2 percent of their net sales to their NISA account, and these contributions will be matched by the federal and provincial governments (shared equally). Farmers can withdraw from NISA whenever the farm's gross revenue falls below the previous five year average, or when net income falls below $10,000.

Government subsidies to grain producers are shown in greater detail in Table 12.4. For each commodity, the ratio of government subsidies to total producer value (i.e., PSEs) is reported on an annual basis. Some of the policies affecting producers of the five major grains and oilseeds provide direct transfer payments from the federal government, especially since the mid-1980s. However, the major policies provide indirect support through their impact on prices realized by farmers. For the major crop, which is wheat, the proportional subsidies varied between 19 and 53 percent during the period covered in Table 12.4. Canada's major subsidy scheme has been in the form of a transportation subsidy. Over the period covered by Table 12.4, Canadian wheat producers did not receive quite as much government support as their U.S. counterparts on average, but the difference was not too substantial—only about 4 or 5 percent of crop value.

TABLE 12.4 Proportional Grain Subsidies (PSEs): Canada, 1982-1987 (in percentages)

Crop Year	Commodity				
	Wheat	Barley	Oats	Rapeseed	Flaxseed
1982	19	28	17	19	15
1983	23	26	19	16	21
1984	32	31	16	22	21
1985	39	71	47	35	33
1986	53	91	30	48	47
1987	51	55	21	39	43

PSE = Producer subsidy equivalent (ratio of total producer subsidy to total producer value)

SOURCE: A. J. Webb, M. Lopez, and R. Penn, *Estimates of Producer and Consumer Subsidy Equivalents: Government Intervention in Agriculture, 1982-1987*, USDA-ERS, Station Bulletin No. 803, Washington, DC, April 1990.

Summary

This chapter has been largely a descriptive analysis of the Canadian grain marketing system. It was shown to have many characteristics that make it substantially different from that in the United States. In Canadian grain sales, there is a large degree of government involvement through its agency, the Canadian Wheat Board. The private grain trade is also active in Canada, and its participation is growing.

The history of the Canadian system was outlined as an introduction to a discussion of the three major marketing channels: the CWB, dual, and open markets. Government regulation and support were also described and briefly analyzed.

The Canadian and U.S. marketing systems continue to serve as interesting contrasts given that they have many similar production practices and are often competitors in foreign markets. Their grain-marketing systems reflect significantly different approaches to similar marketing challenges.

Notes

1. C. F. Wilson, *Grain Marketing in Canada*, Canadian International Grains Institute, Winnipeg, Manitoba, 1979, p. 65.

2. C. Carter, A. Loyns, and Z. Ahmadi-Esfahani, "Varietal Licensing Standards and Canadian Wheat Exports," *Canadian Journal of Agricultural Economics*, 34, November 1986, pp. 361-377.

3. Agriculture Canada, Committee on Unprescribed Varieties, "Report of the Committee on Unprescribed Varieties," Winnipeg, Manitoba, July 1986.

4. A. J. Webb, M. Lopez, and R. Penn. *Estimates of Producer and Consumer Subsidy Equivalents: Government Intervention in Agriculture, 1982-87*, USDA, ERS, Station Bulletin No. 803, Washington, DC, April 1990.

5. C. Carter, J. Karrenborck, and W. Wilson. "Freer Trade in the North American Beer and Flour Markets," Chapter 5 in *Freer Trade and Agricultural Diversification*, A. Schmitz, ed., Westview Press, Boulder, CO, 1989.

Selected References

Carter, C., J. Karrenbrock, and W. Wilson, "Freer Trade in the North American Beer and Flour Markets," Chapter 5 in *Freer Trade and Agricultural Diversification*, A. Schmitz, ed., Westview Press, Boulder, CO, 1989.

Carter, C., A. Loyns, and Z. Ahmadi-Esfahani, "Varietal Licensing Standards and Canadian Wheat Exports," *Canadian Journal of Agricultural Economics*, 34, November 1986, pp. 361-377.

13

Grain Marketing
in the European Community

Mark D. Newman

Since the beginning of the Common Agricultural Policy in 1962, the European Community (EC) has shifted from being a major importer of wheat and coarse grains to being one of the world's most important international grain exporters.

The driving force behind the EC grain-marketing system is the Common Agricultural Policy (CAP). The CAP provides high internal support prices for European grain producers and protects them from international competition through a combination of variable levies on imports and through export refunds, which allow EC grain to be sold at world prices on international markets while preserving the higher internal prices.

The European Community

Since 1986, the EC has been a common market composed of 12 independent nations. The unification of East and West Germany in 1990 made the 12 member states a market with 340 million consumers.

The European common market was established by the Treaty of Rome in 1957 by six countries. The EC-6 included Belgium, West Germany, France, Italy, Luxembourg, and the Netherlands (Figure 13.1). The primary objective in setting up the EC was to facilitate trade among member countries and establish a common set of trade policies between the EC and the rest of the world's trading nations. Some of the founders of the EC hoped that the economic union among the member nations of Europe would eventually lead to a political union, in many ways similar to the United States.

FIGURE 13.1 The European Community.

In the pursuit of European unity, the EC has been enlarged over time. Denmark, Ireland, and the United Kingdom joined in 1973, forming the EC-9. Greece became a member in 1981, making the EC-10; and Spain and Portugal joined to make the EC-12 in 1986 (Figure 13.2). German unification in October 1990 made former East Germany (the German Democratic Republic) part of the EC. It also enlarged the production base of the EC by 12 million metric tons (mmt) of grain to almost 180 mmt annually.

January 1993 was the official beginning of the "European Single Market" in which goods and people flow freely across national borders. This economic union required adoption of a total of 282 proposals by both the EC and the 12 individual national governments. By February of 1993, most of the legislative work was completed. Only 18 measures had not been agreed upon within the Council of Ministers, and member states had enacted an average of 72 percent of the adopted measures into national law. In practice, there remained specific complaints particularly regarding continual passport checks at internal borders.

In December 1991, as part of the Maastricht Treaty on European Union, the EC member states also agreed to work towards the establishment of an economic and monetary union by 1999, with a single currency to replace the 12 currently used. The treaty also sets forth a process leading toward eventual political union. After an initial setback in 1992, when

EC-6

1957 •Belgium

•Federal Republic of Germany

•France

•Italy

•Luxembourg

•Netherlands

EC-9

1973 •Denmark

•Ireland

•United Kingdom

EC-10

1981 •Greece

EC-12

1986 •Portugal

•Spain

1990 •German unification

FIGURE 13.2   Growth of the European Community.

Danish voters failed to ratify the treaty, the other member countries and the EC are working to continue the unification process.

The EC is also currently negotiating with the seven European Free Trade Agreement (EFTA) countries (Austria, Finland, Iceland, Liechtenstein, Norway, Sweden, and Switzerland) to free trade within the European Economic Area (EEA). Association agreements being negotiated will also give many of the Eastern and Central European countries greater access to EC markets and may pave the way for their eventual EC membership. Association agreements with Poland, Hungary, and Czechoslovakia took effect in January 1993, and talks are under way with Romania and Bulgaria. Other nations that have requested EC membership include Turkey, Austria, Malta, Cyprus, Sweden, and Finland.

Economic unity among EC members has been easier to talk about than to actually put into place. Agriculture is the primary sector in which the EC member nations have made progress toward a single policy. The CAP was set up with three objectives:

1. Creation of a single market among the member nations by reducing the barriers to internal trade within the EC;
2. A preference within the EC for products of member nations;
3. Common financing of policy costs.

Grains were the first commodity subsector in which a common policy was agreed to, beginning in 1962. Since that time, policies (called *regimes*) for a broad variety of agricultural products have been developed, many modeled on the grain regime.

Agricultural Decisions and Market Regulations

The design and implementation of agricultural policies for a community of twelve very different countries are not simple matters. EC member countries each maintain individual currencies and national control of their own laws and broad economic policies. Interest groups in each nation work to get policymakers to look out for them.

Decisionmaking in the EC involves:

*   The 12 national capitals;
*   Brussels, Belgium, the headquarters of the EC Commission, the body that proposes and administers agricultural policy;
*   Luxembourg, where the EC Council of National Agricultural Ministers (often called "the agricultural council") must agree on

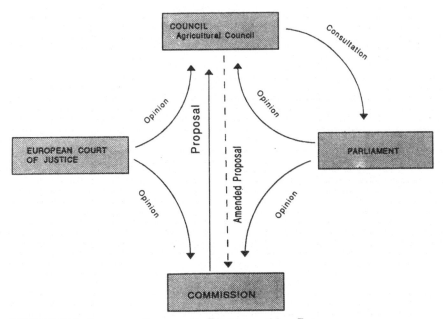

FIGURE 13.3 European Community Decisionmaking Process.

common policy decisions after considering proposals from the commission;
- Strasbourg, France, where the European Parliament meets to discuss policy proposals and the budget required to pay for them;
- The European Community Court of Justice, also in Luxembourg, which reviews the legality of decisions taken by the EC Council of National Agricultural Ministers, the commission, or national member governments, where treaty obligations are concerned (Figure 13.3).

Operation of the Common Agriculture Policy (CAP)

The operation of the CAP differs somewhat for grains and oilseeds. For grains, it provides:

- Minimum support prices as an incentive to produce;
- Protection from imports through variable import levies or tariffs;
- Storage to keep surplus production from depressing internal market prices;

- Export refunds or subsidies that compensate exporters for the difference between internal market prices and world prices, so that high-priced EC grains can be exported.

Because the EC agreed not to charge duties on imported oilseeds as part of the Dillon Round of negotiations under the GATT, the policy for supporting oilseed prices has operated somewhat differently from that for grains. Through 1992, the support program for rapeseed (canola), sunflowerseed, and soybeans involved a crushing subsidy, which varied like the variable levy, and was paid to processors to get them to use EC oilseeds and still pay producers the mandated support prices. The crushing subsidy was set so that processors would use internal EC production before buying imports. In late 1992, the EC agreed to replace this system with a combination of direct support payments to oilseed producers based on the area planted and a payment based on the difference between market prices and a reference price. Producers will have to set aside at least 10 percent of their oilseed acreage to receive the payments.

EC Grain Production and Trade

The CAP was set up for a community that sought to increase food production and decrease dependence on imports. Under the protection of high internal prices, the EC has become much more than self-sufficient in grains and is rapidly increasing its production of rapeseed, sunflowerseed, and soybeans. The EC-10 became a net exporter of wheat in 1974 and a net exporter of coarse grains in 1984 (meaning that its total exports were greater than its imports). By 1991, EC-12 exports of all grains were estimated at 28.4 mmt, with wheat and wheat flour exports estimated at 20 mmt. However, the EC has remained a net importer of about 14 mmt of oilseeds. (The shift from net grain importer to net grain exporter is shown in Table 13.1.)

The EC Grain Price Support System

For an understanding of the operations of the EC grain price support system, explanations of several concepts are critical: the intervention price, the target and threshold prices, the variable levy, and export refunds. Their operation is shown in Figure 13.4.

Intervention, the Support Price

In the EC, grain producers and/or first handlers of grain (elevator operators) may deliver grain to a national intervention agency or its

TABLE 13.1 European Community Net Grain Exports, 1970-1992 (in thousand metric tons)

Year	Quantity[a]
1970	−22,308
1971	−14,502
1972	−13,111
1973	−14,757
1974	−12,066
1975	−11,772
1976	−22,235
1977	−11,983
1978	−6,731
1979	−2,652
1980	3,607
1981	5,200
1982	9,747
1983	10,836
1984	19,374
1985	16,930
1986	19,534
1987	18,251
1988	24,731
1989	24,880
1990	22,449
1991	25,449
1992	27,287

[a]Negative numbers represent imports and positive numbers represent exports. 1970-1986 based on EC-10, 1987-1992 based on EC-12.

SOURCE: USDA.

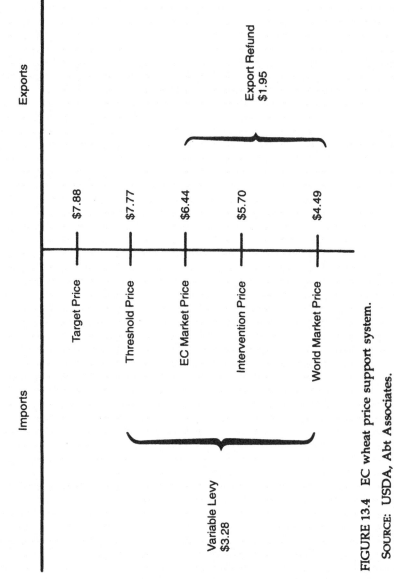

FIGURE 13.4 EC wheat price support system.

SOURCE: USDA, Abt Associates.

licensed representatives and receive the intervention price, or a buying-in price that is closely linked to the intervention price. National intervention agencies perform a function similar in some respects to that of the Commodity Credit Corporation (CCC) in the United States.

Intervention prices are set annually by the EC Council of National Agricultural Ministers ("the council"), generally based on recommendations by the EC Commission. The commission has responsibility for making the CAP work on a daily basis. This accomplishment involves considerable consultation with the Cereals Management Committee, composed of commission and national representatives, generally including delegates from national intervention agencies.

Intervention prices are set for "common" or bread wheat, durum wheat, corn, barley, rye, sorghum, and rice. Prices apply to the grain-marketing year beginning in July and running through the following June 30. Minimum quality and quantity standards are established, with some variation by country. For example, minimum delivery quantities are generally 80 to 100 tons of grain, except for durum wheat, but in the United Kingdom the minimum for bread quality wheat is 500 tons. Maximum grain moisture content also varies by country.

Although grain is actually sold into intervention, the intervention price is like the U.S. loan rate in operating as a price support or market of last resort. Farm-level prices can fall below intervention prices due to the following factors: costs of assembly and transportation; delays in payment for grain delivered to intervention; the coresponsibility levy (which is designed to make farmers share costs of the price support program), and provisions for purchases to actually take place at 94 percent of intervention prices.

Intervention prices for bread-quality soft wheat and rye and for corn are set at the same level; intervention prices for feed wheat and rye and for sorghum and barley are set at a lower level. Durum wheat, used primarily for pasta production, is supported by a higher intervention price as well as a premium for area planted. Oats producers are protected from competition from imports, but there is no intervention for oats.

For the 1991-1992 marketing year, intervention purchases were made only between October 1 and April 30 in certain regions of Italy, Greece, and Spain, where harvests begin early, and between November 1 and May 31 in the rest of the EC. Intervention purchases are made only when market prices less deductions for handling and transportation fall below the intervention price in certain representative markets in France, the UK, and Italy. Then intervention buying is only at 94 percent of the intervention price, adjusted for production levels and monetary distortions. Thus, while the 1991-1992 intervention price for bread wheat was 158.55 ECU/metric ton (mt) or $5.56/bu, the initial buying-in price was 145.04 ECU/mt or $5.09/bu.

Intervention prices are set in European currency units (ECUs), a unit representing a weighted average of member nation currencies (see Table 13.2). Given the nature of the European monetary system, support price changes in ECUs do not translate directly into equivalent changes in the currencies of individual member nations. For example, overall grain support prices for 1991-1992 were cut 0.3 percent in ECUs, but rose an average of 0.2 percent for European farmers in their own money when translated to national currencies through "green" (agricultural) exchange rates. Although a 1990-1991 agreement called for total elimination of these "green rate" adjustments by 1992, this adjustment has not yet occurred.

Intervention prices are increased throughout the season by "monthly increments" to reflect storage and other carrying costs. The idea is to prevent rapid flows of grain into intervention at harvest by providing an incentive for private storage. These increments are applied to the intervention price from November through May. For all grains other than durum wheat, individual monthly increments during 1992-1993 were 1.5 ECUs/mt ($1.93/mt or $.053/bu for wheat). For durum wheat, the monthly increment was set at 2.03 ECUs/mt.

In May 1992 the EC adopted a CAP reform package that is scheduled to go into effect in 1993-1994. The reform program, based on proposals

TABLE 13.2   Value of U.S. Dollars per European Currency Unit (ECU) and Value of French Francs per U.S. Dollar

	Dollars per ECU	French francs per dollar
1985	1.31	8.99
1986	1.02	6.93
1987	1.15	6.01
1988	1.18	5.96
1989	1.10	6.38
1990	1.27	5.45
1991	1.24	5.64
Jan 92	1.28	5.38
Jan 93	1.21	5.52

SOURCE: International Monetary Fund, *International Financial Statistics*, various issues, Washington, DC.

of then EC Agriculture Minister Ray MacSharry, was designed to make European farmers more responsive to international market forces and to make the operation of the CAP more consistent with proposals under discussion in the Uruguay Round of negotiations under the GATT.

The program for grains involves a three-year transition to lower support prices and the use of compensation payments to offset the resulting impact on farm income. Intervention prices for grain other than rice are to be reduced 33 percent between 1992-1993 and 1995-1996, when intervention prices for wheat are to be reduced to 100 ECU per mt ($3.54 per bushel at current exchange rates). The new system will also provide for cash compensation based on acreage and historical yields, which will remain constant over time. A 15 percent set-aside of arable land will be required of those eligible for price and income support, except for those "small farmers" with annual production less than about 3,400 bushels.

Target Prices, Threshold Prices, and Variable Levies

EC intervention prices are generally well above prices at which grain can be purchased on international markets, so the EC uses a levy on imported grain to keep it from undercutting EC grain on internal markets.

The target price used by the EC is the desired price in the community's most grain-deficit area, Duisburg, Germany. While a variety of factors may enter into the determination of the target price, it is theoretically linked to the intervention price. As "normal market prices" are supposed to be above intervention prices, the intervention price at Ormes, France, the area that supposedly has the greatest grain surplus in the EC, is taken as a base. A subjective "market element" is added to the intervention price in Ormes to get the "normal market price." Transportation costs between Ormes and Duisburg are then added to get the target price. Because the derivation of the market element is subjective, target prices are fixed by the Council of National Agricultural Ministers at the same time that intervention prices are set.

The threshold price is intended to be the minimum price at which imported grain can enter the EC, so that it will not be sold at a price that undercuts European grain in Duisburg. The threshold price is derived from the target price by subtracting the cost of transport and transshipment between the most important port of entry, Rotterdam, Netherlands, and Duisburg, Germany, plus a trading margin. For 1990-1991, the difference between target and threshold prices was set at 4.37 ECUs/mt, representing 1.8 ECUs/mt for shipment between Rotterdam and Duisburg, 0.92 ECUs/mt for loading and unloading costs at

Rotterdam, and an importer margin of 1.65 ECUs/mt. For 1991-1992, the target price-threshold price margin was increased to 4.59 ECUs/mt. Threshold and target prices are also increased by monthly increments of 1.5 ECUs/mt, beginning on November 1 and running through the following May.

The variable levy, an import duty, is set with reference to the difference between the threshold price and the lowest price on a delivered (c.i.f.) basis that imported grain is available in Rotterdam, or adjusted for equivalence with Rotterdam. The commission sets levies daily, and they apply to all grain imports, regardless of the port of entry. Levies may be pre-fixed before actually importing a grain shipment. Pre-fixed levies remain valid for 45 days. In addition, a de facto market in levies also exists.

Import levies on grains vary considerably as a result of world prices and exchange rates. They can add anywhere between 75 percent and several hundred percent to the price of imported grain. As mentioned above, there are no levies on imported oilseeds because of an agreement by the EC under the GATT to bind its duties on oilseeds at zero.

Export Refunds

The EC is more than self-sufficient in grain, so something has to be done with the surpluses, or they would be available to depress prices on EC markets. The options are to store, to promote consumption within the EC, or to dispose of surplus grain on international markets.

Government-held grain stocks in the EC grew from 2 million tons in 1978 to 18.6 million tons in 1985. In 1989 they were reduced to about 10.5 mmt by costly export sales. However, by July 1992 intervention stocks had risen to a record 26.4 mmt, an all-time high.

Export refunds (restitutions) are critical to the EC's ability to sell high-cost EC cereals on international markets at prices that are sharply lower than those received by EC producers. These refunds accounted for 9.3 mmt of grain sold out of intervention stocks in 1991-1992. General export refund levels are set daily by the commission and remain valid for the following week. However, these levels are used only for such neighboring countries as Austria and Switzerland. For other countries, restitutions are determined weekly on the basis of tenders awarded to the lowest bidder by the EC Cereals Management Committee. Refunds are paid to traders whose bids are accepted, based on the difference between internal EC prices, prices in importing countries, and transportation and marketing costs. Restitutions remain valid for five months including the month in which they are requested. In exceptional cases, the validity of refunds can be extended to one to two years for large contracts with state agencies.[1]

During 1991-1992, EC export refunds on wheat ranged between about $1.50 and $3.00 per bushel. Although refunds and levies were not intended to be transferrable, a de facto market has arisen for both, so that the traders who actually import or export may not be the ones who are awarded the refunds or levies initially.

Grain Production in the EC

Total grain production in the EC-12 countries (excluding East Germany) has grown from about 80 mmt in 1960, to almost 180.0 mmt in 1991, including former East Germany. Wheat production in EC-12 reached 84.2 mmt in 1992: it was 34 mmt in 1960. Barley increased from 19 mmt to 43.6 mmt; corn production increased from 8 mmt to 26.4 mmt during the same period. As can be seen in Figure 13.5, France, Germany and the United Kingdom account for more than two-thirds of total EC-12 grain production and exports.

Oilseed production has skyrocketed in recent years, going from about 1 mmt in 1971 to 12.2 mmt in 1992, down from 13.3 mmt in 1991. Rapeseed (canola) production accounted for 6.2 mmt, sunflowerseed for 4.3 mmt, and soybeans for 1.3 mmt in 1992. EC oilseed production is expected to fall somewhat over time as a result of the change in support policies in 1993 and the impact of set-asides.

Although there is considerable variability within the EC, on average, grain production is very intensive and highly productive. From 1985 to 1989, wheat yields averaged 78.5 bushels per acre, ranging from 22.3 bu/acre in Portugal to 110 bu/acre in Ireland. Corn yields averaged 107 bu/acre, and barley yielded an average of 75 bu/acre.

Farms are smaller in the EC than in the United States, with an average of only about 30 acres (12 ha) of cropped area per farm in 1984, growing to an average of 57 acres (23 ha) in 1989. In the United States, the average farm was 457 acres in 1989. Average EC farm size in 1989 varied from 16 acres in Greece to 232 acres in the UK.

Grain Use in the EC

Grain use in the EC is divided as follows: about 60 percent is used for animal feed; 26 percent for human consumption; 8 percent for industrial use; and 6 percent for seed use and losses.

Variable levies apply to grain but not to oilseeds, oilmeals, and other nongrain feed ingredients, such as manioc, corn gluten feed, distiller's dried grains, and citrus pulp. As a result, the share of grain in EC feeds has been falling over time. In 1972-1973, cereals provided more than 62

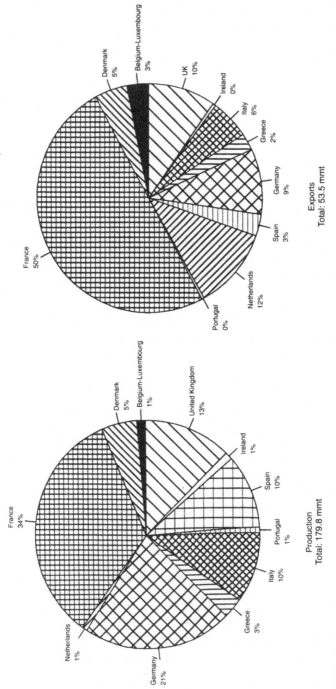

FIGURE 13.5 EC-12 member country grain production and exports, 1991. (Exports include intra-EC trade.)

SOURCE: USDA.

percent of the energy value in EC-10 commercial animal feeds. By 1981-1982, the share of cereals had fallen to 51.3 percent. The absolute volume of grain used in animal feed peaked at 90.7 mmt in 1984-1985, falling to 80.4 mmt for the EC-12 in 1992. About 42 percent of cereals produced for animal feed was used on-farm, with the remainder purchased.

Soft wheat made up about 82 percent of the 32.7 mmt of cereals (other than rice) used for human consumption in the EC-12. Total human consumption of grain has been rather stable.

Industrial use of grains has been increasing, with wheat used for starch and gluten production increasing sharply as prices of wheat have fallen relative to those of corn.

EC Grain Exports

The member nations of the EC sold 61.1 mmt of grain outside of their own borders in 1992. This figure includes intra-EC trade, with a greater portion sold to other members of the EC than is sold on world markets. However, the EC has become an increasingly important competitor in international markets (Figure 13.6). In the 1990-1991 marketing season, EC grain exports (excluding intra-EC trade) were 28.4 mmt. Grain trade within the EC accounted for another 25.1 mmt. The EC share of world wheat trade averages about 21 percent, with exports outside the EC forecast at a record 21.5 mmt for 1992-1993 according to USDA. This share makes the EC the world's third largest wheat exporter behind the United States and Canada. The EC is the world's largest barley exporter, accounting for about half of world trade. Barley exports for 1992-1993 are expected to reach 9.5 mmt.

France accounts for 33 percent of the EC grain production. In 1991, it accounted for about 50 percent of intra-EC grain trade and about 40 percent of EC grain exports to destinations outside of the community (Figure 13.5). France accounted for almost 57 percent of extra-EC wheat and flour exports, and more than 75 percent of corn exports. West Germany and the UK export another 25 percent of EC wheat exports, with Italy exporting over 10 percent of the total, mostly durum wheat.

Costs of the CAP

European Community expenditures for agricultural price-and-income supports have skyrocketed in recent years, going from under $13 billion in 1982 to a budgeted $48 billion for 1992. This budget is based on the

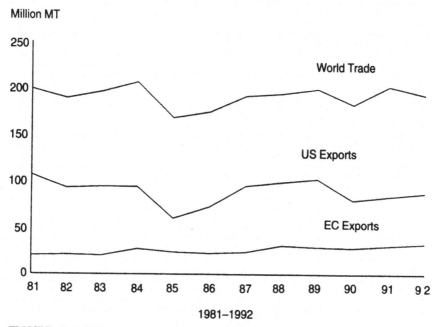

FIGURE 13.6 U.S. and EC shares of world grain trade—wheat and coarse grains.
SOURCE: USDA.

assumption that the current 1991-1992 price package and its reforms
will create the necessary budget savings to keep expenditures below the
spending guidelines set at the EC summit of 1988. In addition, proposals
in the Uruguay Round of the GATT require some modifications of
barriers to trade imposed by the CAP.

Much of the increase in the cost of the CAP in recent years is
attributable to the growing gap between EC support prices and prices on
international markets. This gap has increased as a result of a weakening
U.S. dollar that makes world prices denominated in dollars lower, when
converted to European currencies; reductions in world prices as a result
of the lowered loan rates under the U.S. Food Security Act of 1985;
competition with the United States in markets targeted through the EEP;
and higher world crop production. The gap between intervention and
threshold prices reached 91 ECU/mt for bread wheat in 1991-1992,
compared to 18.30 ECU/mt in the mid-1980s. Under the CAP reform
plan agreed to in 1992, this gap will be reduced to 45 ECU/mt, but
farmers will also receive a flat compensation payment based on area and
historical production.

Grains made up 16.6 percent of the budget cost of the CAP in 1992, up from 12 percent in 1989; oilseeds (including rapeseed, sunflowerseed, soybeans, and flaxseed) accounted for 14.4 percent. These two sectors have been the fastest-growing components of the cost of the CAP, constituting almost 33 percent of its total cost.

In February 1988, the 12 EC heads of state, meeting as the EC Council, agreed to a set of measures for stabilizing the costs of the CAP for grains and other commodities, primarily by reducing support prices for production increases above certain threshold levels, and providing for paid voluntary land set-asides for farmers willing to remove at least 20 percent of their land from production.

The compromise reform program adopted in May 1992 promises some fundamental changes in the way the CAP operates during 1993-1996. However, the costs of the CAP are expected to continue to rise at about 4.5 percent annually, reaching $54 billion in 1997 according to EC forecasts. In December, 1992, EC heads of state agreed to limit agricultural spending to $52.7 billion by 1999 as part of an overall budget agreement. Proposals being discussed in the Uruguay Round of the GATT would also require other reforms of the CAP. A proposal by GATT Director General Arthur Dunkel contained three important features that are likely to affect any agreement reached:

- The first would commit all governments to cut the amount of trade-distorting aid they provide to their farmers by 20 percent by 1999.
- The second would be the "tariffication" of existing trade barriers, replacing all the variable levies as well as quantitative and other restrictions with fixed tariffs.
- The third feature would be to cut the volume of each country's subsidized exports by 24 percent and the value of export subsidies by 36 percent.

In November 1992, the United States and the EC agreed to reduce the quantities of subsidized exports by 21 percent from 1986 to 1990 average levels over six years, once a GATT agreement is reached. They also agreed to a 36 percent reduction in export subsidies on a commodity-by-commodity basis during the same period. Finally, they agreed to reduce overall internal farm support by 20 percent across the board but without subjecting EC set-aside payments to these limits. European farm groups protested the agreement, and as new administrations took over the trade negotiating process in Washington and Brussels in 1993, the ultimate outcome of the Uruguay Round remained uncertain.

Operation of EC Grain-marketing Systems:
The Case of France

Production

France is the largest grain-producing country in the EC, accounting for 34 percent of total 1991 production, 39 percent of EC wheat, 50 percent of EC corn, and 21 percent of EC barley. France was the source of 50 percent of all grains exported by the 12 EC countries in 1991, including intra-EC trade (Figure 13.5).

Grain production is important throughout France, but the Paris Basin is noted as the area with the greatest grain production. Four regions (Centre, Picardie, Champagne-Ardennes, and the Paris region) produce almost 50 percent of the marketed grain in France (Figure 13.7). Grain

FIGURE 13.7   Major grain production regions in France.

area harvested has averaged 23-24 million acres in recent years, almost 33 percent of all the utilized agricultural area and more than 50 percent of all crop land. The average cropped area per farm in France was 107 acres in 1989, up from 72 acres in 1985. This average is well above the 57-acre average for the EC-12, but well below the 232-acre average in the United Kingdom or 457-acre average in the United States.

Large farms, especially in the basin surrounding Paris, are extremely important to French cereal production. According to the Office Nationale Interprofessionel des Céréales (ONIC) data from the mid-1980s, almost 75 percent of marketed grain came from the 20 percent of grain producers farming more than 250 acres. Almost 90 percent of wheat and corn production and 67 percent of barley production are marketed. Seventy-five percent of the producers marketing grain sold less than 100 mt, accounting for less than 25 percent of total marketed grain.

French grain production has ranged from 55 to 60 mmt annually in recent years. Of this total, about 25 percent gets used in the region of production, 25 percent in other regions of France, and 50 percent exported from France, either to other EC member countries or other world markets. Export destinations fluctuate significantly, but most exported corn goes to other EC member countries; barley and wheat are generally exported to markets outside the EC.

In general, the regions with grain surpluses are in the northern half of the country other than Brittany and Nantes, where important livestock production results in regional grain deficits.

Major interregional grain flows within France include movements of corn for industrial uses, starch, and wet milling, from the Bordeaux and Toulouse regions to the north—Lille and Nancy. Marloie estimates that this flow amounts to less than 1 million tons. Milling wheat, accounting for about 2.5 mmt, involves shipments from the Paris basin, centered on Orleans, to all regions of France. Grains for feed use account for about 3.6 mmt traded interregionally, with 50 percent being corn, 26 percent wheat, and 20 percent barley.[2]

Marketing Channels

French farmers market grain through a variety of channels; cooperatives are very important, handling about 75 percent of cereals marketed. Commercial traders and end users assemble the rest. Grain sold to cooperatives or licensed traders is then sold to domestic users, exported, stored, or sold to the intervention agency, Office Nationale Interprofessionel des Céréales (ONIC).

Cooperatives. French cooperatives had their origins in an 1884 law permitting the organization of unions in agriculture and throughout the economy. Out of these organizations grew cooperative credit unions, mutual agricultural savings banks, and agricultural cooperatives, which, by the 1930s, became involved in agricultural marketing, common purchasing of inputs and other supplies, and insurance. More recently, in the 1950s, local cooperatives and their regional and national associations got involved and became owners of port elevators and export capacity.

ONIC estimates that cooperatives control almost 75 percent of commercial grain storage capacity in France, about 21.4 mmt. This capacity includes over 4,900 storage facilities around France. Private traders control another 2,600 storage facilities, with a capacity of 6.8 mmt. On-farm storage provides another 16.9 mmt of capacity.

Among the largest regional grain cooperatives, the Société Coopératif du Syndicat Agricole Départemental d'Eure-et-Loire (SCAEL) is a member of a regional organization regrouping 120 cooperatives that jointly own an elevator at the Port of Rouen. SCAEL also has significant grain origination capability, with a mix of small elevators and terminal facilities.

Overall, cooperatives are extremely important in France, constituting half of the 14 firms in the French food and agriculture sector with sales of more than 3 billion French francs (FF). Dairy cooperatives are the largest group, and the French Grain Cooperatives Union (UNCAC) is among these important firms.[3]

Private Trade. The private and investor-owned grain trade handles about 25 percent of the marketed grain in France at the first-handler level and an increasing share as grain moves to users within France or internationally. Private companies own about 80 percent of French milling capacity, and cooperatives account for the rest. One company, Grand Moulins de Paris, produces 15 to 20 percent of all flour used in France, but the rest of the industry is highly fragmented with many small family-held mills serving neighborhood bakeries. French legislation enacted in 1935 allocated quotas limiting the amount of wheat that mills could grind for domestic markets. Flour exports began to increase in the late 1940s. Throughout the 1970s, milling capacity for export expanded sharply. U.S. and French millers have competed aggressively for markets in North Africa, especially Egypt, with EC export restitutions and the U.S. Export Enhancement Program playing an important role.

About 66 percent of commercial compound feed production comes from private trade, with cooperatives handling the rest. Many small family-owned feed mills are licensed to use formulations and brands by major multinational firms.

ONIC

During the late nineteenth and early twentieth century, Europe in general and France in particular vacillated between periods of trade liberalization and periods of protectionism. After World War I, French wheat producers, finding prices unacceptably low, decided to band together to protect their markets from competition from outside France and its North African colonies. In 1924, this decision led to the creation of the Association Général des Producteurs de Blé (AGPB), the French wheat producers association. With the growth of cooperatives, and continued dissatisfaction with price levels, the government got into the act in 1936 with the establishment of the Office du Blé (wheat board), which later became ONIC, the grain board.[4]

ONIC's role was to bring together the various participants in the grain sector, e.g., producers, merchants, millers, and other users as well as representatives of the Ministries of Agriculture and Finance, and to establish grain prices that were regulated by ONIC's control of licensing of those who assembled cereals, directed purchases and sales of stocks, and had control over exports. End users of grain were no longer permitted to purchase grain directly from producers. They were required to pass through cooperatives and merchants licensed as storage agents by ONIC. In exports, however, the multinational firms that are important in grain trade in the United States are also important in France.

With the formation of the EC, the locus for establishing support price levels and market regulations shifted to Brussels. ONIC became the French Intervention Agency, responsible for implementing EC grain policy in France. As such, ONIC participates in the EC's Cereals Management Committee, which establishes import levies and export refunds. It purchases grain for intervention from its licensed storage agents. It contracts for about 3 mmt of storage capacity used for storing intervention stocks. It assures the availability of financing for intervention purchases, collects levies, and pays export refunds. It also monitors developments in French grain quality and makes available export-grade and origin certification to exporters. As part of these tasks, ONIC monitors developments in markets in France and around the world, gathering and publishing a wide range of data and analysis of importance to the grain sector.

French Grain Exports

France, physically the largest country in Western Europe, covers almost 210,033 square miles (543,965 square kilometers), about four-fifths the area of Texas. Physical size and location close to numerous potential markets give French grain producers important advantages in terms of transportation of grain to be exported. With the bulk of wheat production centered in the Paris Basin, grain export activity has shifted from southern France to the Atlantic Coast. Seven ports account for 90 percent of grain exports by sea, with Rouen, which handled 8.5 mmt of exports in 1989-1990, accounting for about half. Wheat is the primary grain going out of most ports, although Bayonne handles primarily corn, and Bordeaux, through which 2.1 mmt were exported in 1989-1990, about one-half corn and one-half wheat and barley.

Traditionally, inland waterways have made barge transport an important means of moving grain, both within France and for export to neighboring European countries. However, with the introduction of rate structures that encourage use of unit trains, and little change in barge rates for agricultural products, barge transportation has lost market share as traffic has increased, falling from 41 percent of total grain traffic to Rouen in 1976-1977 to 25 percent in 1985-1986.

Transport modes used vary considerably by port. The short distances involved in many cases mean that truck transportation directly to the port is considerably more important than in the United States, accounting for almost one-half of shipments to Rouen. Truck and barge traffic is gradually being eroded by rail shipments, which have increased from only 4 percent of shipments to Rouen in 1976-1977 to 33 percent in the mid-1980s.

The expansion of rail use can be attributed to lower rates for unit trains with a minimum of 1,200 tons of grain and make-a-train rates for 500 to 8,000 tons assembled from different sublines. Rails account for about 50 percent of long-distance grain movements, such as from the Champagne-Ardennes region to Rouen (still short by U.S. standards). Unit train rates between the south of France and Northern Europe have made such shipments more attractive in recent years, offering the potential to cut into volumes shipped through ports. The single European market may further accelerate this trend.

Substantial investments have been made in improving some port facilities in recent years. Rouen, located on the Seine River, can handle grain shipments as large as 65,000 tons on larger vessels, although some ports, such as Bayonne, Port La Nouvelle, and Caen can only handle shipments as large as 5,000 tons. Port storage capacity has grown rapidly. Rouen's port silos reached 420,000 mt of capacity in 1985, up from only 120,000 mt in 1973. Maximum loading rates at Rouen's elevators range from 1,400 to 2,800 tons per hour, with 24-hour loading possible.

Where port draught (depth) and loading facilities do not limit the size of export shipments, limited importer unloading and storage capacity are sometimes constraints. As a result, much of the French grain trade with other European countries passes by barge on canals, the Moselle, Rhine, Rhone, and Meuse rivers, and the Mediterranean Sea.

Grain Export Grading

France has no government agency that exercises control over quantities and qualities throughout the marketing system. Private surveying companies and ONIC provide inspection of grain moving into export channels, and the private companies also provide inspection at locations within France, when called for by contract terms.[5]

Although official export-grade certification has been introduced in France, it accounts for a small share of total exports, with most grades certified by private companies. In 1982, ONIC was given the responsibility for making available grade certification for soft wheat that is exported by sea. In January 1985, ONIC grading laboratories at Rouen and La Rochelle began offering certification of grades and French origin of wheat, as shown in the list below.

	Standard 1	Standard 2
moisture	≤ 14.5%	≤ 15.5%
test weight	≥ 62.4 lbs/bu	≥ 60.8 lbs/bu
broken kernels	≤ 4%	≤ 4%
foreign material	≤ 2%	≤ 2%
Hagberg Falling Number	≥ 220	≥ 180

Grain that does not meet these two standards is graded "outside standard." A higher Hagberg falling number, indicating low levels of alpha amylase enzyme activity resulting from sprouting, is required for fermentation and proper bread loaf volume. The objective in introducing this grading system was to improve the competitive position of French wheat.

French Export Competitiveness

The competitive position of French grain on international markets is influenced by production and marketing costs, support prices, export refund levels, and a range of national government activities. Among these activities is the availability of concessional credit through COFACE, the French export insurance and credit agency. This credit affects the price at which grain can be sold in specific international markets.

Differences in accounting procedures complicate comparisons of production costs, but variations in exchange rates play an important role in determining how costs of production compare among the world's grain exporters. When the U.S. dollar was at its strongest relative to the French franc in 1984 and 1985 (8.8FF/$U.S.), there was little question that the most efficient French producers were highly competitive in production costs with the United States and other major exporters. Short distances from farm to port and from ports to export customers, combined with modernization of marketing facilities, also made marketing costs competitive.

As the U.S. dollar has weakened relative to the franc, down to 5.05 FF per U.S. dollar in early 1993, it is unlikely that even efficient French grain producers could be cost competitive in the absence of government export subsidies. For example, although producer prices for milling-quality soft wheat in French francs have actually fallen since the mid-1980s, the equivalent value in U.S. dollars rose from $3.85 per bushel in 1984-1985 to over $6 per bushel in 1991-1992. This increase has pushed up the subsidies required to sell French wheat on international markets, where prices are generally set in dollars (Figure 13.4, Table 13.2).

EC export subsidies on wheat rose from an average of $17 per ton in 1984 to about $133 per ton in early 1988, and ranged between $67 and $93 per mt during 1991-1992. This subsidy increase has made it possible for wheat that sells at about $6 per bushel on the French market to be exported at competitive prices.

Summary

This chapter has presented an overview of the European Community's Common Agricultural Policy as it affects grain and oilseed marketing. The EC system is different from current policy in the United States in that producer prices are supported at high levels and are protected from international competition by variable levies on imports and export refunds to permit surpluses to be sold on international markets. This system leads to high prices for consumers and costly government support programs. It also means that as long as taxpayers and consumers will pay these high subsidies and prices for the grains, EC grain can compete on international markets without regard to costs of production and marketing. Reforms adopted in 1992 will make producer prices somewhat more responsive to world market forces, while maintaining farm income through direct payments.

The case of grain marketing in France, the EC's largest grain producer and exporter, is described, including some historical developments and marketing channels. Contrasts between U.S. and European marketing systems provide interesting insights into factors affecting competition on international grain markets.

Notes

1. Michel Louis Debatisse, *Cerealexport: Strategies et Techniques d'exportation des céréales Francaises.* ATYA Edition, Paris, 1984.

2. Marcel Marloie, "Les echanges de céréales et d'oleoproteagineux France— C.E.E.—Bassin Mediterraneen—1984," Montpellier, France: INRA—ENSAM, 1986.

3. Jean Paul Charvet, *Les Désordre Alimentaire Mondial: Surplu et Pénurief—Le Scandale*, Hatier, Paris, 1987.

4. Marcel Marloie, "Les Changements dans la Regulation du Marché des Céréales en France depuis la fin du XIXe Siecle" in *Céréales et Produits Céréalières en Méditerranée*. Montpellier, France: Institut Agronomique Méditerranéen, 1986: 220-230.

5. William W. Wilson, and Lowell D. Hill. *The Grain Marketing System and Wheat Quality in France*. North Dakota Research Report No. 110, North Dakota Agricultural Experiment Station, Fargo, 1989.

Selected References

Agra Europe (U.K.), various issues.

Commission of the European Communities, *CAP Working Notes: Cereals and Rice*, Brussels, 1986.

———, The Situation on the Agricultural Markets, EC Commission, Brussels, various issues.

———, The Agricultural Situation in the Community, EC Commission, Brussels, various issues.

CAP Monitor, Agra Europe, Tunbridge Wells, U.K.

Charvet, Jean-Paul, Les Greniers du Monde. Economica, Paris, 1985.

———, *Les Désordre Alimentaire Mondial: Surplus et Pénurief—Le Scandale*, Hatier, Paris, 1987.

Debatisse, Michel Louis. *Cerealexport: Strategies et Techniques d'exportation des cereales Francaises*, ATYA Edition, Paris, 1982.

———, "Hedging and Speculation on the EEC Levy and Restitution Markets," in *Price and Market Policies in European Agriculture: Proceedings of the 6th Symposium of the European Association of Agricultural Economists*, September 14-16, 1983, Newcastle, U.K., 1984, pp. 120-130.

Harris, Simon, Alan Swinbank, and Guy Wilkinson, *The Food and Farm Policies of the European Community*, John Wiley and Sons, New York, 1983.

Home-Grown Cereals Authority, *Cereals Statistics*, London, various issues.

Marloie, Marcel, "Les echanges de cereales et d'oleoproteagineux France—C.E.E.— Basin Mediterraneen, 1984, INRA—ENSAM, Monpellier, France, 1986a.

———, "Les Changements dans la Regulation du Marché des Céréales en France depuis la fin du XIXe Siecle" in *Céréales et Produits Céréalières en Méditerranée*. Institut Agronomique Méditerranéen, Montpellier, France, 1986b, pp. 220-230.

Newman, Mark, Tom Fulton, and Lewrene Glaser, *A Comparison of Agriculture in the United States and the European Community*, USDA, Foreign Agr. Econ. Report 233, Washington, DC, 1987.

Office Nationale Interprofessionel des Céréales (ONIC), "Céréales: Chiffres Cles," Paris, various years.

———, "La Campagne Céréalière: Statistiques Nationales," various years.

———, "La Certification: Un Nouvel Atout pour L'Exportation et la Qualite," Paris, 1986.

Petit, Michel, *Determinants of Agricultural Policies in the United States and the European Community*, IFPRI Research Report 51, International Food Policy Research Institute, Washington, DC, 1985.

Port Autonome de Rouen, *Rouen, European Leading Port for Exports of Cereals*, 1984.

———, *Evolution Des Capacites De Stockage*, Paris, 1986.

Toepfer International, *The E.E.C. Grain Market Regulation*, Hamburg, various issues.

U.S. Department of Agriculture, Economic Research Service, *Western Europe: Agriculture and Trade Report*, various issues.

Wilson, William W., and Lowell D. Hill, *The Grain Marketing System and Wheat Quality in France*, North Dakota Research Report No. 110, North Dakota Agricultural Experiment Station, Fargo, 1989.

World Grain, various issues.

Glossary

Absolute Advantage: The ability of a nation to produce a greater amount of a good from a given set of resources than another nation can produce with a similar set of its resources.

Aeration: The practice of forcing air through bulk stored grain in order to maintain good grain condition.

Agency Tariff: A tariff published by a publishing agent on behalf of two or more carriers participating in all or part of the rates and transport conditions described in the tariff.

Arbitrage: A process of buying a commodity in one market and selling it in another because the prices in the two markets differ by more than the costs incurred in transferring that commodity from one market to the other.

Arc Elasticity of Demand: A measure of the price elasticity of demand between two points on a demand curve.

At-the-Money: A call or put option is at-the-money if its strike price is equal to the current market price of the underlying futures contact.

Back Haul: The use by carriers of their transport space to haul goods on the return trip rather than run empty.

Balance of Payments: A statement showing all of the transactions between residents of one country and all other nations, usually for one year.

Basis: The difference between a cash price and a specific futures price for a commodity, usually the near futures price.

Belief: Our perception of "what is" in regard to the facts of a situation.

Bid: A proposal of price and other terms by a would-be buyer of a commodity.

Bilateral Agreements: Agreements between an exporting and an importing country to purchase or sell a certain amount of products per year.

Blending: The systematic combining of two or more lots or kinds of grain to obtain a uniform mixture to meet a desired specification.

Booking the Basis: An act of entering into a sales agreement that prescribes price in terms of the basis.

Break-even Level: A position that equals the exercise price plus the premium for a call option or the exercise price minus the premium for a put option.

Buffer Stocks: Reserve grain stocks that are available to handle the selling or purchase of grain in order to reduce price fluctuations in the grain market.

By-product: A secondary product that results from processing a commodity to obtain a primary product.

423

c. and f.: Cost and freight to the designated delivery point, paid by the seller.

Call Option: A contract that gives buyers the right, but not the obligation, to assume a long futures position in the underlying futures contract at the exercise price on or before the expiration date.

Canadian Wheat Board: An agency of the Canadian government with monopoly control over the export and domestic sales for human consumption of wheat, barley, and oats.

Carrying Charge Market: A market situation in which each successive future in the market is quoted at a higher price than the previous delivery month.

Carry-over: The amount of grain in inventory at the first (beginning carry-over) or last (ending carry-over) day of a designated crop year; the stock of grain available for consumption but not consumed in the year that it was produced, is, thus, available for consumption in a later year.

Cartel: A group of independent sellers who have joined forces in order to control the production and/or marketing of a commodity.

Cash Forward Contract: A forward contract made in the cash market as contrasted to one made in the futures market.

Cash Grain Merchant: Any person or firm dealing in the buying or selling of grain.

Cash Market: A market other than the futures market, including spot markets and markets in forward contracts.

Certificate Final: A certification of the quality of a shipment of grain being exported.

Certificate of Competency: A Small Business Administration certification in the United States that a party seeking to contract with an agency of the U.S. government is qualified to fulfill the terms of the contract.

Ceteris paribus: Holding some variables constant, while letting specific variables change.

Charter Party: A contract binding ship owners and charterers of ocean freight shipments.

c.i.f.: Cost, insurance, and freight to the designated delivery point, paid by the seller.

Class Rate: A freight rate based on a uniform classification of all freight hauled by rail or motor carriers.

Coarse Grains: A category that includes feed grains (corn, barley, oats, and grain sorghum) and rye, plus millet in some foreign nations.

Commercial Storage: Grain storage space that is commercially provided for a fee.

Commission Merchant: A person who buys and sells grain for others on a consignment basis without taking title to the grain.

Commodity Credit Corporation (CCC): An agency of the USDA created in 1933 to carry out loan and storage operations as a means of supporting prices above the level that would have prevailed in a free market.

Commodity Futures Trading Commission (CFTC): An agency established in 1974 to take over the duties of the Commodity Exchange Authority, with emphasis on regulation and surveillance of futures trading; reports directly to the U.S. Congress.

Commodity Rate: A rate applicable to the specifically identified commodity or commodities only; usually supersedes a class rate.

Common Agricultural Policy (CAP): An act that provides high internal support prices for European grain producers and protects them from international competition through a combination of variable levies, which keep the price of imports high, and export refunds, which allow EC grain to be sold at the going rate on international markets without forcing producers to take lower prices.

Common Carrier: Any person or firm licensed to transport goods, services, or people for a fee, with certain exemptions being granted motor carriers when hauling whole grain or grain products.

Complements: Commodities that are used together because of the additional benefit obtained from using them in combination rather than singly. The computed cross-price elasticity between such commodities is negative.

Comparative Advantage: A situation in which an individual or nation that is relatively superior at producing some goods gains by trading for other goods that another individual or nation is relatively more proficient at producing.

Concessional Sales: Sales of commodities at terms more favorable to the recipient than the going market rate; sales made or subsidized by the U.S. government.

Concurrence: An authorization for a publishing agent to issue a tariff on behalf of the carriers. A letter of concurrence constitutes a power of attorney assignment to the publishing agency in matters identified in the letter.

Conservation Reserve: A U.S. land conservation and supply control program designed to improve farm incomes by removing certain acreages of land from production. Also referred to as the Soil Bank program.

Consignment: A lot or shipment of a commodity that is placed under the control of an agent or broker for custody or sale.

Constructive Placement: An action to place a cargo-carrying vehicle in position for a shipper to load or unload when it cannot be placed at the elevator's rail siding.

Container Freight: Freight that is shipped in sealed containers for loading on rail cars or oceangoing ships.

Corn Competitive Formula (c.c.f.): The Canadian system of pricing grains on the basis of their digestible energy and protein content.

Country Elevator: An establishment that buys grain from farmers; it has facilities for receiving the grain and shipping it by truck or rail.

Country Merchandiser: A person or firm, generally located in a rural area, and buying grain from country elevators and selling it to terminal markets, processors, and exporters.

Crop-Fallow System: A management method that alternates the use of land between cropping and fallowing, especially in arid or semiarid areas, to improve soil moisture and fertility and control plant pests, diseases, and weeds.

Crop Year: The U.S. officially designated production and marketing year for a grain crop. For wheat, the crop year is from June 1 to May 31. For corn and soybeans, it is from October 1 to September 30.

Cross-Price Elasticity of Demand: An index that measures the responsiveness of the quantity of a good, X, demanded to a 1 percent change in the price of another good, Y, *ceteris paribus*.

Crow's Nest Pass Agreement: An agreement signed in 1897 between the government of Canada and Canadian Pacific Railway to build a rail line through the Crow's Nest Pass between southern Alberta and southern British Columbia. The Canadian Pacific Railway agreed to maintain lower freight rates for shipping grain in return for a government subsidy to help build the rail line.

Customs Union: An agreement by a group of nations, such as by the EC member nations, to eliminate tariffs and other restrictions on goods exchanged by those nations and to impose a uniform tariff policy on the exchange of goods with nonmember nations.

Cu-sum ship loading plan: A statistical plan for minimizing variability in quality during the loading of ocean vessels.

Deferred Payment: A payment at a prescribed (later) time after a change of ownership occurs.

Delayed Pricing: A form of cash trading used by country elevators when buying grain from farmers that provides for determining the price of the grain after ownership of the grain has been transferred.

Delivery: The act of transferring ownership or title from a seller to the buyer; it may or may not involve the physical movement of the commodity.

Demand: The quantities of a good or service that will be bought at various prices per unit of time, *ceteris paribus*.

Demand Curve: A graphic representation of the relationship between price and quantity demanded, *ceteris paribus*.

Demurrage: The dollar penalty imposed on a shipper for failing to load or unload a freight-carrying vehicle, following its constructive placement, within the time allotted by the carrier.

Diminishing Marginal Utility: A situation in which the greater the number of units of any given commodity that an individual consumes, *ceteris paribus*, the less will be the amount of satisfaction (i.e., utility) added by each additional unit of the good that is consumed.

Disposable Income: The spendable income remaining to an individual after paying personal and other taxes to the government.

Disposition: The use of products. The terms *utilization, consumption,* and *use* are used interchangeably with the terms *disposition* and *disappearance*. In a broad sense, disposition may refer to domestic and export disappearance; in a narrower sense, it may refer to such specific uses as for food, feed, seed, or industrial purposes.

Diversion and Reconsignment: The selection of alternative destinations for en route shipment.

Diverter-Type Mechanical Sampler: An officially approved mechanical device, operated by electricity or air, that is used to obtain a representative sample (for testing and grading) from a stream of grain being removed from one location to another.

Draft: The vertical distance from waterline to keel of a waterborne vessel.

Dry-processed Products: Products made without the addition of moisture, such as the milling of corn for corn meal or grits.

Dryeration: A process utilizing high speed and a high temperature to remove moisture from grain, with cooling by aeration.

Dunkel Proposal: A proposal in the Uruguay Round of GATT to reduce trade-distorting aid provided to farmers, to "tarrifficate" existing trade barriers, and to cut subsidized exports and the value of export subsidies.

EC-6: The European Common Market established by the Treaty of Rome in 1957 by Belgium, West Germany, France, Italy, Luxembourg, and the Netherlands.

EC-12: See European Community.

Effective Demand: A phrase occasionally used to emphasize that consumers are "willing and able" to buy the quantities indicated by the demand curve.

Elasticity of Demand: *Price* elasticity of demand. An index relating the percentage change in quantity demanded in response to a 1 percent change in the price of a good or service, *ceteris paribus*.

Electronic Trading: The use of electronic communication equipment to enable traders at different locations to exchange bids and offers and enter into sales agreements.

Ellis Cup: An officially approved manual sampling device, designed to draw a sample from grain moving on a conveyor belt. The Ellis cup is used at domestic interior points and not at export points.

Embargo: A government-ordered suspension or prevention of trade with another nation; it may be applied to all trade with that nation or to selected goods and services.

Engel's Curve: A graphic description of the relationship between a family's income and the quantities of goods purchased; based on Engel's law, which states that as a family's income increases, proportionately less of that income is spent on food.

European Community (EC): A customs union with common agricultural and other policies, established in 1957 and signed by Belgium, France, Luxembourg, Italy, West Germany, and the Netherlands. The United Kingdom, Denmark, Ireland, and Greece joined later, thus, the occasional reference, "EC-10"; Spain and Portugal joined to make the "EC-12" in 1986, and with German reunification in 1990, the former East Germany became part of the EC.

European Currency Unit (ECU): A unit representing a weighted average of member nation currencies.

Equilibrium Price: The market-clearing price at which buyers will take the quantity of a good that sellers want to sell.

Exceptions: Statements by individual carriers not wishing to comply with certain provisions of an otherwise governing rate tariff.

Exempt Carriage: Transport not subject to regulation. Exemptions normally apply to products transported, not to the carrier. Bulk containers are generally exempt from regulation when transported by barge. Unprocessed agricultural products are exempt when transported by trucks. Private transport by truck or barge also is exempt.

Exercise or Strike Price: The price at which a call (put) option holder may buy (sell) and a call (put) option writer is required to sell (buy) on the holder's demand.

Expiration Date: The last day that a holder of a put or call option may exercise the right to buy or sell the underlying futures contract at the option exercise price.

Ex-Pit Transaction: A special type of trade allowed under the rules of a futures market involving the exchange of a futures contract for a cash commodity.

Exports: Domestically produced goods and services that are sold abroad.

Fair Average Quality (FAQ): The quality of a shipment of grain is at least equal to the quantity of all such grain shipped during a specified period.

Farmer-Owned Grain Reserve: A grain reserve authorized by the U.S. Food and Agriculture Act of 1977 to support farm prices. Farmers retain ownership of the grain and store it on their farms or in commercial facilities for periods of not less than three years nor more than five years.

f.a.s.: Free alongside ship, a term specifying that the seller delivers goods to the port elevator or dock at the specified location, and the buyer pays for loading the ship and for ocean freight.

Federal Grain Inspection Service: An agency of the USDA established by the Grain Standards Act of 1976. This agency is responsible for maintaining uniform measures of grain quality and for grading grain at ports of export.

Federal Maritime Commission: An agency of the U.S. government responsible for regulating ocean and Great Lakes shipping.

Feed Grains: Grains that are used primarily for animal feed: corn, barley, oats, and grain sorghum.

First Handler: A merchant or processor who buys farm products directly from farmers.

Fixed Costs (of Storing Grain): The costs of facilities ownership. Fixed costs include depreciation on facilities, interest on invested capital, insurance, and other costs that remain constant whether the facility is used for storing grain or is left empty.

Flagout: Omission of rate changes on specific items of a class of freight for which rate changes have been authorized.

Fleeting: The assembly of barges into larger or smaller towing units in response to differing channel conditions.

f.o.b.: Free on board, a term specifying that the seller loads the ship or other conveyance at the specified delivery point, with the buyer paying freight charges.

Food Grains: Grains used primarily for (or in products for) human consumption. Wheat, rye, and rice are classified as food grains.

Formula Pricing: An agreement to set the price for a trade by using a special formula or rule usually based on a reported price to be observed in the future.

Forward Contract: A sales agreement calling for delivery during a specified future time period.

Forward Pricing: An agreement between buyer and seller that sets the price and other terms of trade and provides for the transfer of ownership at a later date.

Forward Trading: Trading in forward contracts.

Fourth-Section Departure: A rule in Section 4 of the Interstate Commerce Act prohibiting higher rates for a longer haul than for a shorter haul of the same commodity over the same line. Departures from this rule are allowed under certain conditions.

Free Market: An economic system in which decisionmakers (people and firms) are able to buy and sell in their own best interests, with a minimum of governmental restriction of their activities; also referred to as an "open" market.

Freight Forwarder: An assembler of freight shipments who arranges shipping details. Ocean container shipments are frequently arranged by a freight forwarder.

f.s.t.: Free on board, stowed, and trimmed; the term specifies that the seller loads the ship and pays for stowing and trimming the load at the specified delivery point, while the buyer pays the ocean freight.

Fumigation: The process of exposing grain to the fumes (vapor) of a chemical agent to kill pests.

Fundamental Analysis: An effort to explain and predict price movements by using the concepts of supply and demand.

Futures Contract: A standardized forward contract that is traded under the rules of an organized exchange.

General Agreement on Tariffs and Trade (GATT): An international code of tariffs and trade rules that became effective on January 1, 1948, following signature by the 23 participating nations, intended to foster the growth and conduct of international trade by reducing tariff barriers and eliminating import quotas and other discriminatory treatment between trading nations. GATT membership currently exceeds 80 nations.

Grain Marketing: Performance of all business activities that coordinate the flow of goods and services from grain producers to consumers and users.

Grain Grades and Standards: Specific standards of grain quality established to maintain uniformity of grains from different lots and to permit the purchase of grain without the need for visual inspection and testing by the buyer.

Grain Merchandiser: A person or firm buying and selling grain; a middleman.

Grain Reserve: Stocks of grain withheld from the market and stored for use in times of critical shortages.

Grain Warehouse Receipt: A legal document by which a warehouseman formally acknowledges that grain has been received for storage.

Grits: Coarsely ground cereal grain used for human consumption.

Hectare: A land area measurement of 10,000 square meters that is used in all countries except the United States and Liberia, and is equal to 2.47 acres.

Hedging: Buying or selling a futures contract as a temporary substitute for an anticipated cash transaction. Hedging is used to protect the firm or individual from losses caused by price fluctuations.

Horizontal Integration: This process occurs when a single management gains control, by voluntary agreement or ownership, over two or more firms performing similar activities at the same level or phase in the production or marketing sequence. An example is the merging of two country elevators.

Identity Preservation: Segregation of a commodity from one point to the next in the marketing system so that the initially identified commodity is delivered to the next point in the marketing system without being mixed with other units of the same commodity during handling and shipment.

Imports: Purchases of foreign-produced goods and services.

Income Elasticity of Demand: An index that measures the relative response of quantity consumed to a change in income.

Integration: The economic linkage of two or more firms under a single management through voluntary agreement or ownership. Examples of such integration are *forward* (one step closer to the final marketing stage), *backward* (one step farther away from the final marketing stage), *vertical* (two or more different stages in the marketing process), and *horizontal* (two or more firms in the same stage of the marketing process).

Interstate Commerce Commission (ICC): A U.S. federal agency, established by the Interstate Commerce Act of 1877, responsible for the economic regulation of railroads, barges, and interstate trucking.

In-the-Money: A term stating that a call (put) option is in-the-money if its exercise price is below (above) the current price of the underlying futures contract.

In-Transit Stoppage: The right of a shipper, when permitted by the rate engaged, to stop a shipment of goods before it can be delivered if the one to whom the shipment is consigned is unable to pay for the goods.

Intrinsic Value: A component of the option premium that is measured in a dollar amount (if any) by which the current market price of the underlying futures contract is above the exercise price for the call option and below the exercise price for the put option.

Inverse Carrying Charge: A charge in which distant futures are quoted at lower prices than near futures; ordinarily occurs when grain supplies are short relative to demand.

Inverted Market: A market situation for a storable commodity where the price for near-term delivery exceeds the price for later delivery during the same marketing year.

Joint Rate: A single rate involving two or more interconnecting carriers.

Land in Farms: The land area under the control of farm producers and in use or available for their use in producing agricultural products.

Land Retirement: A term that usually refers to the removal of land from production according to the objectives of a farm program.

License: A formal permit from a state or federal authority to operate a grain warehouse or other regulated business.

Line Elevator: Two or more grain elevators owned by a grain company, usually located along a railroad line.

Local Tariff: A tariff describing rates and transportation conditions when all origins and destinations are served by a single carrier.

Long: An individual or firm that owns a commodity or holds fixed price agreements to buy the commodity in excess of any fixed price sales agreements.

Malthusian Theory: A theory postulated by Thomas R. Malthus (1766-1834) that the world's population has the biological capability of increasing at a more rapid rate than food supplies can be increased.

Manifest: An invoice listing the cargo aboard ship.

Marginal Rate of Transformation: The rate at which products substitute for one another along a production possibilities curve.

Margin Pricing: In grain merchandising, the practice of setting one's bid price by subtracting a fixed amount (or margin) from a buyer's bid price.

Market: Any group of buyers and sellers who have the ability to communicate with one another.

Market Channels: The agencies and institutions through which products are moved from their original producers to the final consumers.

Marketing: The performance of all the business activities that direct the flow of goods and services from producers to the final consumers.

Marketing Board: A government or quasi-government agency with exclusive marketing privileges for a commodity.

Maturing Future: A futures contract during the last few weeks that it is traded—for example, the March corn contract between March 1 and the last day of trading in that contract.

Metric Ton: 1,000 kilograms of weight, equal to 2,204.6 pounds.

Moisture Content: The amount of water in grain; measured by the weight of water in grain as a percentage of the total weight of that grain.

Multinational: A business firm with subsidiaries in more than one country.

Most Favored Nation: A provision in a treaty between two nations that grants each signatory to the treaty the same tariff rates as the most favorable rates that either may grant to any other nation.

Near Future: A term stating that for a particular commodity or exchange, the futures contract currently being traded has the earliest maturity date.

Net Position: The amount of a commodity that a trader owns plus the amount of fixed price agreements to buy minus the amount of fixed price agreements to sell the commodity.

Nonrecourse Loan: A loan in which the lender, such as the CCC, has no recourse beyond the physical commodity itself in satisfaction of the loan.

Nontariff Barriers: Government regulations that reduce the free flow of commodities in international trade.

Offer: A proposal of price and other terms of trade by a would-be seller.

Off-Farm Storage: Commercially provided facility used for grain storage at a fee.

Oilseeds: Seed crops including soybeans, sunflowers, rapeseed (canola), flaxseed, peanuts, cottonseed, copra, sesameseed, safflowerseed, castor beans, and palm kernel, which are grown both for their oil and high-protein meals.

On-Farm Storage: Farmer-owned grain storage facilities.

Opportunity Cost: The value of a sacrificed alternative.

Option: A contingent contract between two parties to buy or sell a *particular* underlying commodity under specific conditions.

Organization of Petroleum Exporting Countries (OPEC): A cartel of petroleum exporting nations, organized in 1960 to increase their oil revenues.

Out-of-the-Money: A term stating that a call (put) option is out-of-the-money if the exercise price is above (below) the current price of the underlying futures contract.

Parity Prices: The price that gives a unit of product the same purchasing power as it had in a specified base period.

Pelican: An officially approved sampling device that is swung or pulled through a falling stream of grain. The pelican is a leather pouch attached to a long pole, and may be used for sampling grain at domestic interior points but not at export points.

Per Capita Consumption: A simple average, derived by dividing the total amount of a good consumed by the total population.

Perfectly Competitive Market: A type of market that assumes a large number of buyers and sellers; freedom of entry and exit; perfect knowledge of demand, supply, and prices; and rational behavior among participants.

Pink Grading Certificate: A certificate attesting only to the accuracy of the grade of a sample of grain and not to the sampling method or that it is a representative sample of the lot from which it was taken.

Place Utility: The utility created by transporting commodities to locations where those goods are more highly valued.

Policy: A plan by which a government, firm, or individual expects to achieve a given objective by specific actions.

Port Terminal: A grain-handling firm with facilities to load oceangoing ships.

Premium: A negotiated cash payment made by option buyers to option sellers.

Price Determination: The interaction of the forces of supply and demand to establish price in the marketplace.

Price Discovery: The process by which, through exchanging bids and offers, the market clearing price is found that equates supply and demand.

Probe: An officially approved sampling device (a compartmentalized metal tube, 5 to 12 feet long), designed to be pushed into a bin or load of grain and simultaneously draw a sample from several different depths in the lot. The probe is the only approved method for obtaining samples from stationary lots of grain.

Processor: An individual or firm that processes a raw farm product to its desired form.

Production Possibilities Curve: A curve showing all the combinations of two goods that an individual, firm, or nation can produce during a given time period with full utilization of all of the resources and the technology available to produce those goods. When resources are subject to diminishing marginal productivity, the production possibilities curve is drawn concave to the origin to reflect increasing opportunity costs.

Projection: An extrapolation that forecasts a future situation on the basis of historical relationships.

Proportional Rate: A common or uniform balance of a through rate paid beyond a rate-break point.

Public Law 480 (PL-480): The basic legal authority for sharing the surplus agricultural production of the United States with developing countries that have food deficits. Since its inception in July 1954, it has evolved from a temporary, surplus removal measure into a major tool in the world struggle for freedom from hunger and an effective instrument to stimulate economic development and support of U.S. foreign policy goals.

Public Policy: A special kind of group action designed to achieve certain aspirations held by members of society.

Put Option: An option that gives buyers the right, but not the obligation, to assume a short futures position in the underlying futures contract at the exercise price on or before the expiration date.

Putting on the Crush: A term that specifies locking-in a crushing margin, accomplished by buying soybean futures and selling equivalent amounts of soybean meal futures and soybean oil futures.

Quota: A restriction on the absolute quantity of a commodity that may be imported.

Rail Tariff: A legal document published by carriers or carrier associations, showing applicable rates, rules, and regulations governing service, routings, special services, demurrage, and other related matters.

Rate-Break Point: A point in the rail rate system from which incoming rail billing applies to a shipment to qualify it for a lower (proportional) outbound rate in lieu of the higher (flat) rate.

Reciprocal Trade Agreements Act of 1934: An act that expanded the markets for U.S. products by authorizing the president to enter into mutually advantageous, bilateral trade agreements with other nations. Freer world trade was promoted as countries agreed to lower their tariff barriers.

Round Turn: The purchase and sale of one futures contract.

Scale Ticket: A written document that indicates the weight of a load of grain and may also contain such other information as test weight, moisture, grade, and price.

Seasonality: A fluctuation in price, quantity, or other variables within a year or producing season.

Short: An individual or firm holding fixed price sales agreements for a commodity that exceed the amount of the commodity owned plus the fixed price purchase agreements held.

Shrink: In grain, the loss of volume or weight that occurs during drying.

Speculation: An act of carrying unhedged cash grain inventories; buying or selling futures (or commodities), without an opposite cash market (or futures) transaction, on the basis of anticipated price changes.

Splits: Pieces of soybeans that are otherwise undamaged.

Spot Market: The market in sales contracts for immediate delivery, or delivery within a few days.

Spoutline: A vertical core of grain fines that accumulates in the spaces between grain kernels at the peak of the growing pile of grain as it is being poured into a storage bin.

Storage: The marketing function that holds grain from one time period to a later time period when the grain is more highly valued.

Substitutes: Commodities that can replace one another in production or consumption. The coefficient of cross-price elasticity between substitute commodities is positive.

Subterminal elevator: An establishment that buys and sells grain in large quantities and operates facilities for receiving and shipping grain that are not located at a terminal market.

Supply: The quantity of a good or service that producers are willing to sell in the market at various prices per unit of time, *ceteris paribus*. The supply of grain in any one year includes the current year's output, plus carry-over and imports.

Target Prices: Prices established in farm programs that will be supported by government assistance.

Tariffs: Taxes levied on commodities as they cross a nation's boundary; also refers to the schedules of charges for warehouse services, or the schedules of rates charged by carriers.

Technical Analysis: Efforts to predict short-term price movements that are based on observed regularities or patterns in a price series.

Tender: A formal written offer or proposal of price and other terms of trade by a would-be seller.

Terminal Agency: Wholesale dealers, car lot receivers, commission merchants, and brokers. They facilitate the assembly and distribution of grain but do not have the physical facilities for handling grain and may or may not take title to the commodity.

Terminal Elevator: An establishment that operates facilities for receiving and shipping grain in large quantities at a terminal market.

Terminal Elevator Company: A firm involved in cash grain merchandising and in operating terminal and subterminal elevators for storage as well as for merchandising.

Terminal Market: A major assembly and trading point for a commodity. Some of the major U.S. terminal markets for grains are Kansas City, Chicago, Minneapolis, Toledo, Portland, St. Louis, New Orleans, and Houston.

Terminal Market Merchandiser: A person or firm, located in terminal market areas, and buying grain from country positions to be shipped to terminal markets.

Terms of Trade: The specific provisions of a sales agreement, including the price, quantity, quality, time and place of delivery, and method of payment.

Test Weight: A measure of grain density determined by weighing the quantity of grain required to fill a one-quart container, and converting this to a bushel (2150.42 cubic inches) equivalent.

Threshhold Price: The minimum price at which grain can enter the EC so that it will not be sold at a price that undercuts European grain in Duisburg, Germany. It is derived from the target price by subtracting the cost of transport and transshipment between the most important entry port at Rotterdam, Netherlands, and Duisburg, plus a trading margin.

Time Charter: A contract between shipowner and shipper in which the shipper "rents" the ship for a specified period of time.

Time or Extrinsic Value: The amount by which the premium exceeds the intrinsic value of an option.

Time Utility: The utility created through the storage function that makes goods and services available to consumers at the time they are more intensely desired.

To Arrive: A term of sale, specifying that a commodity will be moved to its delivery point at a given time in the future.

Transit Balance: The balance of a through rate from a specific origin to a specific destination that is paid beyond a designated transit point. Differs from a proportional rate because it applies to specific origin and destination points.

Transportation: A marketing function that moves grain (or other goods and services) from one location to another.

Turning: A process of emptying a bin, elevating, rebinning, and possibly blending grain to maintain its quality. Before the advent of aeration, turning was the means used to ventilate or aerate grain.

Two-Price Plans: Farm programs that attempt to take advantage of differences in the price elasticities of demand in domestic and foreign markets to increase total revenue from the sale of agricultural products.

Uniform Commercial Code: A code adopted by the American Bar Association to facilitate the easy transfer of warehouse receipts. Under the code, a warehouse receipt is a certificate of title for agricultural products that are stored in a warehouse.

Uniform Straight Bill of Lading: A uniform shipping contract between a shipper and a for-hire common carrier. The Straight Bill of Lading serves as a nonnegotiable receipt for goods that have been delivered to the carrier.

Unit Train: An assembled train loaded at one location by a single shipper, moving to a single consignee at a single location; used most commonly in shipping coal, grain, and soybeans.

U.S. Warehouse Act: An act passed by Congress in 1916 to protect the interests of producers and other grain owners who store commodities with public warehousemen.

Utility: The satisfaction that people derive from consuming goods and services.

Values: An individual's or group's perception of what "ought to be" regarding a situation.

Variable Costs (of Storing Grain): Costs that are incurred only if grain is stored, including insurance, taxes on the grain, interest foregone on the grain, shrinkage, and quality maintenance.

Variable Levy: An import tax set with reference to the difference between the threshhold price and the lowest price on a delivered (c.i.f.) basis that imported grain is available in Rotterdam, or adjusted for equivalence with Rotterdam.

Vertical Integration: A process that occurs when a firm combines two or more activities that are sequentially related but different phases in the production or marketing sequence. Examples are a country elevator merging with a subterminal or terminal elevator, or a grain producer purchasing a country elevator.

Voyage Charter: A contract between a shipowner and a shipper for a specific shipment or voyage.

Wet-processed Products: Products that are made from soaked grain, such as starch, dextrin, and syrup from corn.

White Grading Certificate: A document issued when the sampling and grading have been performed by licensed employees of an official inspection agency.

Woodside Sampler: An officially approved mechanical grain-sampling device used at domestic interior points. This sampler was often used at export points prior to the invention of the diverter-type sampler.

Working Stocks: Sometimes called "pipeline" stocks. The grain inventory needed to keep a plant operating efficiently throughout the year.

Yellow Grading Certificate: This certificate may be issued by the elevator if it is equipped with a diverter-type mechanical sampler that is operated by a licensed operator. The certificate indicates the accuracy of both the sampling and grading procedure, and the grain must be graded by an official agency.

Acronyms and Abbreviations

AAA	Agricultural Adjustment Act (U.S.)
AGPB	Association Général Des Producteurs de Blé (France)
AID	Agency for International Development
ASCS	Agricultural Stabilization and Conservation Service (U.S.)
BCFM	broken corn and foreign material
bu	bushel
c. and f.	cost and freight
CAP	Common Agricultural Policy (European Community)
CARE	Cooperative for American Relief Everywhere
CBT	Chicago Board of Trade
CCC	Commodity Credit Corporation (U.S.)
c.c.f.	corn competitive formula (Canada)
CFTC	Commodity Futures Trading Commission (U.S.)
CIDS	Computerized Information Delivery System (U.S.)
c.i.f.	cost, insurance, and freight paid by seller
COFACE	French export insurance and credit agency
CPEs	centrally planned economies
CPR	Canadian Pacific Railway
CRS	Catholic Relief Services
CSA	Central Selling Agency (Canada)
CUSTA	Canadian-U.S. Trade Agreement
CWB	Canadian Wheat Board
CWRS	Canadian Western Red Spring (wheat)
cwt	Hundredweight
DTN	Data Transmission Network
EC	European Community
ECU	European currency unit
EEA	European Economic Area
EEC	European Economic Community
EEP	export enhancement program
EFTA	European Free Trade Agreement

ERS	Economic Research Service (U.S.)
FAQ	Fair Average Quality
FAS	Foreign Agricultural Service (U.S.)
f.a.s.	free alongside the ship
FCM	futures commission merchants
FF	French franc
FGIS	Federal Grain Inspection Service (U.S.)
FM	foreign material
f.o.b.	free on board
f.s.t.	free on board, stowed and trimmed
GAFTA	Grain and Feed Trade Association (UK)
GATT	General Agreement on Tariffs and Trade
GRIP	Gross Revenue Insurance Program (Canada)
GTA	Grain Transportation Authority (Canada)
ha	hectare
HFCS	high-fructose corn syrup
HRW	hard red winter (wheat)
ICC	Interstate Commerce Commission (U.S.)
ITM	International Trade Organization
km	kilometer
LTA	long-term agreement
mmt	million metric tons
mt	metric ton
NAEGA	North American Export Grain Association
NASS	National Agricultural Statistics Service (U.S.)
NFA	National Futures Association (U.S.)
NISA	Net Income Stabilization Account (Canada)
ONIC	Office Nationale Interprofessionel des Céréales (France)
OPE	option pricing error
OPEC	Organization of Petroleum Exporting Countries
OTM	out-of-the-money
PCP	posted county price
PIK	payment in kind
PL	Public Law (U.S.)
PSEs	producer subsidy equivalents
ROW	rest of the world
SCAEL	Société Coopératif du Syndicat Agricole Départemental d'Eure-et-Loire (France)
SEC	Securities and Exchange Commission (U.S.)

SGP	special grains program (Canada)
UCC	Uniform Commercial Code (U.S.)
UGSA	Uniform Grain Storage Agreement (U.S.)
UK	United Kingdom
UNCAC	the French grain cooperatives union
URSA	Uniform Rice Storage Agreement (U.S.)
USDA	United States Department of Agriculture
USITC	United States International Trade Commission
USSR	Union of Soviet Socialist Republics
USTR	United States Trade Representative
WGSA	Western Grain Stabilization Act
WGTA	Western Grain Transportation Act (Canada)

Conversion Factors

1 acre = 0.4047 hectares
1 hectare = 2.471 acres
1 kilogram = 2.2046 pounds
1 metric ton = 2,204.622 pounds

1 bushel Barley = 48 pounds or 21.8 kilograms
1 bushel shelled Corn = 56 pounds or 25.4 kilograms
1 bushel Oats = 32 pounds or 14.5 kilograms
1 hundredweight milled Rice = 100 pounds or 45.4 kilograms
1 bushel rough Rice = 45 pounds or 20.4 kilograms
1 bushel Sorghum = 56 pounds or 25.4 kilograms
1 bushel Soybeans = 60 pounds or 27.2 kilograms
1 bushel Wheat = 60 pounds or 27.2 kilograms

About the Book and Editors

The world grain industry affects our daily lives in ways both large and small. It influences what we consume for breakfast, lunch, and dinner and provides about 40 percent of the world's food supply. The U.S. and world grain industry affects our income, our investments, and global politics. As world population and therefore global demand for grain grows, the volume handled by the U.S. grain industry will continue to expand, demanding not only improvement in crop yields but also continued efforts to compete in increasingly sophisticated international markets.

This newly revised, fully updated text provides a practical, comprehensive overview of grain marketing that is useful to both the upper-level undergraduate studying agricultural marketing and the professional working in the industry. *Grain Marketing* blends several approaches to the study of commodity marketing, combining the institutional, functional market structure with the analytical and behavioral systems approach to grain marketing. The book includes basic background information for newcomers to the subject of agricultural marketing as well as more rigorous treatment of advanced subjects in agricultural marketing. The book's overall plan allows the student to follow the movement of the major grains—corn, wheat, and soybeans—from farm production to final consumption. Along the way, it provides a detailed description of the worldwide system, encompassing local and multinational corporations, state agencies and boards, national trade and agricultural policies, and the cash and futures markets that serve this industry.

Gail L. Cramer is L. C. Carter Professor of Marketing and Policy, and, Eric J. Wailes is professor of agricultural economics—both in the Department of Agricultural Economics and Rural Sociology at the University of Arkansas.

About the Contributors

Colin A. Carter is professor of agricultural economics at the University of California, Davis.

Dennis M. Conley is associate professor of agricultural economics at the University of Nebraska, Lincoln.

Gail L. Cramer is L. C. Carter professor of agricultural economics at the University of Arkansas, Fayetteville.

Reynold P. Dahl is professor of agricultural and applied economics at the University of Minnesota, St. Paul.

Walter G. Heid, Jr., formerly with the Economic Research Service, USDA, is now a private consultant specializing in grain marketing and economic development activities. His headquarters are in Harrisonville, Missouri.

Richard G. Heifner is an agricultural economist with the Economic Research Service, USDA, Washington, D.C.

Lowell D. Hill is the L. J. Norton Professor of Agricultural Marketing at the University of Illinois, Urbana/Champaign.

Bob F. Jones is professor of agricultural economics at Purdue University, West Lafayette, Indiana.

Donald W. Larson is professor of agricultural economics at the Ohio State University, Columbus.

Dean Linsenmeyer is former professor of agricultural economics, University of Nebraska, Lincoln.

Mark D. Newman is director of agribusiness consulting for Abt Associates, Inc., Bethesda, Maryland.

Robert L. Oehrtman is professor of agricultural economics at Oklahoma State University, Stillwater.

L. D. Schnake is an agricultural economist with the Federal Crop Insurance Corporation, Kansas City, Missouri.

L. Orlo Sorenson is professor of agricultural economics at Kansas State University, Manhattan.

William W. Wilson is professor of agricultural economics at North Dakota State University, Fargo.

Bruce H. Wright is an agricultural economist with the Economic Research Service, USDA, Washington, D.C.

Index

Contents

Acknowledgments

The quotation from J.-B. Pontalis is reproduced from *Windows*, translated by Anne Quinney: the English translation is copyright 2003 by the Board of Regents of the University of Nebraska; the text was originally published as *Fenêtres* and is copyright 2000 by Editions Gallimard.

The quotation from Simone de Beauvoir is from *The Second Sex*, published by Vintage, and reproduced by permission of The Random House Group Ltd. ©2015.

The quotation from Meir Shalev is reproduced from *My Russian Grandmother and Her American Vacuum Cleaner: A Family Memoir*, published in 2011 by Schocken, with permission from Penguin Random House LLC.

The quotation from Key Redfield Jamison is reproduced from *An Unquiet Mind: A Memory of Moods and Madness*, published in 1995 by Vintage, with permission from Penguin Random House LLC.

1

Why Do I Write about Sleep Disturbances in Women?

"We go to bed. He's asleep before his head hits the pillow. I can't believe that the heated argument we had two hours ago isn't keeping him awake. I am still seething from the argument, and am also going over many other things that happened during the day, and I'm really unable to fall asleep. Remembering the phone call with my mother, that made me so sad, thinking about my son in the army who hasn't been in touch for 2 days, worried about a lecture I'm giving tomorrow and the few corrections I still need to make. And he? He's already snoring deeply. I prod him so that he makes less noise. He stirs but doesn't wake up. I think to myself: how can he be sleeping so deeply, without a worry in the world, when it will take me hours to clear my mind before I can fall asleep?"

I hear so many variations of this story from patients that I have seen with sleep disturbances. I listen, and I always give the same answer: "Yes, there are significant differences between men and women, even in the very neglected area that we call sleep."

Gender differences between men and women regarding sleep are primarily recognizable after puberty. They are a combination of many factors, including:

- The effect of hormones related to fertility in a woman's life cycle.
- Differences in male and female sleep structures.
- Multiplicity of roles in women's lives which affect stress and sleep levels.
- Tendency to depression and anxiety, which is twice as common in women as in men.
- Women's physical and mental diseases affecting sleep (Figure 1.1).

My field of expertise as a psychiatrist is reproductive psychiatry, and it deals with the connection between hormonal and mental changes in the course of a woman's life—that is, mental changes during the menstrual cycle, pregnancy, postpartum, and menopause. I was introduced to this field combining gynecology and psychiatry, which was new in psychiatry at the time, during a 2-year fellowship program in New York, and it fascinated me. I felt I had found my place and calling after many years of uncertainty about my chosen profession.

At a stand of professional books at a conference I attended in the United States, the book *Depression in Women* (2001) by psychiatrists Meir Steiner and Kimberly Yonkers, which deals with the relationship between hormonal and mental changes caught my eye. The book addresses all the stages in the life of a woman, from puberty to menopause. It touches on both normative and non-normative mental changes and

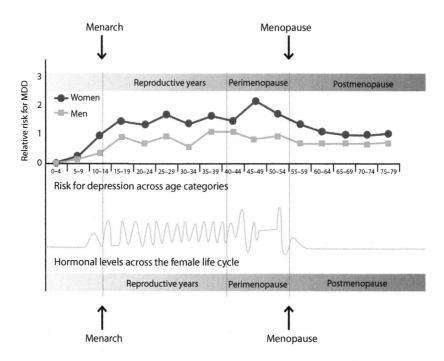

FIGURE 1.1 Risk for depression across age categories.

explains their causes and ways of treatment. I purchased the book, and after reading it, I decided to write a similar book in Hebrew.

My book, *Women, Mood and Hormones*, published in 2005, was based on the experience I gained working as a psychiatrist in hospital gynecological wards in Israel. It describes the mental vicissitudes experienced by women throughout their lives: during the menstrual cycle, pregnancy, postpartum, and menopause. It addresses normative changes that do not require treatment, as well as less normative changes that cause emotional and functional disturbances and need to be treated. The book offers guidelines for treatment ranging from psychological therapy and medication to nutritional changes and alternative remedies.

The book was intended for physicians who specialize in women's health. Surprisingly, it attracted a female readership that included women of all ages, professions, and educational and cultural backgrounds; women who were thirsty for the knowledge I presented and curious about the links between the hormonal changes they experienced and their state of mind. I resolved that my next book should also address those in need of this information—women themselves.

In the following years, a new area within my field of expertise came into focus—sleep disturbances in women. The more knowledge I gained about the subject through treating afflicted women, the more aware I became of how little this common disturbance is understood and, consequently, of the limited scope of viable solutions. I knew this would be the subject of my next book.

While working on the book, I told people close to me about its subject. Their reactions, as expected, were varied. Some wondered if there really is a difference between female and male sleep. Others thought it was an excellent idea, considering that so little is known about sleep. Almost every woman who heard about the content of the book told me that she, or a friend or relative, suffered from sleep disturbances. Everyone agreed on one thing: It is an important subject and very timely for a period in which women are asserting their place in the world and letting their voices be heard.

When I began to collect information and data, I found very little professional literature on sleep disturbances in women. Most of the existing material is in English, highly professional, and aimed at physicians, but not easily accessible for women. Indeed, it is not easy to simplify professional knowledge into a readily comprehensible language, yet that is the challenge I took upon myself.

In my role as a psychiatrist specializing in women's health, I encountered women of various ages and circumstances, who told me of their distress concerning sleep disturbances, whether or not related to hormonal changes. Psychiatry treats sleep disturbances as part of depression or anxiety, but does not pay attention to gender differences, or take into consideration the hormonal changes that affect sleep.

The more my help was sought by women who suffered from sleep disturbances, the more intrigued I became by its causes and by methods of easing them. It is well-known that women are more likely than men to seek medical help, especially in my profession as a psychiatrist of women's health. Yet, there is no doubt (and this is supported by the professional literature) that sleep disturbances are more common in women than men, especially in the current era in which women fill diverse roles in their lives and have an ever-decreasing number of hours of sleep.

Is the medical world dedicating resources to study and treat sleep disturbances in women in light of the above? By no means! Even today, sleep disturbances are not taught in medical schools, except in relation to other disturbances.

Clinical trials, and subsequently diagnoses and treatments, have focused on male subjects, since hormonal changes in women introduce unwanted complexity. The general assumption was that what worked for men would also work for women. We now understand that this is not the case, and we are witnessing the development of gender-specific medicine.

The first time that the U.S. Food and Drug Administration (FDA) addressed gender differences in relation to medicine dosage was in 2013. According to a statement published in medical journals, women were to be given half the recommended dose of pills containing zolpidem (Ambien, Zodorm, Stilnox). In women, these pills can impair morning alertness, cause uneasiness, and at times enhance daytime anxiety. This is only one instance of many gender-oriented differences found in the medical treatment prescribed to men and women. Most have yet to be studied and published. The FDA statement was undoubtedly a milestone in gender medicine and marked the beginning of a new era.

Women tend not to share their sleep problems with those around them. They treat it as a sentence they must accept or, at most, combat with sleeping pills.

Irit, one of my patients, told me that my business card was in her wallet for a year before she contacted me. When she eventually arrived, she told me: "I thought to myself: why should I go to a psychiatrist? The thought terrified me. I remember that my mother took sleeping pills for years and lived with it, so I thought that I would also do that until the problem sorted itself out. I had been suffering from insomnia and early waking since my periods stopped when I was 52. I thought it might be connected with my husband's retirement and the effect it had on our relationship. In conversations with friends of my age, they complained more about hot flashes that interfered with their sleep, but that's not what I was suffering from and therefore I didn't connect my sleep disturbance with menopause."

Women of Irit's mother's generation tended not to talk and not to complain, and the means to help them were more limited. I explained to Irit that today there is no reason to suffer, and that menopause is one of the most common causes of sleep disturbances in women, with or without hot flashes. Irit wanted to know why the problem can't simply be solved with sleeping pills. The answer to this is complicated, and therefore I will dedicate an entire chapter to the medical treatment of sleep disturbances.

My need to write about sleep disturbances in women came from my experience in this area as part of my job, but primarily from the distress I encountered in women close to me and in my patients. I want to encourage women to understand that it is not a sentence they have to accept. Accurate diagnosis of the sources of sleep disturbances can lead to a custom-made treatment for each woman and provide her with the help that will improve the quality of her sleep.

2

What Actually Is Sleep?

"Sleep is the best meditation."

The Dalai Lama

Sleep is an orderly process regulated by the brain. The central aspects of the process are a decrease in awareness of external signals and of physical activity, renewal of the brain's sources of energy, and the processing of acquired information.

Physicians indicate three fundamental elements that are key to good health: sound nutrition, regular physical exercise, and sleep of satisfying quality. The average person spends a third of their life asleep. Ten days without sleep are all that is required for the average human body to collapse. Sleep experts and researchers link sleep deprivation to a variety of medical problems: decrease of cognitive faculties, degenerative diseases of the brain, high blood pressure, weight gain, heart and blood vessel diseases, weakening of the immune system, mental stress, depression, and more.

A research study undertaken in Singapore in 2014 of 66 people over the age of 55 showed that the brains of participants who slept fewer hours aged far more quickly. Magnetic resonance imaging (MRI) scans and cognitive tests of participants who reported stable sleep habits showed better results than those who did not maintain balanced sleep patterns.

Sleep Habits among Various Populations Worldwide

A 2017 study by researchers at the University of Chicago and Monash University, Australia, analyzed sleep habits of a billion users classified by country, based on hours of internet use. The data revealed that, on average, Argentinians sleep the most—10 hours and 16 minutes on average per night; Americans sleep an average of 8.5 hours; and Israelis are situated in the middle of the list, between Brazil and Romania, with an average of slightly under 9 hours of sleep per night. Japanese sleep the least—only 7 hours and 16 minutes on average.

A 2017 survey by the Israel Health Ministry among 460 Israelis found the following:

- 53% of participants in the survey reported they sleep 5–6 hours a night
- 38% reported they sleep 7–8 hours a night
- 4% reported they sleep over 9 hours a night
- 5% reported they sleep up to 4 hours a night

How Many Hours of Sleep Are Recommended?

"The day and the night consist of twenty-four hours. It is enough that a person sleeps a third of this period, which is eight hours. But it should terminate at the end of the night, so that from the time he commences his sleep till sunrise eight hours will have passed. As a consequence, thereof he will arise from his bed before dawn."

Maimonides, Mishnah Torah Human Dispositions 4

The current American Academy of Sleep Medicine recommendations for the daily sleep requirement according to different age groups are as follows:

- Adults up to the age of 65 should sleep 7–9 hours.
- People over 65 should sleep 7–8 hours nightly.
- The best hours for sleep are between 11 p.m. and 7 a.m. (Figure 2.1).

These are the hours in which the secretion of melatonin, the sleep hormone, is at its peak. A detailed explanation appears later in this book. Secretion of melatonin begins at nightfall and stops before dawn. When the sun rises, the human body begins to secrete a hormone called cortisol, which encourages activity. Night sleep is the most

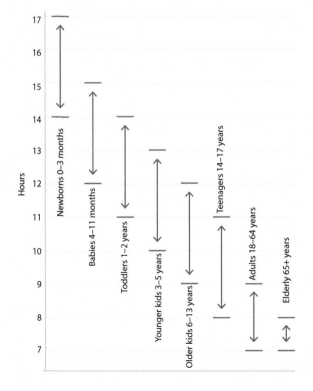

FIGURE 2.1   Recommended sleeping time by age.

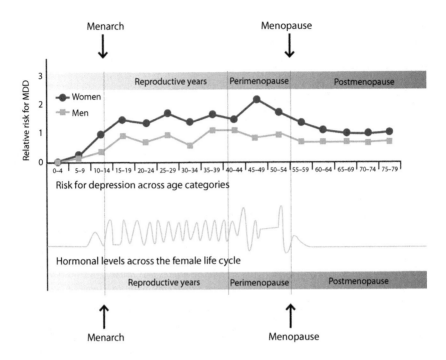

FIGURE 1.1 Risk for depression across age categories.

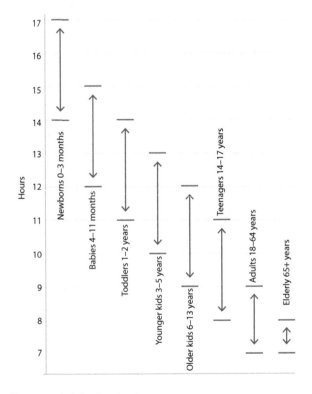

effective and continuous. People who work night shifts know all too well that daytime sleep does not provide the same sense of rest as night sleep.

The required amount of sleep varies with age and is highly individual. I meet women who tell me that if they sleep less than 8 hours a night, they feel they lack energy during the day. On the other hand, there are women who do not require more than 5 hours of sleep. They feel refreshed and function quite well. A baby sleeps during most of the day, and as it grows, the hours of sleep decrease.

Another piece of data emphasized by the American Academy of Sleep Medicine is the importance of the quality of sleep, not just the quantity. I will elaborate on the quality of sleep in Chapter 11 "Treatment of Sleep Disturbances."

Thus, humans spend one-third of their lives sleeping. Despite that, and quite surprisingly, we know very little about the role of sleep. We accept this cycle of sleep and wakefulness as a clear fact and know little about the questions linked to the subject. What is sleep? What causes tiredness? What happens if we don't sleep or sleep too little? What happens to us when we sleep?

Before the nineteenth century, sleep was considered a form of reversible death. The assumption was that when a person sleeps, he or she is dead; when they awaken, they return to life. In 1830, Robert Macnish, a Scottish doctor, surgeon, and author, published *The Philosophy of Sleep*, in which he wrote that sleep is an intermediate state between wakefulness and death. The change in the perception of sleep only came at the beginning of the twentieth century with the discovery that the brain is active during sleep. Gradually, it was revealed that there is a very structured sleep architecture related to the electrical activity of the brain.

What Is Sleep Architecture?

During sleep, there is electrical activity in the brain, and the brain alternates between two distinct states during the night, creating a defined architectural pattern (Figure 2.2): Rapid eye movement, or REM (sleep with rapid eye movement is the stage in which we dream; the brain appears to be as active as when awake, and the eyeballs move quickly right and left; however, muscle tension is low and the body in fact lacks mobility) and non-REM (NREM) sleep.

NREM Sleep

Slow-wave sleep is characterized by a decrease in the frequency of the electrical activity waves of the brain, along with increase in wave intensity, and by the relaxation of the body while maintaining muscle tension and closed and quiet eyelids. This state is divided into three phases, corresponding to the frequency of the waves. The slowest waves are the stage of the deepest sleep.

Stage 1: Falling Asleep (The Stage of Transition from Wakefulness to Sleep)

Characteristics: A sense of floating and incoherent thoughts, limpness of the neck muscles, slowing down of brain activity and eyelid motion.

Sleep quality: A person can be easily woken at this stage.

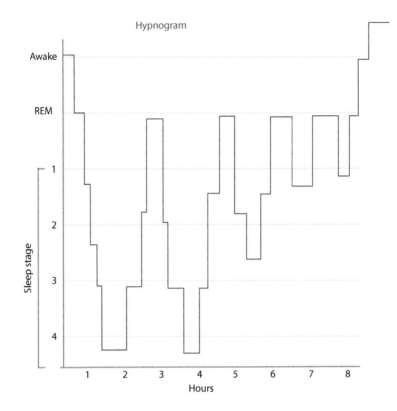

FIGURE 2.2   Architecture of the sleep cycle.

Stage 2: Light Sleep (The Stage in Which We Are in Fact Sleeping)

Characteristics: Awareness of external stimuli is reduced, body heat decreases, eye movements are reduced, there is little brain wave activity.

Sleep quality: Sleep is still light and superficial.

Stage 3: Deep Sleep (The Stage That Provides the Body with the Sense of Rest)

In fact, this is the combination of two different stages previously called stage 3 and stage 4.

Characteristics: Additional decrease in the frequency of the brain waves, in body heat, in breathing rhythm and in blood pressure; there are no eye movements whatsoever. In this stage, bedwetting can appear in children. Others may experience sleepwalking or nightmares that cause wakefulness and fright.

Sleep quality: Characterized by a difficulty in waking up. Those who wake up at this stage will feel confused for several minutes.

Role: This stage appears mostly in the first cycles of sleep. It is in this stage that the largest amount of growth hormone is released; hence its importance for children and adolescents. This stage is extremely important for brain energy renewal and the balancing of many body systems: temperature regulation, immune system function, blood pressure equilibrium, sugar levels, and insulin-release balance. Undoubtedly, this is one of the most important sleep stages.

REM Sleep

The sleeping stage in which we dream takes place periodically after NREM sleep. This stage is characterized by rapid eye movements, an increase in heart rate and breathing, and an increase in sexual arousal. The muscles of the limbs are paralyzed as a protective mechanism to prevent movement during dreaming. Brain wave activity resembles that of wakefulness, even though the sleep is very deep; therefore this stage is also called "paradoxical sleep."

The REM stage plays a critical role in the correct activity of the brain: researchers assume that this is the time when the brain processes the enormous amount of information it has absorbed during the day. Indeed, studies have shown that if a person is prevented from reaching the dreaming stage of sleep, the body will compensate by shortening the time it takes to arrive at this stage. The REM stage lengthens with the duration of sleep. There are those who feel brief wakefulness at the end of this stage. The dreams last between 5–20 minutes, even though we may think we have dreamed all night long. (Dreams are discussed in Chapter 12 "On Women and Dreams.")

Each night comprises an average of 4–6 sleep cycles, with each cycle lasting 90–110 minutes. Each cycle has a segment of NREM sleep followed by a REM segment. The duration of REM sleep increases in each cycle toward morning, whereas the duration of NREM decreases.

Surprisingly, part of the process of normative sleep includes short bouts of waking up for a few seconds during the night at the end of each sleep cycle. For the most part, we do not remember these moments. In cases of various aches and pains, or in states of tension and anxiety, the awakening can be full and is accompanied by difficulty in falling asleep again.

People who use sleeping pills on a regular basis tend to suffer from an impairment of their sleep architecture, especially of the REM stage, and consequently will report that they do not dream. In addition, a large part of their night sleep occurs in the second stage of NREM sleep, in which sleep is still light and superficial, and less in the third stage, in which sleep is deeper.

Within the sleep structure, different roles are attributed to the different stages of sleep. For example, the REM stage, the dreaming stage, is the stage in which—according to many researchers—our memories are stored in the central nervous system.

Why Do We Sleep?

"No small art is it to sleep: it is necessary for that purpose to keep awake all day."

Friedrich Nietzsche

Most experts agree that sleep is more essential to humans than food. Every animal with a nervous system needs sleep.

Conclusions drawn from the physiological changes that occur due to lack of sleep have led researchers to believe that sleep plays a central role in processes of learning and memorizing. The general consensus among researchers is that sleep is the time in which the brain repairs itself, goes through a reboot of sorts, and processes the vast amount of information it has absorbed during the day.

Occasionally, a new scientific publication causes us to stop and rethink the concepts that seem most natural to us. Such was the neurological study conducted at the University of Rochester in 2013, which offered a possible new and interesting explanation as to why we sleep. Maiken Nedergaard, a professor of neurosurgery who led the team of researchers, presented evidence that sleep may possibly be a mechanism that evolved to allow the brain to rid itself of toxins.

In her article "Metabolite Clearance from the Adult Brain," published in the leading medical journal *Sleep Science* in October 2013, Professor Nedergaard writes that sleep is a very unusual phenomenon from the point of view of survival and even a dangerous one. "For an animal like a mouse sleep is very dangerous, because it is then much more vulnerable to the cat that could eat it." Nedergaard uses the example of mice because they were used in the study; however, this is also true of all other animals.

The question then is what is so essential in sleep, that humans and most other animals have risked their lives for it during millions of years of evolution? To discover what makes sleep essential, Nedergaard and her team examined what happens in the brains of sleeping mice. They discovered that the cerebrospinal fluid (CSF)—a fluid that fills the skull and in which the brain floats—serves to evacuate toxic waste.

It turns out that when we are awake, toxic proteins are formed in the brain that can be dangerous at high dosages, and these accumulate among the cells. According to the research, the brain has developed a self-purifying mechanism that occurs during sleep: the CSF is streamed into channels formed between the cells and cleans out the toxic proteins. This process of circulation and rinsing is possible during sleep because the cells shrink, and this allows the fluid to pass between them with ease. In contrast, during the hours when the brain is alert, the cells expand and movement among them is more limited. "It is like opening and closing a tap," says Nedergaard, "that (the difference between sleep and wakefulness) is how dramatic it is."

The toxin clearance mechanism active during sleep is known as the "glymphatic system," like the lymphatic system that clears toxic matter throughout the entire body. To date, the mechanism has been seen in mice and baboons; however, researchers have good reason to assume that the same process also occurs in human brains. One of the proteins that accumulates in the brain during the hours of daily activity is amyloid beta, known to cause Alzheimer's disease. Alzheimer's and other dementia diseases have also been found to be connected to sleep disorders. This has led scientists to assume that Alzheimer's and other neurological diseases are caused by unsatisfactory functioning of the toxin-cleansing process during sleep time. This can explain why we do not think clearly after a sleepless night, and why a lack of sleep over a period of time can cause death. So, if brain cleaning is a lifesaver, or at least a life preserver, why does it occur only during sleep? Why not perform toxin cleaning more frequently? Professor Nedergaard suggests that the cleansing process demands a huge amount of energy: "It is probably impossible for the brain to self-clean and at the same time to be aware of the surroundings, to talk, move, etc."

A clear mind is a creative mind, refreshed and functioning. A clear state of mind represents the difference between tranquility/calm and mental stress, and in the long run—between life and death. If indeed the research findings also apply to humans, this means that every time we perform the everyday act of sleeping without a second thought, we increase the length of time that the brain will function, as well as the quality of its function. According to this research, all these advantages are simply the result of some cleaning and tidying. The findings of this research study make us rethink the role of sleep in our lives and give a positive meaning to the expression "brainwash."

What Causes Lack of Sleep?

In contrast to the lack of clarity that still exists with regard to the active mechanisms during sleep, we know far more about the consequences of lack of sleep.

Anyone who has experienced sleep disturbances knows how it affects daily functioning: tiredness, lack of concentration, anger, low-frustration threshold, tendency toward unbalanced eating (especially of sugar and other carbohydrates), increased smoking. Naturally, all these increase the danger of becoming involved in accidents.

Moreover, it has been proved that chronic disruption of the biological clock, that is, chronic discrepancy between our lifestyle and our biological clock, is linked to sleep disturbances and also to increased danger of depression, bipolar disorders (manic depression), changes in cognitive functioning, memory problems, neurological diseases, cancer, degenerative diseases, infections, and metabolic problems.

The meaning of the Japanese term *inemuri* is to be so exhausted that one could fall asleep during a meeting. In 2015, the American Sleep Disorders Association published a study on the connection between lack of sleep and maladies such as diabetes, heart attacks, obesity, high blood pressure, and, of course, the effect on mental health. All this means that lack of sleep can be a matter of life and death.

With the advances in knowledge about sleep, we learn that the number of hours of sleep are important, but the quality of sleep matters more; that is, refreshing sleep without impairment of its architecture, is of crucial importance.

Occasionally, we sleep for 7 hours at night and do not understand why we wake up tired or irritable. It appears that the quality of sleep was inadequate due to medication that interfered with the sleep architecture, or various health problems or moments of wakefulness that might have disrupted the continuity of the sleep.

Anat, an architect, works with her sister in their joint practice. Anat is an energetic woman in her mid-forties. "I don't recall ever suffering from sleep disturbances," she told me. "On the contrary, I was always so tired that I fell asleep with a book or with my computer in bed sitting down. I never suffered from depression or anxiety. The world of psychology is foreign to me."

Anat came to me because she was suffering from an acute sense of fatigue during the day and functional difficulties due to frequent awakening during the night. "Something wakes me up and I am not sure what it is. I do fall asleep again quickly, but despite sleeping 7 hours, I feel very tired during the day and need a short nap. It interferes with my daily functioning at work and makes me very irritable."

Anat could not connect her frequent waking up with any mental or physical distress. The attempt to find the reasons for her frequent waking up at night included a test of her hormonal state, which was in the normal range, and a sleep-lab test, which also did not reveal any specific problem. Blood tests revealed a lack of iron, vitamin B12, and low levels of hemoglobin, which caused enhanced daytime fatigue and nightly sleep disruptions. A focused treatment of the problem improved the quality of her sleep and alleviated her distress.

How Is Sleep Regulated?

In other words, how does the body know when it is time to go to sleep and when to wake up?

Since life on earth began about 4 billion years ago, plants, animals, and humans have adjusted to the hours of day and night by creating an internal mechanism (biological clock) with electrical activity that has cycles of about 24 hours (circadian) and correlates physiological activities with the daily cycle of day and night. Most of our bodily systems undergo changes during these 24 hours—changes in blood pressure, hormone release, heartbeat, and more. These changes are called *circadian rhythms* and they include regulation of sleep and wakefulness. Normally, in humans, sleep coincides with nighttime and wakefulness with daytime (Figure 2.3). In nocturnal animals, of course, it is the opposite.

The regulation of sleep and wakefulness is managed by two systems:

- The Circadian Process (C-P)
- The Homeostatic Process (H-P)

The C-P

The circadian process controls our biological clock. The word "circadian" comes from the Latin *Circa*, meaning approximately, and *Dies*, meaning day; that is, the cycle is approximately 24 hours.

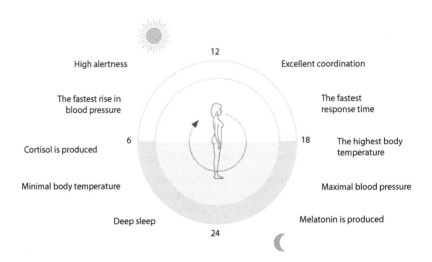

FIGURE 2.3   The effect of the biological clock on the human body.

The biological clock of humans is situated in a group of brain cells—the hypothalamus—that acts as a daily pacemaker for the activity of our brain and other bodily organs and for the regulation of wakefulness and sleep. At times it is also affected by temperature and food, although to a lesser extent.

The H-P

The homeostatic process: the origin of the word "homeostatic" is Greek, meaning a state of equilibrium. A necessary condition for our existence is the preservation of an internal equilibrium in all our bodily systems, for example, maintenance of body temperature, blood sugar levels and electrolyte levels. Equilibrating mechanics is also active in the process of sleep. If there is not enough sleep, a "debt" is created, which is usually solved by this regulatory process by means of longer and deeper sleep the following night.

Wakefulness causes accumulation of the homeostatic component and the creation of a cumulative need for sleep. The biological clock encourages daytime wakefulness and enables sleep at night. Both processes work together to enable satisfying states of sleep and wakefulness. The sum of the two processes is that the night is mainly dedicated to sleep.

According to an article published in a German daily newspaper, the spa town of Bad Kissingen decided to become "Chrono City," that is, a town that allows its residents to live according to their biological clocks, with individualized daily working hours, smart street lighting, and exposure of children to natural light.

Professor Till Roenneberg, an expert in chronobiology (a scientific field that studies cyclic phenomena in living organisms and their correspondence with the rhythms of the sun and the moon,) who is involved in the project, explains, "Internal time is highly individual, and all the bodily functions operate with highs and lows throughout the day and night. There are people who, if they could, would go to sleep at 8 p.m. and wake up at 4 a.m. and others who would go to sleep at 8 a.m. and wake up at 3 p.m. The problem is that society forces us all to live according to external conventions."

It is important to be aware of our internal biological clock and, as far as possible, to be attentive to the hours of sleep and wakefulness that are right for us.

Yael, a longtime patient of mine, asked me to meet her 16-year-old daughter, Gal, who suffers from sleep disturbances. Gal wanted us to meet alone while her mother sat in the waiting room. I found a very articulate and independent teenager. Gal told me that she was awake for most of the night and that she fell asleep only around 4, or even 5, a.m. As a result, she was late for school, tired during most of the day, irritable, and had problems concentrating.

After a few meetings with Gal I did not detect any mental problem or trauma that could cause sleep disturbances. I asked Gal to allow her mother to join us. In conversation with her mother, it transpired that Gal has had difficulties with her sleep since she was a baby. The situation became even more acute during puberty. During school holidays, Gal went to sleep at 3 a.m. and awoke at 10 a.m. She felt that these were the hours that were good for her.

Undoubtedly, Gal's biological clock, since infanthood and even more so during puberty, is different from the average. Throughout her life, Gal will need to adjust her lifestyle to her biological clock as far as possible. Until such time, treatment with melatonin can help her adjust her current lifestyle to her unique biological clock. (See Chapter 11 for further information on melatonin.)

Disturbances in the timing of sleep are found in 7% of teenagers. Usually, their appetites behave accordingly and they tend to eat very little during the morning hours. Their best hours for concentration are at night. Indeed, many teenagers study for exams at night, rather than during the day. They only eat in the late afternoon and at night and have difficulty functioning in the early morning hours. Many of the teenagers who suffer from the symptom of sleep-timing disturbance are mistakenly diagnosed as suffering from attention-deficit disorders, since they are tested in the morning, a time when they have difficulty concentrating.

Today, it is clear that disruption of the biological clock directly impairs health. It has been proved that night shifts cause damage to the immune system and that the proportion of sufferers from heart disease, diabetes, and cancer is higher among shift workers.

It is interesting to note that the Nobel Prize in medicine or physiology was awarded in 2017 to American researchers Jeffery C. Hall, Michael Rossbach, and Michael Yang for the discovery of the molecular mechanisms that control the circadian rhythm. The biological clock is designed to adapt the activity of living organisms to the regular rhythm of the day–night cycle.

Disruptions of the Biological Clock

As with every other physiological process, disturbances can also occur in the functioning of our biological clock, impairing its ability to function properly.

For some of us, the biological clock advances too slowly, and as a result, we are tired only at the late hours of the night. Conversely, for others, the biological clock advances too quickly, causing tiredness in the early hours of the evening. It is important to know that in either case, this is not a disease; however, if one's lifestyle does not comply with one's biological clock, this creates a problem.

Delayed Sleep Phase Syndrome

For one affected by delayed sleep phase syndrome, the sense of tiredness arrives only between 1 a.m. and 3 a.m., if not later. As a result, it is difficult to wake up in the morning for work or school. One feels tired throughout most of the day and has a deep desire to sleep. The sleep deficit is overcome on weekends.

Of course, if possible, it is desirable to find a job that suits a person's biological clock. Recommended treatments include use of melatonin, exposure to sun or artificial light, and treatment of seasonal affective disorder, especially in places where there are few hours of natural sunlight, mainly in the winter.

Advanced Sleep Phase Syndrome

On the other hand, there is also the phenomenon of tiredness and the need to go to sleep earlier than average, advanced sleep phase syndrome. Thus, for example, there are women who feel the need to go to sleep as early as 9 p.m. and wake up at 4 a.m. They often seek medical advice thinking they have a problem, as they wake up while most other family members, and people in general, are still asleep. These women have no problem functioning at work during the day but are tired when they go out for various activities at night.

Many people with this type of biological clock choose professions that suit them, including medical professions and agriculture, in which the work day begins early in the morning and ends by afternoon.

The era in which we live does not take the biological clock and its relation to physical and mental health into consideration: there are workplaces that are active 24 hours a day, airports that never close, places of entertainment that are open at all hours, hospitals in which it is impossible to separate day from night, television that is available around the clock, and computers and internet that are accessible at all hours.

The Link between Women's Health and Sleep Deprivation and Many and Unusual Work Hours

Over the years, many studies on this subject have been published. The first, in 1993 by McGill University in Montreal, indicated a connection between women who work shifts and abortions, depression, and alcohol abuse. Later studies showed a link between women working shifts and risk of breast cancer, disruption of the monthly cycle, and heart disease. It is important for every woman to be aware of the health risks and to decide whether shift work agrees with her personal biological clock. There are women for whom shift work is highly suitable; they do not feel any disruption of their biological clock, and therefore their physical and mental health is not at risk. On the other hand, there are women for whom this can be highly damaging.

Rachel arrived at my clinic late in the evening after many weeks in which she had tried to arrange a meeting and had difficulty finding a suitable time. The psychologist who had been treating Rachel for many years had encouraged her to come to me, as she was suffering from mood swings, frequent bursts of anger, and panic attacks.

Rachel, 30 years old, married, and mother of a 7-month-old baby, presents a daily morning television show. Her job, according to her, is the most important thing in her life. Her husband is a journalist who works in a European country and arrives home on the weekends. "This arrangement suits us," she told me. "I am not prepared to live in a different country, where I would just be a housewife pushing a baby carriage around town."

Rachel's job requires her to wake up every morning at 4:30 a.m. As a young woman who likes going out at night, she does not get to bed before midnight and sleeps about 4 hours per night. "I am constantly tired, irritated, and tense," she told me. "I have no patience for my baby, and especially not for my husband who comes once a week and wants all my attention." Rachel understood that this situation, in which she slept very little, is what was making her tired and irritable, but my attempts at convincing her to go to bed earlier fell on deaf ears. She finds it hard to compromise—she wants it all. I tried explaining to Rachel the consequences of lack of sleep and especially the effect of the disruption of her biological clock on her health, yet due to her youth and desire to drink in the whole world, she was not impressed by what I said. All she wanted was "a small pill" to help her be less stressed and irritable.

I refused to give Rachel "a small pill," because it seemed to me that Rachel was leading an incorrect lifestyle that disrupted her sleep and affected her mental state during the day, and therefore I referred her for psychological treatment to try and help her solve her basic problem—change her lifestyle and perhaps also understand how to create better priorities.

What Do Sacred Texts Tell Us?

Sleep is addressed in many places in the Bible, in different contexts.

Sleep is perceived as a state of weakness and danger

- The prophet Isaiah describes the heroism of the Assyrians in that they do not need sleep: "None is weary, none stumbles, none slumbers nor sleeps, not a waistband is loose, not a sandal strap broken" (Isaiah 5:27).
- Samson lost his potency when he slept on Delilah's lap: "And she made him sleep upon her knees; and she called for a man, and she caused him to shave off the seven locks of his head; and she began to afflict him, and his strength went from him" (Judges 16:19).

Giving up sleep in favor of dedication

- Jacob boasts that he watched over Laban's sheep instead of sleeping: "There I was: by day the heat consumed me, and the cold by night, and my sleep fled from my eyes" (Genesis 31:40).
- King David also prides himself for not having slept until he found the place on which to build God's Temple: "I will not give sleep to my eyes, or slumber to my eyelids" (Psalms 132:4).
- This is also a greatness of God: "Behold, he who keeps Israel will neither slumber nor sleep" (Psalms 121:4).

The sage in the Book of Proverbs advises men to sleep little, especially in cases when there is need for much work

- "How long will you lie there, O sluggard? When will you arise from your sleep? A little sleep, a little slumber, a little folding of the hands to rest…" (Proverbs 6:9–10).
- "Give your eyes no sleep, and your eyelids no slumber" (Proverbs 6:4).

However, there are cases in which sleep can be had quietly and without limit: when a person is connected to God and is protected from the dangers of sleep

- "I lay me down, and I sleep; I awake, for the Lord sustained me" (Psalms 3:6).
- "If you lie down, you will not be afraid; when you lie down, your sleep will be sweet" (Proverbs 3:24).

And Another Look at Sleep

The psychoanalyst Jean-Bertrand Pontalis dedicates a chapter to sleep in his book *Windows*, where he also relates to gender difference regarding sleep.

How I would love to be able to write in praise of sleep, like a poet, or, better still, of the sleeping woman who abandons herself to sleep in a bed after making love, in a field after harvest. Could this calm abandon result from the certainty that she will never be abandoned? Or from the anticipation of a miracle that would satisfy her? I think about the marquise of O. About the secrets she kept to herself. And I think of Proust's Albertine, the sleeping one who was "enclosed in her body" and who "became like a plant."

How I pity them, those I consider really sick, those men—always men—who claim that sleeping is a waste of time! ...

How I suffer with insomniacs, both the nighttime and the daytime kinds. Daytime insomniacs do exist. They are entirely focused on their agendas, incapable of dreaming and panicked by what they cannot control. As for those nighttime insomniacs with their gnawing worries—and everything becomes a worry to them—they think of themselves as lucid, convinced that they see reality as it is, stripped of meaning, bad and deadly. Even though they fail to realize then that, in their hatred of life, of others, of themselves, reality, like love, is something to invent and that each is given this power—except, of course, insomniacs.

Jean-Bertrand Pontalis, Windows, (2003) translated by Anne Quinney

Arianna Huffington, a Greek-American writer, journalist, businesswoman, and founder of the *Huffington Post*, has written several books about women. Her book *The Sleep Revolution* (2016) is dedicated entirely to the topic of lack of sleep and its consequences in modern times. Huffington, for whom an 18-hour workday was the norm, fainted one day as a result of fatigue and broke her cheekbone. This caused her to think about her sleep habits and the culture in which we live, and led her to write the book.

Huffington reveals, among other things, that there are languages that contain a specific word for death from overwork and sleep deprivation—"karoshi" in Japanese, "gwarosa" in Korean, and "guolaosi" in Chinese. English and Hebrew have yet to come up with a word for this phenomenon.

In her book, Huffington dedicates a long chapter to the history of sleep. She describes an interview given by Thomas Edison, inventor of the electric light bulb, to the *New York Times* in 1914, in which he envisioned that "in the future man would spend less time in bed. Man used to go to sleep with the sunset. A million years from now he won't go to bed at all, Edison prophesied, and added that sleep is a waste of time, a bad habit" (Arianna Huffington, *The Sleep Revolution*).

Napoleon, Huffington writes, was among the first to treat sleep in relation to gender, and he too regarded it as a waste of time. When asked how many hours of sleep are needed, he said, "Six for a man, seven for a woman, and eight for a fool." Napoleon also saw sleep as a serious limitation: "If the army can be kept active for many hours, with little sleep, a great advantage is achieved."

Today it is known that important neurological activity occurs during sleep, yet despite this, Huffington argues that modern society still perceives sleep as a waste of time. This is expressed in such popular sayings as "I'll sleep when I'm dead," or, in the Jewish mother's version, "I'll rest in my grave."

3

Women and Sleep

"I slept, but my heart was awake."

Song of Songs 5:2

I am surprised time and again when I talk to a man who has come to me seeking help for his sleep disorder. Although I treat mostly women, sometimes men will also see me, following in their wives' footsteps—and they too need help.

Women usually report that they began suffering from sleep disruptions following births. I have yet to hear a similar report from a man (even though the children were born to him and his partner at the same time). Women mention sleep disturbances as part and parcel of their lives and complexities: the burdens they have to endure, waiting up for a child to come home from a trip or an evening out, pregnancies and births, menopause, a quarrel with a friend, concerns over aging parents, a large overdraft. Certainly, there are men whose sleep is disturbed for many of the same reasons, yet all the men I have met to date attributed their sleep disorders to a single source—work.

Sleep disorders are among the most common complaint among patients who visit family doctors. Studies have shown that 35%–40% of the population suffer from sleep disturbances. Women suffer twice as much (especially from insomnia) compared with men, and with advancing age, the gap between the sex's quadruples (Figure 3.1). When we think about the era in which we are living, a dynamic period with almost daily discoveries in various fields, it is amazing to discover, time and again, how such a fundamental subject as sleep in general, and women's sleep in particular, is still neglected and dismissed.

We all sleep, yet only in recent years have we, as physicians, begun to understand the complexity of the sleep process. So far it has not received the deserved attention, and it appears to fall through the cracks. Very few medical students are trained in the subject, and certainly not specialists in various fields. Naturally, if the subject of sleep disturbances is an orphan, the field of gender within it is almost completely unknown.

A few years ago, I asked professor Nava Zisapel, a neuroscientist at Tel Aviv University, who was responsible for the development of Circadin (a prolonged-release melatonin tablet; details appear in Chapter 11), if her research on the drug took gender into consideration. My question surprised and intrigued her, since until I had asked her, it had not crossed her mind.

As a result of my reputation in this field, I was invited in 2014 to give a talk at a one-day conference of the Israel Sleep Research Society. This society was established in 1997, in continuation of a meeting between academic sleep researchers and practicing physicians who treat this new and developing branch of sleep medicine. My lecture on sleep disturbances among women was new to the society, which, until that time, had not thought that the subject of gender deserved special consideration.

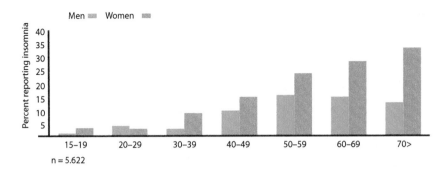

FIGURE 3.1 Insomnia by age and gender.

The extent to which this area is neglected can also be adduced from the fact that before 1993, sleep apnea—a disorder in which breathing stops during sleep, increasing the risk of heart disease and is even possibly fatal—which is considered common in men, was not even diagnosed in women. Women suffering from this disorder were misdiagnosed as suffering from depression and were treated with antidepressants. We now know that 2% of women suffer from sleep apnea. This example reveals that not only do misdiagnoses or non-diagnoses of sleep disturbances damage the quality of life—they can also put lives at risk.

Why Do Women Suffer More Than Men from Sleep Disturbances?

The issue of gender within sleep disturbances raises many questions. Why is it that women suffer more from sleep disorders? Do women complain more than men about sleep disturbances? Do diseases related to sleep manifest differently in women than in men? Do hormonal changes experienced by women affect the structure of sleep? Why is there no correlation between women's complaints about sleep disorders and objective findings in sleep clinics? Should the treatment of sleep disorders be different for men and for women?

I am always surprised to find that women, who have been suffering from sleep disturbances for many years, regard the condition as a fact of life, which they have to accept. I am even more surprised to discover how hard it is for them to change their lifestyles and habits, which are the causes of these sleep disruptions.

As part of my initial evaluation of a woman who comes to me for treatment, I ask various questions, including the following: when do you finish working? When do you eat? How much time passes between the end of work and bedtime? Is there a period of relaxation before bedtime? Do you take your computer from work and continue working at home as well? To my surprise, I discovered that women therapists (psychologists, physicians, social workers), who work with patients until the late evening hours, get into bed with their computers and read articles, write papers, or reply to emails. No wonder they have trouble falling asleep. One could rightfully assume that the same would apply to men. Yet men seem to suffer far less from sleep disturbances, even if they are worried about problems at work, work long hours, or continue to work from home.

Rita is a 64-year-old psychologist, a university lecturer, and has her own practice. She is a vital and active woman who is invited to participate in workshops in her area of expertise around the world. Rita meets her first patient at 7 a.m., has a 2-hour lunch break, and continues to work until 7 p.m. After a short supper, she works on her lectures and usually does not stop working before 10 p.m.

Rita came to me after she began suffering from sleep disturbances. "I'm a morning person," she told me, "In the past I would wake up at 5 a.m. full of energy and only switch off toward 10 p.m. However, in recent years I have stopped sleeping. It takes me 2 hours to fall asleep, and sometimes I am so worried about not being able to sleep, that I take a sleeping pill even before I get into bed. It takes half the following day for me to return to myself again."

Rita's sleep-wakefulness cycle changed with age, yet she finds it difficult to change her lifestyle accordingly. I suggested to Rita to begin working at a later hour and to end earlier, so that in the event that it should take her a long time to fall asleep, she would not have to worry too much about it. Yet it was very difficult for me to convince Rita; despite the suffering caused her by her sleep disturbances, she considered changing her habits as involving far greater suffering.

Women suffer more than men from sleep disturbances during their working lives, and this suffering grows with age. The reasons for this are complex. Some are related to hormonal changes linked to women's fertility and affect sleep throughout life. These are changes that occur during the menstrual cycle, pregnancy, postpartum, and especially during menopause and the end of the menstrual cycle.

How clear the connection between sleep and hormonal changes really is can be seen from studies that proved a link between shift work and disturbances in the monthly cycle, between chronic sleep disturbances and fertility problems, and between polycystic ovaries (a common gynecological condition that will be explained below) and sleep apnea.

Another reason that women suffer more from sleep disturbances than men is the fact that depression and anxiety are twice as common in women than in men, and sleep disturbances are among the symptoms of depression and anxiety.

Additionally, women fulfill many roles in their lives, and despite the many changes taking place in modern society, and in the traditional role division between men and women, it is usually women who must juggle home and work at the same time. This can make it difficult for them to relax, especially before going to bed, when they are busy thinking about the events of the day and those awaiting them in the morning.

4

Types of Sleep Disturbances in Women

When discussing sleep disorders, the tendency is to think about difficulties in falling asleep, tossing and turning or waking up frequently. I admit that until I wrote this book, I too did not think too much about the different types of sleep disturbances, especially those from which women suffer, for the same reason I mentioned above: The tendency among physicians and therapists to treat sleep disturbances as a single disorder, without paying attention to many other details and without investigating them further.

Since I began studying this subject, I have noted different types of sleep disturbances experienced by women. With the right questions, I can distinguish among different sleep disorders, and this enables me to propose the most appropriate and accurate treatment for each woman. Had I not delved more deeply into the subject, I would not have been able to help women in the same way.

In this chapter, I will describe the common types of sleep disorders from which women suffer:

- Insomnia
- Restless legs syndrome (RLS)
- Obstructive sleep apnea (OSA)
- Excessive daytime sleepiness (daytime hypersomnia, including narcolepsy)
- Parasomnia

Insomnia

Insomnia, or sleep disruptions, is the most common sleep disorder: about a quarter of the world's population suffer from this condition at some stage in their lives.

Who among us has not experienced periods of difficulty in falling asleep or of waking up during the night with troubling thoughts about the past day and the chores of the day up ahead? And who among us is not familiar with the term "insomnia loop" when often, after a night of two of difficulty in falling asleep, one becomes obsessed with the inability to fall asleep, and these thoughts about the difficulty prevent the desired sleep.

Characteristics of insomnia include the following:

- Difficulty in falling asleep
- Frequent awakening during the night
- Early waking in the morning

- Waking up in the morning with a sense that the sleep was of poor quality
- Each of these characteristics can occur without relation to the others, accompanied by feelings of distress throughout the day (lack of concentration, fatigue, low energy levels)

It has become more and more common to argue that insomnia is closely linked to our modern lifestyle and the rise in life expectancy. In bygone eras when people rose with the sun and went to sleep at sunset, they did not suffer from insomnia. Studies have shown that the effect of exposure to screens—including smartphone screens—on the brain is similar to exposure to sunlight, and this interferes with the activity of the human biological clock.

Not every difficulty in falling asleep can be described as a sleep disturbance. There is *temporary insomnia*, which lasts for a few days and goes away, without any connection to stressful situations or hormonal or physical problems. There is *short-term insomnia*, which can last up to a month and can be tied to mental or physical states. *Primary insomnia* has no known cause. *Chronic insomnia* refers to a disturbance that occurs at least 3 days a week over 3 months and can be related to mental, physical, and hormonal disturbances.

What Is a Difficulty in Falling Asleep?

There are women who fall asleep in 5 minutes, those who need 15 minutes, and those who fall asleep only after 45 minutes. That is, the time needed to fall asleep is individual, and therefore any deviation from the particular norm of each woman is considered a sleep disturbance. The measure to determine the existence of insomnia is a change in previous habits, or a decrease in daytime functioning because of fatigue.

Many women tell me that they have always loved sleeping and have had no problem falling asleep, and yet now they have difficulty falling asleep or they wake up every hour. Others tell me that they always found it difficult to wake up in the morning, and suddenly they find themselves up at a very early hour. The emphasis in the complaints of the women suffering from insomnia is the deviation from their familiar sleeping habits.

Family physicians report that insomnia is the most common reason for the complaints of the women they treat. It is a symptom—an indication of a problem and not a disease—that may have many causes. Therefore, one of the most important challenges when examining women is finding out the cause of the insomnia, and not necessarily a swift solution in the form of sleeping pills.

In many cases, physicians arrive at a quick conclusion that if a woman complains about insomnia, she is necessarily suffering from depression or anxiety and therefore should be sent to a psychiatrist for a suitable medication. Unfortunately, many psychiatrists, who do not specialize in women's health, may tend to accept this rather than looking for the sources causing the insomnia.

How common is insomnia in women? In one of the books dealing with sleep disorders in women (*Sleep Disorders in Women*, Attarian and Saltzman, 2013), it was reported that almost twice as many women complain about insomnia compared with men. Gender difference appears after puberty, until that age, there is no difference between males and females who suffer from insomnia. From this, it can be deduced that a large part of this disorder in women is related to hormonal changes following menstruation. Indeed, women tend to experience insomnia a few days before menstruation, during pregnancy, after giving birth, and during menopause, when the frequency of insomnia is 2.5–3 times higher compared with the previous period.

The Risk of Suffering from Insomnia

With advancing age, the risk of suffering from insomnia too increases, especially for women. A study conducted in the United States in 2002, found that 58% of the population over 65 suffered from insomnia several nights a week. Of these, 63% were women. Another study, conducted in Japan in 2000, showed that more unmarried women, widows, and unemployed women tended to report insomnia. The men who suffered from insomnia were older than the women who suffered from insomnia, and many of the men were divorced.

Common Causes of Insomnia

Many factors can cause insomnia: hormonal causes (menstrual cycle, pregnancy, postpartum, menopause), distressing events (relationship breakup, moving), worries about parents, financial woes, environmental factors (noisy environment, snoring partner, room too warm/cold, penetrating light), medical reasons (diseases that impair sleep, side effects of medications), lifestyle changes, and psychiatric reasons (depression, anxiety, and post-trauma). Antidepressants such as Lexapro (escitalopram) and Prozac can induce vivid dreams, which may lead to insomnia. (The links between insomnia and physical and mental diseases are dealt with at length in Chapters 9 and 10, respectively.)

Anat, 37 years old, single, is an energetic woman who has been working for many years for a computer company; she likes her job and invests a great deal in it. Anat came to me after she began suffering from insomnia.

"I always got home very tired, ate something, and even before my head hit the pillow, I was fast asleep," she told me. "I am a very energetic women and very optimistic. I have never had problems, and never saw a psychologist or psychiatrist. Everything changed 5 months ago, following the breakup of a long-term relationship that ended very suddenly and painfully. Since then, it takes me about 2 hours to fall asleep. I am very tired, but the thoughts run through my head: perhaps if I had not pressured my boyfriend to move in together, we wouldn't have broken up; perhaps I should try to convince him to get back together. I am becoming afraid of getting into bed, because I know the thoughts will begin racing in my head and I won't be able to fall asleep. Sometimes I'll have a glass of wine before going to bed, to help me fall asleep faster, but that doesn't always happen."

Anat developed a conditional (Pavlovian) reaction to the problem of falling asleep. Even before she got into bed, she was already worrying about the thoughts that would interfere with her being able to fall asleep. As a result, she developed a sleep disturbance, insomnia. The main treatment focused on changing this acquired reaction and helping her deal with the painful breakup, which was causing her anxiety.

The longer insomnia persists, the worse its consequences will be: tension, irritability, anxiety, depression, decreased functioning levels at home and at work, dependence on sleeping pills or antianxiety medication, high risk of car accidents, and sometimes a rise in blood pressure (Figure 4.1).

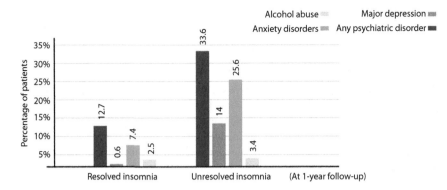

FIGURE 4.1   Persistent insomnia is a risk factor for future psychiatric disorders.

Treatment of Insomnia

The correct treatment of insomnia begins by obtaining the patient's account of her medical and mental history and a description of her lifestyle in order to help understand the cause of the insomnia. As a psychiatrist, it is always important for me to probe the root of the problem and to take into consideration the woman's age and any medical or mental problems from which she may suffer. From here, the route to treatment will be easier and more accurate.

Treating insomnia is a complex challenge. It is true that the easiest way to go about it is to prescribe sleeping pills, which were developed to deal specifically with sleeping disorders. However, this is not the route I recommend. The treatment I believe in includes a deeper understanding of the woman's lifestyle, such as her eating habits, working hours, sleep time, and mental or physical problems, as well as the sleeping conditions in her bedroom. Only after I have all this information will I begin treating the source of the insomnia, with an emphasis on preserving the quality of sleep.

I will elaborate the options for treating insomnia in Chapter 11 on the treatment of sleep disorders. Since it is such a widespread disturbance, I find it important to dedicate a special chapter to therapeutic approaches both with and without the aid of medication.

Restless Legs Syndrome

Restless legs syndrome (RLS) is a known disturbance that inflicts 14% of women and some claim even more. In 50% of the cases, there is a family history of the problem, which indicates a genetic basis. Even in these cases, women are twice as likely to suffer compared with men, and the frequency rises with age.

Characteristics of RLS

RLS is characterized by an uncomfortable sensation at the ends of the legs during rest and by significant relief when the legs are moved. There is a constant need to move, stretch the legs, move them, and go from place to place. The condition worsens in the evenings and impairs sleep and, as a result, leads to fatigue and the desire to sleep during daytime.

Criteria for Diagnosis of RLS

- The woman reports a strong need to move the legs, accompanied by discomfort in the legs.
- The discomfort in the legs worsens when sitting or lying down.
- The discomfort in the legs appears primarily in the evening or at night.
- Significant relief occurs when moving by walking or stretching the legs.
- The phenomenon is unrelated to other sleep disturbances, physical or mental diseases, or medication.

At times, RLS appears along with pregnancy, menopause, diabetes, rheumatic diseases, diseases linked to the thyroid gland, fibromyalgia (muscle pain, chronic widespread pain along with fatigue), anemia with iron deficiency, kidney failure, and more. Additional findings that help with the diagnosis are family history—close relatives who suffer from the condition—and relief following treatment with medication that raises levels of dopamine (a neurotransmitter).

Causes of the Condition

The causes of RLS are not yet clear, there are only conjectures.

Most studies in recent years report a link between the condition and a dopaminergic deficit in the central nervous system (the major part of the body's nervous system, which includes the brain and the spine). It is known that medication that increases dopamine in the central nervous system helps alleviate the symptoms, whereas medication that works on the dopamine in the peripheral nervous system does not provide relief. The peripheral nervous system is situated outside the central nervous system, beginning in the brain and spinal cord and connecting them with the various body organs and the limbs.

Another mechanism attributed to the condition is a deficit in reserves of iron and folic acid. Since symptoms of RLS appear at night, the link between iron and dopamine was examined in relation to the cycle of sleep–wakefulness. Iron levels decrease by 50% at night, and it is known that iron is vital for the production of dopamine, which increases at night. This can explain part of the reason that the condition mostly occurs at night.

Pregnant women suffer from this condition two to three times more than nonpregnant women, especially in the last trimester. The condition disappears after giving birth. This has led some to assume that in addition to the deficit in folic acid, iron, and dopamine, there may be additional factors associated with the female hormones, such as estrogen, progesterone, and prolactin. A deeper understanding of the subject requires more study. The risk of developing this condition during pregnancy increases with the number of pregnancies; in other words, a woman who has had three pregnancies has a risk of RLS that is three times greater than that of men and of women who are not pregnant. The risk increases with age.

Meeting with friends at a restaurant late in the evening, I couldn't help noticing that one of my friends was constantly moving her legs under the table. At one point, she also got up from her chair and began moving around. When I asked her why she was doing this, she told me that she had begun suffering from an uncomfortable sensation in her legs during her first pregnancy with her only child, 25 years earlier.

"Throughout my pregnancy I suffered from iron deficiency," my friend told me. "Toward the third trimester, I hardly slept due to the uncomfortable feeling in my legs and the constant need to get out of bed and move them. This ended after my daughter was born, and in fact, I completely forgot about it. I continued to suffer from lack of iron, which is why I take supplements. When my period stopped, 4 years ago, I suddenly began noticing the same sensation of restlessness in my legs, especially in the evenings, and particularly after a long period of sitting down, like now."

It is vital to obtain a clear description of the symptoms from which the woman is suffering, because not every case of restless legs that cause sleep disturbances is RLS. There are conditions that seem very similar, and so a differential diagnosis is required. Some women have difficulty describing what they feel. Some describe the sensation as tingling, burning, bugs crawling under the skin, or heat under the skin. Their difficulty in falling asleep and their need to move their legs can awaken their partner and cause sleeping problems for him as well. Another hallmark of the problem is a bed that looks like a war zone in the morning.

Women suffering from RLS are at times misdiagnosed with depression because they are depressed due to the sleep disturbance, tend to sleep during the day, are irritable, and function less well. Treatment with antidepressants can worsen the problem instead of relieving it.

Differential Diagnosis between RLS and Similar Phenomena

- Women who have diabetes and suffer from a neural disturbance in the legs (neuropathy) and from leg pain: The disturbance appears during the day as well, and can become more severe at night.
- Leg cramps can appear during the day and become worse at night and usually become worse with walking. These are cramps and not the need to move the legs.
- Arthritis causing leg pain: The pains appear during the day and worsen with activity. The pains are felt in the joints and not just in the legs and are accompanied by rigidity in the joints.
- Additional similar phenomena: Attention-deficit/hyperactivity disorder (ADHD) and akathisia, disquiet and the need to particularly move the legs, as a side effect of antipsychotic drugs.

Treatment of RLS

The aim of the treatment is to primarily reduce the intensity of discomfort in order to enable the woman to have a pleasant night's sleep. In cases of iron deficiency, obviously the treatment would involve correcting the deficit by added iron through pills, and in cases of extreme deficiency, through intravenous infusion.

Nonmedicinal treatment: Acupuncture; massaging the feet before bedtime, taking vitamins, avoiding coffee, alcohol, nicotine, antihistamine medication, antidepressants, and antipsychotics (which usually worsen the situation)—except bupropion, an antidepressant and antianxiety medication that can provide relief.

Medicinal treatment: In extreme cases, when nonmedicinal means do not provide relief, the following medications can be used:

- Medications that increase dopamine levels (dopamine receptor agonists), such as Levodopa, a pill used for treating Parkinson's disease.
- Tranquilizers from the benzodiazepine group, such as oxazepam and diazepam.
- Medication given to epileptics (anticonvulsant agents), such as gabapentin.

It is important to remember that each medication is accompanied by undesirable side effects. It is necessary to find a medication that will provide relief and cause the least side effects.

Obstructive Sleep Apnea (OSA)

This is a sleep disorder in which breathing is briefly and repeatedly interrupted during the night, accompanied usually by snoring, a tendency to sleep during the day, and of course fatigue. Women suffering from this condition often complain of waking up with a dry mouth, morning headaches, irritability, mood swings, and impaired function.

The condition is more common in men—two to three times more than in women—particularly, in overweight men. In the past it was thought to be very rare in women. Today, it is estimated that 24% of men suffer from OSA and 9% of women. A central risk factor for both sexes is obesity, especially in the upper body (including the neck area). About 90% of women suffering from sleep apnea are not diagnosed.

Because, in the past, this condition received no attention in women, and most research studies on sleep disturbances were conducted on men, women who suffered from sleep apnea were diagnosed as suffering from depression and were treated with antidepressants.

Women tend to complain of sleep disturbances at all ages, especially during menopause, so that the family physician to whom the women usually turns will usually not consider OSA in the diagnosis, whereas men complain less of sleep disturbances or depression and therefore OSA is taken into consideration in their case.

Often, people who complain of tiredness and sleep problems, and especially women who suffer from them, are treated as lazy people whose lives are empty. This may also be the reason why, in the past, OSA was not diagnosed in women who complained of fatigue. Indeed, women who suffer from OSA see many doctors before they are correctly diagnosed.

However, it is hard to lay the blame for misdiagnoses solely on the physicians. It becomes clear that there is also a gender difference in the way in which women report sleep disturbances compared with the way men do. Men report snoring, frequent wakening during the night with a choking feeling, tiredness, and a tendency to sleep during the day. Women usually report morning headaches, insomnia, tiredness, anxiety, and depression.

Different descriptions of the symptoms of the same disease are common in medicine, yet only in recent years, with the development of gender medicine, is attention being drawn to this aspect of patients' accounts. For example, men experience a heart attack differently from women and therefore will describe it accordingly. They complain of pain in the heart area that radiates to the arm, whereas women will complain more of pain in the jaw and disturbances in the digestive system. Women will express depression in the classic manner of low moods, depressive thoughts, early rising, or oversleeping, whereas for men, depression can be expressed in workaholism, attacks of rage and anger, need for excessive activity and risk taking, or alcohol and drug abuse, as well as insufficient sleep.

Undoubtedly, when it comes to OSA, gender differences are evident. Some of the reasons for the gender differences are related to differences between women and men with regard to the anatomical structure of the upper respiratory system and the distribution of fat.

When Is the Condition More Common in Women?

- Women who suffer from polycystic ovaries, who are also known to overproduce the male hormone, testosterone
- Women who drink a large quantity of alcohol
- Pregnant women: 27% of women will suffer from this condition in their last trimester
- Women undergoing menopause, who do not use hormonal therapy (women who take hormone therapy suffer far less)
- Overweight women

Because the risk of suffering from OSA rises after the end of menstruation, there is no doubt that a hormonal element plays an important role in this condition.

Effect of Progesterone on OSA

Progesterone is as effective as the benzodiazepine group (see Chapter 11) and known to aid breathing. Therefore, its increase during pregnancy helps breathing and so too during the luteal phase (the second-half of the monthly cycle). On the other hand, during menopause, there is a decrease in progesterone, which explains the rise in frequency of nighttime respiratory problems.

Effect of Estrogen on OSA

Estrogen indirectly improves breathing through its effect on the progesterone receptors. The weight gain that occurs in menopause, especially in the upper half of the body, the neck and stomach, is attributed to a decrease in estrogen levels during this period and an increase in the release of testosterone (the male hormone) from the ovaries. The increase of fat, especially in the neck area, can affect respiratory difficulties and at times cause OSA in women.

The Risks of OSA

Sleep apnea increases the risk of high blood pressure, heart disease, and diabetes. Often it can cause further problems with concentration and cause depression, and therefore it significantly increases the risk of being involved in a car accident.

OSA Diagnosis

OSA is diagnosed by a sleep clinic test, polysomnography, which documents various functions of the body during sleep.

As mentioned in Chapter 2, sleep can be divided into two stages: REM and NREM. During NREM, three different sleep phases occur (seen in examinations of brain activity) that change according to the depth of the sleep.

Sleep tests evaluate sleep cycles and sleep patterns by measuring the flow of oxygen during breathing, the level of blood oxygen, and the position of the body; documenting brain waves; and evaluating respiratory effort and breathing rate, electrical muscle activity, eye movement, and heart rate.

In a sleep test, the frequency of events where breathing stops for 10 seconds (apnea) or a partial blockage of breathing (hypopnea) is traced, as well as activity of brain waves and muscle movement. In most subjects, brief occurrence of pauses in breathing or partial blockage of breathing are observed. The Apnea-Hypopnea Index charts the number of these incidents during the sleep test and serves as the diagnostic tool for detecting sleep apnea.

Treating OSA

Initial treatment of OSA focuses on changes of habit and lifestyle, such as losing weight, guiding the patient to sleep on his/her side so as to prevent the tongue from falling back and clogging the respiratory system, avoiding alcohol, avoiding heavy meals before bedtime, and avoiding tranquilizers and sleeping pills.

In cases of nasal congestion caused by allergies or other factors that can also cause OSA, it is important to solve the problem with antihistamines, steroids, or, in more extreme cases, by operating on the nasal septum.

The main treatment in extreme cases of OSA is with a CPAP (continuous positive airway pressure) machine, an instrument with a compressor that sends high-pressure air to the nose through a mask placed on the face, so that the airways remain open all the time. The CPAP does not solve the underlying problem of apnea but does prevent it from impeding function.

Today, a wide variety of CPAP machines are available and can be suited to the individual, with different masks and air pressure, to facilitate maximum comfort. Instructions for treatment with CPAP is given by the physician in cases defined as medium and above in terms of their severity.

Recent years have seen the introduction of the mandibular advancement splint, a plastic device worn in the mouth that is attached to the teeth during sleep. The device moves the lower jaw far from the base of the jaw and it also moves the base of the tongue. This expands the jaw opening and allows room for the passage of oxygen. The device offers a good and comfortable solution to OSA sufferers who have difficulty adjusting to the CPAP.

Another solution, in extreme cases of OSA when the previous measures have not provided relief is a surgical procedure where excess tissue in the throat area is removed and the oxygen channel is reformed, so it is no longer be blocked: UPPP (uvulopalatopharyngoplasty). This approach may not always be helpful.

More on OSA

- Most studies have been conducted and published from 1975 onward, even though there are many prior indications and examples of the condition—these all relate to men.
- In Charles Dickens's *Pickwick Papers* (1836), he describes a young man named Joe, who suffers from OSA (although of course Dickens did not use that term): He snored all night long and was tired all day long. The medical profession diagnosed the condition only 120 years after Dickens described it.
- William Howard Taft, the twenty-seventh president of the United States, who held office during the years 1909–1913, suffered from OSA, although his physicians did not diagnose it. Taft suffered from snoring and was very tired during large parts of the day. When his term in office ended, he lost weight, he was appointed as a law professor at Yale University, and later served as Chief Justice of the Supreme Court. It can be assumed that there was a marked improvement in his functioning with the loss of weight that restored his health.

I have known Noa, a 45-year-old single mother of 3-year-old twins, since she was 25. She came to me then with severe anxiety attacks and an alarming feeling that she was about to die any moment. Noa was treated at the time with a medication from the SSRI group (medications that increase serotonin levels and affect the mood).

Having failed to find a suitable partner, Noa decided to get pregnant by means of sperm donation, as she feared she would miss out on motherhood. Noa works in the family fashion business; she is financially secure and has a lot of support from her family. I was surprised when she called me asking to meet urgently.

"Everything is alright, the children are great, I even met a lovely partner recently," she reassured me when we met, "but I am constantly tired. My eyes close at work and all I can think about is when I can go to bed."

Considering Noa's age, and after having ruled out depression and anxiety, my diagnostic direction was hormonal changes related to the end of menstruation. However, Noa informed me that her cycle was still regular. Her diet was balanced, and blood tests affirmed that everything was fine. In a few conversations I had with Noa, I tried to examine other factors in her life, which could explain the fatigue she experiences during the day but could find nothing unusual.

It was only when I was writing this chapter that I realized that Noa might suffer from OSA, causing her daytime fatigue: Noa is a tall, overweight woman, and because she lives alone, there is no one to inform her of her snoring or breaks in breathing. I sent Noa to a sleep lab, where it transpired that my diagnosis was correct. Noa was diagnosed as suffering from light to mild OSA and therefore her treatment included dietary changes in order that she lose weight, avoidance of alcohol, and, particularly, avoidance of ant-anxiety pills (such as clonazepam and lorazepam), which she used frequently.

Sleep Apnea and Polycystic Ovary Syndrome (PCOS)

Polycystic ovary syndrome is a common gynecological condition, first described in 1935 by the American gynecologists Irving Freiler Stein and Michael Leventhal, who noticed the combination of large ovaries, excessive body hair, weight gain, and fertility problems were caused by irregular ovulation. It is known that polycystic ovaries increase the risk of diabetes, heart disease, and cancer. Therefore, early diagnosis is very important. It is estimated that 5%–10% of women suffer from this problem. With age, there is an improvement in the regularity of the menstrual cycle, presumably because of a decrease in the secretion of the male androgen hormone.

In recent years, polycystic ovaries have also been found to increase the risk of OSA. This sleep disorder is indeed more common in men; however, women with polycystic ovaries have a higher rate of OSA compared to the general female population. A study conducted on 23 women with polycystic ovaries, who were also overweight, showed that 69.9% suffered from OSA.

There is still not enough information on the combination of the two conditions. More studies are required, and physicians must be more attentive to a possible link between OSA and polycystic ovaries, especially in overweight women.

Excessive Daytime Sleepiness (Daytime Hypersomnia, Including Narcolepsy)

This is a group of several sleep disturbances that are characterized by excessive sleepiness during the day. Those suffering from this disturbance have difficulty maintaining alertness during the day.

Causes of Hypersomnia in Women

- *Neurological disturbances*: The most common among this group of sleep disorders is *narcolepsy*.
- *Depression and anxiety*: Excessive daytime sleep is common in women suffering from depression and anxiety. Women are known to suffer from both more than men.
- *Hormonal factors*: Excessive sleep appears after puberty, especially in women with severe premenstrual symptoms and during menopause. One rare condition is *menstrual-related hypersomnia*.

What Is Narcolepsy?

Narcolepsy is a chronic neurological disorder that is unrelated to mental illness or psychological or hormonal problems and is usually caused by the brain's inability to regulate the sleep and wake cycles because of a lack of cells that produce the neuropeptide orexin (related to wakefulness). Sufferers of narcolepsy fall asleep uncontrollably and suddenly during the day. The disorder usually begins gradually in their 20s or 30s, although it is also known to appear in childhood. It is a rare disorder, with rates varying from one country to another and is most common in Japan and in the Scandinavian countries. In the United States, it is estimated that 1 in 2,000 people suffer from this disorder. In Israel, the number is 1 in 50,000.

The main characteristic of narcolepsy is excessive daytime sleeping (EDS), even after a satisfying night's sleep. In fact, those afflicted with the disorder constantly live in a state of extreme sleep deprivation. Daytime naps can occur suddenly and are impossible to withstand physically. The naps can occur several times per day and are usually refreshing, yet only for a few hours. Lacking any other apparent dramatic symptoms, EDS allows a normal lifestyle on a certain level. Many become used to living in a state of sleep deprivation, to the point that they do not notice how unusual their condition is.

Apart from the tendency to sleep during the day, narcolepsy can be accompanied by other events that are quite serious and complex:

- *Cataplexy episodes*: The loss of muscle tone, ranging from light weakness, such as a slacking of the neck and facial muscles and weak knees, to the inability to speak clearly, and full-body paralysis.

  These episodes usually occur following sudden emotional reactions such as laughter, anger, surprise, or fear and can last a few seconds or several minutes. In some cases, cataplexy can seem like an epileptic seizure involving the paralysis of voluntary muscles. Usually during such an episode, speech and vision may be impaired (double vision or inability to focus). On the other hand, hearing and cognition remain at high-functioning levels. There are also accounts of enhanced sensual experiences of taste and smell.

- *Sleep paralysis*: Transient inability to talk or move upon waking up, or—less frequently—when falling asleep. This symptom can last a few seconds or minutes and at times may be accompanied by hallucinations. Though it is a frightening experience, it is not dangerous.

- *Hypnagogic hallucinations*: Vivid and realistic dreams that occur in the hypnagogic state between wakefulness and sleep.

Cataplexy episodes occur only in those suffering from narcolepsy, while the other phenomena can also appear in different situations.

The main symptom of narcolepsy is extreme daytime fatigue. The other symptoms may appear, alone or in combinations, months to years after the daily attacks of sleepiness. The development, severity, and order of the appearance of cataplexic episodes, sleep paralysis, and hypnogogic hallucinations appear in great diversity in those suffering from the disorder. Only 20%–25% of those diagnosed with narcolepsy will experience the entire range of symptoms. Excessive daytime tiredness usually lasts throughout life, whereas the hypnogogic hallucinations and sleep paralysis can be transient.

The symptoms of narcolepsy, especially EDS and cataplexic episodes, are at times so severe that they impair the private, social, and professional lives of those suffering from them.

What Causes Narcolepsy?

In 1998, it was discovered that the hypothalamus secretes a substance known as hypocretin (also known as orexin), a small protein that regulates the transition from sleep to wakefulness. The hypothalamus is a region of the brain that functions as the main control center for the autonomous nervous system by regulating sleep cycles, body temperature, appetite, and so on. In fact, when hypocretin is released, we wake up. Experiments on mice and dogs found that if the function of this system is irregular, the animals suffer from narcolepsy. A disturbance in the production of hypocretin is usually tied to a specific area on chromosome 6 known as the HLA complex.

In 2000, the brains of narcolepsy sufferers were examined postmortem. Researchers found that cells related to the production of hypocretin had disappeared. This is an autoimmune disease, in which the body attacks itself and kills only those cells. Scientists have yet to discover why this happens. The disease does have a genetic aspect—if one family member suffers from narcolepsy, the chance of another family member to suffer from it is slightly higher; however, it is not a classic hereditary disease.

Michal, 32 years old, married, and mother of a 5-month-old baby, came to see me after the birth of her second son following daytime episodes of tiredness and severe sleep disturbances at night.

"During the day, I carried an awful sense of guilt about not being able to function and being a bad mother," she told me. "In the months following the birth, in addition to the sleep disturbances, I started experiencing bouts of muscle weakness, triggered by stress, crying, or laughter. I've come to you because I don't know what is going on with me. Am I depressed? Am I losing my mind? One moment I'm alert and sharp and a moment later I find it difficult to stand straight, and I can't think or speak clearly."

Michal was in great distress. Only when the episodes of muscle weakness increased, was it clear she was not suffering from postpartum depression, but from symptoms of narcolepsy, which was finally diagnosed by a neurologist to whom I had referred her.

"The medication is very helpful," Michal told me, "but it does not make the symptoms go away completely. I still hardly go out in the evenings and avoid driving altogether. It is very limiting, but I am afraid of taking risks. I make time for naps during the day, especially after lunch. I make sure not to be alone with my baby, and during the day, until my husband comes home from work, I am helped by a nanny."

Narcolepsy is indeed a difficult disorder that limits the lives of those afflicted, especially those of young women with small children. It is important, as was done in Michal's case, to add supportive psychological therapy to the medical treatment, to have help at home, and to provide an explanation and guidance to all family members.

How Do We Treat Narcolepsy?

- *Maintaining sleep hygiene*: This entails adopting habits that will ease sleep, such as avoiding rich meals, coffee, nicotine, and physical activity before bedtime and creating a comfortable sleep environment: a comfortable bed, dark room, adequate temperature. Taking care of sleep hygiene will help maintain a continuous and satisfying night's sleep. Occasionally, a short nap during the day can be added.
- Stimulating medications, such as Ritalin (a medicine for treating attention-deficit disorder).
- Treatment with modafinil, a new, effective stimulating medication that was recently developed and is gaining a central role in treating narcolepsy.
- Treatment of cataplexic episodes is also carried out with SSRI (selective serotonin reuptake inhibitor) drugs, which are used for treating anxiety and depression and are also helpful in treating narcolepsy thanks to their ability to suppress REM sleep.

Narcolepsy and Pregnancy

Because this disorder is characteristic of younger ages, there is a high chance that women suffering from it will have to deal with the problem during pregnancy and need to know how to treat it. It is critical that pregnant women with narcolepsy make sure to take iron and folic acid supplements to prevent the appearance of RLS, which can aggravate the problem. Pregnant women should be sure to sleep during the day, even at the cost of disturbing their everyday lives, and to avoid medication. There are countries in which women suffering from narcolepsy who are not medicated (which can happen when they are pregnant, due to avoidance of medication) are forbidden to drive.

Immediately after giving birth, it is important to resume the medication for narcolepsy. The effect of the medication on breastfeeding is as yet unknown. This needs to be examined at a teratological center (the provider of information on the safety of drugs during pregnancy and lactation).

Menstrual-Related Hypersomnia

This very rare condition was first described in 1975 by the French neurologist, Michel Billiard: A 13-year-old patient slept 14 hours a day for 10 days following ovulation. After menstruation, her sleeping hours decreased and she returned to her normative hours of sleep. An examination showed that this was a cyclic phenomenon related to her menstrual cycle. An examination of hormone levels did not indicate deviation from the standard normal values. Treatment with contraceptive pills solved the problem.

Following the publication of this case, other similar cases were published showing a link between excessive sleeping and the menstrual cycle, which began mostly after ovulation. Treatment with contraceptive pills, which neutralize ovulation, solved the problem.

This problem is common, especially with young girls during puberty, and appears shortly after the first menstruation. In 1993, the case of an older woman, 42 years old, who suffered from the condition was published. In her case, it was tied to high levels of prolactin (the hormone released from the pituitary gland responsible for milk production in women after giving birth).

In extreme premenstrual dysphoric disorder (PMDD), excessive sleeping can be observed as part of the general condition. Women suffering from it, as I elaborate in Chapter 5, which discusses sleep disorders during menstruation, tend to suffer for 2 weeks each month from a difficult mental state and from sleeping disturbances as part of their mental state. It is not always easy to diagnose clearly whether the excessive sleep is only linked to the hormonal changes, or whether it is a worsening of a condition related to an unstable mental state.

Hadas, 38 years old, has been treated by me for many years and suffers from periods of depression that began following the birth of her third child. Hadas is treated regularly with medication, because attempts to stop treatment caused the depression to reappear. Despite the medication, every few years, there is a worsening in Hadas's condition, which is mostly expressed in excessive sleep.

"I go to bed at 11 p.m. and at 9 a.m., I barely manage to wake up," Hadas told me. "I sleep at noon—because I feel that if I don't, I will not be able to continue to function—and sleep a deep sleep for 3 hours, from which I manage to wake only with the help of an alarm clock."

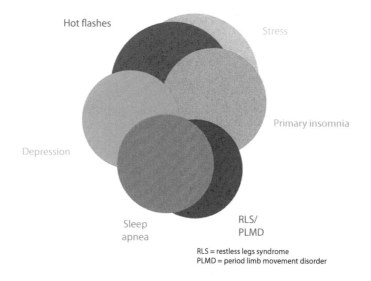

RLS = restless legs syndrome
PLMD = period limb movement disorder

FIGURE 4.2   Co-occurring causes of sleep disturbance.

Over the years, Hadas and I came to realize that excessive sleeping is one of the apparent symptoms indicating the return of her depression, which worsens in the second half of the monthly cycle. Therefore, in Hadas's case this is not excessive sleep as part of a premenstrual syndrome, but a constant depression that worsens during the second phase of the monthly cycle (Figure 4.2).

Parasomnia

For most people, sleep represents the calm and relaxed part of our life. We look forward to the moment when we can rest our head on the pillow and let everything go. Yet, at times, sleep can become threatening and stressful and turns into an uncontrollable period of time that induces worry and unrest. To those who suffer from the sleep disturbance parasomnia, sleep time is indeed a very unsettling, and even dangerous, time.

Parasomnia is a sleep disorder characterized by behaviors or experiences that occur in the transition to sleep, during sleep, or when waking up. It refers to such phenomena as wailing, eating, nightmares, talking and yelling, dissociative states, confusion, and more. This sleep disorder can cause physical and mental damage, which affect both the woman suffering from it and her partner.

Parasomnia may appear at any age. There is even an assumption that it may already be present at the fetal stage and is expressed in the kicks the mother feels—movements that the fetus is probably making while sleeping.

There are types of parasomnia that occur in the deep-sleep stage (NREM) and those that occur in the dreaming stage (REM). Certain phenomena of the disorder show a genetic link. There is a gender difference concerning the various types of parasomnia. It is important to obtain vital information from the partner of the sufferer to help diagnosis and determine the correct treatment.

Types of Parasomnia in Equal Frequency in Women and Men

- Sleepwalking
- Confusional states
- Sleep terrors

These manifestations of parasomnia appear in the deep-sleep phase (NREM) and with the same frequency in men and women. They are more frequent in children. Rates of incidence in adults are 4%–10%.

Sleepwalking usually occurs 15–20 minutes after the onset of sleep but can appear throughout the night. A person gets out of bed and walks, seemingly awake, yet without any recollection of the occurrence the following day. This is a dangerous condition that can involve bumping into objects and leaving the house.

Confusional states occur when a person does not get out of bed, yet is active—talking, for example with the partner—and will have difficulty discerning wakefulness from sleep. The person will have no memory of the episode the following day.

Sleep terrors are usually expressed with sudden screams. They appear mostly in the first half of sleep and cause great anxiety. At times, they are characterized by getting out of bed. They are more common in children.

The latest information on these phenomena in adults indicates a genetic tendency. In half of the cases, there is a history of depression or anxiety. In addition, stressful situations can cause parasomnia to appear.

How Do We Treat Parasomnia?

Treatment recommendations range from adopting various relaxation techniques to creating a safe environment in the bedroom (moving potentially dangerous objects) to the prescription of antidepressants, especially those that increase serotonin levels.

Types of Parasomnia Most Common in Women

Sleep-related eating disorders are an almost unstudied phenomenon for which information is very sparse. It is assumed that 1%–3% of the population suffer from this condition, especially women aged 20–40. Surprisingly, recently, as I was writing this book, two women suffering from the disorder came to see me.

It is important not to confuse this disorder with that known as nocturnal eating syndrome, which is also characterized by night eating, but is done in full awareness and is common with people suffering from eating disorders. Some researchers claim that the difference between these two disorders is not clear and that they are in fact the same phenomenon.

Yasmin, 40 years old, works as a computer programmer and suffers from weight gain. She constantly tries various diets and feels that she herself could write a book on diets. During the day, she is very busy, hardly feels any hunger, and eats very regularly; nighttime is the source of the problem.

"I half wake up at about 2:00 or 3:00 in the morning," she told me, "go to the fridge, eat, go back to bed, and in the morning remember almost nothing, but the fridge preserves the evidence of the binge. It began a few years ago, after the birth of my second son, when

I was up at night with the baby. Since then the habit of night eating while sleeping has not left me."

Yasmin has a vague memory of the nighttime eating, which leaves her with a feeling of an automatic, uncontrollable act. Yasmin was treated with topiramate, a pharmacotherapy for treating epilepsy and eating disorders.

Characteristics of the Phenomenon

This disorder is the opposite of our human physiological process, which is characterized by fasting during the night and hunger and eating during waking hours. The disorder is more common in women and is related to other disorders: sleepwalking and restless legs syndrome.

Approximately 50% of the women suffering from sleep-related eating disorder get up to eat at least once a night, and about 25% get up to eat up to five times a night. The eating is uncontrolled and includes a large amount of carbohydrates and fats. At times, the woman may eat even foods she knows she is allergic to. As a result, the disorder can be deleterious to health: weight gain, diabetes, high blood pressure, sleep apnea, dental caries, and depression caused by a sense of despair.

Some of the women suffering from this condition have a vague memory of getting up at night to eat, whereas others may not recall the episode at all. The medical literature indicates a connection between the lack of memory of the night eating and the ingestion of sleeping pills and various psychiatric medications.

In the book *Sleep Disorders in Women* (Attarian and Saltzman, 2013), the authors devote a detailed chapter to the sleeping pills that contains zolpidem, which might cause this disorder. The side effects of the medication are well-known, and warnings have been published. In Chapter 1, I mentioned that because of the side effects to women, among which are parasomnia, there is a difference in the recommended dosage of sleeping pills containing zolpidem for men and for women.

Liora, 55 years old, an independent book editor, was diagnosed 2 years ago with breast cancer and underwent difficult treatments for a year. As Liora suffers from anxiety and has trouble falling asleep, her oncologist recommended that she take the sleeping pill Stilnox.

The pill worked well and Liora began to sleep for entire nights. Yet a few months after she began taking Stilnox, she felt a strong urge to get up and eat sweet foods shortly after falling asleep. Liora remembers getting out of bed and walking toward the fridge, even though she felt she was not completely awake. The need to eat was uncontrollable, and only after eating a package of ice cream or a bar of chocolate would she calm down and return to bed. Liora's distress was acute regarding the weight she gained and this new nighttime behavior.

Treating Sleep-Related Eating Disorders

Diagnosis of the source of the problem—medication the woman is taking, sleep apnea, or sleep disorder—is important. Medications that have proven to be efficient in treating sleep-related eating disorders are those that increase dopamine and medications used for treating epilepsy, such as topiramate.

Sleep-Related Dissociative Disorder

This is a phenomenon different from those described above with regard to the time at which it appears: This disorder appears at the onset of sleep or a few minutes after waking up. The phenomenon is common in women and can appear in early childhood. It is a difficult and chronic disorder that can cause self-inflicted harm while in the dissociative state.

The characteristics of the disorder are that the woman is in a state of detachment and feels as if she were inside a dream. She is confused, aggressive, fighting with an imaginary figure, shouting at it, and she may try to harm herself or her partner. Most women suffering from this disorder experience episodes of dissociation during the day as well, and a large percentage have a history of physical or sexual abuse, attempted suicide, severe depression, and psychiatric hospitalizations.

This disorder is among the most difficult psychiatric disturbances for treatment and usually needs continuous psychiatric hospitalization.

Sleep Disorders Linked to Hormonal Changes

Chapters 5–8, "Sleep Disturbances during the Menstrual Cycle," "Sleep Disturbances during Pregnancy," "Postpartum Sleep Disorders," and "Sleep Disturbances and Menopause," respectively, address sleep disorders that are linked to hormonal changes (Figure 4.3).

The hormones related to women's fertility, the sex hormones, affect many of their organs, including the brain. In periods of hormonal changes—such as during the menstrual cycle, pregnancy, postpartum, and especially menopause—changes are felt in mood, in sleeping, in appetite, and more. It is important to know that each woman will have different reactions to hormonal changes, and the same woman can experience different phenomena during the menstrual cycle at different ages and throughout life.

Sleep disorders related to hormonal changes are less well-known and, in particular, less well-studied. Sleep disturbances are usually ascribed to states of stress and anxiety and less to physiological and hormonal changes.

The basic assumption concerning the link between hormones and a woman's mental state, including sleep, is that the hormonal fluctuation is the cause rather than the level of the hormone itself.

There is a group of women whose sensitivity to hormonal changes is more acute than usual. In the not too distant past, women who suffered from symptoms related to hormonal changes underwent many tests to check hormone levels (estrogen,

FIGURE 4.3   Hormonal effects of sleep disturbance.

progesterone, prolactin), under the assumption that their hormone levels differed from those of women who did not experience hormone-related mental symptoms. However, the results showed that the hormone levels of these women were similar to those of women who did not suffer from sleep disorders or mental disturbances a few days prior to menstruation or postpartum. The difference in these women is mostly connected to their differing sensitivity to the changes in hormone level rather than to the hormone levels per se.

5

Sleep Disorders during the Menstrual Cycle

Sleep disorders during the menstrual cycle are common, especially before menstruation. Nonetheless, there has not been much research on this subject, and therefore the information comes from various reports of women who describe symptoms such as difficulty falling asleep, oversleeping, fatigue, nightmares, recurring dreams, and, consequently, difficulty in concentrating.

The term "baby brain" is familiar to most women. I find it hard to understand why the term "menstruation forgetfulness" has not yet been invented!

Between the ages of 12 and 52, or thereabouts, namely, over a span of 40 years, women undergo the same hormonal process every month, and each month they are surprised anew when they experience the mental processes related to the menstrual cycle. We hear comments like "you are so irritable today, is your period due?" and yet we are completely unaware of this connection and are angry with the person who is reflecting on our emotional state. And then, surprise: the monthly bleeding arrives, and we understand that the irritability we felt, or the sleepless nights we experienced, are indeed related to our menstruation. The same surprise will occur again the following month.

The menstrual cycle operates in three stages. Sleep disturbances during the menstrual cycle occur particularly when hormonal changes are acute, such as during ovulation and a few days before menstruation.

- *The follicular phase*: The phase in which follicles in the ovary mature into an egg and the levels of estrogen gradually rise. Prior to ovulation, sleep disturbance may occur in connection with the effect of hormone levels on the brain. In sleep recordings, this period manifests itself as an increase in REM sleep (the dreaming stage).
- *Ovulation*: The egg is released from the follicle, usually around the 14th day of the monthly cycle.
- *The luteal phase*: Preparation of the endometrium for the implantation of the fertilized egg. If this does not occur, the hormone level declines with the shedding of the endometrial lining and bleeding. From ovulation to menstruation, there is an increase in the level of progesterone, and a few days before menstruation, both progesterone and estrogen levels decrease. Sleep disturbances such as difficulty falling asleep, repeated episodes of waking up, and lack of feeling of refreshment in the morning are common during this particular stage. The sleep disturbance passes with menstruation (Figure 5.1). In the late luteal phase, there is an increase in subjective sleep complaints, a decrease in sleep efficiency, and a decrease in sleep-onset latency.

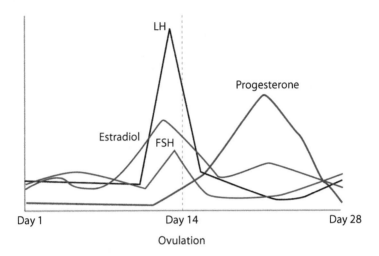

FIGURE 5.1   Sleep and the menstrual cycle.

There are two known premenstrual conditions that also involve sleep disturbances: The most familiar is premenstrual syndrome (PMS); the less well-known is premenstrual dysphoric disorder (PMDD). I will elaborate a little more on these two syndromes in order to clarify the issues involved.

Premenstrual Syndrome (PMS)

Approximately 70%–75% of women suffer from premenstrual symptoms, which appear about a week before menstruation. The symptoms are both physical and mental. The more common ones are abdominal bloating, breast tenderness or swelling, craving carbohydrates, sleep disruptions, lack of concentration, irritability, angry outbursts, and distress.

Sleep disturbances during PMS usually appear 2–3 days before menstruation, most often in the form of insomnia (difficulty falling asleep or frequent episodes of waking up during the night and consequent tiredness during the day). Often, the symptoms may be the exact opposite, hypersomnia (i.e., oversleeping).

PMS does not necessarily occur with every monthly cycle, and it is known to get worse with age. Usually, there is no need for treatment, and it is enough to be aware that these are symptoms that will disappear with menstruation. Charting the monthly cycle (there are apps that do this) helps to anticipate when the symptoms may appear and prevents the element of surprise.

During this period, reducing intake of salt, caffeine, alcohol, and chocolate is recommended. Any means that can induce tranquility and aid in relaxation are recommended: yoga, reflexology, aerobic exercise, and so on.

To prevent sleep disturbances, maintaining correct sleep hygiene is recommended, namely, to avoid heavy meals and physical activity before bedtime and to sleep in a dark room and in a quiet environment.

Premenstrual Dysphoric Disorder (PMDD)

Approximately 5%–8% of women suffer from a more severe type of premenstrual syndrome, which begins more or less after ovulation and lasts until menstruation.

This syndrome is characterized by a depressed mood, heightened anxiety, irritability and outbursts of anger, loss of appetite, sleep disturbances, and a reduction in the ability to function in all facets of life. The physical symptoms associated with PMS can also appear here; however, in this case, it is the mental aspect that predominates. The symptoms disappear with menstruation. PMDD is almost always present with every monthly cycle and is exacerbates with age.

Sleep disturbances during PMDD last about 2 weeks each month. They consist of insomnia expressed as sleeplessness, frequent episodes of waking up, or waking up in the early hours of the morning. The ongoing disturbance leads to fatigue, to a tendency to fall asleep during the day, and to a decline of functional capacity in all areas of life. As this suffering is prolonged, women often use tranquilizers or sleeping pills to help them fall asleep. In rare cases, women report hypersomnia, that is, oversleeping, having difficulty waking up, and experiencing nightmares.

Before the syndrome was diagnosed, women who visited their general practitioner or gynecologist did not receive much help. At times, they were even confronted with disparaging reactions both from their physician as well as from members of their family. Only recently, did PMDD enter the *Diagnostic and Statistical Manual of Mental Disorders* (DSM-V) as one of the conditions of depression. Today, the phenomenon is more widely known, yet still not enough, and in most cases, women do not receive the appropriate help.

Here is what a patient who turned to me for help wrote:

"I read the pamphlet explaining PMDD and how to deal with premenstrual syndrome which you published on your website. I must say I was moved. Wow ... at long last someone understands what I'm going through. Word by word.

I was particularly drawn to the following words: 'Two different women in a month.' This is what I feel during the days before my periods. It's as though a bad animal has entered into me and taken over. And she is not me. I have been dealing for some years with the difficulty of my life, that affects my work, my family, my social ties, and what not. Had I not been so strong, energetic and powerful, I would have given up on it ages ago. But reading your pamphlet made me believe that perhaps there is a chance for a treatment and for a normative life. Is this so? Do you think that you hold in your hands the key to a balanced, functional, normal life?"

Treating PMDD

From the many years I have spent treating women suffering from this problem, I have learned how easy it is to help and what a long route these women take, unfortunately, until they are correctly diagnosed. I always tell patients that in this case, I feel like a surgeon giving a quick-fix solution, in contrast to most of the problems I treat as a psychiatrist. The frustration and agony of women suffering from PMDD is so enormous that often they react by bursting into tears when I diagnose the problem and tell them that the treatment is easy and simple. "Why did I have to suffer for so many years and why did no one believe me?" they say in anger.

Women suffering PMDD need treatment, because for a long period in their lives, they have difficulty functioning. The prevailing understanding is that women with PMDD probably suffer from a low level of serotonin, which affects mood. The most efficient treatment is antidepressants from the SSRI group only, such as Lexapro, Prosak, Paroxetine, and Zoloft, at a lower dosage than that given for cases of depression. The main innovation is that in most cases, the treatment is given only from the middle of the monthly cycle, from the time of ovulation to menstruation. While in other cases, the medication begins working after 4–6 weeks, in the case of PMDD, the medication needs only from a few to 24 hours to take effect.

Dana is 35 years old, married for a decade, mother of two young children, and a graphic designer in a large firm. Dana told me that she always suffered from long and difficult periods. She felt that her reactions during the month were different from those of her friends: after menstruation, she was energetic, in a good mood, and performed all her tasks quickly and efficiently. This situation lasted for about 10 days.

Afterward, quite suddenly, Dana began feeling a great sense of sadness. Every task at home seemed difficult and sleep became a nightmare: she would wake up almost every hour, peek at the clock and calculate how many hours remained for sleep. In the mornings, she often had difficulty getting up and organizing herself for work. She even considered resigning from her job. Her craving for carbohydrates, especially in the evening hours, caused her to add 2 kilograms to her weight.

Dana's husband raised the possibility that her mood swings might be related to her monthly cycle. Dana was offended by the remark. The stressful relations between the couple, which were also hampered by Dana's mental state, led them to couples counseling. When Dana turned to the family doctor, he figured she was suffering from depression and recommended treatment with antidepressants. Dana refused, fearing the side effects. Her gynecologist suggested treatment with contraceptive pills; however, Dana suffered from blood clot issues and thus was not allowed to take them.

Dana had never heard the term PMDD before. She burst into tears of relief mixed with pain upon hearing that there was a name to the condition from which she suffered, as well as a quick treatment. Dana was treated from the time of the ovulation to menstruation with an antidepressant from the SSRI group. At first Dana was worried about the treatment, yet her acute suffering led her to understand that she had no choice. After two monthly cycles, she came for a checkup and told me about the change that had taken place in her life. Two days before menstruation she still suffered from irritability and slept lightly, yet the rest of the month her symptoms had vanished. "If I had known that life could be so different and good," she said, "I would have run to get treatment a long time ago."

Despite the openness that exists these days with relation to psychiatric pharmacotherapy and the increase in their use, when I offer patients the possibility of treating PMDD with antidepressants, they almost always hesitate. When I explain that it will only be for half of the month, they feel relieved in the sense that "it doesn't mean that I'm crazy." In the United States, the reluctance of women to use antidepressants was handled by giving a different name to Seroxat (Paroxetin), indicating its use as a treatment for PMDD with a different dosage indication. In fact, the U.S. drug Brisdelle is the same as Paroxetin, but the treatment dose is 7.5 mg compared with the standard dosage of Paroxetin for treating depression or anxiety that starts at 20 mg.

I am always asked—both at lectures on this topic and by patients—why the medication is effective in the case of PMDD within a few hours, whereas in treating other problems it becomes effective only after a few weeks. Studies have given rise to various theories, but there is no clear answer to the question apart from the fact that in medicine, we know that drugs intended to treat a certain disease can also be effective in the treatment of other diseases or symptoms, and apparently act in a different way.

Conclusion

Women suffering from PMS or PMDD will suffer from sleep disturbances in most cases. Very few studies on sleep disturbances during menstruation have been conducted, therefore our knowledge is very limited. In the case of PMDD, antidepressants given at a different dosage and for a limited time can be very helpful. An interesting fact is that in sleep clinic tests of women who suffer from these syndromes, no differences in sleep architecture were observed compared to the women in the control group.

6

Sleep Disturbances during Pregnancy

Pregnancy is a complex developmental process that involves significant physiological and psychological changes for each woman. During this period, women experience changes in all aspects of their lives, ranging from bodily changes and emotional experiences to preparation for a new role in life: the role of motherhood. This meaningful role represents the end of her existence in the world as an independent, separate person and her transition into a new existential experience, one that is irreversible, within a new structure of mother–child.

The ability of a woman to go through the entirety of changes involved in pregnancy, birth, and motherhood depends on psychological and biological factors, including the woman's characteristic disposition (level of maturity, ability to deal with both mental and physical changes), her family situation, the level of support she receives, and her socioeconomic situation.

The hormonal changes during this period are the most extreme changes a woman will experience in her life, and therefore they can affect her mental and physical state. In some cases, pregnancy can be a catalyst for the onset of mental diseases for the first time in a woman's life.

The Influence of Hormonal Changes during Pregnancy on Sleep

The increase in the release of estrogen and progesterone during pregnancy affects both sleep and wakefulness. Estrogen influences a decrease in the stage of REM sleep (the dreaming phase); progesterone affects a rise in the stage of NREM sleep (the deep-sleep phase) and has a calming effect. In the first trimester of pregnancy, women feel an increased desire to sleep, both during the day and at night, and in the last trimester, there is a decrease in the hours of sleep and an increase in the frequency of waking up at night. The hormonal and physical changes of pregnancy also affect breathing and, consequently, sleep.

The expansion of the womb causes pressure on the diaphragm and breathing difficulties, in addition to those caused by the natural weight gain during pregnancy. The increase in the release of estrogen affects the mucous membrane of the respiratory system and impedes the entry and exit of air, a phenomenon that occurs especially in the last trimester and may cause snoring; 45% of pregnant women snore at night compared with 4% of age-matched fertile women who are not pregnant.

Common Sleep Disturbances during Pregnancy

Sleep disturbances during pregnancy are normative and familiar phenomena. The frequency of sleeplessness changes dramatically in the course of the pregnancy.

In the first trimester, it is 13%; in the second, 19%; and in the third trimester, it rises to 66%–97%.

Just as the field of women's sleep suffers from a lack of adequate research, so too has the study of sleep disturbances during pregnancy been neglected. Very little research has been conducted, and there is insufficient information regarding safe medical treatment of sleep disorders. A questionnaire given to 2,427 pregnant women in a public hospital in California, asking about their sleep during pregnancy, indicated the following disturbances: most of the women, 95%, suffered from frequent episodes of waking up during the night; 75% complained of a reduction in sleep quality; 38% felt they had insufficient hours of sleep; and 49% complained of fatigue and the desire to sleep during the day.

The first trimester: The first part of pregnancy is characterized principally by enhanced tiredness and a marked desire to sleep, both during the day and at night, apparently due to the rise in the level of progesterone. Morning sickness can sometimes cause early rising and contributes to fatigue in the morning.

The second trimester: Sleep disturbances are linked to the growth of the body and uterus and the difficulty of finding the right position. Another difficulty is caused by stomach acidity, which increases mainly at bedtime and during the night as a result of pressure of the womb on the stomach. Restless legs syndrome (RLS)— the need to move the legs and cramps in the legs—can also disrupt sleep during this period.

The third trimester: During this period, several problems that can affect sleep may appear, including RLS, nightly respiratory problems caused by pressure of the womb on the diaphragm, back pain, movement of the fetus, difficulty in finding a comfortable position for the growing body, and the need to toilet frequently.

Mirit, 32 years old, in the 35th week of her pregnancy, came to see me upon the recommendation of her gynecologist. He asked me to see her urgently, after discovering that she had barely been sleeping for the last 2 months. Mirit was married, and this was her first pregnancy. She described herself as a healthy woman who loved life and who had never sought psychological therapy. Her pregnancy was received with joy and she felt quite well, apart for some morning sickness at the beginning.

Gradually, Mirit noticed that she had trouble falling asleep. Almost 2 hours would go by before she could fall asleep, and during the night, she would wake up several times due to bladder pressure. In the morning, she naturally had difficulty waking up. In reply to my question why she waited 2 months before sharing this with her gynecologist, Mirit replied: "I thought this is how it's supposed to be in pregnancy."

Her gynecologist's urgent referral to me led Mirit to understand that her sleep disturbance was not normative for pregnancy, especially not during the third trimester, and of course I agreed with him.

After a thorough investigation with Mirit, I understood that her sleep disturbance was mainly related to anxiety about the birth process itself. Mirit needed to control every aspect of her life, and the issue of the birth bothered her extremely. For the first time, she told me, she was confronting a situation over which she had no control. At night, she would picture the birth process over and over, obsessively, incessantly walking through it in her mind in an attempt to gain control over it.

Because Mirit was very close to her due date, I helped her with antianxiety medication, which may be taken during pregnancy for a few days. I also referred her to a psychologist who treated her with guided imagery for childbirth. This combined treatment led to a gradual improvement in her sleep quality up to the birth, as well as afterward.

Some researchers claim that women sleep less and less the closer they get to giving birth as preparation for the little sleep they will have after the baby is born.

Research conducted by a professor of psychology, Jodi Mindel, and a professor of gynecology, Barry Jacobson, both of Philadelphia, showed that even though most women do not sleep well toward the end of their pregnancy, only a third thought they were suffering from a sleep disorder. They seemed to think, as did my patient, that this is the way things were supposed to be.

Treatment of Common Sleep Disorders during Pregnancy

Treatment of normative sleep disturbances during pregnancy requires attention to the causes of the disturbance: these may involve changing diet, reducing fluid intake ahead of bedtime, adding pillows to support the side of the body to reduce back pain, relaxing before sleep, and refraining from work before going to bed.

Tranquilizers or sleeping pills are prescribed only in rare cases. As mentioned, there has been very little research on this subject, and therefore our knowledge is sparse. Studies conducted on the use of tranquilizers of the benzodiazepine family during pregnancy did not indicate problems for the fetus, especially when they were taken in a low dosage and not in the first trimester.

Before beginning treatment with medication, it is recommended to seek advice from a teratological center (a center that provides information on the impact of environmental exposures and guidance concerning safety or risk to the fetus as a result of exposure to various factors, including medication).

The Most Common Causes of Sleep Disturbances during Pregnancy (in Percentages)

	Weeks 2–14	Weeks 28–42
Need to urinate	51	47
Leg cramps	13	45
Joint pains	4	23
Heartburn	33	51
Nightmares	3	5
Anxiety	3	1
Fetal movement	0	5

Mental Health Problems during Pregnancy

Sleep disturbances during pregnancy can also be related to mental health problems, which may sometimes require treatment. It is important to remember that it is natural and normative to experience mental changes during pregnancy, such as anxiety, oversensitivity, tendency to cry, and sleep disturbances. That being said, it is important to define what the normative changes are, and to diagnose changes that indicate problems which require psychological treatment.

In the past, the common assumption of women as well as professionals was that the process of pregnancy protects women from developing mental problems. Reality proves that this assumption is mistaken.

Which are the mental phenomena that require treatment during pregnancy?

Major Depression

Approximately 10% of pregnant women experience depression. The incidence of depression changes during pregnancy: Its frequency rises especially in the first trimester; in the second trimester, there is a reduction; and toward the third trimester, it may increase again. Depression during pregnancy is a risk factor for many problems, such as the use of unsafe substances by the pregnant woman (alcohol, nicotine, nonprescription drugs), various sedatives, and deficient nutrition. All these present risk factors for the normal development of the fetus. Moreover, depression during pregnancy increases the risk of postpartum depression, which will hamper the connection between the baby and mother, and in severe cases, may lead to suicidal thoughts or to harming the baby.

Not every bad mood is considered to be depression. What do we mean by depression? Depression is a condition that lasts for at least 2 weeks and includes a very bad mood, sadness, changes in appetite, lack of energy, diminished concentration and memory, feelings of guilt, low self-esteem, and at times suicidal thoughts. Sleep disturbances that accompany depression can be expressed as difficulty falling asleep or frequent episodes of wakefulness during the night, bad dreams, and at times the opposite— excessive sleep and difficulty getting up in the morning.

Diagnosis of Major Depression

   a. The symptoms continue for at least 2 weeks.

   Five of the following symptoms should exist, where at least one is number 1 or 2:

   1. Sense of sadness and despair
   2. Diminution of sense of joy and interest in life
   3. Change in appetite and weight
   4. Sleep disturbances: Insufficient or excessive sleep
   5. Lethargy or psychomotor slowness
   6. Low energy levels and tiredness/fatigue
   7. Feelings of lack of value or guilt
   8. Difficulty concentrating
   9. Suicidal thoughts or thoughts about death

b. The symptoms are unrelated to any other mental illness.

c. The symptoms affect the ability to function in all other areas of life.

d. The symptoms are not caused by a physical disease or the use of medication.

e. The symptoms are unrelated to grief.

Panic Attacks

These are brief attacks that are accompanied by physical symptoms such as sweating, racing pulse, shortness of breath, fear of loss of control or madness, dizziness, shaking, and nausea. The attack can reach its peak within 10 minutes and then pass. At times, a sense of panic is felt throughout the entire pregnancy, with attacks appearing every so often.

Panic attacks can lead to frequent waking during the night and especially early in the morning, waking with a sense of breathlessness and fright. Night sleep can be accompanied by nightmares, which have a characteristic motif of suffocation and difficulty breathing.

Diagnosis of Panic Attacks

Sense of fear or uneasiness, accompanied with at least four of the following symptoms and reaching its peak within 10 minutes:

1. Fast pulse, palpitations
2. Sweating
3. Shaking
4. Sense of breathlessness and suffocation
5. Sense of pressure in the chest and uneasiness
6. Nausea and stomach queasiness
7. Dizziness and sense of fainting
8. Sense of self-alienation (derealization) or environmental alienation (depersonalization)
9. Fear of losing control or madness
10. Paresthesia and shuddering

Obsessive-Compulsive Disorder (OCD)

OCD is characterized by obsessive thoughts accompanied at times by the need to perform repetitive actions in order to reduce anxiety. The thoughts and acts cause anxiety, take up time, and interfere with daily functioning. Repetitive rituals are characteristic of the time before going to bed. One of my patients told me that 2 hours before she goes to sleep, she embarks on a regular ritual in which she checks three times that she has locked the door and turned off the gas. She straightens the carpets in all the rooms, straightens the sofa pillows, stretches the bedding in the children's rooms, and finishes off with a long and precise bathing ritual. Often, even after she has completed all these rituals and has gotten into bed, she remembers one more thing that she did not check and gets up to do it. Only when her anxiety level drops at the end of the entire routine, can she relax and fall asleep.

The importance of the right treatment during pregnancy, as I always explain to my patients, is not only the need for relief during that period, but also to prevent the worsening of the anxiety after the birth, when the risk of deterioration of the mental state is greater. The correct treatment during pregnancy reduces the risk.

Diagnosis of Obsessive-Compulsive Disorder

    a. Unpleasant repetitive thoughts that cause anxiety:
- The thoughts are unrelated to everyday worries.
- The woman tries to replace these thoughts with others or with acts but does not always manage to do so.
- The woman is aware that these are her thoughts.
- Compulsive acts that the woman feels she must do in order to reduce the anxiety.

    b. The woman is aware that her thoughts and acts are irrational.

    c. The thoughts and acts induce more anxiety, take up time, and interfere with her life.

    d. If there is another disorder, such as an eating disorder, the content of the thoughts and acts will not necessarily be related to it.

    e. The disorder is not related to another disease or to the use of medication.

Yael suffered from separation anxiety since childhood. She found it hard to separate from her parents when they took her to kindergarten, did not go on school trips, and when she enrolled in the army, she asked to serve close to home. As she got older, she developed obsessive rituals in the morning hours before leaving the house, rituals that at times went on for almost an hour and included making sure she had switched off the gas and closed the windows and doors.

Yael came to me in the 20th week of her first pregnancy. As her pregnancy progressed, she suffered from disturbing thoughts about everything she encountered: the food she ate, the sofa on which she sat, someone who shook her hand, a door she opened in an unfamiliar place. Yael also developed acute anxiety over infections that might pass on to her fetus and kill it. In an attempt to reduce the anxiety, she developed obsessive rituals that revolved around hand washing, up to 20 times a day, and cleaning the house every morning. These thoughts and acts took up her time and energy and made her very exhausted, which was mostly expressed in difficulty falling asleep at night.

In the past, Yael had avoided medication, only occasionally taking a sedative. However, the obsessive-compulsive disorder worsened during her pregnancy, and the risk of additional deterioration after the birth was high. Treatment with Sertraline, a medication from the SSRI antidepressants group, along with cognitive-behavioral therapy (CBT), brought great relief.

Other Sleep Disturbances during Pregnancy

In addition to the physiological sleep disorders that are common during pregnancy and are surveyed here, other disturbances may also appear during this period, which may present a risk factor for postpartum depression, pregnancy-induced hypertension, gestational diabetes, and delayed fetal development. Therefore, it is vital to diagnose the problem and treat it properly without delay to prevent risk to the fetus during both the pregnancy and during lactation.

There are four main common sleep disorders (see Chapter 4 for more details):

Insomnia: This includes difficulty falling asleep and staying asleep during the night, waking up too early, or feeling tired during the day despite having had a reasonable number of hours of sleep during the night. Insomnia worsens as the pregnancy progresses, especially toward the third trimester. In fact, there is hardly a pregnant woman who will not suffer from sleeplessness. A study conducted on 370 pregnant women showed that 73% suffered from varying degrees of insomnia. The causes can be physiological (nausea, frequent need to urinate, back pain) or psychological (i.e., anxiety or depression) during the pregnancy.

Ronit, 30 years old, at the beginning of her first pregnancy, came to me because she had difficulty falling asleep at night. Since the beginning of her pregnancy, she had had difficulty falling asleep before the early morning hours, thinking of her fears about the process of pregnancy and especially about what would happen after the birth. The pregnancy was very much wanted, but Ronit told me she had always experienced difficulties in periods of changes and transitions in her life, and every such change would induce a period of anxiety expressed for the most part in sleep disturbances. In the past, she would occasionally take antianxiety pills for short periods, and these helped reduce the anxiety very efficiently.

Ronit was very concerned about taking medication during her pregnancy. I suggested that she first try a focused treatment of CBT, which did indeed reduce her anxiety to a certain extent. However, already in the first trimester, she showed signs of depression: avoidance of social connections, difficulty functioning at work, episodes of crying, and weight loss. On my recommendation, she began treatment with antidepressants. Significant improvement in her depression was apparent within 4 weeks. Ronit requested to continue the treatment even a year after giving birth, as she felt that her mental state was much better than it had been previously.

The main dilemma in treating insomnia during pregnancy relates to treatment with medication. Most women who have not taken any type of medication prior to becoming pregnant try to avoid it, out of concern for the well-being of the fetus. Women who have taken medication for treating depression, anxiety, or sleeping disorders find it easier to accept treatment with medication, especially when it is prescribed under sound medical supervision.

It is always recommended to begin treatment with an examination of the sleep hygiene (noise, room temperature, snoring of partner, lighting in the room, heavy meal before going to bed, working at a computer before bedtime, etc.) and to seek psychological treatment to reduce stress. This is not always sufficient, and since pregnant women who suffer from sleep disturbances during pregnancy are at risk of depression, of harmful influences that could impair development of the fetus, and at risk of postpartum depression, it is important, where indicated, to add appropriate medication under supervision.

Restless legs syndrome (RLS): RLS manifests as an uncontrollable urge to move the legs, especially in the evening and during sleep. The feeling is very uncomfortable, although moving the legs does afford some relief. The incidence of the condition increases with the advancement of the pregnancy: 13% in the first trimester, 18% in the second, and 23% in the third trimester.

The good news is that most RLS symptoms resolve with the end of the pregnancy. About 87% of women who suffered from the condition reported it had disappeared a month after birth. Some pregnant women who suffer from RLS fear they have contracted a neurological or psychiatric disease and feel great relief when the condition is given a name, especially when they understand it is a condition known to occur during pregnancy and it passes when pregnancy ends.

An interesting fact, which unfortunately is characteristic of women, is that they do not often share their feelings with their physicians. They fear that their physician may think they are "slightly crazy," or else they are confident the condition will pass after the birth.

RLS causes sleep disturbances and tiredness during the day. Women suffering from RLS complain of difficulty falling asleep, difficulty remaining asleep, rising early in the morning, and tiredness during the day. Most of these women have low levels of iron reserves (ferritin) and folic acid before they became pregnant and during pregnancy. The main complications are high blood pressure and premature birth.

RLS is more common in women—twice as common as in men—and the frequency increases two- to threefold during pregnancy. The reason for this is not clear, and there are several hypotheses that link RLS to a decrease in the production of dopamine (neurotransmitter) in the brain. Medications that increase dopamine levels were found to be effective, yet these are not recommended during pregnancy. Risk factors for this condition are family background, a mother or sister who suffered from RLS, its appearance in previous pregnancies, being overweight, smoking, and snoring in the first trimester.

None of the recommended drugs for treating RLS are safe during pregnancy, or there is not enough information about their level of safety. Because it is assumed that the condition is linked to iron deficiency, one recommended treatment is to take iron supplements, which has been shown to provide some relief. The problem is that iron only helps those women who lack iron. Adding folic acid has also proved helpful for many women. Other recommended treatments include abstinence from coffee, alcohol, and smoking; avoiding long spells of sitting, which can act as a catalyst for the condition. Yoga, massage, various relaxation techniques, and exercise are suitable for pregnancy.

In severe cases, when none of the means described above provide relief, medication from the dopaminergic agonist group (drugs that increase dopamine levels in the brain) can be used in low doses during pregnancy and lactation.

Obstructive sleep apnea (OSA): OSA is characterized by several episodes of breath cessation during the night, which last for 10–30 seconds each, and at times longer. The frequency of this disorder in pregnant women is 10%–27%, and rises according to the woman's age and if she is overweight.

What indicates the existence of OSA? The main symptoms are snoring during the night, accompanied with fatigue the next day, a tendency to fall asleep during the day, and a decrease in the level of functioning. The complications related to this can be many and include high blood pressure, early birth, gestational diabetes, complications after caesarian section and giving birth to a low-weight baby. Snoring during the night, weight gain, and tiredness during pregnancy are all common features of pregnancy, making the diagnosis of OSA rather complex. The diagnosis should always be done in a sleep lab, where the night sleep process is recorded, and breathing pauses characteristic of OSA can be detected.

There is no special treatment for pregnant women suffering from sleep apnea. The standard treatment for sleep apnea is a special device, a CPAP machine, which consists of a compressor that pumps air into the nose under pressure via a hose attached to a face mask; there is no reason to avoid using this device during pregnancy. In addition, it is recommended to maintain good weight, to sleep with the head raised on a pillow, and to avoid sedatives.

Narcolepsy: This is a chronic neurological disorder, unrelated to mental disease, psychological problems, or hormonal problems, which is usually caused by the inability of the brain to regulate the cycles of sleep and wakefulness. Those suffering from narcolepsy fall asleep uncontrollably. Diagnosis of this disorder follows defined criteria: it must be present for over 3 months and show interference in the dreaming phase in a sleep-lab examination.

In this case, women who suffer from narcolepsy did so before they became pregnant; this is not a disorder that began during the pregnancy. The disorder has severe effects on everyday functioning and increases the risk of being involved in road accidents.

Women who were diagnosed with narcolepsy prior to becoming pregnant and are receiving medication are advised to stop taking their medication if it is unsafe for pregnancy. The common medications for treating narcolepsy, such as amphetamines, are not safe to use during pregnancy. If need be, antidepressants from the SSRI or serotonin-norepinephrine reuptake inhibitors (Cymbalta, Effexor) (SNRI) families can be used. These are effective in maintaining alertness and can be used safely during pregnancy.

Actually, the main treatment is behavioral and includes instructions to nap at defined intervals during the day as much as possible, to avoid driving, and to work at the most convenient hours. In cases in which treatment with medication cannot be avoided, it is important to conduct more frequent ultrasound examinations than is usual, and to make sure to take folic acid to prevent fetal defects.

Conclusion

- Pregnant women sleep an hour less due to frequent episodes of waking up.
- Women in their first pregnancy suffer more from sleep disorders compared with women who have already had a baby.
- Complaints about difficulty falling asleep are not characteristic of pregnancy and require investigation. They can indicate anxiety, depression, marital problems, restless legs syndrome, and obstructive sleep apnea.
- Women who had difficulties falling asleep prior to pregnancy will feel relief during pregnancy due to the calming effect of progesterone.
- Heavy tiredness during the day and a tendency to fall asleep demands examination to diagnose or eliminate the sleep disorders obstructive sleep apnea and narcolepsy.

7

Postpartum Sleep Disorders

There is no clearer milestone in a woman's sleeping habits than the period immediately after childbirth.

"When did you start suffering from sleep disturbances?" I always ask women who see me for complaints of difficulty in sleeping. In most cases, they reply, "Since the children were born, I don't sleep the way I used to." Many patients tell me that their sleeping habits changed right after the first time they gave birth, when they came home with their baby. Since then, they say, they wake up with every sound of a crying child, or just to check the children's breathing. And when these children grow up and go out, they don't fall asleep until they come home. "Only on vacation without the kids can I sleep a whole night through," many will tell me.

As mentioned, the literature on sleep disorders in women is very sparse. I was surprised to discover that in the book *Sleep Disorders in Women* (Attarian and Saltzman, 2013), the chapter dedicated to postpartum sleep disorders is short and touches only lightly on sleep disturbances in the first year after childbirth. In my estimation, this has a number of possible explanations. The common assumption is that sleep disorders after childbirth are normative for a period in which waking up from sleep is related to the baby and its care. Another explanation is that the young mother may not complain about her sleep disorder to her physician because she assumes it to be a normative disturbance. The immediate environment reinforces this view, and therefore physicians receive very little information regarding sleep disorders following childbirth.

However, the prevailing assumption may not necessarily be correct. After childbirth, mental disorders may develop. One of the most pronounced symptoms is sleep disorder.

The postpartum period is a special time, which is unique in a woman's life. During this period, she experiences, simultaneously, both significant hormonal changes and immediate and stressful mental-developmental changes. Many physicians have difficulty distinguishing between normative postpartum mental adjustment and disturbances that are non-normative and require treatment. It seems to me that the difficulty encountered by physicians, and therefore by women as well, stems from the fact that until recently, postpartum mental disorders did not earn special and separate consideration within either the field of medical or clinical research. It is only recently, as part of the growing interest on the subject of gender, that postpartum mental disorders have come under more intensive consideration, both from a research and clinical aspect.

The History of Postpartum Mental Disorders

The fourth-century BC Greek physician Hippocrates was the first to describe postpartum mental disorders. Hippocrates described the case of a woman who gave birth to twins, and on the 6th day following the birth, was restless and had difficulty sleeping at night. On the 11th day, she fell unconscious, and 17 days after the birth, she died.

Until the nineteenth century, there are no clear reports of postpartum mental phenomena. At the beginning of that century, French physicians Jean Esquirol and Louis Marce developed an interest in the subject and conducted research studies on women who suffered from mental disturbances during their pregnancy, immediately after childbirth, and 6 months or more later. Esquirol's study was the first to be published and included 90 women. Both physicians mentioned that many women suffered mental phenomena in these periods of their lives and were afraid to tell anyone about them, fearing they would not be understood and would be banished. Marce conducted a study on a larger group, about 300 women, and was among the first to raise the question whether postpartum depression is the same as general depression, or whether it is a separate and different condition. He concluded that mental disturbances during pregnancy are not different from mental disorders diagnosed in nonpregnant women, whereas postpartum mental disorders have their own unique characteristics. He linked the disorders to the physical changes that occur in the body following childbirth but had difficulty identifying them. His findings served as the foundation for understanding postpartum mental disorders, both by treating these disorders as unique and different, and by assuming that there exists a link between body and soul.

In the early twentieth century, when psychiatry was accepted as part of the medical profession, postpartum mental disorders were still not included in medical textbooks under a separate title. Surprising as this may sound, until the 1980s, terms such as postpartum depression and psychosis did not appear in medical books at all.

In 1980, the British psychiatrist Dr. Ian Brockington organized an international conference on postpartum psychiatric illnesses, which for the first time brought together physicians and researchers from around the globe. They discovered that their studies of postpartum mental disorders led them to the same conclusions reached by the French physicians in the early nineteenth century.

The American *Diagnostic and Statistical Manual of Mental Disorders* (DSM-V) broadened its definition of postpartum depression by relating to the existence of symptoms of depression during pregnancy as part of the correct diagnosis of postpartum depression.

Postpartum Mental Disorders (Figure 7.1)

- *Postpartum blues*: Common among 80% of women.
- *Postpartum depression*: Common among 15% of women. Begins 4–6 weeks after childbirth.
- *Anxiety disorder*: Can appear during the first year after childbirth.
- *Postpartum psychosis*: A rare phenomenon in 0.1%–0.2% of new mothers. Appears immediately after childbirth.

One of the earliest and clearest symptoms in all these phenomena is insomnia.

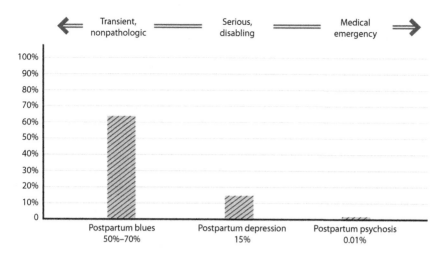

FIGURE 7.1 The spectrum of postpartum mood changes.

A close friend, a physician, asked me to meet Tamar urgently. Tamar was the daughter of his friend from medical school. He told me that his friend was very worried because his daughter had given birth 3 months earlier, and since then had been in a very difficult mental state. I agreed to his request and asked Tamar to contact me. To my surprise, in our phone conversation Tamar sounded very aggressive and angry, whereas I was expecting to hear someone lacking energy and with a very weak voice.

Tamar, a 32-year-old woman, carefully dressed and well-groomed, sat throughout our entire meeting on the edge of the armchair and evinced irritability. She is a mother of two, a 3-year-old boy and the 3-month-old baby. Tamar told me that after she got married, she chose not to work, "Because it is important for me to be at home and raise my children." After the birth of her eldest son, she suffered some sort of depression, "But I function very well, and no one can tell that I'm not my usual self. That's very important to me." She told me that during her second pregnancy, she was looking forward to having a daughter and was disappointed to discover she was carrying another son. "I even thought about having an abortion, but my husband wouldn't hear of it, and I came to terms with the situation. Throughout my pregnancy, I was obsessed with thoughts about the decision to go on with the pregnancy, which made me restless and lose sleep. Immediately after the birth, I had difficulty falling asleep and I stay awake during many nights. I do have a night nurse, but despite this, I hardly sleep."

At the recommendation of her family physician, Tamar began using sleeping pills, yet she was worried about becoming addicted to them. She gradually developed severe panic attacks and started to feel depressed. This was accompanied by obsessive thoughts about her not being a good enough mother to her sons, and she even had suicidal thoughts.

There is no doubt that Tamar's mental state had already been impaired during her pregnancy and had become more severe after the birth. Tamar required antidepressant treatment combined with psychological therapy. The sleep disorders she experienced were the first indication of the anxiety and the depression, which developed subsequently. The clearest indication that mental problems were the source of her sleeping disorders was Tamar's inability to fall asleep during the day, when the baby slept. The correct treatment is not sleeping pills, but treatment of the source of the sleep disorder, namely, of her mental state. Once this improved, there was also an amelioration of her sleep disorder.

When a woman comes to therapy after childbirth for evaluation of her mental state, it is important to ask her two questions: How long does it take you to fall asleep? Are you able to sleep during the day when the baby is sleeping? Reports of difficulty in falling asleep and difficulty sleeping during the day while the baby is sleeping are clear indications of mental distress.

- Early stages of postpartum depression are difficult to detect because all women are tired after childbirth and are not sleeping well.
- Diagnostic questions for the diagnosis of postpartum depression:
  - How long does it take you to fall asleep at night?
  - Are you able to sleep during the day when the baby is sleeping?

Causes of Postpartum Mental Disorders

These can include a number of factors:

- Hormonal changes after childbirth: A sharp decrease in the levels of progesterone and estrogen
- Changes in functioning of the thyroid gland, which are common after childbirth (approximately 6%) and can be expressed as changes in mental state
- Changes in the release of neurotransmitters—substances secreted from the ends of nerve cells that serve to pass information among nerve cells—such as serotonin, noradrenalin, and dopamine, which affect the mental state
- Changes in levels of cortisone
- History of psychiatric disorders
- Mental disorders in previous pregnancies
- Mental disorders during the current pregnancy
- Personality traits, such as a tendency to anxiety or obsessiveness
- Lack of a supportive environment
- Problematic relations with partner
- Unplanned pregnancy
- Problematic relations with mother
- Medical and gynecological problems during the pregnancy
- Multiple and frequent pregnancies
- A baby that is difficult to take care of

There is no evidence of a connection between postpartum mental disorders and a mother's mature age, socioeconomic status, low IQ, and whether or not she breastfeeds.

Professor Shaila Misri, from Vancouver, a leading psychiatrist in the field of postpartum depression raises the following question in her book *Shouldn't I Be Happy?* (1995): If the main assumption is that postpartum mental disorders occur, primarily, because of biochemical changes, why do these disorders not show the same frequency in different cultures? Misri also assumes that postpartum biochemical changes are only one factor among many that influence the ability of women to cope mentally after the birth.

A number of anthropological studies on postpartum mental disorders find a connection between the way women are treated after childbirth in different cultures and the development of mental disorders. These studies show that when the new mother is considered to be in a "healing period" and enjoys the help of her mother, her partner, and other family members, the risk of developing postpartum mental disorder is reduced.

In Western society, the period in which the new mother is treated in this manner corresponds to the time she spends at the hospital following the birth, usually from a few days to a week. In contrast, other cultures have developed special traditions and rites around the birthing mother that are based on the assumption that after childbirth, a woman requires special protection and treatment.

In the rural areas of Guatemala, it is customary to watch over the birthing mother for 8 days after the birth, protecting her from the need for activity and, in particular, from the cold. The belief is that after childbirth, women are more sensitive to cold and that this sensitivity can affect the milk and cause the baby to become ill. In the Philippines, for 8 days after the birth, it is customary to give birthing mothers a special diet of remedial plants to prevent them from engaging in strenuous tasks, and to protect them from the cold. In Nigeria, women after childbirth are brought to a separate place called the "Fattening Room" in which they are looked after by older women. The treatment comprises mostly feeding the mother and the baby and ensuring that the mother sleeps. One anthropologist called this custom "Mothering the Mother." After a period of rest, the new mother leaves the "Fattening Room" and is welcomed with the baby at a party, in which the father introduces his wife and the baby and plants a tree in their honor.

Shaila Misri, who was born in India and spent most of her adult life in Canada, describes the treatment she received from her mother after she gave birth to her sons. After she gave birth, her mother arrived in Canada for a period of 6 months, accompanied by a woman who specialized in treating postpartum women. For a month, Shaila Misri and the baby received daily massages and baths. Misri was fed a special diet to enrich her milk and reduce the baby's stomachaches. Special emphasis was placed on the rest and sleep of both mother and child, both during day and at night. For 6 months, Professor Misri enjoyed a supportive environment that understood her new needs and allowed her to heal and to rest.

In Western society, in contrast, no special attention is given to a woman's needs following childbirth. She is mostly alone, without any support, and even if she does enjoy support and help, it is usually for only a brief period. The irony is that it is specifically during this period, in which the woman begins her own existence as a mother, that she needs to feel the mothering from her own mother, who is not always there at her side.

Postpartum Blues

This is a very frequent phenomenon—experienced by about 80% of mothers who give birth—that appears immediately after childbirth. The symptoms are sleep disturbances (difficulty falling asleep), emotional turmoil, the tendency to alternate between crying and feeling happy, angriness and irritability, sadness, feeling the need to talk a lot, and a tendency to withdraw from surroundings—that is, a myriad of fluctuating mental phenomena, that appear from the first or second day after the birth, peak during the first week after the birth, and then disappear. These are normative reactions to the hormonal and mental changes that take place after childbirth. It is important to know of these phenomena in advance, to calm the new mother, to enlist support, and especially to allow her to sleep well and to rest.

In 25% of the cases, the phenomena last longer than 2 weeks and can develop into mental phenomena that require treatment and help.

Postpartum Depression

This is depression that develops in 15%–20% of women about a month to 6 weeks after birth, but that can also develop throughout the first year. It is more common in women after their first birth. However, postpartum depression can also appear in women for whom this is their third or fourth birth and who have not experienced postpartum depression in the past.

The characteristics of postpartum depression are the time at which it occurs and the depressive features that are usually connected to functioning as a mother, such as guilt over the inability to take care of the baby appropriately. As mentioned previously, one of the main symptoms of postpartum depression is the inability to sleep when the baby sleeps, both during the day and at night.

Postpartum depression can also be a depressive stage within the framework of bipolar disease, manic depression, which generally appears for the first time only after childbirth and is characterized by periods of depression that alternate with manic episodes (the opposite of depression: elated mood, disquietude, difficulty falling asleep, and more). A more detailed explanation of bipolar disorder appears in Chapter 10 on sleep disturbances related to mental disorders.

As of July 2012, women in Israel are given the Edinburgh Postnatal Depression Scale (EPDS) questionnaire after childbirth at the clinics that care for mother and baby, to help identify women suffering from postnatal depression. The questionnaire comprises 10 questions, with a choice of four possible answers that examine the woman's mood and functioning.

The treatment of postpartum depression depends on its level of severity. The main key for help is through sleep, lack of which aggravates the mental state. It is vital to let the new mother sleep at least one or two full nights, with the help of her partner, a night nurse, or any other person who can take care of the baby on her behalf. If the professionals to whom the woman turns suspect she is suffering from depression, it is essential that they alert the family members, because in many cases, the woman tends to conceal her mental state, fearing she will disappoint those around her who expect her to be happy. Family members for their part tend to think that the depression is temporary and will pass on its own accord. In cases of severe depression (sleep disorders, heavy mood, low level of function, guilt feelings, suicidal thoughts), it is important to begin medication with an antidepressant. The risk of postpartum depression is twofold: It affects both mother and baby. Therefore, it demands a quick medical intervention and close monitoring. In extreme cases, in which there is a fear that the mother will harm herself or the baby, the need for hospitalization should be considered.

This was Ifat's third birth, a baby girl after two boys. The joy was immense. Ifat, 35 years old, had been working for many years at a computer company, which is also where she met her husband. "I was born to be a mother," she told me, "It was easy and natural for me to raise the two boys."

The pregnancy with her daughter was difficult as she was hospitalized a few times because of bleeding and was also busy renovating their new house, to which they planned to move around the time of the birth. In addition, during the pregnancy, Ifat was transferred to a new and more demanding role in the company, which on the one hand made her very happy, but on the other hand increased her workload.

Ifat had her daughter by caesarian section, which left her in pain for a month. The baby was very restless, and for the first month, Ifat hardly slept. However, Ifat, according to her own admission, is a woman who does not easily let herself off the hook. "I breastfed the baby as I had the boys," she told me, "One week after the birth I was already back at the computer, the baby in my arms, and I returned to work from home."

Ifat came to me only a year after the birth and in a state of severe depression. She had lost 22 pounds, had difficulty functioning at home as well as at work, and slept on average 3–4 hours per night. Despite that, she insisted on handling her depression alone and considered that getting help from an external source was a failure. Her boss, with whom I was acquainted, was the one who eventually managed to convince her to come see me.

Ifat's story emphasizes the accumulation of a number of stress factors that preceded the birth, together with her personality. The combination of these factors led to the postpartum depression after the third birth. My treatment of Ifat included enlisting the help of those around her in order to reduce the stress; an instruction to sleep continuously for 6–7 hours a night; to acknowledge the fact that she was suffering from depression, which she could not cure by herself; and treatment with medication accompanied by psychological therapy.

Postpartum Anxiety Disorder

Postpartum anxiety disorders are quite frequent after childbirth, and one should remember that not every anxiety that a woman experiences after childbirth should be defined as an "anxiety disorder." Anxiety is natural among new mothers, especially after the first birth, and is a normative process, particularly in the first month of adjusting to their new role. Persistent anxiety, which disrupts function and especially sleep, is an indication of the existence of a disorder. Feelings of anxiety can be manifested in different ways. Some of the most common include difficulty falling asleep, frequent episodes of waking up to check if the baby is sleeping, difficulty in resuming sleep after feeding the baby, elevated heart rate, frequent use of the toilet, appetite disruption, increased stress over the baby's development, dizziness, feelings of constant panic, and every so often—panic attacks. Another manifestation of anxiety is obsessive thinking, sometimes accompanied by compulsive acts. These can scare the woman and lead her to think that she is losing her mind.

Women who have gone through difficult births often experience post-traumatic phenomena, accompanied by recurring dreams linked to the birth experience that can persist and accompany them for many years after the birth.

Postpartum Psychosis

This is a rare disorder with an incidence of one to two in every 1,000 births. In all my professional career, I have seen only isolated cases of women who suffered postpartum psychosis after giving birth. Two were hospitalized in a psych ward where I was doing my internship.

Anna gave birth to her eldest daughter under ideal conditions, just as she had envisioned: a private midwife accompanied her and her husband, she was given an epidural to prevent pain, and within 6 hours, her daughter, Chen, burst into the world. Anna could imagine no greater joy.

On the first night, Anna could not fall asleep because she was so excited. However, in no time at all, she also began to feel a sense of unease, a need to walk around the ward, and insatiable hunger. She ate everything she could lay her hands on. Thoughts began racing through her mind, and turned into voices that spoke to her: "You won't be able to function as a mother; give the baby away; you have no future; the baby is deformed." Anna began reacting to the voices and speaking simultaneously with both anger and fear. Her speech became muddled, she laughed and cried, entered the rooms of the other mothers, and became a nuisance.

Anna was diagnosed as suffering from postpartum psychosis and was hospitalized for a month in a psychiatric hospital for immediate treatment, since she posed a threat to herself and to her baby daughter. During her hospitalization, she was given antipsychotic medication. After she was discharged from hospital, she continued to take the medication and was under supervision.

Anna came to see me for a consultation when her daughter was 2 years old, after having recovered. The cause of her postpartum psychosis was not completely clear, and Anna was nervous about stopping the medication and especially of having another pregnancy. In view of her situation, she decided to wait before deciding whether or not to have another child, since the trauma was still fresh in her mind and quite threatening.

Postpartum psychosis also manifests with an extreme sleep disturbance, which is one of its dominant symptoms. In this case, it is accompanied by extreme agitation, confusion, rapid speech, irrational thoughts (thoughts that have no grounding in reality, yet the woman is convinced of their validity and acts accordingly); as well as hearing voices that command her to act in various ways—to harm the baby, for example (auditory hallucinations); and frightening and threatening visions that do not exist in reality (visual hallucinations). The moods swing markedly in accordance with the different object in the psychotic state: from joy to suspicion, irritation, fear, and worry.

Postpartum psychosis occurs a few days after birth and is considered a dangerous disorder because of the high risk of suicide or of harm to the baby. It is rare for postpartum psychosis to erupt without a history of mental disorders in the woman's past, usually bipolar or manic depression disorder.

The need for psychiatric hospitalization causes a difficult problem because of the separation of the baby from the mother in the days that are critical for their bonding. Some countries have hospitalization units for both the mother and baby. Attempts of this kind were made in Israel, yet the need for a diverse staff of experts posed a heavy financial burden and they were terminated.

Conclusion

Mental changes in pregnant women and especially after birth are frequent and normative. They are caused by several factors and are not the result of a single cause. One of the principle expressions of these changes is sleep disorders, and treating them aids recovery. I always make sure to talk with pregnant women about their sleep during their pregnancy and, in particular, to begin to prepare them for sleeping well after the birth as it is an important factor for maintaining her mental health.

In most women, the hormonal changes during pregnancy and postpartum will not lead to mental disorders that require treatment. Also, today women are still ashamed of their feelings and attempt to conceal them. The title of Shaila Misri's book *Shouldn't I Be Happy?* reflects the cultural expectations of our society. Any other reaction, which does not comply with these expectations, leads to concealment and shame.

In the past, pregnancy was considered to be a period that conferred protection from mental disorders, yet today it is clear that there is no truth to that assumption and that ignoring states of depression or anxiety during pregnancy can affect the development of the fetus as well as the mental well-being of the pregnant woman and increase the risk of postpartum depression. The principle of benefit versus risk should be at the center of all considerations regarding treatment with medication of pregnant and lactating women who suffer from mental disorders.

In the postpartum period, the risk of the appearance of mental disorders is high compared with every other period in a woman's life. Every mental disorder left untreated poses a risk for the mother–baby bond. Here too, it is vital to maintain the benefit versus risk principle while considering treatment with medication.

8

Sleep Disturbances and Menopause

I have always been curious as to how professionals choose their area of specialization. Why have physicians, like me, chosen a particular specialization; why have historians delved into one particular period and not another; why have psychologists decided to probe one type of trauma rather than a wide array? From the distance of many years of observation, I have discovered that most choices were not meaningless or random, but stemmed from a personal affiliation with a certain subject.

And so too did mine. During my years at medical school as well as during the years of my internship in psychiatry, the subject of menopause was hardly given any consideration. My focus on the connection between mental and hormonal changes throughout a woman's life began in my mid-thirties. Since menopause was still far away for me personally, I didn't take much interest in the subject. Only when I reached that age did the subject attract me.

We live in an era that does not respect aging or the changes brought by age, especially in women. In the field of women's health, there are very few physicians who treat menopause. Most physicians choose to focus on young women in their fertile stage. Following the rise in life expectancy, women live some 30–40 years after menstruation ends. This is a group of women who are in the midst of their lives in terms of their professional and personal capabilities. It is really unfortunate that society and the medical world do not relate to this group with more respect.

I am still amazed at various conferences to hear the derogative attitude toward menopause, and to find this is the case even for women who come to me seeking treatment, expressing a cultural attitude deeply embedded in society and reflecting the prevailing wisdom concerning this period in a woman's life. I very much admire the title of the book *The Wisdom of Menopause*, by gynecologist Christiane Northrup, who chose to emphasize the positive aspects and the life experience, and not the loss of youth.

Menopause

Mental Symptoms

- Irritability, anger, agitation
- Despair or depression
- Extreme mood swings
- Loss of sexual desire

Physical Symptoms

- Hot flashes
- Sleep disorders
- Vaginal dryness
- Urinary infection
- Tiredness
- Weight gain

Historical and Cultural Characteristics of Menopause

In the past, many women did not reach the age of menopause, yet it is still possible to find historical references to this period. In the Book of Genesis, we read that Abraham was aware that his wife had ceased menstruation from his deliberations on the possibility of bringing a child into the world in his old age and that of his wife Sarah.

The British anthropologist Margaret Luk raised the question—from a developmental point of view—Why does menopause exist? Why is the fertility of women limited, in contrast to that of men? What advantage is there to life beyond the period of fertility? Anthropologists developed the "grandmother theory": society needs women past the age of fertility to serve as teachers, to care for children, to train younger women, and to gather food. The elder women filled an important role in society as partners in the tribal decisions due to their life experience; the loss of fertility was not perceived as a tragedy, but as a change in their tribal role.

In Celtic cultures, the young woman was considered to be a flower; the mother, a fruit; and the grandmother, a seed. And why a seed? Because this is the part that contains within it the knowledge and power encoded in all the other parts. The role of the grandmother, the postmenopausal woman, was to again sow in the tribe the seed of truth and wisdom.

Different studies emphasize the cultural differences in attitudes to women in menopause. For example, women from northern India, who are meticulously separated from the men during their fertile years, are less restricted during menopause and are allowed to move around the village with more freedom. These women experience menopause as a period in which their quality of life improves. A study conducted in 1944 in a Thai village found that women happily accepted the period that liberated them from pregnancies. They regarded the menopausal symptoms as temporary, did not seek medical help, and allowed nature to take its course.

It is always fascinating to read the French philosopher, writer, and feminist Simone de Beauvoir's thoughts on gender differences in her book *The Second Sex*, written in 1949: because a woman is trapped in her female functions, her history depends much more on physiological destiny, which is more erratic and discontinuous, with dangerously abrupt transitions from one stage to another; "*The male grows older continuously, the woman is brusquely stripped of her femininity; still young, she loses sexual attraction and fertility, from which, in society's and her own eyes, she derives the justification for her existence and her chances of happiness.*" She concludes that women's disorders come less from the body itself than from their anxiety about it.

The Medical Profession's Approach to Menopause

Modern medicine, which offers excellent solutions for different illnesses, has, at times, difficulty accepting a physiological process and allowing nature to run its course. The medical system regards menopause as a deficiency disease. Since there is a decline in the ability of the ovaries to produce estrogen, the common medical perception is that women at the age of menopause are experiencing a hormonal-deficiency disease, and they relate to it in the same way as to any other hormone deficiency, or as the gynecological literature defines it: "ovarian failure."

In 1966, the American gynecologist Robert Wilson published the book *Feminine Forever*, which aroused much enthusiasm and constituted one of the groundbreaking revolutions in medicine. In his book, Wilson describes menopause as a process of chemical castration due to the lack of estrogen and suggests, for the first time, treatment with estrogen to "continue and remain feminine forever." Wilson writes that menopause causes an emotional revolution and he also ties the mental aspect to the deficiency in estrogen.

A year later, in 1967, Paul Rhodes, an American physician specializing in geriatrics, published an article in the *Journal of the American Geriatrics Society*, in which he called menopause a tragedy. "Many women lead active and fruitful lives when this tragedy attacks them," he wrote.

In 1985, members of the International Menopause Society met in France and formulated a milder opinion: "Menopause occurs around the age of 51 and presents a transition period from a state of fertility to a state of infertility." This definition aroused disagreement among many in the medical community, who saw it as too simplistic and lacking the tragic dimension. Prominent among them was the American gynecologist, Wulf Utian, who claimed that menopause is endocrinopathy (glandular disease). A study conducted in 1986 in a menopause clinic in a general hospital in Los Angeles found that 53% of the women considered menopause to indeed constitute a medical problem that requires medical intervention.

The attitude of the medical world to menopause as a disease had practical consequences. The medications offered to women carry the name "hormone replacement therapy," namely: hormones that replace those that are no longer in the body. Over the years, it transpired that treatment with estrogen alone increases the risk of uterine cancer, whereas adding progesterone reduces this risk.

Jerilynn C. Prior, a Canadian endocrinologist and researcher, writes, "Our culture finds it easy to blame the women's reproductive system as being responsible for the disease. Drawing a connection between the change in reproductive ability due to the end of menstruation and aging, the transformation of menstruation into a point in time and not a process, and naming it an estrogen deficiency disease—all these reflect an unscientific and harmful perception by medical science."

In July 2002, the *Journal of the American Medical Association* (*JAMA*) published a comprehensive study conducted by the National Institutes of Health (NIH) among 17,000 women who ranged in age from 50–80, with an average age of 63. The aim was to examine the benefit/damage of hormonal therapy after menopause, especially in respect to prevention of heart disease, brain aging, and risk of breast cancer due to the use of hormone replacement therapy (HRT).

The study was stopped after about 5 years, before the planned date, as a result of findings that showed that the risk of treatment with hormones outweighed the benefit.

An increase was found in the risk of contracting heart and blood vessel diseases and in the risk of developing breast cancer. The findings of the study sent shock waves through the community of gynecologists and led to a drastic cessation of hormonal replacement therapy and to much confusion among women. Much criticism has been directed at the study since that time, and physicians have gradually returned to hormonal replacement therapy with clear guidelines, in small doses, and with a limit on the duration of the treatment.

My numerous years of experience with the subject indicate to me without a doubt that there are women for whom hormonal deficiency causes difficult symptoms and significantly impairs their quality of life. Consequently, in these cases, hormone therapy markedly improves their lives. That said, not all women react similarly to the cessation of hormone secretion, and therefore the attitude and the treatment should be individually tailored to each woman.

I recall a case of a woman who came to my office and said she was in menopause and did not suffer from any symptoms; however, she had understood from reading women's magazines that she needed treatment. I had difficulty explaining to her that only women suffering from symptoms need treatment. There are women with no symptoms or only a few, and therefore they do not require any medical intervention or need specific pinpointed intervention. I will elaborate on this point in Chapter 11, on treatments for sleep disorders.

What Is Menopause?

The term menopause derives from Latin: *meno*, the monthly moon, and *pause*, that is, the end of the monthly cycle. I was very happy to read the following sentence in Meir Kryger's book (*A Woman's Guide to Sleep Disorders*, 2004): "Menopause is not a disease but a normative physiological process."

Menopause (Figure 8.1) includes the following processes:

- *Perimenopause*: A period that begins around age 48–52 years, characterized mostly by irregular periods, sleep disturbances, hot flashes and sweating, and mood swings. This period usually ends a year after the last menstruation.

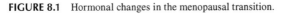

FIGURE 8.1   Hormonal changes in the menopausal transition.

- The ovaries gradually cease their activity until hormone secretion ceases completely.
- *Menopause*: The time at which the last menstruation takes place. Only a year afterward can it be known for certain that the woman is no longer in the perimenopause stage.
- *Postmenopause*: The period that begins a year after the last menstruation. The ovaries are no longer active, and therefore there is no secretion of estrogen and progesterone from the ovaries and as a result no more menstruation.

The most difficult period as far as the symptoms are concerned is perimenopause, because this is a period of hormonal fluctuations and our body does not like instability.

Sleep disturbances in menopause are often the first symptom to appear while menstruation is still regular. Both women and physicians find it difficult to relate the sleep disturbances to the slight hormonal changes, which are not detectable in blood tests, at a time when menstruation is still regular.

Sleep Disturbances during the Period of Menopause

Restless Legs Syndrome (RLS)

Characteristics: RLS includes an uncomfortable sensation at the tips of the legs while resting, with marked relief when the legs are moved. There is a constant need to move, to stretch the legs, to move them, to walk from place to place. It is most common among pregnant women.

The literature on this subject has failed to provide clear explanations for the condition itself as well as for the rise in its frequency during menopause. Some attribute the disorder to a deficiency in iron and to a connection to the neurotransmitter dopamine. Medications that increase the level of dopamine are known to be helpful.

Obstructive Sleep Apnea (OSA)

Characteristics: OSA consists of episodes of sleep disruptions during the night, associated mostly with snoring, uneasy sleep with several breathing pauses, a tendency to sleep during the day, and, of course, severe fatigue. Usually, this disorder is more common in men, yet its frequency, which increases in women in menopause, led to the assumption that the decrease in reproductive hormones has a negative influence on breathing at night.

Many studies that compared women in menopause who took hormone replacement therapy (HRT) to women who did not take hormones showed that those using hormones suffered less from breathing problems at night. This raises the question of whether or not it is right to give hormone replacement therapy as a protection against breathing disturbances at night.

The answer is not conclusive. In the last decade, we have learned that hormone replacement therapy is not given automatically as in the past, because of its adverse side effects (such as rise in the frequency of breast cancer). In addition, many studies have shown a much higher frequency of breathing disturbances at night among overweight women compared with women who are not overweight. Weight gain is a frequent occurrence in menopause. Additional studies have been conducted aiming to explain why premenopausal women are more protected from breathing disturbances at night than women in menopause, when there is no doubt that there is a connection to the hormonal changes that occur at this age. Unfortunately, these studies were conducted on a very small scale, and the precise mechanism has not yet been determined.

Insomnia

Characteristics: Insomnia includes difficulty falling asleep, frequent waking during the night, waking up very early in the morning, or a feeling during the day that the sleep was not a restorative sleep.

This is the most common sleep disorder in menopause: approximately 40%–61% of women will suffer from insomnia during menopause compared with 33%–36% of women before menopause.

I hear many complaints from women in menopause about sleep disturbances. It is always the same story: Most have no difficulty falling asleep, yet after several hours, between 2:00 and 4:00 in the morning; they awaken and find it very difficult to fall back asleep.

One of my patients, who approached me as a result of severe sleep disturbances and depression that appeared with the end of her menstruation, wrote me a letter expressing chiefly frustration and anger at the lack of an appropriate and viable solution for women in menopause. Her frustration accumulated after she had to go through a long ordeal until she obtained the long-desired treatment; here is the text in her own words:

In recent years there has been an organizational shakeup in the health services. Everything has become efficient, measured, driven by inputs, outputs, tables, and statistics. Among all this, we, the patients, have somehow been forgotten. To provide a good medical response one doesn't only need to know how to write a prescription. One also has to look the patient in the eye, to answer her, to listen and to try to understand what really bothers her.

Women in menopause are a large segment of society, and with the changing times their needs are changing too, the dilemmas they face, and their life conditions. If once, women experienced menopause and clenched their teeth, today hormones have "fixed" the problem. Then came the news that the hormones cause cancer and women stopped taking hormones and began to seek alternative ways to solve the problem. Because menopause is a problem. A problem on an alternating axis: a mild problem, a physical problem, a mental problem, or a severe problem—or all of the above.

This is a chapter in our lives that needs a solution. The point is that there is no formulaic answer. There is need to determine the correct dosage for each woman or, in other words: each woman deserves her custom-made outfit. Not something mass-produced. As someone who has gone through the stage of mass-produced outfits, I can attest that it is problematic. The health clinics give you an appointment and allow you the option of a gynecologist—hormones, or a psychiatrist.

"What did I have? God knows. To me personally, it was unclear," wrote the poet Nathan Alterman. However, his poem speaks of love. For me it was depression. Depression has many sides. With me it began with lack of sleep, with waking up at four in the morning and with the inability to go back to sleep. With a heavy sense of distress from the moment I opened my eyes that stayed with me throughout the day. I lost my appetite, and I certainly like eating; I tried to maintain an organized life, tried to uphold routine and life habits, yet I was in deep distress. No, I didn't want to die, but I didn't see a way out of the ongoing nightmare my life had become. I had hot flashes but as friends told me: "If you don't have pools of sweat at the roots of your hair, you're not suffering from hot flashes."

So, the hot flashes were relatively light but the mental disorders were heavy. Few are the physicians in general, and family practitioners in particular, who can give an answer to a woman in the despair of menopause. In the age of modern reproduction, we need many such physicians, who have accumulated knowledge from experience and from specializing in the needs of women in menopause. Many gynecologists who are aware of the combination of gynecology and psychiatry. Many psychiatrists need to understand the significance of the hormonal aspect. It is not either-or, but a wise combination based on specialization and experience in this unique field called menopause.

Public health services must understand that this is not a subject that can be solved with diet, herbal medicine or a visit to the gynecologist. Various ideas from a variety of fields should be mixed together with the knowledge of how to diagnose and to notice each woman's symptoms. There is no single prescription. There should also be conversation and listening. There is listening to a woman's needs and then there is a solution, which may be incorrect, but at least there is a listening ear ready to change.

This cannot be done in the 10-minute time slot the health service allocates each patient. It demands more time but saves a lot of money later and much misery and physical and mental pain. And more than anything, it gives back to us—mature, beautiful, strong and creative women – the sparkle in our eyes. We are opinionated women, at an age that is a significant turning point in our lives, and we want to be treated with respect.

Had I not been referred to the right physician, would I have been sentenced to a long period of depression, distress, anxiety and anguish? To terrible sleeplessness?

I want—yes, I want, because I deserve it—the physician sitting opposite me to be attentive, aware of the connections between the body that is going through the physical process and the mind that is oscillating as a result of chemical changes. Because no woman deserves to suffer unnecessary pain and we deserve our pains to be eased and that a solution be found for the distress in which we find ourselves and to be treated appropriately and respectfully."

The reasons for insomnia are numerous and intertwined, affecting one another. The causes are connected to the hormonal changes, physiological age-related diseases, mental illnesses that worsen with age, change in the circadian cycle, and various life experiences.

I will address the most frequent causes:

- *Hot flashes* that cause frequent episodes of waking during the night.
- *Depression and anxiety* that are common in menopause.

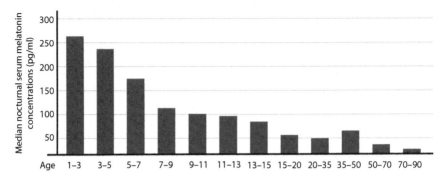

FIGURE 8.2   Melatonin secretion decreases as a person ages.

- *Changes in the cycle of sleep and wakefulness (circadian rhythm)*, especially changes in the release of melatonin (the hormone that is secreted in darkness and affects the circadian cycle; Figure 8.2).

Hot Flashes

These can appear by day or by night, and the more frequent they are, the more sleep is disturbed.

Hot flashes were described in the seventeenth century and are the most well-known sign of menopause. They are caused by a change that occurs in the temperature-regulating center of the brain, yet the exact mechanism that generates them is not completely clear to this day.

Around 80% of women in menopause suffer from hot flashes. Women report a feeling of heat that begins at the top part of the body, spreads to the neck and face, and ends in sweating. Hot flashes vary in their frequency from woman to woman, and some women do not experience them at all, even though they go through the same physiological process. Usually, without hormone replacement therapy, there will be a decrease in the frequency of hot flashes 3–6 years after they first appear, until they disappear almost entirely. However, a small percentage of women will continue to suffer from them 10 years after their first appearance and perhaps even longer.

I meet women who enter my office with flushed cheeks, dressed in summer clothes even in winter, their faces covered in sweat, and they are in constant agitation, which is connected to their hot flashes. Their sleep is disturbed due to frequent bouts of waking up as a result of the hot flashes, and therefore during the day, they may suffer severe fatigue and a decrease in their ability to function. Hot flashes may cause embarrassment in social circumstances, at professional presentations, in front of a crowd, and consequently may cause a change in lifestyle, such as a change in clothing style and makeup to suit the situation.

Many studies indicate a decrease in the flow of blood to the brain during hot flashes, which explains the difficulty to function when they appear. Additional studies have shown a link between hot flashes and the risk of heart disease, high blood pressure, and diabetes. For this variety of reasons, it is extremely important that hot flashes are taken seriously and treated accordingly.

Hormone replacement therapy (HRT): Treatment with HRT eases the hot flashes results in a marked improvement in the quality of sleep.

HRT is an addition of the estrogen hormone, with or without progesterone. Until the summer of 2002, this treatment was given almost automatically to any woman who was diagnosed by her physician as suffering from a decrease in the function of her ovaries, even if she was not suffering from any other problems. The widely held perception was that replacement hormones help to prevent heart disease and hot flashes and prevent a decrease in bone density, as well as improving mood and the quality of sleep. A comprehensive study conducted by the American National Institutes of Health showed that the use of hormones increased the risk of heart attacks, intracranial hemorrhage, and breast cancer. As a result of the publication of this study, there was a marked decrease in the use of replacement hormones, although it was later proved that the population that was tested was too old and was not representative of younger women who were entering menopause.

Other treatments of hot flashes: Treatment with antidepressants, especially venlafaxine or paroxetine at low doses, has been proven to reduce hot flashes and consequently also sleep disturbances. The treatment can be beneficial for women who cannot, or refuse to, take hormones. There are reports that acupuncture and natural remedies may help to reduce hot flashes, however, the research in this field has not yet presented conclusive results.

Japanese women suffer far less from hot flashes, and there are those who attribute this to their diet, especially to soy products, which are popular in Japan. Indeed, some of the natural products developed for easing hot flashes are based on soybeans (such as Femarelle) or on the black cohosh plant (such as Cimidona, an herbal medication).

Women who, for various medical reasons, have their ovaries removed (surgical menopause) will usually suffer from acute hot flashes and insomnia, because their body was not prepared for this situation as well as for the cessation of hormone secretion, which happened abruptly because of the surgery. The treatment of sleep disturbances in these women is rather complex, since it is usually not recommended to treat them with hormone replacement therapy. Treatment with antidepressants can provide some relief both to the hot flashes and to their mental state and usually has a positive effect on sleep quality.

Women who have had breast cancer will also suffer more from hot flashes, depression, and sleep disorders (linked to the treatments they are given), and they too will not be helped with hormone replacement therapy, but with the other treatments described previously.

A very interesting fact I came across was that many women who have had breast cancer began suffering from sleep disorders or depression only after the end of their treatment (operation, radiation, chemotherapy) and not during treatment. What is the reason for this? The answer is that during treatment, these women are supported by physicians, family members, and friends, but when their treatment ends, they remain alone, because seemingly everything is behind them. However, it is particularly during this period that the awareness of the disease begins to seep into their cognition, together with the fear that the disease may recur. This awareness causes fears, anxieties, and thoughts about death. All these make it very hard to sleep and may cause depression. A young patient, who had breast cancer and had to deal with fears and anxieties, once said to me: "I always leave our sessions with a smile, because you make me feel that I am normal." I will never forget this sentence.

Depression and Anxiety

Women suffer from depression twice as much as men, and during menopause, the frequency leaps to almost four times in comparison with men. Approximately 30%–35% of women in menopause will suffer from depression and anxiety. One of its main characteristics is insomnia. There is a subgroup of women who are sensitive to hormonal changes and suffer more from premenstrual symptoms or postpartum depression. These women could suffer from depression and anxiety at a higher frequency considering the hormonal changes brought about by menopause. I would like to add that I also meet women who do not have a history of mental symptoms and come for treatment for the first time in their life due to mental problems that began at menopause.

Yael, 52 years old, had been my patient for 2 years now, due to a worsening in the depression from which she has been suffering for many years. Over the past year, her periods have been irregular, and in particular, she experienced deterioration in her mood and sleep disorders, expressed in frequent waking during the night.

Yael lost her mother when she was 10, and after her father remarried, about 2 years after the death of her mother, she moved in with her grandmother. She had felt depressed ever since then. This was manifested especially in excessive sleeping, excessive eating, weight gain, and a feeling of apathy toward life. Yael finished her high school education with difficulty and was released from the military service on the grounds of difficulties adjusting. She met her husband at the law office where she worked as a secretary.

After the birth of her second daughter, Yael was overcome by severe postpartum depression, and there was no choice but to hospitalize her for a 3-month period because of suicidal thoughts and risk to her life. "The depression became part of who I am," she told me, "but I learned to live with it and not be alarmed by it." When Yael reached menopause, her therapist turned to me with a request for help.

Nothing in Yael's well-kept appearance bore witness to the depression from which she had been suffering for so many years. "It is very important to me that nobody knows or notices anything from the way I appear, and therefore I make a lot of effort to dress well, to use makeup, and to behave as if all is well. I function extremely well when I'm at work, but getting up in the morning is very difficult for me. During the day, I have no appetite whatsoever. At night I wake up every hour."

The hormonal change experienced by Yael demanded a reevaluation of her antidepressant medication. Over the years, she was treated with sertraline, a member of the SSRI drug group, which elevates serotonin levels. Various studies have shown a higher efficacy of treatment with drugs from the SNRI family, which raise the levels of both serotonin and noradrenalin (an activating neurotransmitter) among women over the age of 50. Indeed, Yael reacted well to treatment with Cymbalta, which is a drug from the SNRI group.

Figure 8.3 illustrates just how difficult it can sometimes be to differentiate between symptoms of depression and symptoms of menopause. Sometimes, as in Yael's case, the symptoms may coexist. However, despite this, it is possible to find some distinguishing features: In depression, there are features connected with mood: Thoughts about death, feelings of purposelessness, and lack of energy. In contrast, in menopause, we observe more features related to the irregularity of menstruation until it ends: Hot flashes,

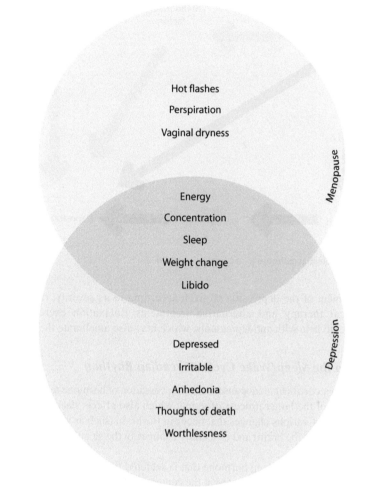

Hot flashes

Perspiration

Vaginal dryness

Menopause

Energy

Concentration

Sleep

Weight change

Libido

Depression

Depressed

Irritable

Anhedonia

Thoughts of death

Worthlessness

FIGURE 8.3 Depression or menopause?

sweating, and vaginal dryness. Nonetheless, both depression and menopause can be associated with sleep disorders, loss of appetite, difficulties concentrating, and decrease in sexual drive.

Causes of depression during menopause:

- Direct effect of the decline in the level of estrogen
- Consequences of the decline in the level of estrogen: hot flashes that cause sleep disorders, which can lead to depression
- Effect of the decrease in estrogen on serotonin in the brain, which affects mood
- Environmental–psychological factors related to menopause, such as aging, the difficulty of accepting mature age, the empty nest syndrome due to departure of the children, and relationship problems that can arise as a result of this (Figure 8.4)

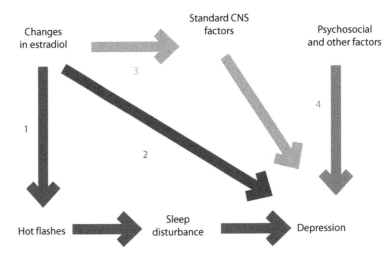

FIGURE 8.4   Potential pathways to menopausal depression.

The treatment of the depression is given according to its severity, ranging from psychological therapy and alternative treatments (relaxation exercises, yoga, acupuncture) to help with antidepressants, which may also ameliorate the hot flashes.

Changes in the Sleep/Wake Cycle (Circadian Rhythm)

The change that occurs in menopause following cessation of hormone release from the ovaries is part of the larger process of aging, which also affects sleep. The effect on sleep is a result of various changes that occur in the brain (such as atrophy, reduction of the blood flow to the brain) and especially changes in the sleep/wake cycle and the release of melatonin.

Melatonin is a natural sleep hormone that is secreted from the pineal body, a very small gland the size of a pea, situated at the center of the brain. Melatonin is secreted every time we are in the night part of the biological clock and in the dark (which is why it is also called the "darkness hormone"). It breaks down quickly after its production and is not present when we are exposed to light.

Melatonin was discovered in 1958. The first reports on induction of sleep with the aid of melatonin appeared in the 1970s. Later, many studies proved that there is a link between melatonin and the timing and induction of sleep in both youngsters and adults.

During the day, the pineal body is inactive. When darkness falls, it begins to produce melatonin and to release it into the blood. Melatonin serves as a physiological signal, informing the brain and other organs of the body of the hours of darkness in each 24-hour day–night cycle and, indirectly, also the seasons of the year. Melatonin levels in the body rise in the evening, after nightfall, and reach their peak between midnight and 2:00 or 3:00 a.m. before daybreak. Thereafter, there is a gradual decrease of the hormone until the complete cessation of secretion occurring at daybreak, with the first rays of light. The maximum duration of melatonin secretion is around 12 hours (Figure 8.5).

FIGURE 8.5 Melatonin secretion starts soon after darkness, peaks in the middle of the night, and declines during the remainder of the night.

It is interesting to note that most of the complaints of insomnia which I hear from women relate to waking up between 2:00 and 3:00 a.m. before dawn, which is the time when melatonin levels begin to decrease.

The increase in the level of melatonin at night causes a sense of tiredness and sleepiness and a reduction of alertness. Consequently, taking a drug that contains melatonin will decrease the time needed to fall asleep. In parallel, during the day, when melatonin levels are low—after a night with high melatonin levels—we feel more refreshed, have an improved mood, and our ability to concentrate increases. Treatment with melatonin has also been proved to be effective for depressive states, so that it may be effective both for sleep disorders and for depression, which are intertwined.

In addition, melatonin also helps treat metabolic changes that appear with age, such as an increase of fats in the blood, increase of sugar levels, changes in blood pressure, and a decline in the functioning of the immune system.

Bear in mind: The secretion of melatonin decreases as a result of a stimulus from light rays—not only sunlight. Artificial light, produced by electric bulbs, a flickering television screen, or a computer or mobile phone screen, also decreases the production and secretion of melatonin. Many women who suffer from sleep disorders get into bed opposite a computer screen, a TV, or a mobile phone. Because melatonin secretion begins around 9:00 p.m., it is recommended to avoid screens from this hour on, especially screens held near the eyes (such as cell phones and laptops) which are normally illuminated with a blue–green light.

What Happens to the Secretion of Melatonin with Age?

Melatonin secretion decreases with age due to shrinking of the pineal gland body. Recent studies that aimed to examine whether there are gender differences in the secretion of melatonin indicate an enhanced age-related decrease in melatonin in women. Indeed, one of the explanations for insomnia in women is also connected to a significant decrease in secretion of melatonin. This becomes more evident particularly after age 55.

> In the 1990s, commercial products containing melatonin appeared in the United States for use as food additives, without any medical guidelines and without a recommended dosage. Consumption of these products is the responsibility of the users themselves, as the FDA does not monitor products defined as food additives. There is no quality control of these products or of their ingredients. They have never been clinically tested to prove their efficacy and safety, as in the case of medications. These food additives usually act by releasing melatonin and lead to a rapid increase of the hormone in the blood, which peaks within about an hour and returns to basic levels after 2 hours. Namely, these food additives do not mimic the form of the physiological secretion of the hormone.

To my great joy, in 2007, EU countries, to be followed by Israel as well as some other countries, approved the use of a drug called Circadin developed by Professor Nava Zisapel, the founder of the drug company Neurim. Circadin is a tablet that releases melatonin in a way that mimics the physiological secretion of the hormone into the blood. The hormone level peaks 3 hours after ingestion, remains stable for 3 more hours, and decreases gradually to basic daytime levels about 9 hours after it is taken. According to comprehensive research studies, Circadin has been found to be effective in treating insomnia in people aged 55 and over.

And why did I write "To my great joy?" Over and over again, I hear from women in menopause about waking up between 2:00 and 3:00 a.m. and finding it difficult to go back to sleep. The thought that passed through my mind was that this was probably a case of a deficiency in melatonin. I avoided offering the American food additive, for the reasons I mentioned before, even though many of my patients did use it. When Circadin came onto the market in 2008, I began prescribing it to women and was glad to witness the improvement that occurred in the quality of their sleep. Since then, I have gained much experience using this particular medicine. Treatment with Circadin is a breakthrough for me in helping women who suffer from insomnia, especially if the insomnia is not related to hot flashes or to depression.

The downside of Circadin is the time that it takes before it becomes effective (about 4–6 weeks). However, during this period, until it begins to exert its effect, the Circadin can be supplemented with a low dose of an antidepressant (mirtazapine, amitriptyline, trazodone), which improves sleep quality.

So, How Do We Treat Insomnia during Menopause?

The correct treatment—as I emphasize over and again in this book—is to deal with the cause or causes of the insomnia. This sounds very simple, on the face of things, yet it is not always so. In very simple terms, Figure 8.6 describes the phases of treatment:

- If the cause of insomnia is hot flashes, the most effective treatment will be hormonal replacement therapy, which will reduce the hot flashes and alleviate the sleep disorder. Not every woman can or wants to use hormones, and in that case, the solution is antidepressant drugs, which can reduce the hot flashes, although to a lesser degree in comparison with hormones.

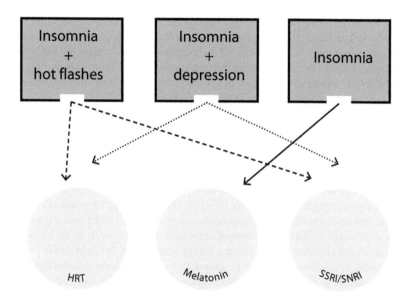

FIGURE 8.6 Treatment of insomnia during menopause.

- If the cause of insomnia is depression or anxiety, treatment with antidepressants or hormone replacement therapy, or both combined, can improve the mental symptoms and, as a result, can also improve the quality of sleep.
- If the sleep disorder is not caused by hot flashes or depression, then it is recommended to add melatonin by taking Circadin.

"Do you think I will ever go back to sleeping the way I used to in the past, or have I lost forever the ability I had to sleep well?" I am asked this question over and over again by women suffering from insomnia. My answer is "This is a process that if treated correctly, will allow the body to adapt to its new state, and it's reasonable to assume that with time, sleep will return to normal." Indeed, in most cases, sleep disorders exist as long as the hormone fluctuations are at their peak. When they end, there is calm, and in most cases, normal sleep can be enjoyed once again.

Keep in mind: When treating sleep disorders at every age, and especially during menopause, it is vital to check the source of the problem and not to solve it by prescribing sleeping pills or sedatives. Although these do help temporarily, they do not treat the source of the problem and often aggravate the sleep disorder.

When I finished writing about sleep disorders in the different phases of a woman's life, I thought about how I am moving from stage to stage and I wondered about the transitions in our lives. I recalled the book by American writer Judith Viorst, *Necessary Losses* (1986), which discusses periods of stability (where we put together a structure for our life) alternating with periods of transition (where we challenge the premises of what preceded, terminate the previous structure, and explore new possibilities).

The hormonal changes women experience demand endless coping with transitions, both physical and mental. A woman's sexual maturation is the beginning of a new era that marks a separation from childhood and transition into womanhood. Each month, menstruation leads women to face a process that ends and begins anew. Pregnancy and childbirth, among the most significant processes in a woman's life, are marked by an end, a separation and a new beginning—motherhood.

The end of the fertile period—or by its other name: the entry into menopause—is a period that unfortunately, we women still experience as an end only and not as a transition to another, new period, with new beginnings. This is a time in which we tend to experience the end and the loss rather than the advantages of age, the wisdom, the experience, the freedom of choice, and the liberation from the significant roles we bore and which are now already behind us.

I was also delighted by *Lighter as We Go* (2014) by Mindy Greenstein and Jamie Holland, whose lecture I heard at a conference in New York. The writers, a psychiatrist and a psychologist, share their experiences from life and work at an oncological hospital, which afforded them a broader perspective on age, life, and death.

The book, as is intimated by its title, attends in fact to the advantages and lightness of older age and not to its burden, hardship, and loss. Holland, an 85-year-old psychiatrist, writes that at her age, she has already completed all the assignments life presented her with, such as finding a partner, raising and nurturing a family, acquiring a profession, professional promotion, and economic well-being, and therefore everything seems easier and simpler to her. Greenstein, a psychologist in her 50s who had breast cancer, looks at her maturing process from a completely different angle compared with her viewpoint prior to the disease, and she wishes only to reach a ripe old age.

It is important for me to end this chapter with the main message I have tried to convey throughout: Menopause, which for many women symbolizes old age, loss, and nearing the end of life, is just another stage in our life as women. It is definitely not the end. The correct way of coping with this stage depends on the course we have taken in life up to that point: The experiences we have accumulated, the wisdom, the ability to cope, and our self-acceptance.

"When one door of happiness closes, another opens; but often we look so long at the closed door that we do not see the one which has been opened for us."

Helen Keller

9

Sleep Disturbances Related to Physical Diseases

Gender medicine, a field still in its infancy in Israel and the world at large, is dedicated to the study of the different frequencies of diseases and the differences in the way the same disease manifests itself in each gender. I hope that an understanding of these differences will extend into all areas of medical care, to allow each field of medicine to achieve a more comprehensive and nuanced approach than what has prevailed up to now.

In the past, most clinical tests were conducted solely on male patients, and female patients were excluded; the diseases and their characteristics were written in an identical way and applied to both sexes. As Professor Marek Glezerman, founder of the Israel Society for Gender and Sex Conscious Medicine writes in his book *Gender Medicine* (2016): "Gender medicine is the abandonment of the medical point of view which sees men as representative also of women, and their bodies as capable of representing the woman's body with slight weight adjustments. Gender medicine offers a new viewpoint, which sees the physiology and pathophysiology of men and women and places the emphasis on the differences between them."

Why do I emphasize the subject of gender medicine in this part of my book? My involvement with the issue of sleep in women is a gendered look into the medical world, and my engagement with gender in the context of physical diseases that are more frequent in women highlights this issue.

Unfortunately, physical diseases increase with age, and so too do sleep disorders. It is difficult to think of a single disease that does not impair sleep quality, considering that the anxiety and depression that frequently accompany physical diseases, and the different medications used to treat them, can also disturb sleep.

In this chapter, I will discuss a number of problems and physical diseases, which occur more frequently in women, and that contribute to sleep disorders:

- Pain
- Headaches
- Fibromyalgia
- Parkinson's disease
- Thyroid gland imbalance
- Urinary tract infections
- Multiple sclerosis

Pain

Just as sleep disorders can be defined as accompanying symptoms of various diseases, so too can pain, a symptom that occurs in many different diseases. Pain can be acute or chronic and is determined according to the report of the person who suffers from it. Namely, there is no objective index to measure the extent of the pain. The index is solely subjective. That said, over the years, there have been developments in measuring pain by MRI scans, which allow areas of the brain that react to pain to be identified and examined, but obviously people are not sent for an MRI scan every time they complain of pain.

What do we know about gender differences in relation to pain?

The pain threshold of women changes over the years in relation to hormonal changes. Women are more sensitive to pain in the first half of the menstruation cycle (until ovulation) than in the second half. Women also suffer more from headaches before menstruation. Pregnant women are less sensitive to pain, and one proof of this is that women who suffer normally from migraines suffer less during pregnancy. The assumption is that the body, preparing for the pain of childbirth, reduces the woman's sensitivity to pain by elevating progesterone. Progesterone levels at the end of the pregnancy are very high, especially in comparison with their levels during the menstrual cycle and, more particularly, before ovulation. Women suffer more from joint pains with the decrease of hormones during the menopausal period.

In men, the pain threshold changes and decreases over the years, apparently due to a decrease in the level of testosterone.

Shifra, 60 years old, married and a mother of two daughters, is a clinical psychologist who spends many hours sitting with patients. "My sleep was never easy," she told me. "I get up at least twice a night to use the toilet and fall back asleep immediately. It's been some months now that I have not been sleeping well. It is difficult for me to explain exactly what's happening to me. I am tired during the day and have even fallen asleep in front of a patient. That was, of course, very embarrassing and it had never happened to me before. Yes, there are occasionally stresses in my daily life and worries, but nothing out of the ordinary. I do not even like taking pills for headache as I don't want to put unnecessary chemicals into my body. Yet I've reached a state where I no longer have a choice. I'm tired all the time and feel that I look really bad."

Shifra's menstrual cycle ended almost a decade ago, and she passed through menopause without any sleep disorders. In our conversation, I could not detect mental issues or medications that could explain the problem. About a quarter of an hour before the end of our meeting, when I had already begun thinking about the possibilities of treatment with medication, Shifra mentioned, in passing, that in recent months she had been suffering from pain in her shoulder. This pain worsens at night and interferes with her sleep.

I wondered why she hadn't begun her account of her difficulty sleeping by mentioning her shoulder pain and asked her about this. "Because you're a psychiatrist," she answered, "I thought shoulder pains had nothing to do with you." However, the information Shifra provided did help me treat her: in addition to the treatment she received for the shoulder pain, I gave her the drug, amitriptyline, which is widely prescribed by psychiatrists and physicians specializing in pain. The medication helped her sleep and reduced the severity of the pain.

We know that more women than men visit their family physician with complaints of pain (including menstrual pain), and women use more medication than men, including painkillers. As a result, physicians tend to disregard women's complaints of pain. In contrast, they place considerable importance on men's complaints of pain, assuming that if a man, who in general tends to complain less about pain, makes the effort to go and see a physician, it is likely to be because of something "serious." When physicians do not take women's complaints about pain seriously and dismiss them with sentences like "it isn't something serious," women too (such as Shifra) tend to agree and do not demand a more thorough investigation and an adequate solution.

Who are more sensitive to pain, men or women?

It may sound surprising, but women are far more sensitive to pain than men. The prevailing assumption that women have more tolerance to pain is grounded mainly on the experience of childbirth, during which women suffer excruciating labor pains. However, several new studies reveal that indeed women do complain more of pain throughout their lives. These studies also show that women experience the pain in more areas of the body, at a higher frequency rate, and for a longer duration compared with men. In one research study, men and women were asked to place their arm in a bath of hot water and then move it to a tub of ice water. The researchers documented the moment at which the women first felt the pain and the moment it became intolerable and discovered that women have a lower pain threshold than men.

Researchers at Bath University in England have suggested that a combination of genetic and hormonal mechanisms as well as psychological factors are responsible for the differences between the pain tolerance in men and women. The researchers also suspect that because women concentrate on the emotional aspects during pain and at times develop anxiety, which enhances the feeling of pain; while men tend to concentrate on the physical sensation alone and therefore are better able to tolerate the pain.

Pain-relieving medications are less effective in women, and they are prone to suffer more from the side effects, such as nausea and vomiting. Morphine, one of the most effective drugs for treating pain, takes longer to exert its effect in women (especially when administered intravenously) and dissipates faster. Nonsteroidal anti-inflammatory drugs, such as ibuprofen (Advil) and aspirin, are also more effective in men.

Any kind of pain will cause sleep disturbance. We must not, as women, underestimate its role as a factor in the disruption of our sleep and must find the best way to treat it, knowing that painkillers (like all other drugs) have been studied primarily in men, and thus their influence on women will differ.

Headaches

Headaches are more frequent in women after puberty. Until that time, their frequency is identical in both genders. There are different types of headaches:

- *Migraine*: The source of the word is the Greek word *hemicrania*, which means "half of the head." Migraines are almost three times more frequent in women than in men: about 17% of women suffer from migraines compared with about 6% of men. Migraine is a strong pain limited to one side of the head that can last between 2 hours to 3 days and is accompanied by nausea, sensitivity to light, smells, and noise. Usually, about 30 minutes before the onset of the migraine, precursory signs appear, such as light flashes, paresthesia in the limbs, and often even impairment of speech and half-body paralysis. Migraines are chronic and have a genetic origin. They can be aggravated by hormonal changes, but with age, and particularly when menstruation ceases, a significant improvement can be felt. Migraine attacks influence sleep and vice versa: often lack of sleep may trigger the onset of a migraine attack.

- *Tension headache*: This is caused, as its name indicates, from states of tension, when contractions of the muscles of the forehead and nape of the neck occur. This headache is more frequent in women from the age of 20 and often appears before menstruation.

- *Neuropathic headache*: This originates from a damaged nerve and is twice as frequent in women than in men. This type of headache may appear in women who suffer from diabetes, multiple sclerosis, and various malignant diseases.

- *Cluster headache*: This is a more rare type of headache that is felt around the eye on one side of the face. It is the most severe type of headache, more common in men and therefore often not diagnosed in women. As its name attests, the headache attacks arrive in clusters or one after the other with short gaps of time. After a period of attacks, there can be a pause that may even extend for several years. Cluster headaches begin mostly during sleep, usually in the dreaming (REM sleep) stage and cause the sleeping person to wake up.

Fibromyalgia

The meaning of the term is pain in the connective tissue and the muscles. The name of the condition is a combination of three Greek words: *fiber* (the structure of the tendon fibers, ligaments, and muscle envelope), *mus* (muscle), and *algie* (pain). Fibromyalgia is a chronic disease that affects 2%–8% of the population and is seven times more frequently in women. The disease is usually diagnosed between the ages of 20 and 50, although it can also appear in children. The most surprising fact is that it is estimated that 75% of fibromyalgia sufferers are not diagnosed. The syndrome appears to have a genetic background, and the risk of contracting it increases if a close family member suffers from it. Diverse traumatic events, such as physical trauma and mental stress, can cause the outburst of fibromyalgia. The syndrome highly impairs the functioning of sufferers in many aspects, personal, professional, and social.

Characteristics of the disease are as follows:

- *Chronic pain (that continues longer than 3 months)*: Pain in all muscle systems, joints, and skeleton, especially in the neck, back, legs, and arms.

- *No relief in muscle ache when resting*: Pain may even be more severe during the night.

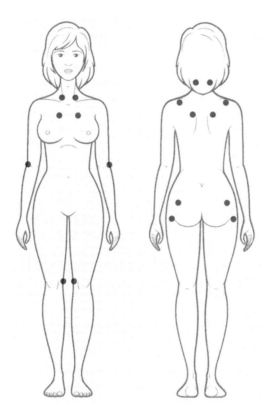

FIGURE 9.1 Sensitive points on the body.

- *Marked sensitivity to light pressure over many points on the body (18 such points have been defined)*: Most of the sensitive points are on both sides of the body (as illustrated in Figure 9.1).
- *Sleep disorders*: The disorders include difficulty falling asleep and frequent wakening through the night. Therefore, sleep quality is impaired, and the patients suffer from daytime fatigue. Around 90% of fibromyalgia patients suffer from mild to acute fatigue, accompanied by a sense of lack of energy, reduction of endurance during physical activity, or fatigue of the type felt when one is ill with the flu or suffering from lack of sleep. In many cases, the fatigue causes more suffering than the pain. Studies have shown that disturbances in deep sleep impair bodily functions as well as the process of pain evaluation. Therefore, fibromyalgia sufferers are treated with medications that improve sleep and relieve pain.
- Stomachaches and irritable bowel syndrome (stomachaches, alternating diarrhea and constipation, bloating of the stomach).
- Symptoms of anxiety or depression, and at times, also difficulties in concentration and in memory.
- Need to urinate frequently.

Even though a general physical examination will not reveal pathological findings and patients seem healthy, a thorough examination of their muscles will pinpoint areas that are sensitive to pain. Because the complaints of patients are general and resemble those of other medical disorders, many patients undergo a complex series of tests before they are diagnosed with fibromyalgia. Often, because the disease is not noticeable in the patients, and many test results are normal, family, friends, and physicians may doubt the validity of the complaints—and more so because most fibromyalgia sufferers are women. This increases the sense of isolation, guilt, and anger among patients.

Most fibromyalgia sufferers that I have seen were women aged 50 and above. Indeed, the professional literature indicates a higher incidence of the disease among women in menopause. The women complained of pain at all hours of the day, great fatigue, and a marked decline in their quality of life. Fortunately, as a psychiatrist, I can offer them help with the antidepressant medicine Cymbalta (duloxetine), known to be effective in treating depression and anxiety, but it is also effective for treatment of fibromyalgia. Another effective medication in reducing the pain is Lyrica (pregabalin), which is used for treating severe chronic pain. Physiotherapy and physical exercise—walking, swimming, bike riding—are also an important part of the overall treatment.

Parkinson's Disease

I met Aviva when she was 45 years old, and she came to see me with severe depression that manifested itself mostly as sleep disorders, especially problems in falling asleep. Aviva, married and mother of a son, immigrated to Israel from Germany about a decade ago. I found it easy to treat Aviva's depression as she responded rapidly and well to a low dose of antidepressant medication.

Aviva came to see me again when she was 51, after being diagnosed with Parkinson's disease, which was expressed by a slight tremor of her left hand and a very slow gait. Her mother was diagnosed as suffering from Parkinson's at the age of 50. It transpired that the depression Aviva had experienced in the past—a common phenomenon among Parkinson's sufferers—was the onset of the disease. Today, she is treated with Parkinson's disease medication combined with an antidepressant.

Even though Parkinson's disease is more common in men, it is important for me to relate to the sleep disorders associated with the disease because, as a psychiatrist, I meet quite a few women in their 50s who are coping with Parkinson's disease and suffering from sleep disorders.

Parkinson's disease is named after the British physician James Parkinson, who was the first to describe its symptoms. It is a perturbation in the central nervous system (the largest portion of the nervous system, which includes the brain and the spinal cord). The outbreak of the disease occurs because of a deficiency in the neurotransmitter dopamine, which, in turn, results from the degeneration of the group of black-matter cells in the brain. Patients with Parkinson's disease suffer a host of symptoms, including reduction and deceleration of gait, muscle rigidity, tremors both when at rest and during times of stress, balance problems, lack of stability when walking, and sleep disorders, which is what I focus on.

Around 70% of Parkinson's patients suffer from sleep disorders, such as disruption of continuous sleep, daytime sleepiness, difficulty falling asleep, and sleep disturbances in the dreaming stage (REM sleep).

Some of the main reasons for sleep disorders in Parkinson's disease patients:

- *Obstructive sleep apnea (OSA)*: One of the frequent causes of excessive daytime sleepiness in Parkinson's patients.
- *Restless legs syndrome (RLS)*: Common in older people including Parkinson's patients. This condition gives rise to repetitive and slow movements of the legs and feet and to unpleasant jumpy sensations in the legs, which consequently contribute to sleeplessness.
- *Sleep disturbances linked to REM sleep*: Often appear a few years prior to the outbreak of the disease itself.
- *Depression*: Very common in Parkinson's patients and is often the first symptom of the disease. Sleep disorders are part of the signs of the depression.
- *Medication used for treating Parkinson's*: May interfere with sleep. In this case, the type of medication can be changed as well as the time of its administration. All drugs should be taken before 8:00 p.m., since often dopaminergic remedies (which raise dopamine levels) lead to sleep disorders.

Men and women with Parkinson's disease suffer from the same symptoms. Studies show that in men, the disease appears at a younger age and that men suffer from more severe symptoms. Women, on the other hand, have a longer life expectancy. The reasons for this have yet to be understood.

Do the hormonal changes that women undergo during their life affect the disease? Physicians who treat women with Parkinson's detect changes in the severity of symptoms during the monthly menstruation cycle. In different studies, most women reported that they experienced a deterioration in their situation in the days prior to their menstruation and that their medication was less effective during those days. The researchers could not determine whether these reports were due to a physical-motor deterioration or due to the women's emotional state in the days before menstruation, as only a small number of the women who were examined showed an objective deterioration in relation to the low levels of estrogen present in the days before menstruation. Thus, it is not clear whether treatment with replacement estrogen during menopause reduces the risk of developing Parkinson's or provides relief to those women who already have the disease. The few studies that have been conducted have shown contradictory results.

The treatment of women with Parkinson's consists of medication with drugs that raise dopamine levels together with antidepressants, combined with treatments which include physical activities such as sports, dancing, Feldenkrais, massage, and physiotherapy.

Thyroid Gland Imbalance

Problems in the functioning of the thyroid gland are five times more common in women than in men. I always ask women during their initial consultation to bring recent blood test results, particularly the test results of the thyroid gland function. Disorders in the function of the thyroid gland very often resemble various mental disorders, and especially symptoms linked to menopause.

The thyroid gland is one of the largest endocrinal glands (glands that synthesize and secrete hormones into the blood circulatory system) in the human body. It is shaped like a butterfly and is located in the neck, on both sides of the windpipe, just under the skin.

The thyroid gland secretes hormones, the most important of which are T3 and T4. These hormones affect the metabolic rate of the body and exert a strong influence on the rate of growth and the function of many systems in our body. Another hormone produced by the thyroid gland is calcitonin, which plays an important role in the calcium economy in the body.

Women who have never suffered from any problem with their thyroid gland may see a problem develop during pregnancy or after childbirth ("postpartum thyroid gland"). The problem usually appears in the first 6 months after childbirth, and in most cases, the thyroid regains its correct balance without any intervention. Women who suffer from the problem should continue to be monitored.

The most common issues include:

- *Hypothyroidism*: This can appear at all ages. However, it is particularly common in women over 50. The assumption is that up to the age of 50, one in every 10–12 women suffers from a certain degree of hypothyroidism. At the age of 60, the incidence rises to one in every five to six women. Hypothyroidism has many symptoms because of the effect of the thyroid gland on so many body systems. The symptoms usually develop gradually and therefore it is often quite difficult to notice them. Many women tend to ignore the symptoms because some resemble the symptoms of menopause, such as sleeplessness, fatigue and exhaustion, irritability, depression, heart palpitations, and brain fog. With time, additional symptoms that are more unique to an underactive thyroid gland may also appear. These include an increase in sensitivity to cold, constipation, dry and pale skin, hoarseness, rise of blood cholesterol levels, unexplained weight gain, muscle or joint pain, muscle weakness, changes in the menstruation cycle, fragile nails, and hair loss.

- *Hyperthyroidism*: This usually appears between the ages of 40 and 60. The symptoms include unexplained weight loss, fast and irregular pulse, increased appetite, irritability, anxiety or uneasiness, trembling (mostly in the hands and fingers), perspiration, changes in the menstruation cycle, increased sensitivity to heat, increase in frequency of bowel activity, and muscle weakness. In some cases, the eyes protrude, the neck swells, and sleep difficulties appear. Hyperthyroidism may lead to difficulties falling asleep and to a feeling of hyperactivity and excessive energy.

Urinary Tract Infections (UTIs)

UTIs are more common in women, especially in young women who have become sexually active. Urinary tract infections are far more common in women than in men.

The reason may be that the urinary tract is closer to the anus in women, which makes it easier for germs to reach the bladder.

Approximately 30% of women experience at least one isolated incident of urinary tract infection during their lives, 6%–10% contract recurring infections, and at advanced ages, the frequency of the inflammations rises at a rate of 2%–3% each decade.

The connection between sexual activity and urinary tract infections in women is unequivocal. Women who have not suffered from urinary tract inflammations in their childhood begin suffering recurring inflammations when they become sexually active, and most inflammations appear within 24 hours of the sexual act.

Another age group in which the frequency of inflammation is high is women in menopause. About 20% of postmenopausal women develop urinary tract infections. One possible cause may be the decrease in the level of estrogen and, as a result, changes in the natural population of microbes in the vagina, as well as in its level of acidity. In addition, during this period, structural changes occur in the area of the vagina, such as cystocele (fallen bladder) or uterine prolapse following births, which are liable to disrupt proper emptying of the bladder.

Inflammation of the bladder is expressed in a frequent and urgent need to urinate, a sensation that the bladder has not properly emptied itself even after having urinated, a strong stinging sensation and pain during urination, pain in the lower abdomen, cloudy urine with a bad smell, blood in the urine, and sometimes also fever.

In every age group, inflammation of the bladder will disrupt sleep and cause urgent episodes of waking up with a need to urinate. In menopausal women in particular, there is an increase in the frequency of waking up at night with the need to urinate, which is linked to recurring inflammations or irritable bladder (overactivity of the bladder muscle).

Multiple Sclerosis (MS)

In 1868, the French neurologist Jean-Martin Charcot was the first to diagnose multiple sclerosis as a disease in itself. The three signs of multiple sclerosis that he described— scanning or staccato speech, intention tremor, and nystagmus (involuntary, rapid, and repetitive movement of the eyes)—are known as Charcot's neurologic triad.

The meaning of the word "sclerosis" is "scar within healthy tissue." In MS, scar-like areas are formed in the central nervous system. This sclerosis is "multiple" because the sclerotic areas are prevalent in all parts of the central nervous system.

MS is an autoimmune disease, and similar to other autoimmune diseases, the reasons for its outbreak are unknown. The commonly accepted assumption is that the disease results from a combination of several factors. The principal factors are the genetic background of the person and various environmental influences. The disease usually affects people aged 20–40 and is twice as common in women. It is not known why the disease is more prevalent in women than in men, but it is known that in general, women are more susceptible to autoimmune diseases than men.

MS is a chronic disease of the central nervous system that impairs the regular function of the nerve cells by damaging the myelin nerve fibers. The myelin assists chiefly in the transmission of electric currents between nerve cells. The damage to the transmission of the currents impairs the normal function of the brain (which is based on transmitting electric currents) and this, in turn, affects the functioning of the entire body.

The symptoms vary from individual to individual and depend mainly on the damaged area of the nervous system and on the severity of the disease. The symptoms include, among others, limb paresthesia and weakness, impairment of vision, urge and frequency of urination, tremors, speech impairment, fatigue, and loss of balance. There are patients who suffer only from minor symptoms and whose routine life is hardly affected and others who suffer acute symptoms that deteriorate over the years and can lead to severe handicaps.

In most cases, the disease is characterized by periods of attacks and intervals between them. Often, especially in the early years of the disease, symptoms such as impairment of vision may appear only during an attack and later disappear altogether. However, at times there is also neurological damage that will not disappear completely after the attack although it may improve partially.

Apart from the direct symptoms related to the disease, MS patients are at a high risk of suffering from additional diseases such as diabetes, intestinal diseases, depression and anxiety, and sleep disorders such as OSA and insomnia, which cause a state of constant fatigue. Over the years, women with MS have come to me seeking treatment for dealing with depression and anxiety, which also affected the quality of their sleep.

Irit, in her early 40s, a mother of two sons, worked with her husband in the family food-importing business. At 28, a few months after the birth of her first son, she began feeling the need to urinate frequently during the day and at night would wake up for this reason at least four times. The gynecologist she went to see assumed she was suffering from urinary tract infection, even though her urine tests were normal. A year after the birth of her son, Irit began suffering from blurred vision. After comprehensive tests, she was diagnosed with MS. The symptoms of blurred vision and frequent urination disappeared over time and her second pregnancy proceeded without any problems.

Irit and her husband often go abroad for work and visit many food exhibitions, something that involves a considerable physical effort. Over time, Irit began suffering from loss of balance and even fell a few times. During all these years, Irit was not monitored by a physician, nor was she treated. "I erased the disease from my life," she told me. "In fact, I did not believe the diagnosis the physician gave me. What did that have to do with me? I was sure it was tiredness and other phenomena characteristic of the postpartum period."

The family physician to whom Irit turned after she began experiencing problems with losing her balance reminded her of her disease. With time, other symptoms appeared—inability to control urination and difficulties walking—yet Irit still had difficulty coming to terms with the fact that she suffered from MS. Her husband also refused to comprehend that his wife was coping with a severe disease, and was even angry with her for not keeping up with the work load. The couple continued their frequent work trips.

Irit came to me at the urging of a close friend. "I am in fact doing her a favor because she keeps nagging me," she told me, "but I really don't know what I have come here for." Irit was suffering from depression, which manifested itself mostly in great sadness, loss of weight, sleeplessness, decrease of functioning, and guilt feelings about not being able to function as in the past.

Irit's treatment focused mainly on her understanding the essence of her disease, and accepting it. Only after she and also her husband accepted the fact that Irit had MS was it possible to help her introduce changes into her life that were compatible with her current level of functioning. She was also referred to a center for the treatment of MS.

Conclusion

I have described a few diseases that are more common in women and that affect the quality of sleep. There are many other diseases that comply with this definition, including the following:

- *Irritable bowel syndrome*: Twice as many women suffer from this condition as men.
- *Chronic fatigue syndrome*: Appears at the ages of 40–50 and is four times more frequent in women than in men.
- *Asthma*: From the age of 30, it is twice as frequent in women. Coughing and breathlessness may interfere with a night's sleep.
- *Heart disease*: Can impair sleep and vice versa: sleep disorders are a risk factor for heart disease.
- *Breast cancer and ovarian cancer*: Some side effects of the treatment of the disease can cause sleep disorders and tiredness. The distress and anxiety that accompany the disease can also impair sleep quality.

As I stated at the beginning of this chapter, it is hard to think of a physical disease that does not impair sleep. Correct diagnosis of the disease and precise treatment can help to attain good sleep as well.

10

Sleep Disturbances and Mental Disorders

Sleep disorders are one of the most common complaints of patients who see a psychiatrist. Indeed, it is a known fact that mental disorders almost always include sleep disorders. In some cases, sleep disorders are intrinsic to the psychiatric disorder itself, yet it often transpires that the sleep disorders are due to side effects of medication. We know from professional literature—and I know this from my own experience as a psychiatrist—that sleep disorders may be the first indication of a mental disease, before the appearance of any other symptoms.

Sleep disorders in psychiatric patients can manifest themselves in a wide variety of symptoms, such as great difficulty in falling asleep, inability to sleep continuously through the night, inability to get out of bed in the morning, and sleeping for many hours during the day. There are psychiatric patients who eat in their sleep, sleepwalk, or suffer from abnormal transitions between the different stages of sleep. Sleep disorders may also cause excessive daytime tiredness, difficulty in concentration, and a decrease in alertness.

There is a reciprocal link between sleep and mental disorders: Sleep disorders increase the risk of developing mental disorders or diseases, and conversely, sleep disorders as part of a given mental disorder or disease indicate the severity of the mental state and the need for more complex treatment. Sometimes treatment of the sleep disorder alone, without treating the underlying mental problem, can result in an improved mental state.

A woman suffering from prolonged sleeplessness can develop depression over time as a result of the sleep disorder, and conversely, a woman suffering from depression will also suffer from sleep disorders. I am not referring to a mild depression that lasts a few days but to clinical depression, known professionally as major depression. I always begin treatment by focusing on the sleep disorder, and usually in a matter of days, sleep improves, and consequently, the general mood improves as well. Only after sleep quality improves do I begin treating the mental state, which in the case of depression or anxiety, requires a longer period of time to achieve an improvement.

Differences between Men and Women in Mental Disorders

The correlation between mental problems/disorders and sleep disorders is very clear in both sexes. You do not have to be a psychiatrist in order to notice the link between sleep disorders and mental distress. Nonetheless, there is a marked gender difference in the way emotional experiences, mental problems, and mental illness affect the sleep of men and women. Women suffer more from depression and various forms of anxiety, eating disorders, type 2 bipolar disorder (more on that in the next section), and personality disorders (borderline, dependent, histrionic), whereas men suffer more from addictions, from behavior which is associated with risk taking, from a tendency to commit suicide, and from a variety of type 1 personality disorders (narcissistic, schizoid, antisocial) and the autistic spectrum.

Common Mental Disorders among Women

- Major depression
- Bipolar type 2 disorder (rapid cycling)
- Anxiety disorders (phobias, panic attacks, obsessive-compulsive disorder, post-trauma)
- Eating disorders
- Personality disorders (borderline, dependence, histrionic)

Common Mental Disorders among Men

- Various addictions (to work, alcohol, drugs, and more)
- Suicidal tendencies
- Personality disorders (narcissistic, schizoid, antisocial)
- Autism

The explanation of the gender difference is based on hereditary genetic diseases, which differ between men and women, from hormones that affect mental problems in women and also from the different gender-based cultural behavioral norms. Despite the progress we have made toward equality between the sexes, men are still expected to "appear strong" and to not externalize emotions. This is why many men deal with depression or anxiety by repressing their problems, by addiction, and by a tendency for risk taking. In contrast, there is much less expectations that women "be strong," and "weakness" is accepted as a normative part of the female identity. It is accepted by society that a woman can be depressed or anxious and often turns to a psychologist or a psychiatrist for help. Consequently, some would claim that in fact there is no gender difference between men and women regarding depression or anxiety, but there is only a difference in the expression of the symptoms. For example, a man suffering from depression will be aggressive, abuse alcohol or drugs, and work long hours, whereas a woman will have difficulty functioning, will externalize her difficulties by sharing them with those around her, and will turn to professional help. Moreover, since men tend to seek help less than women, the available data on depression and anxiety that comes from physicians and clinics refer mostly to women.

And what about sleep? Is it possible that the difference is not a gender issue but a culture issue? Namely, is it that men do not complain about sleep disorders and do not seek treatment but self-treat themselves by using alcohol or drugs or by working long hours? Unfortunately, the medical literature does not deal with this issue. This is not surprising considering that the subject of gender medicine is still in its early stages.

The Biological Link between Sleep Disorders and Mental Disorders

When mental disorders are present, for various reasons, the normal functioning of the biological clock is disrupted. (The biological clock is a mechanism in the central part of the brain that regulates and synchronizes a large portion of our daily functioning.) Many biological processes may be disrupted, among these the secretion of cortisol (stress hormone), the secretion of melatonin (sleep hormone), and the secretion of neurotransmitters (noradrenaline and serotonin), which control moods (Figure 10.1).

The disruption of the balance in the circadian rhythm is tied to several mental disorders, including depression, manic depression, seasonal affective disorder, and schizophrenia. Over 90% of patients with clinical depression suffer from sleep disorders, abnormal secretion of hormones, heart function irregularities, and the regulation of body temperature. Similar disturbances appear also in those who suffer from manic depression.

In this chapter, I will address three common mental disorders in women, which are accompanied by sleep disorders:

- Anxiety disorders
- Major depression
- Manic depression, bipolar disorder

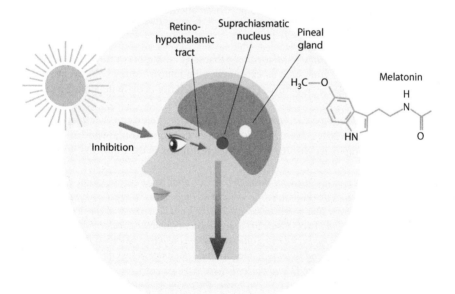

FIGURE 10.1 Neurotransmitters and the biological clock.

Anxiety Disorders

Anxiety disorders are the most common mental disorders. These are a group of phenomena during which certain states or objects are experienced as extremely threatening and dangerous, causing emotional distress, avoidance of action, or impairment of functioning. It is important to know that anxiety is a normal component of life, and at times, it alerts us to danger. However, for some people, this is an ongoing problem that interferes with their daily functioning.

Like many mental diseases, the specific reason for the onset of anxiety disorders is not entirely understood. The general causes are numerous and may include genetic tendencies, disruption of the brain's chemical balance, and experiences encountered during life (such as traumatic experiences). At times, anxiety derives from a background related to a medical condition. In some cases, the symptoms of anxiety are the first symptoms that indicate the presence of a medical disease.

Anxiety disorders, such as depression, are almost twice as frequent in women. It is estimated that between 5% and 10% of the population suffer from anxiety disorders of different intensities at any given moment.

There are several types of anxiety disorders:

- *Generalized anxiety disorder (GAD)*: Continuous and exaggerated worrying about almost every occurrence in life.
- *Panic attacks*: Short and strong attacks characterized by an acute sense of fear, loss of control, and at times even the fear of a heart attack or of going crazy. In addition, physical symptoms such as accelerated heartbeat, sweating, and trembling may appear.
- *Phobia*: Excessive fear of an object or a situation such as animals, flights, elevators, speaking in public, crowded places, and heights, which leads to avoidance.
- *Obsessive-compulsive disorder*: Obsessive thoughts and uncontrolled repetitive behavior, such as extremely long hand washing to prevent contracting a disease, repetitive checks of locked doors or switched off ovens.

In his book *My Russian Grandmother and Her American Vacuum Cleaner: A Family Memoir* (2011), Meir Shalev describes Grandma Tonia who suffered from an obsessive-compulsive disorder regarding cleanliness in the following way:

First, she insisted that people enter through the back door and never the front, because if a guest came in through the front door, he would find himself in the preserved and forbidden part of the house. Every time someone knocked at the front door, her resolute cry issued from inside the house "Around back! Come in the second door" and the guest would have to circle the house—without letting his foot step onto earth from the paved pathway, thereby bringing in mud or dust—only to find that there, too, he could not enter unless he was a particularly important and special visitor. Grandma Tonia liked having guests, but her hospitality did not include actually allowing guests into the house, she preferred entertaining outside.

- *Social anxiety*: Excessive discomfort or avoidance of certain social situations, such as speaking in front of an audience, eating in company, or social situations in general. It is interesting to note that the actress and singer Barbra Streisand suffered from stage fright, after having forgotten the words of a song during a concert in Central Park in 1967. She then avoided appearing on stage for 27 years.

- *Post-traumatic stress disorder (PTSD)*: This is a situation following a traumatic event, such as an accident, a terror attack, or a sexual attack that can lead to anxiety, over-alertness, avoidance, dreams, or intrusive flashbacks of the event.

Sleep and Anxiety

The link between sleep disorders and anxiety seems obvious, and in fact, one of the most genuine signs of states of anxiety are sleep disorders, such as difficulty falling asleep due to disturbing thoughts, general anxiety that prevents calming down and entering into sleep mode, intermittent sleep, waking up with panic attacks, and being unable to fall back to sleep. All these are among the most frequent occurrences in women suffering from different types of anxiety. Furthermore, after nights without sleep, the anxiety of not being able to fall asleep is added to the general anxiety. Many women who use sleeping pills or sedatives regularly see me with the intention to stop taking medication, out of fear of dependency. Yet at the same time, they are very anxious about any change in their medications, fearing their anxiety will increase or that they will once again suffer from sleepless nights.

Patients suffering from anxiety may experience frequent episodes of waking up due to overactivity of the sympathetic (autonomic) nervous system. The sympathetic nervous system innervates different organs and is active mostly at times of emergency, stress, and sudden threat. Consequently, the autonomic nervous system causes physical changes such as elevated heartbeat rate, increased supply of blood to the muscles, reduced bowel activity, constricted blood vessels, and elevated blood pressure. All these, naturally, affect sleep.

This is how my friend, Adi, described, in her poetic language, the sleeplessness that she's been suffering from for many years due to anxiety:

My eyes begin to shut. I begin to crash in front of the computer screen and my head droops, I attempt to hold on for a bit more, not to reach the battle scene too early, the place where there will be a war of attrition between my tiredness and my thoughts. My war over sleep.

The computer moves to "sleep." I envy it. I drag myself to the shower and to bed, plump up my pillows so I'll be comfortable, sniff the intoxicating scent of the fresh sheets. Within the coziness I've created, I close my eyes and imagine myself floating in a dark pool, moving slowly with the flow of the water, trying to skip between the stages of sleep. And then into this calm quiet a cruel dialogue between tiredness and thoughts intrudes.

I am almost asleep, and it arrives: "Hi, I'm here—your intruder, I have a lot to say. Let's choose a subject. I have a wonderful suggestion: your father will fall and break some vital organ in his body and your entire parental system will crash. What do you say, shall we elaborate on this?" I don't manage to think too deeply about it and she continues with a host of ideas: "Do you remember the night we spent together, when your brother left the physical world? Let's create another thought disaster and continue to meet every night. And it's not that I'm threatening, just giving you food for thought."

The anxiety scares me, and I try to repress the distressing intruder that brings it to me and to return to the imaginary pool in which I tried to float a few seconds ago. This is where I want my thoughts to take me.

I think I should get up and go to the toilet. Maybe that will spare me waking up in the middle of the night. My legs feel cold and I am cold under the blanket. I look for the hairdryer on the cabinet by the bed, but it's loaded with things and covered by the clothes I took off earlier. I find the relevant electrical cord and pull it free, all the pile that includes chargers, clothes, important papers, reading lamp and the hairdryer falls on the dog, who goes into a session of scratching. I hold the hairdryer in my hand, I turn it on to the maximum under the duvet. A lovely warmth begins to enter and warm my legs but after a few seconds the hairdryer chokes and stops working.

I am tired. Yearning for some sleep and I cannot stop thinking about it.

The cat, who escaped from my bed, lands on the keyboard on the desk and wakes up the computer. Now we are three.

Treating Anxiety and Sleep Disorders

Treatment is provided, of course, after a comprehensive evaluation of the possible sources of the anxiety.

- *Psychological therapy*: Mainly cognitive behavioral therapy (CBT), short-term therapy that focuses on thought patterns and is very effective in states of anxiety; biofeedback (therapy via a system that allows patients better control over their body, as well as over their physiological and emotional reactions in situations of pain, stress, and anxiety); and psychotherapy.
- *Medication*: As with every mental disturbance, it is recommended to begin by treating the sleep disorder with sedatives or sleeping pills (detailed in Chapter 11 which deals with treatment of sleep disorders).
- *Soothing means*: Yoga, guided imagery, relaxation, aerobic sports, and acupuncture.
- In cases of severe anxiety disorders, it is important to consider administering an antidepressant or antianxiety medication such as sertraline or Lexapro, in addition to the other therapeutic means mentioned before.

Major Depression

Many define "major depression" as the epidemic of the twenty-first century, and various data show that cases of depression are constantly on the rise. Major depression is not a state of a passing depression or temporary sadness. Major depression is characterized by a constant feeling of very low mood, deep sadness, hopelessness, and despair. Mood swings that occur in major depression last for more than 2 weeks and sometimes even longer. At times, this may even last for months. Some people might suffer from a major depression once in their lifetime; others might experience this situation several times during their lifetime. As mentioned previously, the incidence of depression in women is two times higher than in men, and with age, this ratio increases to almost four times.

Symptoms of Major Depression

- Bad mood, profound sadness, despair
- Exhaustion, lack of energy during most the day and almost every day
- Feeling of low self-esteem or guilt most of the day and almost every day
- Inability to concentrate and make decisions
- Insomnia or excessive sleep almost every day
- Loss of interest in things that were once a source of interest and enjoyment
- Restlessness or slowness
- Frequent thoughts about death and suicide
- Significant changes in body weight

Sylvia Plath, the American poet and author, who committed suicide after suffering from depression for many years, wrote,

> *Dying*
> *Is an art, like everything else.*
> *I do it exceptionally well.*
>
> *I do it so it feels like hell.*
> *I do it so it feels real.*
> *I guess you could say I've a call.*

> *Excerpt from "Lady Lazarus"*

Sleep and Depression

Sleep disorders will appear in 75% of the cases of depression. They are one of the risk factors of suicide. Sleep disorders usually manifest themselves as difficulty in falling asleep, frequent waking up, or waking up early in the morning. Sleep disorders during depression can also be expressed as excessive sleep while turning the night into hours of wakefulness and the day into hours of sleep.

The brain activity of people prone to depression is different from that of healthy people. During the day, signals that encourage alertness are sent from the circadian system (our biological clock) and help us stay awake. These signals alternate at night with signals that encourage sleep. Our brain cells also act in a cyclical manner and are especially sensitive to stimuli in times of alertness and relatively indifferent to them at times of sleep. Among those prone to depression or manic depression, these cyclical fluctuations are attenuated or even nonexistent. Depression is also characterized by changes in the daily cycle of hormone secretion and in the regulation of body temperature (Figure 10.2).

I have been treating Rachel, in her mid-fifties, for 20 years, since the birth of her eldest son. She came to me due to postpartum depression, which I treated with antidepressants. Since that time, Rachel has suffered recurring episodes of depression and therefore continued the antidepressant medication throughout the years, with slight changes in dosage.

One of the pronounced features of Rachel's depression was the insomnia from which she suffers. Over the years, she managed to fall asleep only toward 4:00 or 5:00 a.m. and woke up

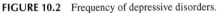

FIGURE 10.2   Frequency of depressive disorders.

at 4:00 in the afternoon. Rachel, educated as a lawyer, worked at translating legal documents from English to Hebrew and therefore could allow herself flexible working hours. However, the change in her sleeping hours, which was in contradiction to her biological clock, led to disturbances in concentration and memory. Eventually, she was fired from her job.

Rachel found it very difficult to change her sleep habits. For those suffering from depression, the morning hours are the most difficult hours, and in the evening, there is some relief. Therefore, Rachel deals with her depression by escaping into sleep during the morning hours. Rachel's treatment, which continues to this day, is a combination of medication and behavioral therapy aiming to ease her depression and to change her sleeping hours. Unfortunately, it is hard for me to say that the treatment is producing good results. There are short periods in which Rachel manages to get herself out of bed toward noon, yet for the most part she persists in her sleep until the late afternoon hours.

In chronic depression, such as that experienced by Rachel, daytime hours are a threat. "In the evening and at night no one expects me to function," so tell me many patients who are sufferers of depression, "and therefore I find it much easier to cope during these hours."

Treating Depression and Sleep Disorders

- *Short-term cognitive behavioral therapy (CBT)*: This can be useful, especially for treating the negative thought patterns and sleep difficulties.
- *Medication*: As with every mental disease which is affected by sleep and which affects sleep, it is recommended to treat the depression (with antidepressants) as well as the sleep disturbance. As I mentioned at the beginning of this chapter, alleviating the sleep disorder will ease the severity of the depression. The use of antidepressants that also help sleep disturbance (amitriptyline, mirtazapine, trazodone) should be considered. Sometimes antidepressants cause sleep disturbance and therefore it is important to pay attention to the hour they are taken and to find the best suitable time for each patient individually, rather than adhering to the pharmaceutical company's guidelines.
- *Light therapy*: The beneficial effect of light on human health has been known since ancient times. Studies show a link between light and depression in general, not only seasonal affective disorder (SAD) (a tendency of people, especially women, to fall into depression with the changes of seasons). The insight is that in a state of depression, there is a disruption in a number

of biological rhythms including light and darkness (day and night). Light treatment in modern psychiatry developed only in the 1980s as a quick and effective therapy, especially in cases of SAD. Multiple studies indicate that exposure for 30–60 minutes to clear white light (daylight, a standard light bulb) in the early hours of the morning can significantly improve mood in a matter of days.

- *Sleep deprivation*: Studies have revealed that lack of sleep actually has a positive effect on people who suffer from major depression.

A team of researchers from the University of Pennsylvania, led by Philip German, analyzed studies from the last 30 years and found that partial lack of sleep (sleep for 3 or 4 hours, after which the patient is kept awake for 20–21 hours) and complete lack of sleep (preventing sleep for 36 hours) are effective in reducing the symptoms of depression. It must be emphasized that in all these studies, the deprivation of sleep was done under the supervision of physicians and during hospitalization. In all the studies it was found that in 40%–60% of patients, there was a marked improvement in the depression symptoms within 24 hours of treatment, in contrast to antidepressant medications that began to show an effect only after a few weeks.

Professor Stin Stingrimson, a psychiatrist at the Swedish University Hospital, offers this explanation for treatment through sleep deprivation: "Depression is identified with changes in the daily cycle of hormone secretion and in the regulation of body temperature. People in deep depression do not react in the same way to an increase in levels of melatonin (hormone that promotes sleep) in the evening, and their cortisol levels (hormone that promotes activity of many body systems, including digestion and stress) remain high even at night." It seems that sleep deprivation corrects the sleep–wake cycle.

Treatment for sleep deprivation can indeed help those suffering from depression, yet most studies show that the relief is short term; improvement may be felt for a few days or a week, after which the symptoms return. The preferred treatment is a combination of medications combined with other therapies.

Bipolar Disorder

Key Redfield Jamison, an American psychologist and writer who suffers from manic depression, writes in her book *An Unquiet Mind: A Memory of Moods and Madness* (1995):

I have often asked myself whether, given the choice, I would choose to have manic-depressive illness. Strangely enough I think I would choose to have it. It's complicated. Depression is awful beyond words or sounds or images.

So why would I want anything to do with this illness? Because I honestly believe that as a result of it I have felt more things, more deeply, had more experience, more intensely, loved more, and been more loved, laughed more often for having cried more often, appreciated more the springs for all the winters, worn death "as close as dungarees" appreciated it—and life—more, seen the finest and the most terrible in people, and slowly learned the values of crying, loyalty and seeing things through.

But normal or manic I have run faster, thought faster, and loved faster than most I know. And I think much of it is related to my illness—the intensity it gives to things and the perspective it forces on me. I think it has made me test the limits of my mind (which, while wanting, is holding) and the limits of my upbringing, family, education and friends.

Manic depression, also known as bipolar disorder, is a mental disorder characterized by extreme mood swings. A person suffering manic depression moves between poles of euphoria and hyperactivity, called mania or hypomania, and periods of depression expressed as low mood, lack of interest and enjoyment, and pronounced disruption of functioning. The duration of these periods ranges from a few days to a few months. Manic depression tends to appear between the ages of 15 and 30.

It is estimated that about 1% of the general population suffers from manic depression. A different estimate suggests about 6% of the population suffers from the disease, as there are scientists who hold that manic depression occurs on a spectrum and many additional people suffer various forms of the disorder.

Manic depression is ranked by its severity:

- *Type 1 bipolar disorder*: A syndrome that includes manic episodes during which the person experiences psychotic symptoms, may lose control over their behavior, and in some cases may have to be hospitalized.
- *Type 2 bipolar disorder*: A syndrome that includes hypomanic episodes, without psychotic symptoms, and behavior that is usually not dangerous.

The depressive episodes in both types of bipolar disorder can be similar. What differentiates them is the existence or not of a psychotic state.

- *Rapid cyclic bipolar disorder*: This is the most severe form of the disease. The patient undergoes four or more episodes of depression, mania, hypomania, or a mixture of all four within a year.
- *Combined states*: During a manic-depressive attack, the patient will experience symptoms of mania and of depression at almost the same time.

Type 1 bipolar disorder occurs with the same frequency in women and in men; however, type 2 bipolar disorder and rapid cyclic bipolar disorder are more common in women. More episodes of depression than manic states are observed in women, whereas in men manic states are more common. The first appearance of manic depression, or an increase in the severity of the disease, coincides with two milestones in a woman's life: postpartum (postpartum psychosis) and after menopause. The appearance of manic depression after menopause is a phenomenon that has not yet been studied enough.

What Are the Causes of Manic Depression?

- *Biological factors*: Those suffering from manic depression are characterized by changes in the biochemistry of their brain activity, changes in the hormonal balance of the body, and changes in sleep patterns.

- *Genetic factors*: The risk of contracting the disorder increases if is another family member has been known to suffer from it.
- *Psychological factors*: The psychodynamic approach, which attends to reciprocal activity of the conscious and the unconscious, has tried to find explanations for the disorder. The most famous one is the "manic protection theory," according to which the mania is a protective mechanism from self-harm, depression, and despair.

Varda, 55 years old, married for the third time and mother of a daughter, is a well-known international artist. A happy, energetic woman, who enjoys alcohol and entertainment and needs very few hours of sleep. She is always at the center of attention and frequently travels around the globe.

Varda suffered in the past from several episodes of depression and was treated at intervals with antidepressants. She stopped the medication, against the advice of her psychiatrist, because she felt that the medication made her "not myself." She felt that the drugs were harming her creative abilities and particularly her main identity—her professional identity.

At 45, following a total hysterectomy, she begun to suffer from hot flashes, sweating, and sleep disorders. A few months later, other symptoms appeared, including rapid speech fluency, reckless spending of money, need for very little sleep, extremely happy and elevated mood, and especially delusions that she was about to win the Nobel Prize for her creative work. Varda was suffering from a manic state, which appeared for the first time with the onset of menopause and needed treatment. However, it was very difficult to convince her of that, since she claimed that she had never felt better. When her situation deteriorated, there was no alternative but to hospitalize her for a short period of treatment with antipsychotic medication, which helped restore the balance of her mental state.

Researchers at the Karolinska Institute in Sweden found a connection between activities in the creative, artistic, or scientific professions—for example, dancers, photographers, designers, writers, and researchers—and manic depression. In an article published in October 2012 in the *Journal of Psychiatric Research*, they provided some explanations for this link, among them the known connection between creativity and a tendency for associative thinking, which is also characteristic of various mental disorders, especially schizophrenia, manic depression, and autism.

"The findings give us reason to think again about attitude toward mental disorders," says PhD student Simon Kyaga of the Karolinska Institute, the lead researcher of the study. "If we consider that the disease might be beneficial to the person, it opens the way to a new approach to treatment. In these cases, the physician and patient need to arrive at an agreement as to what will be treated and at what cost."

Indeed, I always face a dilemma regarding treatment when dealing with a patient who suffers from manic depression, whose creativity is expressed chiefly in hypomanic states. She comes for treatment when she is in the depressed state and needs help, yet fears that the treatment will harm her creativity. Usually, the suffering created by the depression leads patients to use medication, yet the moment there is an improvement and at times transition to a hypomanic stage (which may be the result of medication), the patients stop the treatment and disappear.

The British writer Virginia Woolf, who committed suicide at age 59, suffered her entire life from cycles of physical and mental symptoms of disease: Feelings of anxiety, great sense of loneliness, despair, guilt feelings, hallucinations, threatening voices in her mind, loss of appetite, weight loss, severe headaches, fever, and rapid heartbeat. Many researchers are convinced that the changes in her mood, which swung between severe depression and the excitement of mania with episodes of psychosis, would today be diagnosed as manic depression. In her diaries and letters, Virginia Woolf described the strange link between the attacks of "madness" and the attacks of creativity, as in this fascinating paragraph: "I believe that these diseases, in me are—how do I express it?—mystical in part. Something happens to my spirit. It refuses to continue to hold images within. It closes. It becomes a chrysalis…." (diary entry, February 1930).

Sleep and Manic Depression

One of the most characteristic signs of manic depression is changes in sleep; these are often the first to appear. Of those suffering from the disease, 69%–99% report a decrease in their need to sleep during manic or hypomanic episodes. People in manic or hypomanic states feel full of energy and vigor and do not feel the need to sleep at all. A woman in a manic state may recount that at night she answers emails, cooks, cleans, and tidies the house and does not even feel that something is wrong. With the transition to the depressive state of the disease, there can be a sudden change to excessive sleep or very low-quality sleep, marked with many episodes of waking up during the night and pronounced daytime tiredness. Often, the state of lack of sleep can be a catalyst for the outburst of the manic state.

In recordings of the sleep architecture in manic-depressive patients, a shortening of the dream stage (REM sleep) is observed. The sleep disorder will persist, at a lower intensity, even when patients are balanced, namely, in neither a manic, hypomanic, or depressive state. This fact has led researchers to hypothesize that those suffering from manic depression may have a problem with their melatonin production (the hormone that affects sleep).

Treating Manic Depression and Sleep Disorders

The treatment of the disease is complex. The most effective treatment is with medication.

- In manic and hypomanic states, treatment with antipsychotic drugs (such as Zyprexa and Seroquel) will affect both sleep and the manic states. At times, there is a need to add minor tranquilizers.
- In depressive states, antidepressants will help (such as Lexapro and sertraline) together with antianxiety pills or sleeping pills for immediate help with sleeping.
- Mood-balancing medications (such as Lamictal and lithium) are recommended for equilibrium and maintaining a balanced mood over time, together with other drugs.
- Addition of melatonin is possible, especially under the assumption that manic depression involves a decrease in melatonin production.
- Cognitive behavioral therapy (CBT).

And Something to End With

In her book, *A Beautiful Mind*, a biography of the mathematician John Nash, who was diagnosed with schizophrenia, Sylvia Nasar writes about a meeting between Nash, hospitalized in a psychiatric ward, and a colleague who visited him and asked how he—a mathematician, a man dedicated to logic and logical truth—could believe that aliens were sending him messages. Nash's answer was that his ideas about supernatural beings came to him in the same way as the mathematical ideas and therefore he treated them seriously.

"Certain people see things others are unable to see, and they are right. We call them 'creative geniuses.' Certain people see things that others cannot see, and they are wrong. We call them "mentally ill.' And some people, such as John Nash, are both."

11

Treatment of Sleep Disturbances

Treatment of sleep disorders appears easy and simple: Why do we have sleep medication if not to help us get a good night's sleep? Many of us are familiar with the most common sleep medications—Ambien, Stilnox, and others—just as we know the names of sedatives, such as clonazepam and alprazolam, that can also help us sleep. One pill and here comes the long-desired sleep.

However, treating sleep disorders is not so simple. It usually requires a complex solution that takes into consideration both a change in life habits including sleep habits, and also consideration of the side effects of the medication prescribed to induce sleep.

I encounter many women who suffer from extended periods of sleep disorders, and by the time they see me (one doesn't go to a psychiatrist immediately as it's too frightening), they're already suffering from prolonged sleep disturbance, are taking various medications and suffering from their side effects, and most importantly: They have already developed behavioral disorders which are connected to their chronic sleep deprivation problem. By this I mean faulty sleep habits, anxiety about sleeping, night eating, working at night, sleeping during the day instead of at night, and a belief that has become embedded in their mind that "I won't fall asleep unless I take a pill."

The first step in treating sleep disorders is to attempt to understand the source of the disorder and to treat this in parallel with the underlying disease. As I repeatedly point out, treating the sleep disorders can also ameliorate the underlying disease that is causing the disorders. After understanding the source of these sleep disorders and treating them, it is always important to examine the sleep hygiene and to instill a change in sleep habits.

Sleep Hygiene

The theory known as "sleep hygiene" determines that just as we take care of the cleanliness of our body, so too can we concern ourselves with "clean sleep," and from this, the term "sleep hygiene" is derived. This term may seem a little strange, but what we mean is to be very meticulous about a number of rules before going to bed (just as we are careful to always brush our teeth). These rules are completely logical and can also considerably improve the quality of our sleep.

- *Hours of sleep*: Each of us has a natural biological clock that makes us feel sleepy when we are ready to sleep. It is important to maintain regular bedtime and waking times every day. Keeping to a routine in cycles of sleep and wakefulness helps maintain a balanced hormonal cycle and regular biological clock. Even during times of change in our daily routine, for example, weekends and holidays, it is important to try to go to bed and to get up as closely as possible to our usual hours. For some reason, we tend to think that sleep is not an integral part of our daily schedule, and if something has to be sacrificed for the completion of some task during a period of time pressure, we are only too willing to give up on our sleep.

- *Sleep ambience*: It is important to make sure that the bedroom exudes an atmosphere that is pleasant and soothing: A comfortable bed, a pleasant temperature, and a room that is sealed off from noise and light with no clock in sight. Women tend to object when I ask them to move the clock away from the night table or to remove of the bedroom altogether. I explain that the presence of the clock increases stress when one is having difficulty falling asleep or when one awakens during the night.

- *Prebedtime habits*: It is recommended not to perform any stimulating activities before going to bed (a competitive computer game, watching an exciting television program, sport, or an argument with one's partner). For a few hours before sleep, it is important to engage in tranquil activity, such as reading a book or watching a pleasant television program, and to avoid physical activity. One should also avoid eating a heavy meal close to bedtime, and it is best to keep away from alcohol about 6 hours before bedtime. Alcohol may help one to fall asleep, but it has several detrimental effects: It causes premature awakening and disrupts one's sleep by making one get up to go to the toilet. Sex before sleep is certainly recommended: Sexual activity induces a pleasant feeling and helps get a good night's sleep.

- *Coffee before bedtime*? Drinking a cup of coffee in the evening contradicts the natural activity of our biological clock. Researchers have found that caffeine slows down the function of the body's biological clock. A double espresso, 3 hours before bedtime, will delay the production of the sleep hormone melatonin by 40 minutes and lead to difficulty falling asleep.

"If you are tired and drink coffee at night in order to stay awake, it is a bad idea, because you will then have far more difficulty falling asleep and you will not enjoy sufficient hours of sleep," explains John O'Neil of the Laboratory of Molecular Biology at Cambridge. Study results indeed prove that it is better not to consume caffeine during the late hours of the day, yet for some things, such as long airplane flights from east to west, it can be helpful. "If you consume caffeine at the right time of the day," O'Neil explains, "it could reduce the time it takes to overcome jetlag."

- *The bed is not an office* ... The brain makes connections between places (bedroom) and events (sleep). The habit of getting into bed with the smartphone or laptop, and to run one's entire life using these, causes wakefulness and interferes with sleep. In addition, the screen light, held close to the eyes, interferes with the secretion of melatonin, which is secreted in response to darkness.

- *Get out of bed or insist on falling asleep?* One of the dilemmas when having trouble falling asleep is whether to get out of bed and risk losing the sleep momentum, or to turn over and hope that in one second, sleep will arrive. The problem is that the more effort we make to fall asleep, the higher our level of alertness, and sleep slips away. It is recommended to get out of bed, leave the bedroom, and find something to do that will calm one down and induce sleepiness, such as reading a book.

- *Afternoon nap?* Sleeping in the afternoon is a common custom in hot countries, in order to escape the intolerable afternoon heat. In Spain, the "siesta" is taken seriously, and the term has found its way into other languages. The western lifestyle has made the afternoon nap much rarer.

Professor Giora Piller, a pediatrician and expert on sleep disorders from the Carmel Medical Center in Israel, imposes some order on the various studies:

"There are a few nonorganized studies that showed that an afternoon nap is not recommended, especially for adults, because it raises the risk of their dying from heart disease. However, other studies show the exact opposite, that sleeping in the afternoon contributes to the health of the heart. In conclusion, each person [...] must decide for himself or herself if sleeping in the afternoon is good for them or not."

Cognitive Behavioral Therapy (CBT)

CBT is a short-term treatment during which therapists who specialize in this type of treatment attempt to understand the source of the problem—in this case the mechanism that prevents sleep—and, together with the patients, to alter the thought and behavioral patterns that interfere with their sleep. Thought patterns that have become firmly established in the mind of a patient who is having trouble falling asleep lead to a situation whereby the bed becomes a "battlefield" and going to bed is perceived as a threat. CBT is very useful for identifying these thought patterns and for creating alternative thought patterns, as well as for providing prebedtime relaxation methodologies to reduce the anxiety associated with going to bed.

CBT is effective in treating many disorders, but in my experience of treating sleep disorders, it is effective by itself only in cases of mild sleep disorders. When it comes to prolonged sleep disorders, it should be combined with medication.

Hila, 55 years old, a piano teacher, married and mother of two grown children, is a very energetic women who works long hours at the music conservatory and as a private music teacher at home. Her husband has been unemployed for several years now, and she is responsible for the family's livelihood. Hila is also the only child of two Holocaust survivors and has suffered from generalized anxiety all her life. Hila's sleep is not good: it takes her an hour to fall asleep, and during the night she wakes up almost every hour and checks the clock in fear that she will get up too late, and eventually she does not get more than 4–5 hours of sleep per night. When I met her, I saw, clearly, that it would not be right to offer her medications only, since I was aware of her thinking and behavior before going to bed, which demanded an intervention of a different nature. As soon as she went to sleep, Hila would be flooded with thoughts, about the next day's assignments, her parents' health problems, a pupil she was having trouble teaching well, and so on. During the treatment with CBT (in addition to medication for reducing her anxiety), Hila became more aware of the thoughts that disturbed her sleep, learned presleep relaxation techniques, and changed her habits, including removing the clock from her bedroom.

Pharmacological Treatment for Sleep Disorders

Pharmacological treatment for sleep disorders focuses on two problems: Falling asleep and remaining asleep.

In a survey of sleep-inducing medications, published in 2010 in the Israeli *Journal of Sleep Medicine*, neurologist Dr. Baruch Elad presented a short survey of the history of the development of sleep medication:

"Since ancient times people have been using different means to induce sleep. Apart from relaxation techniques and massages with various oils, prevalent in the Far East, it was common around the world to use herbs such as mandrake bark, lettuce juice, valerian essence, and others. In the 19th century two types of synthetic compounds were added: chloral hydrate, which is still in use for sleep induction, especially in children, and bromide, which was introduced into medical practice initially for treating seizures and later as a sleep-inducing medicine. Toward the end of the 19th century paraldehyde was introduced and its widespread use continued until the beginning of the 20th century. At the beginning of the 20th century barbiturates were developed for the treatment of epilepsy, difficulties falling asleep and sleeping."

The modern history of sleep-inducing medication begins with the introduction of benzodiazepines—at the beginning of the 1960s—which helped induce sleep and also reduce anxiety and were safer than drugs of the barbiturate drug family. Medications of this group include, for example, alprazolam, diazepam, flurazepam, oxazepam, and clorazepate. In the 1980s, another class of drugs was developed, Z-drugs—these are not benzodiazepines. Drugs from this family include Stilnox, Ambien, Nocturno, and Imovane. This class of drugs is considered to have a shorter half-life (the time it takes the drug to be eliminated from the body until its level in the blood decreases by half) and therefore to be more suitable for treating sleep disorders.

The drugs of the benzodiazepine and the Z-drug families are certainly effective in inducing sleep. The main concern regarding their use is that long-term use can cause tolerance (in which case the dose has to be increased), addiction (when it becomes impossible to sleep without taking the pill), episodes of falling (especially among the elderly), effect on breathing (particularly in those suffering from sleep apnea), diminished concentration and memory, and, at times, may also create changes in the mental state. The longer the half-life of the drug, the greater the risk of side effects on the following day.

In the late 1980s, there was a clear tendency to use antidepressants to induce sleep. This was probably based on the assumption that sleep disorders are part of depression and that these drugs do indeed affect sleep as well. Some antidepressants (trazodone, mirtazapine, amitriptyline) do indeed induce sleep as a side effect. In addition, in contrast to benzodiazepines, the prescription conditions for antidepressants do not have a time limit, and there is a lower risk factor for developing dependence and tolerance. Nevertheless, these drugs are not entirely without side effects, such as weight gain (mirtazapine), anticholinergic symptoms (dry mouth, constipation, blurred vision, and urine retention), cardiac side effects (increased heart rate, cardiac arrhythmia), drowsiness, and lowered blood pressure.

At the beginning of 2000, two new products that increase the release of melatonin came into use: ramelteon and Circadin; the latter was developed by the Israeli drug company Neurim. These drugs, which increase the secretion of melatonin, are meant

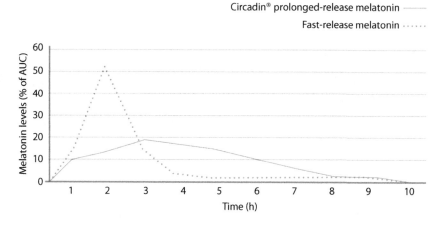

FIGURE 11.1   Circadin vs. immediate-release melatonin. (Data courtesy of EPAR Assessment Report on Circadin.)

mostly to maintain sleep throughout the night. This involves delayed release of melatonin, which imitates the natural secretion of melatonin from the pineal gland in the brain. Circadin is a prescription drug that should not be confused with over-the-counter melatonin, which does not mimic biological melatonin secretion and is sold mainly in the United States.

Circadin is a groundbreaking sleeping pill. Unlike the older generation drugs, which induce sleep by repressing the activity of the cerebral cortex, Circadin mimics the body's natural sleep mechanism. The medication shortens the time it takes to fall asleep and also improves sleep quality and alertness on the following morning. Unlike other sleeping pills, Circadin is nonaddictive and does not affect sleep architecture (Figure 11.1).

Professor Nava Zisapel, the founder and CEO of Neurim, described the process that led to development of the drug in an interview to the economic daily *The Marker* in 2009:

"Sleep always interested me. It is a big and important process, and until the 1980s there was no hypothesis about the way it works. A lecture I heard by chance in the early 1980s, about sex in crabs, inspired my initial idea. The lecturer said that during the mating season crabs release different hormones into the blood in males and females. As a result, the muscles of the male crab begin to stick up, whereas in the female crab the muscles become lax and her body shortens. I suddenly realized that a hormone can perhaps influence nerve conduction and I decided that I must find a hormone that affects nerve conduction at night and that causes sleep. In 1991 the results of the basic research evolved into an idea for a patent and then we decided to turn it into a product."

Other Sleep-Inducing Products

- *Antipsychotic products*: Especially products referred to as "atypical," such as quetiapine and olanzapine, which are used primarily as antipsychotic medications for treating psychiatric patients but can also enhance sleep.

- *Antihistamines (nonprescription)*: Many products are available at pharmacies, for example, Sleep-Aid and Tonight. Their long-term efficacy has yet to be determined.
- *Various herbal treatments*: Valerian (data on this plant are inconsistent), Bach flower essences (Rescue), and essences of passion fruit, melissa, and chamomile.

At a conference on herbal treatment for mental health problems in women I attended in New York, one of the lecturers shared her technique for getting her daughter and friends to sleep when they come over for a pajama party: she takes three chamomile tea bags, soaks them in half a cup of warm water with a teaspoon of honey for 10 minutes and offers them to the girls. According to her, 10–15 minutes later, complete silence reigns, and the girls sleep until morning.

Sleeping Pills and Gender

Just as research on diseases is conducted primarily on men, so too are studies on medications. We now know that women and men do not react similarly to the same medication. Here are a few examples that illustrate the gender difference:

- At times, women and men are prescribed an identical dose of medication, even though in most cases women weigh less. A dose of medication that does not consider the patient's weight results in a larger concentration of the drug in the body. Indeed, women suffer far more from the side effects of a particular drug than men.
- Enzymes in the liver and the wall of the intestine, linked to a system called P450 that is responsible for metabolizing the drug, perform differently in each gender.
- The acidity in the stomach of women is lower than in men, and therefore the rhythm of the stomach depletion is slower. This means that the active components of medications are absorbed in a woman's stomach over a longer period of time, and can therefore affect women more than men.

In 2013, the U.S. Food and Drug Administration (FDA) made history when, for the first time, it pronounced a different dose of medication for women and men. The instruction was made in reference to sleeping pills that include zolpidem (Ambien, Stilnox), and recommended that, women using the drug should take only half of the recommended dose. The decision was made after finding that there were differences in the concentration of the drug between men and women. It was found that the drug in women remained for a longer time in the blood circulatory system, leading to a more pronounced effect. This unusual effect of zolpidem-based drugs caused morning drowsiness and led to a higher risk of car accidents. The producers of zolpidem-based drugs are now required to publish guidelines regarding different doses for men and women.

Today, there is a long list of drugs that have different specifications for dosages and the side effects in men and women. These include statins, aspirin, certain high blood pressure drugs, painkillers, sleeping agents, and drugs for epilepsy.

Treating Sleep Disorders in Women during the Hormonal Changes in Their Lives

Sleep disorders are common throughout a woman's life in relation to the hormonal changes she experiences. At every such stage, it is recommended to begin the treatment of sleep disorders by considering sleep hygiene and CBT. In the event that these do not solve the problem, adding medication should be considered.

- *During the menstrual cycle*: Sleep disorders are frequent a few days before menstruation in cases of mild premenstrual syndrome (PMS), and a week to 10 days in cases of the more severe premenstrual dysphoric disorder (PMDD). Drugs from the benzodiazepine group or Z-drugs can be used for a few days until menstruation, when the sleep disorders usually pass.
- *During pregnancy*: It is not recommended to use sleep-inducing medications and especially not nonprescription drugs, which are erroneously considered safer, but certainly are not, considering that there is not enough information on them. Antidepressant prescription drugs that affect sleep mentioned above can be used. In any case, I always recommend contacting a teratological center (centers that provide information about the safety of using a drug during pregnancy and lactation).
- *Postpartum period*: Sleep-inducing drugs are not recommended for lactating women for fear of passing the drug to the baby via the breast milk. In this case, too, antidepressant prescription drugs that affect sleep can be used under the guidance of a teratological center.
- *During menopause*: My professional experience in treating sleep disorders in women comes especially from treating women during menopause. The challenge of treating women at that stage is very complex since there are usually prolonged sleep disorders. Women whose sleep disorders stem from hot flashes are treated with hormone replacement therapy (HRT), which usually improves sleep. Women whose sleep disorders originate from depression or anxiety are treated with antidepressants. A large proportion of the women I meet do not suffer from either hot flashes or depression, and yet suffer from insomnia.

Why Is Long-Term Use of Sleep-Inducing Pills Not Recommended?

The medical aim of sleeping pills is to break what experts call the "vicious circle," a cycle of lack of sleep that feeds itself with irritability, sleeplessness, impaired daytime functioning, more irritability, and more sleeplessness. Stopping this cycle with sleeping pills may seem easy, yet for many, the path to addiction is short. People with a tendency to addiction require ever-larger doses to maintain efficacy, since their brains usually develop resistance to the drug. People begin with one pill, and when that no longer helps, they increase the dose until they develop a tolerance and dependency. When they try to stop, they suffer from withdrawal symptoms, such as difficulty falling asleep, which was the initial reason for using the drug, along with restlessness and anxiety.

The Ministry of Health does try to constrain the HMOs and pharmacies by regulating the number of the prescriptions that are issued and sometimes also requires explanations, but this doesn't always help. According to data collected by Israel's Ministry of Health from the HMOs, in 2012, the number of people being treated with sleeping pills was 800,000? Dr. Paula Roshka, Director for the Treatment of Addiction at the Ministry of Health, estimates that today this number may be even higher. "According to a national health survey," she says, "Over 30% of people aged 65 and above are treated chronically with sleeping pills that contain benzodiazepine or sleeping pills from the Z-drug group, and the number of people using sleeping pills is ever increasing. This is an epidemic."

Another problem in long-term use of sleeping pills is the disruption of the natural sleep pattern. *Sleeping pills disrupt sleep architecture.* The pills induce chemical sleep, like anesthesia during an operation. The disruption of sleep architecture produces, among other things, a severe disruption of the dream stage, both by reducing the time of this stage and by postponing its onset, which is harmful from a neurological standpoint. This sleep is not restful because it does not include all the stages of natural physiological sleep.

Some physicians are of the opinion that "chemical sleep" is better than no sleep at all. Many experts in the field of sleep believe that sleeping pills help people maintain a normal life routine, yet express reservations about prolonged use of sleeping pills due to the risk of addiction. Professor Yaron Dagan, Head of the Institute for Sleep Medication at Assuta Hospital in Tel Aviv, has a different opinion. He estimates that there is no conclusive proof of addiction to sleeping pills. In his opinion, even if a certain dependency develops among a minority of users, there is no reason to prevent others from using a pill that allows them to maintain their life routine. His conclusion: When using sleeping pills becomes an option, the termination of its use should also be planned in advance.

There is no shortage of examples of people who became addicted to sleeping pills. One example is Judy Garland, who began her career at a very young age and became addicted to sleeping pills and tranquilizers to enable her to work long hours at the film studios with no breaks for sleep. After years in which her body was stuffed with these drugs, it was not at all surprising that Garland developed a dangerous drug dependence. "They would give us amphetamines, and then take us to the clinic and give us sleeping pills … Some hours later they would wake us up and give us more amphetamines. We became lost and we lost contact with the world."

In 1947, Garland had her first nervous breakdown, she was hospitalized, and attempted suicide. Over the years, her addiction to sleeping pills got worse and she also used morphine pills as well as a fair amount of alcohol. In June 1969, Garland's body was found in the bathroom of her London apartment. She was only 47 years old when she died. Her autopsy indicated that her death was due to a drug overdose.

My Approach to Treating Sleep Disorders in Women

"Habit is habit and not to be flung out of the window by any man, but coaxed downstairs a step at a time."

Mark Twain

This quote is the focus of my attitude toward a woman who suffers from sleep disorders and who comes to me for treatment. And why? Women who suffer from sleep disorders (especially toward midlife) see me after a long period of sleep disorders and various treatment attempts. Their sleep habits and sleep disorders have also become a habit or, in other words, a chronic state. Even if the habit causes misery and disturbance, it is already part and parcel of the woman's life. To change the habit requires a slow and gradual management and a combination of several different approaches. Or, as Mark Twain said: To coax it down "one step at a time."

- My first step is always to diagnose the cause of the sleep disorder and to characterize the type of disorder.
- I examine the woman's sleep habits and try to provide her with correct sleep habits, known as "sleep hygiene." It is not at all trivial to change habits, especially in women who have become used to working until late at night, having their laptop in bed, eating in the middle of the night, or turning night into day and day into night.
- If necessary, I refer women to CBT for therapy in order to change behavioral patterns and to adopt various relaxation exercises. Often, people who suffer from insomnia because of an external factor continue to experience sleep difficulties even after the problem has been solved. This is characteristic of people who need a sense of control. The difficulty in falling asleep makes them stressed and then they enter into a "loop," a cycle of tension and thoughts about insomnia, which subsequently leads to insomnia.
- The biggest challenge, as far as I am concerned, is medication. Most of the women that I see are already using sleep medications. They have already tried several medications and they are experiencing side effects, and often they use an overly high dosage or a combination of drugs. The principle that guides me in the treatment is to return the woman's physiological sleep in a slow process, and not via immediate treatment with sleep-inducing drugs; these help sleep, but this sleep would be of poor quality because of the change that is created in sleep architecture.

I try to gradually wean women off sleep medications with an antidepressant that affects sleep, at a low dose, to prevent side effects. One of the medications I mentioned is mirtazapine, which at a very low dose, well below the range in the guidelines for treatment of depression, is very helpful for sleep. After having weaned the woman off the sleeping medication (which she began taking prior to seeing me) with mirtazapine and achieving normal sleep (mirtazapine does not change sleep architecture), I add Circadin (which increases melatonin by slow release, mimicking the natural rise of melatonin during the night) for a period of 6–8 weeks, which is the time that it takes for Circadin to be effective. After 2 months of combined treatment with mirtazapine and Circadin, I gradually stop the treatment with mirtazapine and leave only the treatment with Circadin.

In 2001, together with Professor Nava Zisapel, the developer of Circardin, and her associates at Neurim drug company, I published a research article on this treatment system in the medical journal *Archives of Women's Mental Health*. The article generated considerable attention.

Circadin not only changed my attitude toward treating sleep disorders, it also brought relief to many women who had suffered from sleep disorders for many years. It is true that the process is not always smooth sailing in every case—there are women who do not react well to mirtazapine and suffer from side effects, such as fatigue and weight gain. There are women who do not react well to Circadin and complain of dizziness or headaches, and there are those who need to take Circadin for a very long period of time before achieving its effect. In each case, the principal guideline I adhere to in my treatment is to work slowly by creating a tailor-made process to suit the individual needs of each woman.

In this process, which takes about 2 months, the woman becomes used to physiological sleep, which does not impair her sleep architecture. I believe that our bodies learn to adapt to the new state, when given the opportunity. A multitude of examples of this phenomenon can be found in medicine. Most women, for example, stop having hot flashes once their bodies adapt to the deficiency in sex hormones.

The Oxford Dictionary defines "process" as "a series of actions or steps taken in order to achieve a particular end." Yet we live in an era in which we often seek a quick fix to a problem that requires a longer process. Both patients and therapists are impatient and are not always willing to invest in a process that requires commitment and time. Every process demands patience, persistence, and belief, as well as the ability to endure suffering.

12

On Women and Dreams

The Israeli poet Agi Mishol writes of the difficulty of keeping a dream diary:

"A dream is like a sensitive film in a camera, a film that is overexposed by the light of external impressions ... Every attempt to capture the dream makes it flee for good"

"To Grab the Poem Before the Mind," Dream Notebook

Indeed, dreaming is dreaming and writing is writing and therefore any attempt to tell of a dream, to turn it into words or to write it, imposes order on the dream and makes it lose its very disordered, associative character, which is made up mostly of pictures, experiences, and feelings. Nonetheless, there is no way to convey a dream except in words and in writing, and proof of this is the vast literature on dreams and their significance.

What is the role of dreams in our lives? What is the significance of our dreams? Why don't we all remember the dreams we dreamed? When do we dream during sleep? How many times in one night do we dream? Do the dreams predict the future? Are there differences between the dreams of women and those of men?

So many questions and, yet, very few answers. To this day, scientists are trying to decode the secrets of dreams, their role in our lives, their biological function, and the mechanism that creates them.

From the dawn of history, people were engaged with their dreams and gave them mystical meanings. Ancient civilizations around the world, including the Hebrew, Celtic, Greek, Arabic, Chinese, Japanese, French, and Russian, left images that indicate the important role of dreams, both in the lives of individuals and in shaping the heritage of a people. One of the richest sources of dreams from the ancient world is, of course, the Bible.

Dreams were a source of inspiration for many poets and authors even before Freud, Jung, and their disciples focused on the subject. It is enough to mention Shakespeare, whose work is interwoven with references relating to sleep and dreams.

The Role of Dreams

In 1900, Sigmund Freud's book *The Interpretation of Dreams* was published, in which he presented his viewpoint on dreams. It is based on the assumption that the path to the unconscious is through the understanding of dreams. According to Freud, the unconscious is a sewage cistern of subjects that are not allowed entrance to our consciousness, such as repressed memories and forbidden desires. All those elements that we have difficulty accepting in our everyday life appear in our dreams in the form of symbols.

Swiss psychiatrist Karl Gustav Jung, Freud's student, saw in dreams a mechanism that compensates for elements lacking in a person's life that dare not find expression in everyday life, and the mechanism helps us solve problems. Jung was the first to claim that it is not enough to interpret a single dream, but that many dreams should be analyzed in order to find the universal elements they share. Jung has generated quite a bit of opposition, including psychologists Alfred Adler and Fritz Perls, who claimed that there are no universal symbols that are expressed in dreams, and that everyone has their own system of symbols linked to their individual personality.

In fact, we dream even before we are actually born. Thanks to dozens of studies and to the development of ultrasound, we now know that from the 32nd week of gestation, when the fetus reaches the last stage of its development, it spends about 90% of the time sleeping, in three different patterns: Deep sleep, dream sleep, and intermittent modes. During the stage of dream sleep, the fetus's eyeballs move forward and backward similarly to the movement of the eyes of sleeping adults. This has led scientists to conclude that fetuses dream too.

The importance of dreams can already be noticed during infancy. Babies spend up to 70% of the 24-hour period sleeping; 50% of this time they dream. Taking into consideration that the early years of life are dedicated to extremely rapid development of the memory and learning skills, it appears that dream sleep fills an important role in infant development.

One of the main hypotheses regarding the essential role of dreams is that we need them in order to organize our memory. The assumption is that while dreaming, the brain spills everything onto the table and organizes it. This theory has yet to be proved, but studies have shown that, in people whose dream sleep is lacking, brain function and memory capabilities are inferior to those of people who sleep and dream.

The big revolution in understanding dream processes began in 1929 with the development of electroencephalography (EEG), which records electrical activity of the brain. EEG showed that there are different stages of sleep and that during dream sleep, the brain shows vital activity. This evidence confirmed Freud's theory that dreams play a central role in our lives. The subsequent development of the MRI scanner also brought about significant breakthroughs in our understanding of all that happens during the time that we dream. A new generation of the scanner called functional MRI (fMRI) has allowed us to detect differences in brain blood flow that occur during sleep.

The ability to examine brain waves during the dream stage led to many studies. One study compared two groups of people: People who claimed not to dream at night and those who reported dreaming. Brain scans did not show any difference whatsoever between the two groups with respect to dreaming. In other words, all people dream at night, during the cyclical activity of sleep, but some people remember their dreams and some do not.

William Charles Dement, an American sleep researcher who noticed the eye movements during the dreaming stage, reported additional interesting findings: The longer the night's sleep, the longer the dreaming stage. The first dreaming stage, which occurs about an hour after one falls asleep, lasts about 9 minutes. The second dreaming stage, beginning about 90 minutes later, lasts about 19 minutes. The dreams that occur before waking up are the longest and these are the dreams we usually remember. After waking from a dream, it may seem that the dream lasted only a few seconds; however, recordings show that this is not the case. Moreover, when the duration of the dream is compared with the time that it would take to actually carry out the same activities in reality, no difference is noted.

Another interesting finding from the sleep lab is that examinees whose sleep was disrupted (with the intention of making them recall their dreams) began dreaming at more frequent intervals. This led researchers to conclude that dreams play a role in our mental and physical health and that if a dream is prevented from unfolding correctly, nature will compensate for this with more frequent dreams. When sleeping people were woken up but not during the dream stage, there was no shortening of the gaps between the dream stages, since no dream time was taken and therefore there was no need for compensation.

The psychologist Alan Siegel, author of *Dream Wisdom* (2002), assembled a list of our most common dreams, from childhood to old age: Flying in a plane or crashing; failure or forgetting an exam or forgetting words of a text; appearing naked at a public event; trying to run away from someone or from a threat; sexual experiences (hetero- or homosexual); natural disasters; teeth loss; violence or self-harm or hurting others; abandonment, betrayal, or humiliation; driving in a car or a road accident; missing a bus, a train, or a flight; discovering hidden rooms in a house or a building; finding or losing money, a wallet, or valuable objects; people from the past, alive or dead; and returning to the past, to childhood, to a former house or school.

Gender Differences in Dreaming

I treat mostly women and therefore know more about women's dreams. When I ask a woman to tell me about one of her significant dreams, or a recurring dream, it is rare that I get the answer, "I don't remember my dreams" or "It's only a dream." I do hear these answers from men and sometimes the answer "But it's only a dream." Women are willing to write down parts of their dreams and to come in for treatment in an attempt to figure out their meaning with the therapist. Men, even if they do remember their dreams, are more reluctant to do this for fear of seeming silly or weak or fearing what the therapist will discover about them through the dream. This is, of course, a generalization. There are men who are happy to share their dreams just as there are women who claim that they do not remember their dreams.

Two interesting studies about dreams and gender were published by Michael Schredl and Iris Reinhard from the Central Institute of Research and Mental Health in Mannheim, Germany.

The first study was published in the *Journal of Sleep Research* in 2008 and addresses gender differences in the recollection of dreams. The conclusion of the study is that there is a difference, albeit small, between men and women, especially in that women tend to remember dreams more, particularly in the age group 25–55. The scientists offered several hypotheses yet did not reach a conclusive result beyond what they called sex role orientation, namely, the tendency of women to remember emotional experiences and to translate emotional experiences into words. Women believe that dreams represent and contain an unconscious content, and therefore they are more inclined to try and understand their meaning.

The second study was published in the *Journal of Sleep Medicine* in 2011 and referred to gender differences with regard to nightmares. This refers to bad dreams that begin during childhood and continue throughout life. The nightmares usually occur during the second half of the night, when sleep is not deep. This research study found that women suffer from nightmares more than men (34% compared with 19%), and in this case, the gender difference is not noticed in children or in elderly people. These findings gave rise to different hypotheses but no clear conclusions.

G. William Domhoff, a psychologist and sociologist from the University of California, Santa Cruz and author of *Finding Meaning in Dreams: A Quantitative Approach,* published in 1996, found a number of gender differences in relation to the content of dreams.

- *Men*: Of the characters about whom men dream, 67% are men. The dreams of men include more violent and aggressive subjects (such as chases, beatings, break-ins, robbery, killing). Most of the dreams occur in unfamiliar places, not at their homes, and have to do with competition or an unpleasant confrontation with another man.
- *Women*: Women dream about men and women equally. They mostly dream about familiar people, friends, or close family members. The location of the dreams is usually inside a house or in a familiar place (their home or workplace). If the woman's dream does contain elements of aggression, it will usually be expressed verbally, not physically. Women's dreams contain elements linked to water (usually symbolizes feelings), babies, births, or weddings.

What about sexual content? There is no difference; about 8% of dreams of both genders include sexual activity.

According to Domhoff, gender differences in dreams reflect the cultural gender differences and the different roles of men and women in reality. These are also expressed through the content of dreams.

What Do Women Dream About?

I have known Idit since she was 21 years old. She came to me once in a while to get urgent help for panic attacks from which she suffered. During her university studies, Idit met Yoram and after 3 years they moved in together. Idit was not certain about the relationship for many reasons. In particular she was apprehensive about Yoram's employment instability and his attitude toward her that at times bordered on the verbally abusive. Following a job offer in Europe, Yoram proposed to her. Idit had many doubts. Her panic attacks worsened. In one of our meetings, she told me of a dream she had:

"I am alone in an open space, perhaps a large field, and standing at the edge of a large pit that looks like a deep well. I ask myself whether to go into it or not. I look into the deep well and the thought passes through my mind that if I enter the well I will be trapped and unable to exit, because it is very deep and it frightens me a lot."

Idit woke up from the dream with a pounding heart and a feeling of suffocation and panic. In a matter of minutes, she understood that she had received an answer to her doubts and decided to end the relationship with Yoram. The dream served her as affirmation of her fears of getting into this marriage, which she interpreted as a closed place, a prison that would entrap her.

I write and lecture a lot on gender differences in the sleep between men and women, and people are surprised to learn that these do exist. Eyebrows lift in amazement when

I talk about gender differences in dreams. "Are there really differences in the dreams of women and of men?" I am often asked, and my answer is "Why not? Dreams reflect our feelings, experiences, thoughts and in fact—who we are. The differences between men and women are reflected both in our biology and in our dreams."

Here are some of the main topics of dreams that I have heard from women: In the early stages of pregnancy, dreams about bleeding and miscarriage, dreams about the children (now grown) as youngsters, in which they get lost; betrayal by the partner; returning to a former house or to life with an ex-partner; returning to the childhood home; matriculation exams that still need to be passed; a dead partner comes back to life in the dream and the woman doesn't understand how it happened; and unsuccessful attempts to reach someone by phone.

Patients tell me that they continue to experience the dream many hours after waking up, and if it was an unpleasant dream, they continue to feel the distress and anxiety that accompanied it.

Virginia Woolf, in her essay *A Room of One's Own*, claimed that a woman needs money and a room of her own if she is interested in writing books. The desire for a "room of one's own" appears frequently in dreams of women and I hear it especially from married women.

A patient who married at a very young age and spent most of her time taking care of the family told me that she always wanted to achieve financial independence and have a space of her own—with no success. This desire came up in a recurring dream: "I am in a small flat, perhaps it has one room, I think it is in my hometown. I understand that I am alone in it and that the flat is mine. A sense of joy fills my heart and I cannot believe it has happened at long last. I wake up, understand it was only a dream, and refuse to let go of it."

Another patient, who had always wanted a place of her own in which she could sit and think, read, and perhaps also write, also told of a recurring dream: "I wander around the rooms of my house, go upstairs to the second floor and suddenly discover a large room I was unaware existed. It is a lit room, with two large windows, and my heart is filled with happiness."

Professor Monique Lortie-Lussier of the University of Ottawa, Canada, published an interesting article in 2015, which deals with the development of the dreams of women throughout their lives. The study, published in the journal *Consciousness and Cognition,* was carried out with 375 Canadian women aged 17–85, who were divided into five age groups (12–17, 18–24, 25–39, 40–64, and 65–85). Lortie-Lussier's working premise was that throughout life, the dreams of women would reflect the psychological and social changes they experience. The results of the study affirmed her hypothesis.

Lortie-Lussier examined a number of aspects:

- *Feelings*: Puberty is characterized by a high intensity of feelings (anger, jealousy, love, hate, loneliness, shame, etc.) and so is the content of the dreams of girls of this age. As they grow up, especially after the age of 45, there is a decrease in the intensity of the emotions that appear in their dreams and particularly with respect to aggression and negative feelings.
- *Social ties*: With age, there is a tendency to reduce social interactions, to withdraw inward, and to prefer quality over quantity. Indeed, the dreams of

women of a more mature age showed a decrease in subjects related to social interactions and a decrease in content relating to activities in life.

- *Sexual content*: Here, there was a surprise: there was no age-related decrease in the frequency of sexual content in dreams of women. It appears at about the same frequency in young women's dreams as in those of elderly women.

Dreams, according to Lortie-Lussier, reflect the woman's natural maturation, which over the years achieves greater tranquility and acceptance of the place where she is. At this stage, she seeks social ties as well as activities of a greater quality, and so she is less busy with the types of social activity that characterized her younger life, her feelings are more positive, and she is much more satisfied with her life. All this is reflected in her dreams.

Dreams and Hormonal Changes

When I wrote my first book, *Women, Mood and Hormones*, the link between the hormonal changes that women experience and the mental changes was very clear to me. It is also very clear to me that there is a link between hormonal changes in a woman's life and her dreams. Just as hormones affect our mental states as women, they certainly affect the content of our dreams.

Indeed, I have found some support for this in the professional literature, not much, but enough to confirm that this connection—even if it is very rarely mentioned—does indeed exist.

The Menstrual Cycle

During the menstrual cycle, particularly from the luteal phase (after ovulation) onward, more sleep disturbances occur. The reason is the hormonal changes, especially the effect of progesterone, which begins to rise after ovulation and to decrease a few days before menstruation. The progesterone causes tiredness and a decrease in the frequency of the dream stages (REM). Sleep disturbances near menstruation will manifest mostly in difficulty falling asleep and frequent waking, often during the dream stage. As a result, the memory of the dream will be clear and very sharp. Very often, women tell of strange dreams or even nightmares in the nights before their menstruation.

Dr. Patricia Garfield, a psychologist and dream researcher, has published ten books about dreams. In her book *Women's Bodies, Women's Dreams* (1998), she writes about women's dreams at different stages of their lives based on research studies done on this subject. Garfield states that during the menstrual cycle, recollection of dreams is relatively low. During this period, women dream of floods, of being wounded, and the color red often appears in the dream. Friendly female characters appear and there is an increase in sexual dreams and in aggressive content. In the week after menstruation, when estrogen levels begin to rise, the dreams include journeys in open places and meetings with various characters.

At the time of ovulation, the content of dreams is connected to birth and babies. There are images of eggs, diamonds, and expensive or fragile objects. Sexual desire is at a peak and there is an increase in the frequency of erotic dreams. The figures of males become closer and more friendly, whereas the female characters are distant and less friendly to the dreamer. After ovulation, when the corpus luteum secretes progesterone, there is an increase in the intensity of dreams and in their recollection. The images are relatively calm, and figures of babies and children appear with great frequency. The dreamer has more a more passive role in the dream scenes.

Pregnancy

The most significant hormonal changes in the life of a woman occur during pregnancy, which is an emotionally turbulent period for every woman. The biological processes and hormonal changes that occur in the woman's body lead to feelings that she had never experienced previously. This is an exciting and stressful period, during which she is troubled by many things.

Several studies on dreams during pregnancy show that the dreams that a woman dreams during pregnancy are special and very tangible. Their content is very different from those of nonpregnant women. Often, the dreams during pregnancy make the dreaming woman feel like she is losing her mind, but in fact these dreams are normal and healthy and allow the pregnant woman to express tension and fears in a nonthreatening manner. Many women tell of dreams related to nightmares about the birth and the meeting with the baby and even prophetic dreams. Furthermore, studies reveal that pregnant women tend to remember more bad dreams and nightmares than women who are not pregnant.

The most common topics of the dreams of a pregnant woman include giving birth to something that is not a human baby—cat, doll, or other object; a birth that takes place in a car; encounter with the baby; miscarriage or scenes of troublesome birth; and various deformities found in the fetus. Nearing the birth, the dreams are tinged with even more anxiety and include horror scenarios of hours of screaming and pain, caesarian sections without anesthesia, and intimidating doctors and midwives. In contrast, there are often optimistic dreams of an easy birth or of a baby born clean, combed, and dressed.

Incidentally, the dreams of men during the pregnancy of their partner have almost never been studied. It would be interesting to know whether or not their dreams during this period are also special and express similar themes.

Deborah Jackson, a journalist and writer, wrote in *Eve's Wisdom* (1999) about the cultural attitude to guessing the sex of the fetus through dreams dreamed during pregnancy. In cultures in which the sex of the newborn is of great importance in the eyes of its parents, the symbols appearing in the dreams of the pregnant woman are meticulously examined in search of clues as to the sex of the baby. In many cultures, the sex signs that appear in dreams are divided into two groups: round objects, which, according to beliefs, symbolize a girl and long, phallic, objects that attest to the arrival of a boy.

For example, Nyinba women of Nepal assume that they are carrying a boy if they dream about long vegetables such as a cucumber or eggplant. They also say that cutting tools—traditionally male tools—indicate a boy. Conversely, dreams of girls are characterized by round vegetables or objects, such as prayer beads. When a pregnant Maori woman of New Zealand dreams of a human skull, she will interpret it as a sign that the fetus is a girl. Dreams of feathers are interpreted as a sign of a boy.

Postpartum

In search of predictions of postpartum depression, which could allow early detection of women at risk during their pregnancy, the possibility arises of exploiting the dreams dreamed during the pregnancy. Studies have indicated a way to distinguish between the dreams of women who suffered from postpartum depression and those who did not through interpretation of their dreams during the pregnancy. The aim is to identify women with a high risk of postpartum depression during their pregnancy and perhaps help prevent the outbreak of the depression through preventative treatment.

An Israeli study conducted in 2003 by the psychologists Adi Brosh and Tamar Kron showed that postpartum depression can be predicted by studying the dreams of pregnant women. Kron and Brosh compared the dreams of pregnant women who experienced postpartum depression with the dreams of women who did not. They found that in the dreams of women who did not experience postpartum depression there were more expressions of fear and masochistic elements in comparison with the dreams of pregnant women who subsequently suffered postpartum depression.

The researchers concluded that the ability to express distress and anxiety through dreams serves as a type of emotional release that allows women to prepare for the difficulties that follow birth.

Adi Brosh, a clinical psychologist, explained: "If a woman does the mental work of preparation toward the transition from her being a woman to becoming a mother, which undoubtedly is a frightening and stressful transition, it is most likely that she will not suffer from postpartum depression. Women who dreamed difficult dreams prior to the birth tended to suffer less from postpartum depression compared to women whose dreams were pleasant or those who had no recollection of their dreams at all."

Women who had experienced traumas related to fertility and giving birth tell of recurring dreams related to the trauma, which remain with them for many years afterward:

"I feel pressure and contractions in my stomach and know that the baby is dead," a patient tells me; she experienced many miscarriages before the birth of her daughter and the miscarriage nightmare recurs over and over again in her dreams. "I look, I see heavy bleeding, I shout to make it stop and I wake up crying."

"I arrive at the delivery room and I know that in fact nothing is going to come out," a patient undergoing years of fertility treatments says, sharing her dream with me; "I feel the contractions, I lie on the birthing bed and push hard, but nothing comes out."

"I am pregnant and arrive at the delivery room," says a woman who suffered a miscarriage after the birth of her two children and since then has had many dreams about pregnancies and births. "I know that it is going to be long and hard. Suddenly they tell me that my adored boss, who left the workplace recently, is the 'on-call' midwife. I am already in the delivery room with contractions and he enters to help me give birth."

Menopause

Efrat, 56 years old, has been in therapy for about 5 years. She came to me with severe depression, which was related to the breakup of her marriage after 30 years. Efrat was treated with an antidepressant, which helped her a lot. At some stage, Efrat began complaining about prolonged nightmares, which did not leave her. We tried to attribute the nightmares to the difficult years she had experienced, yet they appeared after there was an improvement in her mental state and after a new partner had entered her life. It was difficult to understand why the nightmares were recurring now. Efrat suspected that they had to do with the antidepressant, and since her mental state had improved, we decided to gradually stop the medication. Despite that, the nightmares did not cease. A sleep clinic examination clarified that Efrat suffered from sleep apnea, which began after the end of her menstruation and caused frequent waking and nightmares. Treating the sleep apnea quickly made the nightmares disappear and brought about a calmer and more continuous sleep.

Very little has been written about the dreams of women in menopause, yet one subject appears in all the articles I read: an increase in the frequency of nightmares during this period. What is the reason for this? Sleep disorders are much more frequent after the end of menstruation, especially as a result of phenomena related to this period, such as hot flashes, sleep apnea, or restless legs syndrome. As a result, women suffer more from frequent waking at night, often during the dream stage, accompanied by an increase in their stress level. The combination of the physiological disorder of sleep and the psychological disorder may lead to nightmares. Sleep perturbation combined with nightmares increases the risk of high blood pressure and heart problems.

Conclusion

I will end, as I began, with words written by the Israeli poet Agi Mishol, who dreams a lot and writes about her dreams. In one interview she was asked if she had ever dreamed a poem and she replied: "Many times, connecting to my dreams is part of my connection with myself. Depicting the dream is part of my morning writing ... I am amazed by the ingenuity and creativity that occurs in me when I sleep. Sometimes it is more fascinating than reality, but it's not the psychological aspect ... What fascinates me is the magic, the surprise, the wisdom of the dream itself."

"A dream is a personal certificate from a person's letter to himself."

Calvin S. Hall, an American psychologist

13

A Few Closing Words

At a wintry evening meeting with my friend Tali, we talked about many things, and Tali asked me, "Why don't you write about women's sleep disorders? You helped me so much in coping with sleep difficulties during my menopause, and I saw what happened to my daughter and her friends after they gave birth, how hard it was for them to resume normal sleep. This problem is not talked about enough. It's time that you wrote about it."

Like everything in life, timing is everything, and Tali's words came at exactly the right time for me. I felt that I was indeed ready to start writing about sleep disorders in women. I thought about the knowledge and experience that I've acquired over the years of studying this subject, lecturing about it, writing about it, and I thought particularly about the women who suffer from sleep disorders and whom I treat. I knew the subject was indeed worthy of attention and that I had a lot to contribute.

The process of writing for women means concurrently writing for myself, as a woman and physician who treats mainly women. As I was writing, I thought of the patients who came to see me in their distress. Some of their stories are included in the book, incognito, needless to say. Along with the scientific presentation, I felt the need to intersperse excerpts of literature and prose relating to sleep and dreams, and written primarily by women.

A deep sense of responsibility accompanied the writing of this book. Anat, the Hebrew editor with whom I chose to work (a very happy choice), followed the writing process closely all the way through. Aside from her professional qualities, I chose her because she is a mature woman who is aware of the importance of the subject. Professor Nava Zisapel added her accurate scientific comments for which I thank her.

With the completion of my book, I came to understand even more how wonderful and complex the sleep process is, the extent to which the importance of good sleep is undervalued in the culture in which we live, and the extent to which correct treatment of sleep disorders is the result of understanding their causes.

And no less important, I understand the extent to which we women, who suffer from sleep disturbances so frequently throughout our lives, are not sufficiently aware of their causes and tend to consider this situation as a problem that we have to accept and to live with. It is important that we know that in some cases, the solution will be quick and in other cases restoring good sleep will require a more lengthy process, but it will always be possible to find a solution to the sleep problem.

And in Albert Einstein's words:

"For each problem there is a solution, otherwise there wouldn't have been a problem."

Bibliography

Allen RP, Picchietti D, Hening WA et al. Restless legs syndrome: Diagnostic criteria, special considerations, and epidemiology. A report from the restless legs syndrome diagnosis and epidemiology workshop at the National Institutes of Health. *Sleep Med.* 2003;4:101–119.

American Psychiatric Association. *Sleep-Wake Disorders/DSM-5.* American Psychiatric Association, 2013.

Antonijevic IA, Murck H, Frieboes RM et al. On the role of menopause for sleep-endocrine alterations associated with major depression. *Psychoneuroendocrinology.* 2003;28(3):401–418.

Attarian H, Hachul H, Guttuso T et al. Treatment of chronic insomnia disorder in menopause: Evaluation of literature. *Menopause.* 2015;22(6):674–684.

Attarian HP, Saltzman MV. *Sleep Disorders in Women: A Guide to Practical Management.* 2nd ed. Springer, 2013.

Baker FC, Driver HS. Circadian rhythms, sleep, and the menstrual cycle. *Sleep Medicine.* 2007;8(6):613–622.

Baker FC, Sassoon SA, Kahan T et al. Perceived poor sleep quality in the absence of polysomnographic sleep disturbance in women with severe premenstrual syndrome. *J Sleep Res.* 2012;21:535–545.

Baker FC, Willoughby AR, Sassoon SA et al. Insomnia in women approaching menopause: Beyond perception. *Psychoneuroendocrinology.* 2015;60:96–104.

Balserak BI, Lee K. Sleep and sleep disorders associated with pregnancy. In: Kryger MH, Roth T, Dement WC, editors. *Principles and Practice of Sleep Medicine.* 6th ed. Saunders/Elsevier, 2011; Chapter 156: 1525–1539.

Baughman KR, Bourguet CC, Ober SK. Gender differences in the association between antidepressant use and restless legs syndrome. *Mov Disord.* 2009;24:1054–1059.

Bentley AJ, Rosman KD, Mitchell D. Gender differences in the presentation of subjects with restless legs syndrome. *Sleep Med.* 2006;7:37–41.

Berger K, Luedemann J, Trenkwalder C et al. Sex and the risk of restless legs syndrome in the general population. *Arch Intern Med.* 2004;164:196–202.

Bixler EO, Papaliaga MN, Vgontzas AN et al. Women sleep objectively better than men and the sleep of young women is more resilient to external stressors: Effects of age and menopause. *J Sleep Res.* 2009;18:221–228.

Blumel JE, Cano A, Mezones-Holguin E et al. A multinational study of sleep disorders during female mid-life. *Maturitas.* 2012;72:359–366.

Blyton DM, Sullivan CE, Edwards N. Lactation is associated with an increase in slow-wave sleep in women. *J Sleep Res.* 2002;11(4):297–303.

Campos-Rodriguez F, Martinez-Garcia MA, Monserrat JM. Gender differences in treatment recommendations for sleep apnea. *Clin Pract.* 2012;9:565–578.

Chung SA, Wolf TK, Shapiro CM. Sleep and health consequences of shift work in women. *J Womens Health (Larchmt).* 2009;18:965–977.

Chuong CJ, Kim SR, Taskin O et al. Sleep pattern changes in menstrual cycles of women with premenstrual syndrome: A preliminary study. *Am J Obstet Gynecol.* 1997;177(3):554.

Dale A, Lortie-Lussier M, De Koninck J. Ontogenetic patterns in the dreams of women across the lifespan. *Consciousness and Cognition*. 2015;37:214–224.

de Beauvoir S. *The Second Sex*. New York: Vintage, 2015.

de Zambotti M, Colrain IM, Baker FC. Interaction between reproductive hormones and physiological sleep in women. *J Clin Endocrinol Metab*. 2015a;100(4):1426–1433.

de Zambotti M, Willoughby AR, Sassoon SA et al. Menstrual cycle-related variation in physiological sleep in women in the early menopausal transition. *J Clin Endocrinol Metab*. 2015b;100(8):2918–2926.

Desai K, Mao JJ, Su I et al. Prevalence and risk factors for insomnia among breast cancer patients on aromatase inhibitors. *Support Care Cancer*. 2013;21:43–45.

Dobkin RD, Menza M, Bienfait KL et al. Ramelteon for the treatment of insomnia in menopausal women. *Menopause Int*. 2009;15:13–18.

Dolev Z. *Women, Mood and Hormones*. Kotarot Publishers, 2005.

Dolev Z. Case series of perimenopausal women with insomnia treated with mirtazapine followed by prolonged-release melatonin add-on and monotherapy. *Arch Womens Ment Health*. 2011 Jun;14(3):269–273.

Domhoff GW. *Finding Meaning in Dreams: A Quantitative Approach*. Springer, 1996.

Dorsey CM, Lee KA, Scharf MB. Effect of zolpidem on sleep in women with perimenopausal and postmenopausal insomnia: A 4-week, randomized, multicenter, double blind, placebo-controlled study. *Clin Ther*. 2004;26:1578–1586.

Earley CJ, Allen RP, Beard JL et al. Insight into the pathophysiology of restless legs syndrome. *J Neurosci Res*. 2000;62:623–628.

Ehlers CL, Kupfer DJ. Slow-wave sleep: Do young adult men and women age differently? *J Sleep Res*. 1997;6(3):211.

Eichling PS, Sahni J. Menopause related sleep disorders. *J Clin Sleep Med*. 2005;1:291–300.

Ford DE, Kamerow DB. Epidemiologic study of sleep disturbances and psychiatric disorders. An opportunity for prevention? *JAMA*. 1989;262:1479–1484.

Frey BN, Haber E, Mendes GC et al. Effects of quetiapine extended release on sleep and quality of life in midlife women with major depressive disorder. *Arch Womens Ment Health*. 2012;16:83–85.

Gabbay IE, Lavie P. Age- and gender-related characteristics of obstructive sleep apnea. *Sleep Breath*. 2012;16:453–460.

Garcia-Borreguero D, Egatz R, Winkelmann J et al. Epidemiology of restless legs syndrome: The current status. *Sleep Med Rev*. 2006;10:153–167.

Garfield P. *Women's Bodies, Women's Dreams*. New York: Ballantine Books, 1998.

Glezerman M. *Gender Medicine*. New York: Harry N Abrams, 2016.

Goldman SE, Stone KL, Ancoli-Israel S et al. Poor sleep is associated with poorer physical performance and greater functional limitations in older women. *Sleep*. 2007;30:1317–1324.

Greenstein M, Holland J. *Lighter as We Go: Virtues, Character Strengths, and Aging*. New York: Oxford University Press, 2014.

Hachul H, Andersen ML, Bittencourt LR et al. Does the reproductive cycle influence sleep patterns in women with sleep complaints? *Climacteric*. 2010;13:594–603.

Hachul H, Bittencourt LR, Soares JM Jr et al. Sleep in post-menopausal women: Differences between early and late post-menopause. *Eur J Obstet Gynecol Reprod Biol*. 2009;145:81–84.

Hale L, Do DP, Basurto-Davila R et al. Does mental health history explain gender disparities in insomnia symptoms among young adults? *Sleep Med*. 2009;10:1118–1123.

Hollander LE, Freeman EW, Sammel MD et al. Sleep quality, estradiol levels, and behavioral factors in late reproductive age women. *Obstet Gynecol*. 2001;98:391–397.

Huffington A. *The Sleep Revolution: Transforming Your Life One Night at a Time.* New York: Harmony Books, 2016.

Jackson D. *Eve's Wisdom: Traditional Secrets of Pregnancy, Birth and Motherhood.* Duncan Baird Publishers, 1999.

Joffe H, Massler A, Sharkey KM. Evaluation and management of sleep disturbance during the menopause transition. *Semin Reprod Med.* 2010a;28(5):404–421.

Joffe H, Petrillo L, Viguera A et al. Eszopiclone improves insomnia and depressive and anxious symptoms in perimenopausal and postmenopausal women with hot flashes: A randomized, double-blinded, placebo-controlled crossover trial. *Am J Obstet Gynecol.* 2010b;202:171.e1–171.e11.

Jordan AS, McEvoy RD. Gender differences in sleep apnea: Epidemiology, clinical presentation and pathogenic mechanisms. *Sleep Med Rev.* 2003;7:377–389.

Kalleinen N, Polo O, Himanen SL et al. The effect of estrogen plus progestin treatment on sleep: A randomized, placebo-controlled, double-blind trial in premenopausal and late postmenopausal women. *Climacteric.* 2008a;11:233–243.

Kalleinen N, Polo-Kantola P, Himanen S-L et al. Sleep and the menopause—Do postmenopausal women experience worse sleep than premenopausal women? *Menopause Int.* 2008b;14:97–104.

Kline CE, Irish LA, Krafty RT et al. Consistently high sports/exercise activity is associated with better sleep quality, continuity and depth in midlife women: The SWAN sleep study. *Sleep.* 2013;36:1279–1288.

Koo BB, Dostal J, Ioachimescu O et al. The effects of gender and age on REM-related sleep-disordered breathing. *Sleep Breath.* 2008;12:259–264.

Kravitz H, Zhao X, Bromberger J et al. Sleep disturbance during the menopausal transition in a multi-ethnic community sample of women. *Sleep.* 2008;31(7):979–990.

Krishnan V, Collop NA. Gender differences in sleep disorders. *Curr Opin Pulm Med.* 2006;12(6):383–389.

Kryger MH. *A Woman's Guide to Sleep Disorders.* McGraw-Hill, 2004.

Krystal A, Attarian H. Sleep medications and women: A review of issues to consider for optimizing the care of women with sleep disorders. *Curr Sleep Med Rep.* 2016;2(4):218–222.

Krystal AD. Insomnia in women. *Clin Cornerstone.* 2003;5(3):41–50.

Kyaga S, Landén M, Boman M et al. Mental illness, suicide and creativity: 40-year prospective total population study. *J Psychiatr Res.* 2013 Jan;47(1):83–90. Epub 2012 Oct 9.

Lee KA, Baker FC, Newton KM et al. The influence of reproductive status and age on women's sleep. *J Womens Health (Larchmt).* 2008;17:1209–1214.

Lee KA, Kryger MH. Women and sleep. *J Womens Health (Larchmt).* 2008;17:1189–1190.

Lee KA, Shaver JF, Giblin EC et al. Sleep patterns related to menstrual cycle phase and premenstrual affective symptoms. *Sleep.* 1990;13(5):403–409.

Mishol A. *Dream Notebook [Hebrew].* Even Hoshen, 2000.

Misri S. *Shouldn't I Be Happy? Emotional Problems of Pregnant and Postpartum Women.* The Free Press, 1985.

Misri S. *Shouldn't I be Happy?* New York: The Free Press, 1995.

Moline ML, Broch L, Zak R et al. Sleep in women across the life cycle from adulthood through menopause. *Sleep Med Rev.* 2003;7(2):155–177.

Mong JA, Baker FC, Mahoney MM et al. Sleep, rhythms, and the endocrine brain: Influence of sex and gonadal hormones. *J Neurosci.* 2011;31:16107–16116.

Mong JA, Cusmano DM. Sex differences in sleep: Impact of biological sex and sex steroids. *Philos Trans R Soc Lond B Biol Sci.* 2016;371(1688):20150110.

Morgenthaler T, Kramer M, Alessi C et al. Practice parameters for the psychological and behavioral treatment of insomnia: An update. An American Academy of Sleep Medicine Report. *Sleep.* 2006;29:1415–1419.

Morin CM, Vallieres A, Guay B et al. Cognitive behavioral therapy, singly and combined with medication, for persistent insomnia: A randomized controlled trial. *JAMA.* 2009;301:2005–2015.

Mullington JN, Haack M, Toth M et al. Cardiovascular, inflammatory and metabolic consequences of sleep deprivation. *Prog Cardiovasc Dis.* 2009;51:294–302.

Nasar S. *A Beautiful Mind.* Simon & Schuster, 2011.

Northrup C. *The Wisdom of Menopause: Creating Physical and Emotional Health During the Change.* Revised ed. Bantam, 2012.

Nowakowski S, Meliska CJ, Martinez LF et al. Sleep and menopause. *Curr Neurol Neurosci Rep.* 2009;9:165–172.

Ohayon MM. Epidemiology of insomnia: What we know and what we still need to learn. *Sleep Med Rev.* 2002;6:97–111.

Okatani Y, Morioka N, Wakatsuki A. Changes in nocturnal melatonin secretion in perimenopausal women: Correlation with endogenous estrogen concentrations. *J Pineal Res.* 2000;28(2):111–118.

Okun ML. Sleep and postpartum depression. *Current Opinion in Psychiatry.* 2015;28(6):490–496.

Oyayon MM. Severe hot flashes are associated with chronic insomnia. *Arch Intern Med.* 2006;166:1262–1268.

Parry BL, Martínez LF, Maurer EL et al. Sleep, rhythms and women's mood. Part 1. Menstrual cycle, pregnancy and postpartum. *Sleep Med Rev.* 2006;10(2):129–144.

Paul KN, Turek FW, Kryger MH. Influence of sex on sleep regulatory mechanisms. *J Womens Health (Larchmt).* 2008;17:1201–1208.

Pavlova M, Sheikh LS. Sleep in women. *Seminars in Neurology.* 2011;31(4):397–403.

Phillips BA, Collop NA, Drake C et al. Sleep disorders and medical conditions in women. *Proceedings of the Women & Sleep Workshop, National Sleep Foundation,* Washington DC, March 5–6, 2007. *J Womens Health (Larchmt).* 2008;17:1191–1199.

Picchietti D, Winkelman JW. Restless legs syndrome, periodic limb movements in sleep, and depression. *Sleep.* 2005;28:891–898.

Plath S. *The Collected Poems.* Harper & Row, 1981.

Polo-Kantola P. Sleep and menopause. *Womens Health.* 2007;3:99–106.

Polo-Kantola P. Sleep problems in midlife and beyond. *Maturitas.* 2011;68(3):224–232.

Pontalis JB. *Windows.* Translated by Anne Quinney. Lincoln: University of Nebraska Press, 2003.

Rahman SA, Kayumov L, Shapiro CM. Antidepressant action of melatonin in the treatment of delayed sleep phase syndrome. *Sleep Med.* 2010;11:131–136.

Redfield Jamison K. *An Unquiet Mind: A Memory of Moods and Madness.* Vintage, 1995.

Redline S, Kirchner HL, Quan SF et al. The effects of age, sex, ethnicity, and sleep-disordered breathing on sleep architecture. *Arch Intern Med.* 2004;164:406–418.

Regestein QR. Estrogen therapy and the cognitive impairments of sleep deprivation. *Menopause.* 2005;12(2):125–127.

Roth T. Insomnia: Definition, prevalence, etiology and consequences. *J Clin Sleep Med.* 2007;3(suppl 5):S7–S10.

Sarris J, Byrne GJ. A systematic review of insomnia and complementary medicine. *Sleep Med Rev.* 2011;15:99–106.

Sarti CD, Chiantera A, Graziottin A et al. Hormone therapy and sleep quality in women around menopause. *Menopause.* 2005;12:545–551.

Bibliography
137

Schredl M, Reinhard I. Gender differences in dream recall: A meta-analysis. *J Sleep Res.* 2008 Jun;17(2):125–131.

Schredl M, Reinhard I. Gender differences in nightmare frequency: A meta-analysis. *Sleep Med Rev.* 2011 Apr;15(2):115–121.

Shalev M. *My Russian Grandmother and Her American Vacuum Cleaner: A Family Memoir.* Schocken, 2011.

Shepertycky MR, Banno K, Kryger MH. Differences between men and women in the clinical presentation of patients diagnosed with obstructive sleep apnea syndrome. *Sleep.* 2005;28:309–314.

Siegel AB. *Dream Wisdom: Uncovering Life's Answers in Your Dreams.* Celestial Arts, 2002.

Smith MT, Perlis ML, Park A et al. Comparative meta-analysis of pharmacotherapy and behavior therapy for persistent insomnia. *Am J Psychiatry.* 2002;159:5–1.

Soares CN, Joffe H, Rubens R et al. Eszopiclone in patients with insomnia during perimenopause and early postmenopause: A randomized controlled trial. *Obstet Gynecol.* 2006;108:1402–1410.

Sowers MF, Zheng H, Kravitz HM et al. Sex steroid hormone profiles are related to sleep measures from polysomnography and the Pittsburgh Sleep Quality Index. *Sleep.* 2008;31:1339–1349.

Steiner M, Yonkers K. *Depression in Women.* Martin Dunitz, 2001.

Thomson J, Oswald I. Effect of oestrogen on the sleep, mood, and anxiety of menopausal women.. *Br Med J.* 1977;2(6098):1317–1319.

Tom SE, Anderson ML, Landis CA et al. Sleep problems after short-term hormone therapy suspension: Secondary analysis of a randomized trial. *Menopause.* 2011;18:1184–1190.

Tom SE, Kuh D, Guralnik JM et al. Self-reported sleep difficulty during the menopausal transition: Results from a prospective cohort study. *Menopause.* 2010;17:1128–1135.

Tranah GJ, Parimi N, Blackwell T et al. Postmenopausal hormones and sleep quality in the elderly: A population based study. *BMC Womens Health.* 2010;10:15.

Valham F, Stegmayr B, Eriksson M et al. Snoring and witnessed sleep apnea is related to diabetes mellitus in women. *Sleep Med.* 2009;10:112–117.

Viorst J. *Necessary Losses: The Loves, Illusions, Dependencies, and Impossible Expectations That All of Us Have to Give Up in Order to Grow.* Simon & Shuster, 1986.

Wilson R. *Feminine Forever.* M Evans and Company, 1966.

Wolfson AR, Crowley SJ, Anwer U et al. Changes in sleep patterns and depressive symptoms in first-time mothers: Last trimester to 1-year postpartum. *Behav Sleep Med.* 2003;1(1):54–67.

Xie L, Kang H, Xu Q, Nedergaard M et al. Sleep drives metabolite clearance from the adult brain. *Science.* 2013 Oct 18;342(6156):373–377.

Xu M, Belanger L, Ivers H et al. Comparison of subjective and objective sleep quality in menopausal and nonmenopausal women with insomnia. *Sleep Med.* 2011;12:65–69.

Young T, Peppard PE. Clinical presentation of OSAS: Gender does matter. *Sleep.* 2005;28:293–295.

Zhang B, Wing YK. Sex differences in insomnia: A meta-analysis. *Sleep.* 2006;29:85–93.

Zhdanova IV, Wurtman RJ, Regan MM et al. Melatonin treatment for age-related insomnia. *J Clin Endocrinol Metab.* 2001;86:4727–4730.

Index